Major Transitions in Vertebrate Evolution

LIFE OF THE PAST
James O. Farlow, editor

Major Transitions in Vertebrate Evolution

Edited by Jason S. Anderson and Hans-Dieter Sues

Indiana University Press
Bloomington and Indianapolis

This book is a publication of

Indiana University Press

601 North Morton Street

Bloomington, IN 47404-3797 USA

http://iupress.indiana.edu

Telephone orders   800-842-6796

Fax orders   812-855-7931

Orders by e-mail   iuporder@indiana.edu

The paper used in this publication meets
the minimum requirements of American
National Standard for Information Sciences—Permanence of Paper for Printed
Library Materials, ANSI Z39.48-1984.

Manufactured in the United States of
America

Library of Congress Cataloging-in-
Publication Data
Major transitions in vertebrate evolution /
edited by Jason S. Anderson and Hans-
Dieter Sues.
     p. cm. — (Life of the past)
Includes bibliographical references and
index.
ISBN 978-0-253-34926-2 (cloth : alk.
paper)   1. Vertebrates—
Evolution.   I. Anderson, Jason S., date
II. Sues, Hans-Dieter, date
QL607.5.M348 2007
596—dc22   2007001167
1 2 3 4 5 12 11 10 09 08 07

Figure D.1. Bob Carroll and Anna Di Turi relaxing in Kejimkujik National Park, Nova Scotia, in fall 2001. Photograph by Brian K. Hall.

This volume is inspired by and dedicated to Robert Lynn "Bob" Carroll (Figs. D.1 and D.2). Bob has spent his distinguished career examining many of the major evolutionary transitions in vertebrate history. Perhaps this was inevitable given his graduate training at Harvard, under the preeminent anatomist and paleontologist Alfred Sherwood Romer, where graduate students attended weekly discussions between some of the leading thinkers on evolution in the twentieth century, including Romer, George Gaylord Simpson, and Ernst Mayr.

Bob has not only worked on many of the major transitions in vertebrate evolution, but he has to a large extent framed subsequent discussions on these topics. His preference for focusing on specific, complex morphological systems and biologically plausible explanations for transformation has proven resilient to changing methodologies in evolutionary biology. Specifically, his hypothesis concerning the constraints imposed by miniaturization driving rapid morphological change in the origin of amniotes and modern amphibians has become widely accepted (but see Reisz, Chapter 6). His interest in macroevolution culminated in the publication of *Patterns and Processes in Vertebrate Evolution* (Carroll, 1997), the first comprehensive treatment of the subject since Simpson's *Tempo and Mode in Evolution* (Simpson, 1944). A comprehensive recent bibliography of Bob's publications can be found in Sues et al. (2003).

Bob was one of the first paleontologists to recognize the importance of recent innovations in molecular developmental biology, and he has worked passionately to incorporate developmental studies into the Modern Synthesis, becoming a leader in the emerging field of evolutionary developmental biology (colloquially known as "evo-devo"). Although many of the chapters in the present volume are primarily concerned with vertebrate paleontology, they have a substantial developmental component, at least in part

Figure D.2. Paleontologist dwarfed by nature; Bob Carroll in front of cross section of a giant sequoia (*Sequoiadendron giganteum*) in Sierra Nevada, California, 1997. The circumference of the tree, using Bob Carroll's head as the reference, is 17.5 m. Photograph by Brian K. Hall.

as a result of Bob's efforts. Most of Bob's current research is devoted to the study of development in extant amphibians in an effort to understand their evolutionary origins.

In addition to being a leading researcher, Bob also has an impressive record as a teacher and supervisor. Most of the postdoctoral fellows, graduate students, and undergraduate students who have passed through his lab have gone on to successful careers in paleontology, biology, and anthropology (Sues et al., 2003). Bob's door has always been open to students, and he treats them as colleagues, often passionately discussing the latest developments in the field. His magnum opus, *Vertebrate Paleontology and Evolution* (Carroll, 1988), remains the text of choice for courses in vertebrate paleontology around the world.

For his research, for his pioneering spirit, and for his dedication to the training of new professionals, we feel it is right and fitting that a volume on the subject of major transitions in vertebrate evolution be dedicated to our colleague, mentor, and friend, Robert Lynn Carroll.

"It's a long road from *Amphioxus,* but we come from there!"

References Cited

Carroll, R. L. 1988. Vertebrate Paleontology and Evolution. W. H. Freeman and Company, New York, 698 pp.

Carroll, R. L. 1997. Patterns and Processes of Vertebrate Evolution. Cambridge University Press, Cambridge and New York, 448 pp.

Sues, H.-D., A. M. Murray, and J. S. Anderson. 2003. Robert Lynn Carroll—an appreciation. Canadian Journal of Earth Sciences 40:469–472.

Simpson, G. G. 1944. Tempo and Mode in Evolution. Columbia University Press, New York, 237 pp.

CONTENTS

Acknowledgments

The production of a volume such as this would not be possible without the assistance of many individuals and institutions. We would first like to thank the contributors to this project; they are the leading researchers in their respective fields, and we appreciate their dedication and efforts in helping to create this wide-ranging overview of current thinking on major transitions in vertebrate evolution. We owe a debt of gratitude to the many colleagues who assisted us by reviewing individual chapters: Larry Barnes, Chris Bell, Chris Brochu, Chi-Hua Chiu-Groff, Jenny Clack, Peter Forey, Jason Head, Philippe Janvier, Zofia Kielan-Jaworowska, John Long, Zhe-Xi Luo, Andrew Milner, Sean Modesto, Kevin Padian, John Scanlon, Moya Meredith Smith, Hans Thewissen, Greg Wilson, Mark Wilson, and four colleagues who requested anonymity. We thank Robert Sloan at Indiana University Press for his patience with our questions and occasional deadline extensions. For their support during the production of this volume, JSA thanks the College of Veterinary Medicine at Western University of Health Sciences (Pomona, California) and the Faculty of Veterinary Medicine at the University of Calgary (Calgary, Alberta), and H-DS is indebted to the National Museum of Natural History (Washington, D.C.). Final stages of the preparation of this volume were supported in part by a Natural Sciences and Engineering Research Council of Canada Discovery Grant (to JSA). This volume is a joint and equal effort on the part of the two editors.

Major Transitions in Vertebrate Evolution

Introduction: Studying Evolutionary Transitions among Vertebrates

HANS-DIETER SUES AND
JASON S. ANDERSON

Historical Background

Comparative biology has its foundation in the work of the Greek philosopher Aristotle (384–322 BC; Russell, 1916). Aristotle first postulated gradual transitions from inanimate to animate objects, from plants to animals, and from quadrupedal animals to man. He applied the principle of continuity to natural diversity (Lovejoy, 1936). Although Aristotle did not (as is frequently claimed) hold that all organisms form a single ascending sequence of forms, he observed that the differences between organisms tended to show many gradations rather than clear-cut distinctions. During the eighteenth century, the influential Genevan natural philosopher Charles Bonnet adopted Aristotle's principle of continuity as a foundation for his natural philosophy (Rieppel, 1988). In an early work, Bonnet (1745) envisioned an uninterrupted succession that began with the four Aristotelian elements and proceeded through minerals to fossils to plants and animals to culminate in man. He was concerned about an apparent gap that existed between minerals and plants, but he ascribed it (as well as other gaps) to imperfections of human cognition and perception. Later, he wrote: "Nature never proceeds by saltations. Everything has its sufficient reason, or its immediate cause. The actual state of an organized body is the consequence of the product of the antecedent state, or, to put it more correctly, the present state of an organized being is determined by the antecedent state" (Bonnet, 1768: vol. I, 4–5, translated from the original French text). Bonnet classified nature along a continuous linear hierarchy, his *échelle des êtres* or the Great Chain of Being, which ascended to increasingly more complex levels of organization. This notion of the Great Chain of Being, with a gradual series of forms, influenced all subsequent work.

Bonnet (1764: vol. I, 242, translated from the original French text) also argued that "the same general design embraces all parts of the earthly creation. A globule of light, a molecule of earth, a grain of salt, a mold, a polyp, a shellfish, a bird, a quadruped, man are nothing but different expressions of this design that represents all possible modifications of matter on our globe." This statement implies the existence of an underlying common structural plan (what German-speaking morphologists subsequently referred to as a *Bauplan*) that can be abstracted from the considerable variety of organic forms on the basis of the principle of continuity (Rieppel, 1988). During the first half of the nineteenth century, the study of comparative anatomy became concerned with the identification of these underlying plans or archetypes, which were believed to represent the essence of each group of organisms (Owen, 1848). Drawing on the work of earlier authors, Richard Owen (1843) formalized the concept of homology to denote the correspondence of organic structures in terms of relative position—"under every variety of form and function" (Owen, 1843:379). For example, the forelimb of a crocodile is regarded as homologous to the flipper of a marine turtle and the wing of a bird.

The publication of Charles Darwin's *On the Origin of Species by Means of Natural Selection* in 1859 was a pivotal event in the history of biology. Darwin introduced the notion of genealogical kinship between taxa as the result of descent with modification from a common ancestor. In this context, the similarities and differences between groups of organisms reflect relationships of common ancestry among them. This led to a novel research agenda that sought to establish the evolutionary connections between and within the major groups of organisms. Unraveling homologies became critical to tracing the course of evolution, and common ancestors came to replace Owen's archetypes. Variation in homologous features could be interpreted as the result of transformation of an ancestral (or primitive) character state into derived (or advanced) character states that could be used to characterize taxa. This new agenda established the research tradition of evolutionary morphology and launched the quest for ancestor-descendant relationships.

Thomas Henry Huxley noted the wide morphological gaps that separate the major groups of present-day vertebrates. For example, although they share numerous anatomical features and other characters in common, indicating genealogical relationship, extant birds and reptiles can be easily distinguished from each other. Although Huxley (1868) observed that the nonflying ratites appear in some respects to be more similar to reptiles than to other birds, this resemblance is insufficient to fill the structural gap between birds and reptiles. Huxley viewed Darwin's theory as having predictive power, permitting inference of the forms that undiscovered intermediate stages (commonly referred to as "missing links") might take. He also came to realize that the fossil record would be critical to the recovery of such intermediates. When Darwin published his "long argument," the fossil record of vertebrates was still inadequately known. Soon afterward, however, the discovery of the small theropod dinosaur *Compsognathus* from the Upper Jurassic Solnhofen Formation of Bavaria (Germany) provided Huxley (1868) with a link between reptiles and birds. Although *Compsognathus* is clearly reptilian and apparently lacked feathers, it closely resembles birds in the structure of its hind limb, especially the distal end of the tibia and tarsus. Huxley (1868, 1870) examined the anatomical similarities between the few dinosaurs known at that time and extant birds in detail. He noted the close resemblance between certain dinosaurs and birds in the structure of the pelvic girdle, hind limb, and feet, and he considered dinosaurs the ancestors of birds. Interestingly, unlike later authors, Huxley paid much less attention to the earliest-known bird, *Archaeopteryx*, from the same provenance as *Compsognathus*. Although he noted its intermediate position between reptiles and birds, he considered it anatomically closer to reptiles than to extant birds. After Huxley's hypothesis of the dinosaurian origin of birds had fallen out of favor for several decades during the twentieth century, Ostrom (1973, 1976) revived and supported it by an impressive suite of skeletal features. Indeed, as a result of numerous discoveries in recent years of feathered nonavian theropods and primitive birds from Jurassic and especially Cretaceous strata, mainly in China, the transition from nonavian theropod dinosaurs to birds now represents one of the best-documented examples of a major evolutionary transition (Chiappe and Dyke, Chapter 8).

The tradition of evolutionary morphology came to dominate investigations of the pattern of vertebrate evolution during the second half of the nineteenth and much of the twentieth century. In the context of Darwinian thinking, the process of evolutionary transition requires explanation of the observed changes in terms of their adaptive significance. Scenarios for this purpose were constructed on the basis of diverse functional-morphological and ecological arguments.

For a long time, the fossil record of vertebrates remained too poorly known to be of much use in reconstructing even the major evolutionary transitions in detail. In his classic works on evolutionary theory, George Gaylord Simpson (1944, 1953) emphasized the rarity of intermediate forms and the apparent speed with which structural and functional changes took place during these transitions. In classical Darwinian tradition, Simpson assumed that all change was continuous, without breaks or saltation, and proceeded at rates that, given the inherently incomplete and biased nature of the fossil record, would make it virtually impossible to identify a series of intermediates. Indeed, it is never possible to establish the exact interval of time during which a transition occurs. The transition is part of a longer continuum, and thus determination of a point of transition must necessarily be arbitrary. It is worth noting that some authors have read breaks in the fossil record more literally, interpreting morphological discontinuities between taxa from successive strata as reflecting periods of rapid evolution (Eldredge and Gould, 1972; Sheldon, 1996). However, even when evolution does occur in bursts, natural selection has been shown to proceed at sufficiently rapid rates that the assumption of continuity of form over time need not be rejected (see Carroll, 1997).

The advent of phylogenetic systematics, based on the pioneering work of Willi Hennig (1950, 1966), replaced the traditional quest for ancestor-descendant relationships with the search for shared derived character states (synapomorphies) to establish the monophyly of groups of taxa. The equivalence of homology with synapomorphy was already implicit in Hennig's writings, but it was made explicit by subsequent authors (Patterson, 1982; de Pinna, 1991; Nelson, 1994). Only the presence of homologies in different taxa is meaningful for establishing relationships between them. Recognition of two features as homologous requires analysis not only of those characters but also of other characters unrelated to those under consideration (Patterson, 1982; Lauder, 1994).

One can establish whether a particular character or character state is present or absent in a set of taxa. This "taxic" approach to evolutionary studies (Eldredge, 1979) is concerned with the discovery of monophyletic groups (clades) and is thus testable (Patterson, 1982; de Pinna, 1991). The statement that a character was absent and then came into being implies that evolutionary change has taken place. Studying such change represents the "transformational" approach (Eldredge, 1979), which is not concerned with grouping and thus, unlike the taxic approach, is not testable. Traditional scenarios of evolutionary transformation relied primarily on inferred series of structural and functional changes, literal reading of the order of appearance of taxa and features in the fossil record, and the fundamental assumption that evolutionary change typically proceeds from simple to complex. Even researchers who had already abandoned the traditional notion of ancestor-descendant relationships between taxa often retained this notion for the relationship of characters—for example, stating that bird wings "evolved" from reptilian arms (Nelson, 1994).

Lauder (1981) proposed a new approach to structural analysis in evolutionary morphology. Its objective is the analysis of generalized historical pathways or tracks of change in form and function. It starts by generating a testable phylogenetic hypothesis with nested sets of structural features, which are used to build a cladogram. This cladogram provides a historical framework for tracking and interpreting changes in form. The second step is the identification of structural, functional, and developmental interactions. These interactions result in an observable hierarchy of relationships between structure, function, and development in organisms (Bock and Wahlert, 1965; Lauder, 1981, 1994). This hierarchy is reflected by constraints on changes that can occur in developmental, structural, and functional pathways. Following Lauder (1981), the network of structural and functional interactions must be determined for at least three taxa and then examined in a phylogenetic context. The cladogram with its nested sets of similarities is used to identify the sequence of appearance of homologous and convergent character states among the terminal taxa and examine the consequences of the patterns of genealogical relationship for the structural and functional changes in a given clade. The phylogenetic hypothesis also permits inferences concerning the historical sequence of structural and functional transformation. Thus, an evolutionary scenario is an inference that depends on the distribution of character-state changes on a cladogram (Eldredge and Cracraft, 1980) rather than a construct from morphological sequences read directly from the fossil record in a phylogeny-free context. In Lauder's approach, the transformational analysis stands and falls with the underlying testable phylogenetic analysis of the taxa under consideration.

By using homologies, extinct taxa can be placed on the cladogram along with extant taxa. Fossils provide three critical elements for the study of evolutionary change: they provide the datum of first appearance of characters, the sequence in which characters appeared, and intermediate character states that are not observed in extant taxa. For example, Reichert (1837) first proposed the homology of the incus and malleus of the mammalian middle ear to the quadrate and articular in nonmammalian tetrapods, but his argument was based only on embryological observations, and it was difficult to envisage intermediate stages of this transformation in an evolutionary context. Subsequently, numerous discoveries of late Paleozoic and Mesozoic nonmammalian synapsids demonstrated that the postdentary bones were already involved in sound reception in basal taxa, became increasingly reduced in size, and eventually became separated from the lower jaw (Allin, 1975). Following Lauder's approach, the successive appearance of these changes can be mapped onto a phylogeny of Synapsida, which is well corroborated by a substantial set of characters from the cranial and postcranial skeleton (Kemp, 1982; Gauthier et al., 1988). Although some evolutionary biologists questioned the importance of fossils for phylogenetic reconstruction during the 1980s, the analysis of tetrapod interrelationships by Gauthier et al. (1988) again underscored their critical importance. Recent decades have witnessed dramatic improvements in the extent and quality of the fossil record and the development of new methods for the study of fossils, resulting in a wealth of available new information.

Another area of investigation that, after a long hiatus, has once again assumed a critical role in the investigation of evolutionary change is the study of development. Although many nineteenth-century morphologists (e.g., Ernst Haeckel) already drew heavily on developmental changes to understand patterns of evolutionary transformation (Russell, 1916), details of the underlying morphogenetic mechanisms remained obscure until recently. As a result, most evolutionary biologists, notably the founders of the Modern

Synthesis, paid little attention to development. Starting in the late 1970s, increasing attention was paid to development as a mechanism to explain large-scale evolutionary changes with the recognition that changes in the timing of developmental events can lead to changes in adult features (Gould, 1977; Alberch et al., 1979); however, the exact nature of the causes underlying such changes remained unknown. In recent years, however, enormous advances have been made in the identification of specific loci and patterns of gene expression, which now make it possible to determine the mechanisms by which structural features are formed. Larsson (Chapter 4) discusses avian feathers as a case study for the development of a complex evolutionary novelty. Prum (1999) compared the early ontogenetic development of feathers in extant birds to that of their hypothesized evolutionary history. He divided feather morphogenesis into five discrete stages, each of which represents an elaboration upon its predecessor. Prum ordered these stages so that each developmental stage in the series is conditionally dependent on its preceding stage, resulting in an ordered developmental series for examining evolutionary change. He then associated specific stages of feather development with evolutionary stages of feathers documented from the fossil record. Subsequently, the sequence of morphogenetic stages has been further corroborated by expression stages of the genes *Bmp2* and *Shh* that apparently support the early stages of feather development (Harris et al., 2002).

Overview

This volume presents a series of reviews of recent advances and unresolved issues in our understanding of major evolutionary changes among vertebrates. It is not intended to be a comprehensive overview of this fascinating subject—such an objective is unrealistic in view of the current rate of research progress. For example, just as this volume went to press, the extraordinary *Tiktaalik roseae*, a fish-like sarcopterygian from the Upper Devonian of Nunavut (Canada) with pectoral appendages that are morphologically intermediate between fish fins and tetrapod limbs, was described (Daeschler et al., 2006; Shubin et al., 2006). Most chapters in this volume were drawn from a symposium organized by the editors in honor of Robert Lynn Carroll (McGill University) and held as part of the Sixty-third Annual Meeting of the Society of Vertebrate Paleontology in St. Paul, Minnesota, in 2003. Several symposium participants were unable to submit manuscripts for this volume, but we were fortunate to recruit other leading researchers to fill in some of the gaps.

Hall and Witten (Chapter 1) begin the volume with a discussion of a key attribute of vertebrates: bone. They explain that the four basic types of skeletal tissues present in vertebrates—bone, cartilage, dentine, and enamel—are frequently considered distinct entities. Vertebrate skeletons, however, also contain a variety of other tissues that are intermediate between two types of skeletal tissue, such as chondroid bone and various kinds of dentine. The authors characterize the four principal types of skeletal tissues and certain intermediate tissues with emphasis on their characteristic properties. They review their phylogenetic distribution among extant and extinct vertebrate taxa as well as their modes of development. Hall and Witten conclude that these tissues reflect the early evolution of highly plastic skeletogenic cells that could modulate their behavior in response to intrinsic and extrinsic signals. Even once formed, skeletal tissues are constantly remodeled and reshaped, and sometimes replaced or even regenerated. The authors demonstrate that the origin of skeletal resorption accompanied the origin of the mineralized skeleton. They provide a series of examples of cellular responsiveness, underlying plasticity, and transition by considering the origin of intermediate tissues by two particular modes of transformation, modulation and metaplasia. Modulation, a temporary change in cell behavior, structure, and/or the type of matrix product(s) produced, occurs in response to changing extrinsic and/or intrinsic conditions. Thus, Hall and Witten view modulation as the cellular equivalent of phenotypic plasticity. Metaplasia refers to the transformation of one differentiated cell type or tissue into another. Differentiation, modulation, metaplasia, and resorption are four phylogenetically ancient processes that have facilitated the development and evolution of the vertebrate skeleton.

Janvier (Chapter 2) reviews the morphological gaps between cephalochordates and vertebrates and between hagfishes and lampreys (collectively referred to as cyclostomes) and jawed vertebrates (gnathostomes), respectively, in the light of paleontological information. The myllokunmingiids from the Early Cambrian of China probably already had a variety of structures derived from neural crest and epidermal placodes, but they exhibit character combinations that suggest they may be stem vertebrates. Along with the coeval yunnanozoans, myllokunmingiids also help bridge the morphological gap between cephalochordates and vertebrates. Although the fossil record of both hagfishes and lampreys is now known to extend back to the Carboniferous, their relationships to extinct groups of jawless vertebrates remain unresolved. Euphaneropids and possibly anaspids display a number of lamprey-like features, but these features may characterize a clade comprising lampreys and gnathostomes. Most authors now regard osteostracans as the closest relatives to gnathostomes, but osteostracans share some unique similarities with lampreys (e.g., structure of the nasohypophysial complex) that are currently interpreted as homoplastic and therefore are ignored in narratives about early vertebrate evolution. Janvier reviews the characters that support current hypotheses of vertebrate interrelationships. Although there are incongruences in the distribution of certain character states, he argues that "ostracoderms" provide a means for establishing the sequence in which characters diagnostic for gnathostomes appeared. Finally, Janvier revisits the long-controversial issue of the origin of jaws in the light of a substantial body of recent developmental data.

Wilson et al. (Chapter 3) review the origin of gnathostomes, which is one of the most important events in the evolution of verte-brates but is still one of the most poorly understood. Among the many features shared by jawed vertebrates with possible precursors among jawless vertebrates ("agnathans") is the presence of paired fins. Among the latter, fin-like structures have been recorded in numerous taxa of anaspids, osteostracans, and thelodonts. Some forms have paired precursors of pectoral fins, whereas others have precursors of pelvic fins. At least one thelodont taxon appears to have both types of fins, sharing this character state with gnathostomes. Some "agnathan" groups likely lost either pectoral fins (furcacaudiforms) or pelvic fins (osteostracans and possibly some thelodonts) that were present in their ancestors. Pectoral and pelvic fins or their precursors differed fundamentally in both position and structure even before the origin of jaws, and within most of the major clades of basal gnathostomes.

Larsson (Chapter 4) argues that the widely used terms *evolutionary novelty* and *modularity* lack properties to be properly applied to the dynamic process of evolution. Therefore, he proposes a novel comparative system to combine definitions of these terms with data from developmental and evolutionary biology. Larsson introduces the concept of modules of developmental evolution (MODEs) to describe instances where a sequence of evolutionary transformations can be causally explained by a sequence of developmental transformations with equivalent structural information (mechanism and pattern). As an example, he analyzes the much-discussed transition from the fins of fish-like sarcopterygians to the tetrapod limb in detail. Larsson reviews the skeletal structure of the fins in a series of known extinct and extant taxa of fish-like sarcopterygians in conjunction with new developmental data for the paired appendages in fishes and tetrapods. He develops a comparative framework to place the evolution of an autopodial field at the dipnoan-tetrapodomorph divergence, the evolution of digits within the autopodial field at the node Tetrapodomorpha, and the evolution of digit identity at the node Tetrapoda. Finally, Larsson proposes a new definition for the autopodium and digits that suggests the metapterygial axis contributes to a single digital anlage within the autopodium. Larsson regards all other digits either as repetitions of the same axis or as representing a novel axis within the autopodium.

Anderson (Chapter 5) argues that the presence of derived developmental states should be coded and included in phylogenetic analyses. Some recent authors have applied developmental data to the classical issue of the monophyly versus paraphyly of extant amphibians (lissamphibians). The pattern of vertebral development was coded and included in the analysis whereas the pattern of cranial ossification was discussed but not included for a number of reasons. Metamorphosis was incorporated as a series of discrete characters, because it is through the cumulative presence of these correlates that metamorphosis can be inferred in extinct taxa. Anderson's phylogenetic analysis supports a separate evolutionary origin for caecilians within lepospondyls. Specifically, it places caecilians as the sister-group to brachystelechid microsaurs and *Rhynchonkos*. Frogs and salamanders, on the other hand, form a clade (Batrachia). In turn, these two groups together are found to be the sister-taxon of the Early Permian *Doleserpeton*, rather than of amphibamid and branchiosaurid dissorophoids, respectively, as previously suggested by other authors. Anderson's phylogenetic hypothesis implies that developmental changes were already important among the late Paleozoic precursors of frogs and salamanders for facilitating exploitation of various ecological opportunities.

Reisz (Chapter 6) presents results of new observations on the cranial structure in the Permo-Carboniferous diadectomorph tetrapods *Limnoscelis* and *Tseajaia* and examines them in the context of cotylosaurian history and relationships. Previous phylogenetic analyses of tetrapods have established the monophyly of Cotylosauria, placing Diadectomorpha as the sister-taxon of Amniota, but the pattern of relationships within diadectomorphs has not previously been examined in detail. Phylogenetic analysis supports the hypotheses that *Limnoscelis*, *Tseajaia*, and Diadectidae form a clade, in which diadectids and *Tseajaia* share a more recent common ancestor than either does with *Limnoscelis*. The pattern of diadectomorph relationships and the sister-group relationship between Diadectomorpha and Amniota result in long "ghost lineages" (undocumented gaps in the fossil record predicted by phylogeny) between the minimum divergence date for these taxa and the earliest-known occurrence of diadectomorphs in the latest Pennsylvanian (Late Carboniferous). Reevaluation of the known fossil record implicated in the anamniote-amniote transition, the known Permo-Carboniferous diadectomorphs, and Carboniferous amniotes indicates that the origin of amniotes remains yet to be resolved.

Caldwell (Chapter 7) examines the role that fossils have played in shaping current thinking about the origin of snakes. The modern controversy over snake origins began when Cope (1869) suggested there was a close relationship between snakes and mosasaurs. This hypothesis spawned a vigorous debate in the late nineteenth and early twentieth centuries, centering on fossil lizards. From the mid-1920s to the late 1970s the focus of the debate shifted from fossils to evidence obtained from extant lizards and snakes, but the description of early Late Cretaceous *Pachyrhachis problematicus* reignited a fossil-based discussion that continues to the present. Caldwell follows the progress of hypotheses generated by fossils though these two phases of investigation, and he concludes by reviewing problems of establishing the homologies of certain cranial bones that have played key roles in the current debate: the jugal, postorbital, postfrontal, and ectopterygoid.

Chiappe and Dyke (Chapter 8) review the diverse lines of evidence in support of the hypothesis that birds evolved from maniraptoran theropods. As a direct result of an enormous increase in the number of discoveries of mostly Cretaceous birds and nonavian theropods in recent decades, it can now be established that many features long considered diagnostic for birds were in fact already present among nonavian maniraptoran theropods. A large suite of skeletal features now unambiguously places birds among maniraptorans. Chiappe and Dyke note that the debate has now shifted toward the search for the identity of the immediate sister-group of birds. A substantial gap remains in the fossil record between the Late Jurassic *Archaeopteryx* and the earliest-known Cretaceous avian taxa. Many new discoveries in the last few years have revealed an unexpected Early Cretaceous diversity of long-tailed birds, which display an intriguing mosaic of features. The earliest short-tailed birds appear rather suddenly in the fossil record, and to date, no morphological intermediates are known between these forms and their more primitive, long-tail avian sister-taxa. Recent discoveries are also improving our understanding of the earliest evolution of the body plan of modern birds, and have begun to provide evidence concerning the pre-Cenozoic radiation of extant avian lineages.

Luo (Chapter 9) characterizes the evolutionary history of Mesozoic mammals as successive diversifications of relatively short-lived clades in a series of radiations. Most constituent clades of a major group are clustered into several episodes of diversification. Constituent clades clustered in a previous episode of diversification tend to be dead-end "evolutionary experiments," frequently without any direct relationship to the emergent family- or order-level taxa in the succeeding mammalian communities. Most of the higher-level taxa of Mesozoic mammals left no Cenozoic descendants. Taxonomic turnovers between mammalian assemblages of the different intervals of geological time and within major groups occur by successive clusters of emergent clades or their constituent groupings. Less common are major groups with a long history of low diversity before significant diversification of their constituent taxa. Current stratigraphic and phylogenetic data for early mammals reveal few examples of rapid divergence and diversification of the constituent taxa of a major group soon after the appearance of that group. Currently available dates for Cretaceous metatherians and eutherians are not in conflict with molecular estimates, suggesting that certain extant mammalian superorders may extend back to the Cretaceous Period. However, they do not fully support recent hypotheses based on molecular data that various long-lived lineages of extant mammals already diversified during the Cretaceous.

Uhen (Chapter 10) notes that the rich and growing fossil record of cetaceans demonstrates that the shared derived characteristics of extant whales, dolphins, and porpoises arose in a mosaic fashion over the course of the first 10 million years of the evolutionary history of this highly distinctive group of marine mammals. Many features traditionally considered diagnostic for extant cetaceans actually appear at many different points during the evolutionary history of this group. The fact that these features were not acquired all at once relates at least in part to the probably gradual adoption of the habit of living in the water. Uhen demonstrates that the earliest documented cetacean synapomorphies are involved with feeding and sensory perception. Features that are more widely shared among marine vertebrates, such as a streamlined body and flipper-shaped forelimbs, are adaptations for underwater locomotion. Uhen notes that although there are many ways for sensory perception of the aquatic environment, feeding in the water, and reproduction in the water, there are few options to evolve a body plan to minimize drag and create efficient control surfaces for underwater locomotion.

References Cited

Alberch, P., S. J. Gould, G. F. Oster, and D. B. Wake. 1979. Size and shape in ontogeny and phylogeny. Paleobiology 5:296–317.

Allin, E. F. 1975. Evolution of the mammalian middle ear. Journal of Morphology 147:403–438.

Bock, W. J., and G. von Wahlert. 1965. Adaptation and the form-function complex. Evolution 19:269–299.

Bonnet, C. 1745. Traité d'Insectologie; ou Observations sur les Pucerons. Two volumes. Durand Librairie, Paris, 228 + 232 pp.

Bonnet, C. 1764. Contemplation de la Nature. Two volumes. Marc-Michel Rey, Amsterdam, 298 + 260 pp.

Bonnet, C. 1768. Considérations sur les Corps Organisés, où l'on traité de leur Origine, de leur Développement, de leur Reproduction, etc. Second edition, two volumes. Marc-Michel Rey, Amsterdam, 234 + 280 pp.

Carroll, R. L. 1997. Patterns and Processes of Vertebrate Evolution. Cambridge University Press, Cambridge, 448 pp.

Cope, E. D. 1869. On the reptilian orders Pythonomorpha and Streptosauria. Proceedings of the Boston Society of Natural History 12:250–261.

Daeschler, E. B., N. H. Shubin, and F. A. Jenkins Jr. 2006. A Devonian tetrapod-like fish and the evolution of the tetrapod body plan. Nature 440:757–763.

Darwin, C. 1859. On the Origin of Species by Means of Natural Selection, or the Preservation of Favoured Races in the Struggle for Life. John Murray, London, 513 pp.

de Pinna, M. 1991. Concepts and tests of homology in the cladistic paradigm. Cladistics 7:367–394.

Eldredge, N. 1979. Alternative approaches to evolutionary theory. Bulletin of Carnegie Museum of Natural History 13:7–19.

Eldredge, N., and J. Cracraft. 1980. Phylogenetic Patterns and the Evolutionary Process. Method and Theory in Comparative Biology. Columbia University Press, New York, 349 pp.

Eldredge, N., and S. J. Gould. 1972. Punctuated equilibria: an alternative to phyletic gradualism; pp. 82–115 in T. J. M. Schopf (ed.), Models in Paleobiology. Freeman, Cooper & Company, San Francisco.

Gauthier, J. A., A. G. Kluge, and T. Rowe. 1988. Amniote phylogeny and the importance of fossils. Cladistics 4:105–209.

Gould, S. J. 1977. Ontogeny and Phylogeny. Belknap Press of Harvard University Press, Cambridge, 501 pp.

Harris, M. P., J. F. Fallon, and R. O. Prum. 2002. *Shh-Bmp2* signaling module and the evolutionary origin and diversification of feathers. Journal of Experimental Zoology, Part B: Molecular and Developmental Evolution 294:160–172.

Hennig, W. 1950. Grundzüge einer Theorie der phylogenetischen Systematik. Deutscher Zentralverlag, Berlin, 370 pp.

Hennig, W. 1966. Phylogenetic Systematics. (Trans. D. D. Davis and R. Zangerl.) University of Illinois Press, Urbana, 263 pp.

Huxley, T. H. 1868. On the animals which are most nearly intermediate between birds and reptiles. Annals and Magazine of Natural History (4)2:66–75.

Huxley, T. H. 1870. Further evidences of the affinity between the dinosaurian reptiles and birds. Quarterly Journal of the Geological Society of London 26:12–31.

Kemp, T. S. 1982. Mammal-like Reptiles and the Origin of Mammals. Academic Press, London and New York, 363 pp.

Lauder, G. V. 1981. Form and function: structural analysis in evolutionary morphology. Paleobiology 7:430–442.

Lauder, G. V. 1994. Homology, form, and function; pp. 152–196 in B. K. Hall (ed.), Homology: The Hierarchical Basis of Comparative Biology. Academic Press, San Diego.

Lovejoy, A. O. 1936. The Great Chain of Being. A Study of the History of an Idea. Harvard University Press, Cambridge, 382 pp.

Nelson, G. 1994. Homology and systematics; pp. 101–149 in B. K. Hall (ed.), Homology: The Hierarchical Basis of Comparative Biology. Academic Press, San Diego.

Ostrom, J. H. 1973. The ancestry of birds. Nature 242:136.

Ostrom, J. H. 1976. *Archaeopteryx* and the origin of birds. Biological Journal of the Linnean Society 8:91–182.

Owen, R. 1843. Lectures on Comparative Anatomy and Physiology of the Invertebrate Animals. Delivered at the Royal College of Surgeons, in 1843. Longman, Brown, Green, and Longmans, London, 392 pp.

Owen, R. 1848. On the Archetype and Homologies of the Vertebrate Skeleton. John Van Voorst, London, 203 pp.

Patterson, C. 1982. Morphological characters and homology; pp. 21–74 in K. A. Joysey and A. E. Friday (eds.), Problems of Phylogenetic Reconstruction. Systematics Association Special Volume 21. Academic Press, London and New York.

Prum, R. O. 1999. Development and evolutionary origin of feathers. Journal of Experimental Zoology, Part B: Molecular and Developmental Evolution 285:291–306.

Reichert, C. 1837. Ueber die Visceralbögen der Wirbelthiere im Allgemeinen und deren Metamorphosen bei den Vögeln und Säugethieren. Müllers Archiv für Anatomie, Physiologie und wissenschaftliche Medizin 1837:120–222.

Rieppel, O. C. 1988. Fundamentals of Comparative Biology. Birkhäuser Verlag, Basel and Boston, 202 pp.

Russell, E. S. 1916. Form and Function. A Contribution to the History of Animal Morphology. John Murray, London, 383 pp.

Sheldon, P. R. 1996. Plus ça change—a model for stasis and evolution in different environments. Palaeogeography, Palaeoclimatology, Palaeoecology 127:209–227.

Shubin, N. H., E. B. Daeschler, and F. A. Jenkins Jr. 2006. The pectoral fin of *Tiktaalik roseae* and the origin of the tetrapod limb. Nature 440:764–771.

Simpson, G. G. 1944. Tempo and Mode in Evolution. Columbia University Press, New York, 237 pp.

Simpson, G. G. 1953. The Major Features of Evolution. Columbia University Press, New York, 434 pp.

1. Plasticity of and Transitions between Skeletal Tissues in Vertebrate Evolution and Development

BRIAN K. HALL AND
P. ECKHARD WITTEN

ABSTRACT

Vertebrate skeletons are composed of four long-recognized skeletal tissues—bone, cartilage, dentine, and enamel—that are often viewed as separate entities. Nevertheless, vertebrate skeletons also contain tissues that are intermediate between two skeletal tissues, examples being chondroid, chondroid bone, cementum, and various types of dentine. We describe the four skeletal tissues and selected intermediate tissues to emphasize their distinctive properties, distribution in extant and extinct taxa, and modes of development, and we conclude that they reflect the early evolution of highly plastic skeletogenic cells that could modulate their behavior in response to intrinsic and environmental signals. Once formed, skeletal tissues are not static but rather dynamic, subject to constant reshaping and remodeling and even replacement or regeneration (teeth and antlers). We document the evolutionary origin of skeletal resorption as concomitant with the origin of the mineralized skeleton. Two themes that emerge are (1) the plasticity of skeletogenic cells and (2) tissues and skeletal tissues as part of a continuum with adjacent cells in the continuum capable of transforming to another skeletogenic tissue. We provide examples of the cellular responsiveness, underlying plasticity, and transition by considering the origin of intermediate tissues by two modes of transformation, modulation and metaplasia (transdifferentiation). Modulation, a temporary change in cell

Figure 1.1. *(opposite page)* Diagram representing the main subcategories of dental (dark-gray shaded semicircles) and skeletal tissues (light-gray shaded semicircle). Within the outer circle of the major cell types, **1** represents cementoblasts/-cytes, **2** represents intermediate stages between osteocytes and chondrocytes, and **3** represents fibroblasts/-cytes. Major categories of resorbing cells are shown in inner ring. Conventional categorization distinguishes dental and skeletal tissues and addresses only four subcategories: *enamel* and *dentine* as dental tissues; and *cartilage* and *bone* as skeletal tissues. Common mesenchymal origin of dentine, bone, and cartilage-forming cells; coincident appearance of bone, dentine, and enamel in early vertebrates; many tissue types intermediate between dentine and bone and between bone and cartilage; and potential of skeletal cells to transdifferentiate reveal inadequacy of any simplistic categorization. From developmental and evolutionary perspectives, dental and skeletal tissues represent a continuum, as indicated by overlapping semicircles. The one exception is enamel, because of its exclusively epithelial origin and complete lack of collagen.

behavior, structure, and/or the type of matrix product(s) produced, occurs in response to altered intrinsic and/or environmental conditions. Modulation is therefore the cellular equivalent of phenotypic plasticity. Metaplasia is the transformation of one differentiated cell type or tissue into another. Differentiation, modulation, metaplasia, and resorption are four ancient processes that facilitated the evolution and the development of the vertebrate skeleton. Skeletal tissues might best be understood by viewing them as dynamic rather than static.

Introduction

In this chapter, we discuss skeletal tissues within the context of several themes, which we organize around several questions. What are skeletal tissues, and how many types of skeletal tissue exist now or have ever existed? Can we recognize transitions between such well-recognized skeletal tissues as bone and cartilage? If so, on what basis do we identify such transitional or intermediate tissues? What are the developmental and evolutionary relationships between the formation and removal or remodeling of skeletal tissues? Did remodeling evolve with formation, or was it a later evolutionary event? Finally, what mechanisms allow us to explain the formation of the transitional/intermediate skeletal tissues? We draw on data and hypotheses derived from extinct (fossil) and extant vertebrates, from normal and abnormal (pathological) tissues, and from descriptive and experimental analyses, and we use genetic, molecular, cell, developmental, and morphological information to inform our topic. We conclude that the four skeletal tissues (bone, cartilage, dentine, enamel) and the intermediate tissues (chondroid, chondroid bone, cementum, various types of dentine) are reflections of the early evolution of highly plastic cells and of the mechanisms to modulate cell behavior among and between these skeletal tissues.

Four Skeletal Tissues

Four classes of mineralized tissues are found in vertebrates: bone, cartilage, dentine, and enamel (Fig. 1.1). We think of cartilage and bone as skeletal tissues and of enamel and dentine as dental tissues, but enamel and dentine arose evolutionarily together with bone as skeletal tissues in the dermal skeleton (exoskeleton) of early vertebrates. Scales and teeth of sharks are examples of dermal skeletal elements that are still composed of the three ancient components—enamel, dentine, and bone (Plate 1). Cartilage, on the other hand, provided the basis for the second vertebrate skeletal system, the endoskeleton (Smith and Hall, 1990; Hall, 1998a,b).

Whereas bone, dentine, and enamel are new characters that appeared with the evolution of the vertebrates, the functional design solution of combining a mineralized exoskeleton with endoskeletal elements made from cartilage, cartilage-like, or chondroid tissues evolved in invertebrates. These tissues have a diverse distribution within the animal kingdom, although cellular and matrix properties

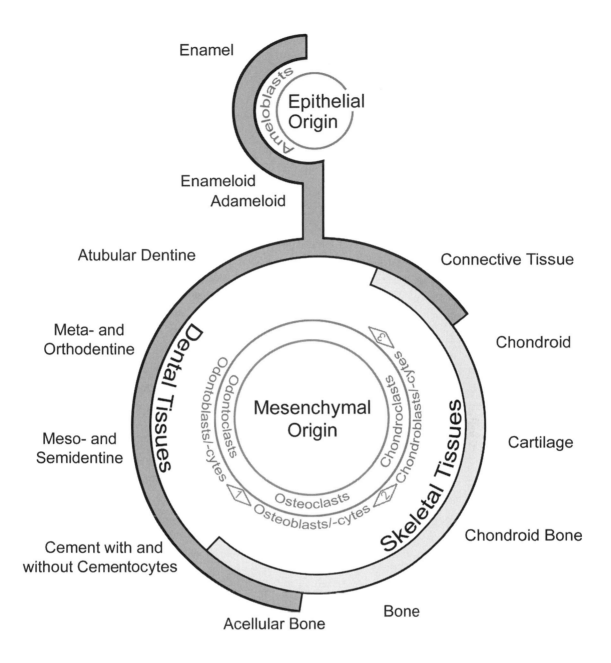

of these invertebrate tissues differ from those of vertebrate cartilage (Cole and Hall, 2004a,b). Also, and in contrast to the calcium phosphate–based vertebrate skeleton, invertebrate exoskeletons are usually composed of calcium carbonate. Nevertheless, some invertebrate skeletal tissues have surprisingly bone-like features. Examples include chondrocytes interconnected by cell processes in cephalopod cartilages (Cole and Hall, 2004a,b), and the calcium phosphate layer in the shells of brachiopods (Rodland et al., 2003). However, neither bone nor mineralized cartilage have been found in invertebrates. See Patterson (1977), Hanken and Hall (1983), Smith and Hall (1990, 1993), Janvier (1996), Donoghue et al. (2000), Franz-Odendaal et al. (2003), Vickaryous et al. (2003), Witten et al. (2004a), and Hall (2005a,b) for overviews of the evolution of skeletal issues, for the delineation of dermal (exo-) and endoskeletons, and for models of dermal-skeletal and endoskeletal evolution.

The four classes of vertebrate skeletal tissues may be further subdivided in ways that reflect the interests of individual skeletal biologists and the scope of their fields. Embryologists use developmental process (intramembranous versus endochondral ossification), anatomists use structure (woven versus lamellar bone; spongy versus compact bone), and pathologists use deviation from the norm (regular versus metaplastic osteogenesis).

A fifth category encompasses those tissues that are recognized as *intermediate* between two of the four mineralized tissues (Hall, 2005a). The best-recognized intermediate tissues are *chondroid* and *chondroid bone,* which are intermediate between connective tissue and cartilage and cartilage and bone, respectively; *osteodentine* and *cementum,* which share features of bone and dentine; and *enameloid,* a range of tissues that share features of enamel and dentine. These intermediate tissues and their modes of formation are discussed later in the chapter. As background for that discussion, we provide an overview of the four classes of vertebrate mineralized tissues (Fig. 1.1). For reviews of the mineralized skeletal tissues, see the chapters in Bourne (1971–1976), Carter (1990), and Bilezikian et al. (1996), as well as the nine-volume series, *Bone,* edited by Hall (1990–1994).

Bone

Bone is a vascularized, supporting skeletal tissue, which also can form ectopically outside the skeleton. Bone consists of cells and a mineralized extracellular matrix that is permeated by canals (*canaliculi*) that contain osteocyte processes (Plate 2). Bone is deposited by bone-forming cells (*osteoblasts*), which, when they cease dividing, transform into *osteocytes.* The first bone matrix deposited is unmineralized and is known as *osteoid* (Plate 3). Subsequently, osteoid is impregnated with hydroxyapatite to form the mineralized tissue we know as bone, a tissue that is modeled, remodeled, and/or removed by mono- or multinucleated osteoclasts (and sometimes by osteocytes). Type I collagen, a product of the *Col1a1* and *Col2a1* genes, is the major extracellular matrix component. The three major noncollagenous proteins of bone matrix are: (1) *osteocalcin* (bone γ-carboxyglutamic acid, Gla, bone Gla protein, BGP), a 5,800 MW, vitamin K–dependent, γ-carboxyglutamic acid-containing, Ca^{2+}-binding protein that recruits osteoclasts or osteoclast precursors to bone for resorption; (2) *osteopontin,* a 66-kDa glycosylated phosphoprotein that enhances survival and migration of osteogenic cells and inhibits mineralization; and (3) *osteonectin* (or SPARC; *s*ecreted *p*rotein, *a*cidic, *c*ysteine-*r*ich), a 32,000 MW extracellular matrix protein that links collagen to hydroxyapatite, serves as a nucleus for mineralization, and regulates the formation and growth of hydroxy-apatite crystals (Jundt et al., 1987; Nomura et al., 1988; Owen et al., 1991; Ayad et al., 1994; Yagami et al., 1999; Hall, 2005a).

As an aerobic tissue with high oxygen consumption, bone as an organ functions to support the body, protect major organs such as the brain and spinal cord, and is a site of attachment for ligaments and muscles. Bone is a storehouse for calcium and phosphorous, is a major site for the metabolic regulation of mineral homeostasis in tetrapods, and houses the hematopoietic tissues of adult tetrapods. Bone is found only in vertebrates; extant jawless vertebrates (lampreys), craniates (hagfishes), and all invertebrates lack bone.

Cartilage

Cartilage is an avascular supporting and articular skeletal tissue. It also functions as the primary endoskeletal support in vertebrate embryos (Plates 3 and 4). Cartilage is deposited by and is composed of *chondroblasts* and *chondrocytes* (Plate 4) separated by an extracellular matrix, which may or may not mineralize depending on cartilage type, age, or taxon (Plates 5 and 6). Like bone, cartilage can arise ectopically outside the skeleton in connective tissue, muscle, and the heart. Unlike other skeletal tissue cells, chondrocytes lack connecting cell processes (Plate 3) and are thus isolated from each other. This and the abundant extracellular matrix make cartilage an anaerobic tissue, which, consequently, has a low oxygen consumption. Most chondrocytes continue to divide throughout life (Plate 7), although in some cartilages—mammalian articular cartilages, for example—the number of dividing cells may be less than 1% of the chondrocyte population. Cartilage resists pressure and has a high fluid content.

The hydrated extracellular matrix of vertebrate cartilage is primarily composed of glycosaminoglycans, notably chondroitin sulfates and proteoglycans (Plate 8). The major collagen is type II, often referred to as cartilage-type collagen. Some types of vertebrate cartilages contain additional collagens—for example, type I in articular, fibro-, and secondary cartilages, and type X in hypertrophic cartilage. In shark cartilage, in addition to collagen type II, collagen type I is a principal extracellular matrix component (Witten and Huysseune, 2005). Lamprey and hagfish cartilages lack collagen type II but have different classes of fibrous proteins (lamprins, myxin) in place of collagen (McBurney and Wright, 1996; Wright et al., 2001). Cartilages of extant "agnathans" do not mineralize in vivo (Langille and Hall, 1993). The extracellular matrix of invertebrate cartilage is composed of glycosaminoglycans and a modified form of type I collagen. No invertebrate cartilages mineralize (Cole and Hall, 2004a,b).

Dentine

Dentine, a tubular mineralized tissue, comprises the bulk of true teeth, which are distinguished from the keratinized structures that function as teeth in anuran tadpoles. Dentine is a primary tissue of the vertebrate dermal (exo-)skeleton, and thus is both an *odontogenic* (tooth-forming) and a *skeletogenic* (skeleton-forming) tissue (Fig. 1.1). Seventy-five percent of the mass of dentine is inorganic, 20% organic (cf. enamel), and 5% liquid. Dentine is produced by *odontoblasts*. Unlike bone (with the exception of acellular bone), where osteoblasts become embedded in the matrix, odontoblasts remain outside the mineralized matrix, which is, however, penetrated by odontoblast cell processes.

Various types of dentine are recognized—orthodentine, osteodentine, and vasodentine are three—primarily on the bases of the shape and patterns of the odontoblasts process within the matrix (Ørvig, 1989; Smith and Hall, 1990; Huysseune and Sire, 1998). Osteo-(cellular) dentine is found in the basal osteichthyan, *Amia* (Moss, 1964b), and in the puffer fish, *Tetraodon* (Ørvig, 1989). As is bone and enamel, dentine is produced in two sequential phases. Synthesis and deposition of *predentine*—an organic matrix composed of glycosaminoglycans and type I collagen—comprises the first phase. Mineralization of predentine by hydroxyapatite to form dentine is the second phase.

Enamel

Enamel is the highly mineralized, hard, prismatic, avascular outer layer of vertebrate teeth and some scales. Enamel is unique in being the only *noncollagenous* mineralized vertebrate skeletal tissue (Diekwisch, 2002). The enamel proteins amelogenin, enamelin, tuftelin, MMP-20, and EMSP-1 form a highly conserved family of matrix proteins (Satchell et al., 2002). Like dentine, enamel is both odontogenic and skeletogenic; enamel covers teeth and dermal skeletal tissues. Because of its very high mineral content, enamel is much more resistant to wear than dentine; some 96% of mammalian enamel is inorganic, and only 0.5% organic.

In contrast to the other three mineralized tissues, enamel is an epithelial tissue produced by *ameloblasts* (Fig. 1.1). As with the formation of all mineralized tissues in vertebrates, enamel formation (*amelogenesis*) consists of two phases: deposition of an organic matrix and subsequent mineralization of the matrix, the latter being a process in which organic matrix is removed to facilitate the precipitation of a more inorganic substance (Nanci, 2003). Enamel is not found in invertebrates. See Glimcher et al. (1965), Miles (1967), Scott and Symons (1974), and Linde (1984) for the structure and function of enamel and dentine, and see the papers in Diekwisch (2002) for reviews of the development and mineralization of dentine and enamel.

Plasticity and Transformation of Intermediate Tissues

Intermediate tissues possess characteristics of two or more of the four skeletal tissues described above. The most recognizable intermediate tissues are *cementum, enameloid, chondroid,* and *chondroid bone,* but certain types of dentine and bone also display characters intermediate between bone and dentine (Fig. 1.1). The existence of such *permanent* intermediate tissues suggests that the four skeletal tissues represent a continuum (or perhaps continua), and not discrete, bounded categories. Some tissues are so intermediate that they cannot be classified as bone, cartilage, dentine, enamel, or as any one of the more readily recognizable intermediate tissues. Some intermediate tissues, but by no means all, are pathological.

The existence of tissues such as chondroid and chondroid bone—whether transitional or permanent—raises fundamental questions about how we identify, define, and distinguish bone and cartilage as tissues, or osteoblasts and chondrocytes as skeletogenic cells, both in development and through evolutionary history. These issues are the topics of the remainder of the chapter. Before going on to discuss chondroid, chondroid bone, cementum, dentine, and enameloid as intermediate tissues, we assess what may appear to be an oxymoron: the widespread existence of acellular bone in teleost fishes, and whether such a tissue should be regarded as intermediate. Further, because it remains an issue whether acellular or cellular bone is the primitive type of vertebrate bone (Witten et al., 2004a), it is appropriate to discuss acellular bone in this section.

Acellular Bone in Teleost Fishes

Kölliker (1859) first described acellular bone in an analysis of teleost fishes. Kölliker called it *osteoid*, which is unfortunate because Virchow (1853) had already used this designation for the initial unmineralized but mineralizable matrix deposited by osteogenic cells. The term *osteoid* should be reserved for recently deposited, unmineralized bone, whether cellular or acellular. Acellular (anosteocytic) bone is characterized by the absence of bone cells and cell processes in the bone matrix (Plate 14). We agree with Ørvig (1989) in regarding acellular bone not as an intermediate but as a derived bone type. Nevertheless, acellular bone shares characters with dentine, acellular dentine (dentine without cell processes that is deposited in first-generation teeth of teleosts; Sire and Huysseune, 2003), aspidine, and osteocyte-containing bone.

In extant vertebrates acellular bone sensu stricto is restricted to teleosts; acellular cementum is restricted to mammals and crocodiles (Huysseune and Sire, 1998; Yamamoto et al., 2004). Moss (1961a) and Enlow and Brown (1956, 1957, 1958) conducted the bulk of the classic histological studies on acellular bone in teleosts.

Acellular bone is the most widespread type of bone in teleosts, which are the most speciose group of vertebrates, whose estimated 24,000 species exceed the 8,600 avian and 3,500 mammalian species combined (Lecointre, 1994). Loss of osteocytes also occurs in mammalian bone, but as a pathological phenomenon in response to weightlessness or prolonged exposure to high pressure, as experienced by divers or those who tunnel underground (Hall, 2005a). To reflect the presence of cells on the *surfaces* of acellular bones, Weiss and Watabe (1978a,b) argued that acellular bone should be renamed *anosteocytic* bone and cellular bone renamed *osteocytic* bone. We sympathize with this view, but because of recognized usage, we continue to use the terms *cellular* and *acellular.*

In his analysis of teleost acellular bone, Moss (1961a) convincingly demonstrated that acellular bone is a character of derived teleostean groups. More basal teleosts, and all other osteichthyans, including tetrapods, have cellular (osteocyte-containing) bone. Moss noted that only one teleost has both cellular and acellular bone, the bonefish, *Albula vulpes,* but Moss did not examine every species of teleost so we cannot say that only one species has both cellular and acellular bone. The pike (*Esox*) was described by Moss as having acellular bone, but it nests phylogenetically within a group of fishes that all have cellular bone. According to Meunier (1989), the entire family Esocidae, as well as allied groups, lack osteocytes. Hughes et al. (1994a,b) suggested that osteocytes are present in the alleged acellular bone of three species of sea bream off the east coast of Australia. Clearly, more work needs to be done, especially through examination of different life history stages and of additional species.

Acellular teleost bone develops through the polarized secretion of bone matrix. According to Ekanayake and Hall (1987, 1988), at no stage or time are cells or processes incorporated in the new bone matrix, although Moss (1961b) suggested that bone cells can initially become embedded into bone matrix and disappear soon after.

From a developmental perspective, there is no transitional tissue lacking osteocytes and cell processes between acellular and cellular bone (Huyssenne, 2000). Acellular bone is present in the dermal skeleton and in the endoskeleton, can be produced by neural crest–derived and trunk mesodermally derived mesenchyme (Hall, 1978, 2005a), and its appearance is not dependent on the presence of teeth. From an evolutionary perspective, the last two characters discriminate acellular bone from other acellular skeletal tissues that evolved earlier, such as basal plates of placoid scales or teeth, bone of attachment, acellular cementum, and acellular dentine (Huysseune and Sire, 1998; Witten et al., 2004a).

Interestingly, advanced teleosts have apparently not lost the capacity to develop osteocyte-containing bone. Hyperostotic bony outgrowth (exostoses, osteoma), which is now believed to be a regular part of late skeletal development in many advanced teleostean species with acellular bone, contains osteocytes. Hyperostosis is not uncommon (Meunier and Desse, 1986; Smith-Vaniz et al., 1995). Hyperostosis in particular is interesting when it occurs in advanced teleosts that redevelop osteocytes, but other vertebrates also develop hyperostotic bone. The ribs of two genera of basal whales (archaeocetes) from the late Eocene, *Basilosaurus,* which was large (15–20 m in length), and the much smaller (5–6 m in length) *Zygorhiza,* show histological evidence of hyperostosis that was interpreted by Buffrènil et al. (1990) as an adaptation to aquatic life.

Maisey (1988) included acellular bone in his phylogenetic analysis of vertebrate skeletal tissues, whereas Smith and Hall

(1990) analyzed the phylogeny of acellular bone and dealt with the perennial question of whether cellular or acellular bone arose first in evolution. Comparative studies on the development and structure of acellular- and cellular-boned teleosts might shed new light on the evolutionary origin of acellular bone in advanced teleosts. In various species examined by Witten et al. (2004a), developing cellular bone in teleosts goes through a phase of acellularity (Plate 3) in which it resembles acellular bone, suggesting that acellular bone might have arisen in advanced teleosts through heterochrony, and that examinations that rely only on sexually mature adults may be inadequate. How would this interpretation affect our view concerning the developmental and/or evolutionary relationships of acellular bone to other acellular or partly acellular skeletal tissues such as aspidin and the various types of dentine?

Chondroid and Chondroid Bone

Chondroid and chondroid bone possess, to different degrees, features of connective tissue and cartilage (chondroid) or of cartilage and bone (chondroid bone) and resemble, to different degrees, cartilage or bone in their development (Plates 8–10). Typically, chondroid and chondroid bone have collagen types I and II in their matrices, although this is not sufficient to identify either tissue (Plates 11 and 12). Shark cartilage, as well as articular and secondary cartilages, also possesses both collagen types, as demonstrated in the articular cartilage of rat tibiae over the first 11 weeks of postnatal life. Beresford (1981) and Dhem et al. (1989) analyzed chondroid, the contribution of chondroid to skeletal growth, and the use of the term in the literature. See Wright and Moffett (1974) and Hall (2005a) for further analyses; Sasano et al. (1996), Witten and Hall (2002), and Witten and Huysseune (2005) for collagen types; and Benjamin (1986, 1988, 1989a,b, 1990), Benjamin and Ralphs (1991), and Taylor et al. (1994) for chondroid and other cartilaginous tissues in teleost fishes.

The matrix of *chondroid bone* stains like regular bone matrix in light-microscopic histological analyses but has nests of cells in lacunae resembling chondrocytes that can be surrounded by cartilage matrix (Witten and Hall, 2002). Especially in mammals, chondroid bone is often a permanent skeletal tissue (Goret-Nicaise and Dhem, 1985; Mizoguchi et al., 1992, 1993). In this sense, chondroid bone can be a stable skeletal tissue, and not necessarily a transitional one. Indeed, primary cartilage, being a transitional tissue replaced by bone, is a less permanent part of the skeleton than is chondroid bone.

The processes of skeletal formation in the development and re-growth of antlers shed considerable light on how the presence of a tissue such as chondroid and chondroid bone modifies the process of endochondral ossification (Hall, 2005a). In the antlers of white-tailed deer, *Odocoileus virginianus*, chondroid is replaced by bone in what Wislocki et al. (1947) called "chondroidal bone formation," a process intermediate between intramembranous and endochondral bone formation. In other situations, chondroid is remodeled and transformed to bone (Enlow, 1962). Examples of the formation of chondroid and chondroid bone are discussed in the last section of this chapter.

Cementum

In a recent and thorough review, Diekwisch (2001:695) defined tooth cementum as "a bone-like mineralized tissue secreted by cementoblasts on the surface of root dentine or, in some animals, crown enamel." Cementum, the supporting tissue that anchors mammalian and crocodylian teeth into their sockets, differs among species and can have features of dentine, of bone, and of calcified cartilage, as well as unique features. Cementum is deposited by *cementoblasts* onto existing dentine. As in bone and dentine, *cementoblasts* produce an organic matrix, the main constituent of which is collagen type I (Bosshardt, 2005), but cementum, dentine, and bone also share a number of important noncollagenous matrix components such as osteopontin, osteocalcin, bone sialoprotein, α_2-HS-glycoprotein, dentine matrix protein, dentine sialoprotein, and dentine phosphoprotein (McKee et al., 1996; Bosshardt, 2005).

Cementum is composed of mineralized collagen fibers—both extrinsic (Sharpey's) and intrinsic fibers—and of interfibrillar matrices. The principal cells are cementocytes, but as with acellular bone, cementum can lack cells, when it is known as *acellular cementum*. Three types of mammalian cementum are recognized: crown cementum, primary root cementum, and secondary root cementum. Cementum equivalent to mammalian root cementum is found in alligators and crocodiles, in toothed birds such as *Hesperornis* from the Late Cretaceous, and in various other groups. The cementum on the coronal surface of the molar teeth in guinea pigs, *Cavia porcellus,* has been described as cartilage (Hunt, 1959), calcified hyaline cartilage (Brunn, 1891), embryonic cartilage (Santone, 1935), and cartilaginous cement (Weidenreich, 1930). Consequently, it has been argued that in all rodents and ruminants, cementum is a form of calcified cartilage. The presence of cementum appears to be related to a thecodont type of tooth implantation, in which the tooth base sits in a deep pocket surrounded by alveolar bone. Other types of tooth attachment are characterized by mineralized tissues other than cementum that mediate the tooth-bone connection; pedicel tissue and attachment bone are two examples. See Yamamoto et al. (2004) for acellular cementum; Miles (1967), Listgarten (1973), Osborn (1981), and Sperber (2001) for rodent and ruminant cementum; and Huysseune and Sire (1998) for other tissues associated with tooth attachment.

Types of Dentine

Dentine attached to underlying dermal bone, covered or not covered by a hypermineralized cap, constitutes the basic unit of the vertebrate dermal skeleton, the *odontode* (Reif, 1978, 1980a,b; Ørvig, 1989; Smith and Hall, 1990, 1993; Huysseune and Sire,

1998; Reif and Richter, 2001; Donoghue, 2002). A substantial variety of dentine types existed in ancient vertebrates and are found in extant vertebrates. Most varieties of dentine display, to a certain degree, features that are intermediate between dentine and bone. Dentine is classified on the basis of the anatomy of structures that are formed by dentine, on whether the cell bodies are inside or outside the matrix, and on the orientation of the cell processes inside dentine matrix and of the collagen fibers (Ørvig, 1951, 1967, 1989; Huysseune and Sire, 1998; Smith and Sansom, 2000).

In his classic treatise, Ørvig (1951) distinguished and established four basic types of dentine in the dermal skeletons of early vertebrates, two cellular and two acellular (Fig. 1.1). *Mesodentine* and *semidentine* are both characterized by cell bodies and cell processes (tubular dentine) located inside the matrix. Mesodentine bears the largest similarity to bone on the basis of the orientation of cells and cell processes. Semidentine displays a clear polarization of cells and cell processes. In extant vertebrates, dentine is usually acellular. Mammals and teleosts possess orthodentine. *Vasodentine* (vascularized dentine) is found on the lingual side of the incisors of the little pocket mouse, *Perognathus lingimembris*. In *metadentine* and *orthodentine,* cell bodies are located outside the dentine matrix (acellular dentine), but cell processes penetrate the matrix inside dentine tubules (tubular dentine). See Smith et al. (1972) and Miller (1979) for tubular dentine, Moss-Salentijn and Moss (1975) for vasodentine, and Huysseune and Sire (1998) for orthodentine.

As with the hyperostotic osteocyte-containing bone that can develop in teleosts from acellular bone, pathological conditions or repair processes can permit the inclusion of odontoblasts into dentine matrix and generate a cellular dentine, usually regarded as osteodentine (Tziafas, 2004), although Ørvig (1951) established the term *osteodentine* because of the osteon-like morphology of dentine structures (denteons), irrespective of the inclusion or exclusion of cell bodies.

The early occurrence of dentine and bone in the vertebrate fossil record raises the question of whether dentine or bone appeared first in the dermal skeleton. Romer (1970) suggested that the first odontodes (dermal denticles) were composed of pure cellular bone and that dentine evolved through the retreat of osteocyte cell bodies into the pulp cavity. Donoghue et al. (2000), dissenting from Romer's view, suggest that bone was primitively acellular, that dentine is more primitive than bone, and that atubular dentine preceded tubular dentine phylogenetically. Carroll (1988) took an intermediate stand when he pointed out that the dermal skeleton of the earliest fossil vertebrates resembles both the bone and teeth of modern vertebrates.

The discussion of whether dentine or bone came first, whether the first bone or dentine was cellular or acellular and atubular, and whether intermediate tissues were present or not could be taken as suggesting that no clear distinction existed between dentine and bone in the early vertebrates. The unsettled question of whether aspidin—the skeletal tissue in the dermal skeleton of heterostracans—derives from bone or from dentine may exemplify the difficulties of distinguishing between bone and dentine; see Smith and Hall (1990:300) and Witten et al. (2004a).

Even in extant vertebrates such a distinction can be difficult. One example is the dentine of the first generation of teeth in teleosts, which is completely deprived of cell processes (Sire and Huysseune, 2003). As Huysseune (2000:315) pointed out, this acellular and atubular dentine is hardy distinguishable from acellular bone or from bone with a low level of osteocytic incorporation. Indeed, dentine and dermal bone are largely composed of the same organic and inorganic constituents (Huysseune, 2000; Nalla et al., 2003), both tissues develop from neural crest–derived mesenchyme, and the formation of both tissues requires epithelial mesenchymal interaction (Graveson et al., 1997; Hall, 1999, 2005a).

Enameloid

Enameloid is a mineralized aprismatic tissue that coats the teeth of teleosts and larval urodeles and is a component of the scales of some taxa. Enameloid is also variously known as meso-enamel, vitrodentine, and durodentine, names that reflect the presence of components typical of enamel and dentine.

Enameloid is deposited by the combined and orchestrated action of ameloblasts and odontoblasts, the bulk of the enameloid being deposited by ameloblasts but penetrated by fibers and cellular processes from the adjacent odontoblasts that are already depositing dentine. The dual contribution of these two cells types is demonstrated nicely in the Ballan wrasse, *Labrus bergylta,* the European eel, *Anguilla anguilla,* and the checkered puffer, *Sphaeroides testudineus,* in all of which odontoblasts produce and deposit collagen while ameloblasts produce and deposit the other proteins of the enameloid matrix. As with enamel, where amelogenin is removed as mineralization proceeds, collagen is removed from enameloid as mineralization occurs. See Andreucci and Blumen (1971), Baratz (1974), Shellis and Miles (1974), Moss (1968b, 1977), Diekwisch (2002), and Sasagawa (2002) for the mineralization of enameloid.

Prostak et al. (1990) showed that enameloid in the little skate, *Leucoraja erinacea,* has giant fibers unique to elasmobranchs, and that odontoblasts deposit the noncollagenous fibers of the enameloid. See Kemp (1985) for a discussion of the ultrastructure of enamel, and enamel versus enameloid in shark teeth. Although enameloid is found in sharks (known also as *adameloid* because of its long and parallel-oriented mineral crystals; Diekwisch et al., 2002), the enameloid of teleosts is a derived character because enamel is found in all basic osteichthyan groups and their terrestrial descendants, the tetrapods.

The teeth of adult urodeles are true teeth capped by enamel (Smith and Miles, 1971). The teeth of larval urodeles contain enameloid, not enamel. In two urodele amphibians, the Mexican axolotl, *Ambystoma mexicanum,* and the smooth newt, *Triturus vulgaris,* ameloblasts produce and deposit enameloid from the *larval* dental epithelium but produce and deposit enamel from the dental epithelium in adults, one of the few examples of a switch in ameloblast activity between two life history stages, reflecting differential activation of regulatory genes before and after metamorphosis. Similar regulatory shifts have been proposed as underlying changes in the production of mineralized tissues during the evolution of the vertebrates.

Skeletal Tissues in Early Vertebrates

Much information is now available on the nature of the skeletal tissues from the earliest vertebrates. An examination of the skeletal tissues found in ancestral vertebrates, even in the earliest chordates, indicates a surprisingly high degree of homology with skeletal tissues in present-day vertebrates.

From histological studies of fossilized skeletal tissues, we can conclude that these tissues in the earliest vertebrates were no less specialized than those of extant vertebrates. Rather than involving major changes in cell or tissue type, evolution worked with adaptive modulation of a highly plastic series of *scleroblasts* (Moss, 1964a) as dictated by local conditions. Furthermore, it is now possible to interpret developmental processes and the functional significance of skeletal tissues during vertebrate evolution.

As summarized in Box 1.1, fossilized skeletal tissues can be processed for light and electron microscopy as well as for chemical analyses (Mathews, 1975; Fietzek et al., 1977; Bada et al., 1999). Osteocalcin, hydroxyproline, and hydroxylysine have been reported to have been isolated (Box 1.1). Pawlicki's (1975a,b, 1977a,b, 1978, 1983, 1984, 1985) extensive studies of dinosaur bone serve as an example of the methods used and the cellular detail that can be obtained from fossil histology (paleohistology). Even cellular processes within dentine can be recognized (Schaedler et al., 1992).

Preservation of All Four Skeletal Tissues

All four major mineralized tissues—enamel, dentine, bone, and mineralized cartilage—are present in representatives of the Ordovician Agnatha as fossilized head shields or elements of the endoskeleton; despite the name, the Agnatha are not a monophyletic group. One of the best samples derives from Early and Late Ordovician vertebrates from Colorado with cellular bone and dentine. Indeed, these specimens provide important evidence for the early association of dentine and bone (Smith, 1991). A specimen, restudied by Smith and Sansom (1997) and named *Skiichthys halsteadia,* was identified as an acanthodian or placoderm with scales of enameloid, mesodentine, odontocytes, and basal cellular bone. Smith and Sansom differentiated *Skiichthys* from two well-known genera, *Astraspis* and *Eriptychius,* both of which have acellular bone and tubular dentine. Whether these jawless vertebrates also possessed

Box 1.1. Summary of techniques applied to, and findings obtained from, fossilized skeletal tissues in the three decades between 1960 and 1990

Histological description and comparison of acellular and cellular bone in teleosts (Moss, 1961b).

Automatic amino acid analysis to show micro-level quantities of amino acids at higher concentrations in Jurassic and Devonian fossils than in the matrix in which they are embedded (Armstrong and Halstead Tarlo, 1966).

Decalcification of dentine in heterostracan "ostracoderms" to analyze the organic content (Halstead Tarlo and Mercer, 1966).

Use of the linear relationship between body temperature and amino acid content in the collagen of extant mammals to estimate the body temperatures of Pleistocene mammals (Ho, 1967a,b).

Macromolecular evidence in a phylogenetic context to analyze evolutionary relationships between invertebrate and vertebrate species (Mathews, 1967, 1975).

The demonstration that fossilized bone of the Pliocene elephant, *Gompthotherium*, Jurassic dinosaur, *Apatosaurus* (*Brontosaurus*), and the Eocene uintathere, *Uintatherium*, is partially depleted of organic material, which has been replaced by calcium carbonate (Biltz and Pellegrino, 1969).

A review of studies on chemical analysis of fossilized skeletons (Halstead, 1969a–c).

Extensive studies by Pawlicki on the dinosaur, *Tarbosaurus bataar*, using scanning electron microscopy to visualize vascular canals with collagenous walls (1975a,b); staining procedures on ground unfixed sections to visualize lipids and acid mucopolysaccharides in the perivascular space of vascular canals (1977a,b); light, transmission, and scanning electron microscopy to identify two cell types and a vascular communication system similar to that found in contemporary animals (1978, 1983); determination that intermediary osteocytes are distributed near blood vessels, with implications for metabolic pathways (1984); identification of osteocytic processes linking osteocytes to one another and two lacunar types and associated collagen (1985).

Light microscopy to demonstrate lamellar-zonal compact bone in Jurassic theropods (Reid, 1981, 1984).

Electrophoresis and chondroitinase digestion to isolate chondroitin sulfate from fossilized antlers (Scott and Hughes, 1981).

Transmission electron microscopy to determine that Pleistocene fossils with collagen profiles typical of modern collagen showed significant peptide fingerprints differences when exposed to proteolytic enzymes, although the change in collagen may have occurred as a result of fossilization (Armstrong et al., 1983).

Scanning electron microscopy/replica technique used for microscopical taphonomy, bone growth/damage, and remodeling analysis in early hominids (Bromage, 1987).

Detection of osteocalcin in bovid fossils between 12,000 and 13 million years old and in 30 million year old rodents with high-pressure liquid chromatography, ion exchange, and size exclusion columns (Ulrich et al., 1987).

Scanning electron and light microscopy to argue that aspidine may not be present in Ordovician heterostracans (Ørvig, 1989).

Amino acid analysis to identify hydroxyproline and hydroxylysine as evidence of preservation of collagen in fossil crocodiles and rhinos (Glimcher et al., 1990).

unmineralized tissues such as cartilage has been a matter of inference for a century (Eastoe, 1970; Hall, 1975; Smith and Hall, 1990).

Given what we know of present-day fishes (Witten, 1997; Roy et al., 2004), the Ordovician skeletal tissues may not have served as a major reservoir for calcium but possibly for phosphorus, as suggested by Tarlo (1964). Given the enormous surface area of the gills and the potential for loss of calcium to the environment from the gills (Simmons and Marshall, 1970; Simmons, 1971), it is unlikely that exoskeletal denticles or plates acted as a barrier against loss of calcium to the aquatic environment. If the dermal skeleton of early vertebrates was unrelated to the animals' mineral metabolism, it may have functioned as protective armor against predators such as large sea scorpions (eurypterids), marine arthropods that ranged in size to over 2 m.

Mechanical support from the skeleton was probably of lesser importance for these early marine vertebrates than for later terrestrial taxa; extant primitive vertebrates such as lamprey and hagfish, and basal osteichthyans such as sturgeon show that the notochord can stabilize the body axis, even with no or only rudimentary vertebral structures present. See Denison (1963), Romer (1963), Ørvig (1968), Miles and Poole (1969), Dacke (1979), Carroll (1988), Smith and Hall (1990), and Janvier (1996) for the function of the early vertebrate dermal skeleton.

During vertebrate evolution, especially with the evolution of jaws and the origin of the jawed vertebrates (gnathostomes), adaptations of skeletal tissues arose in response to new and local environmental demands, utilizing the ability of skeletal cells (scleroblasts) to modulate their synthetic activities and differentiative pathways. There is, however, no evidence of a generalized trend toward progressive specialization of skeletogenic tissues during vertebrate evolution. As examples, the degree of skeletal mineralization and bone density in "ostracoderms" that existed over 500 Ma, in Permo-Carboniferous temnospondyl amphibians, and in extinct and extant whales, dolphins, and porpoises (cetaceans) reflects adaptations for equilibrium, buoyancy, and conservation of temperature and energy, rather than progressive evolution within the groups. Ricqlès (1989) argued that there is no selective advantage to one bony tissue over another in the non-weight-bearing skeletons of aquatic tetrapods.

Although Ricqlès's point is an accurate generalization when comparing skeletal tissues, variation can and does exist from skeletal element to skeletal element at the organism level and from taxon to taxon. Variation includes bone that is very dense, elements that lack cortical bone, and various degrees of loss of bone marrow. The spongy bone of the vertebrae of osteichthyans is *deposited* in response to mechanical stresses, although the biconid *shape* of each vertebra reflects an inductive response to the alternating constrictions and dilations of the notochord and sclerotome, rather than response to mechanical stresses. For detailed discussions of the significance of histological variation in fossil vertebrates, see Enlow and Brown (1956, 1957, 1958), Enlow (1969), Ricqlès (1973, 1974, 1976, 2000), Hall (1975, 1978), Smith and Hall (1990, 1993), and Horner et al. (1999, 2000, 2001).

Intermediate Tissues in Early Vertebrates

The presence of intermediate tissues in early vertebrates suggests that osteoblasts, odontoblasts, and chondroblasts could modulate their synthetic and differentiative activity much as they do today.

Ørvig (1951, 1967, 1989) undertook detailed studies in the earliest vertebrates of three types of dentine, tissues intermediate between dentine and bone, and intermediate between bone and mineralized cartilage. Similarly, Jarvik (1959) saw dentine, bone, cartilage, connective tissue, and tissues intermediate between two or more of these tissues as evidence for transformation among and between these tissues. Acquisition of such adaptable cells so early in their evolution was perhaps a major reason for the later evolutionary success of the vertebrates. See Halstead (1969a–c, 1973), Hall (1975, 1977, 1978, 2005a), Maisey (1988), and Smith and Hall (1990, 1993) for discussions of the significance of intermediate tissues for vertebrate evolution.

Dermal skeletal tissues of ancient "agnathans" fall into two classes, respectively, in two major groups, the osteostracans (cephalaspids) and the heterostracans. Both share dentine, enamel, and mineralized cartilage as skeletal tissues, but the skeletal similarities stop there. "Ostracoderms" have cellular bone and hetcrostracans a kind of acellular bone or aspidine, both aspidine and cellular bone having arisen contemporaneously during the Ordovician. Once this was known, much debate ensued over whether aspidine was bone or dentine. Ørvig (1957, 1965, 1968, 1989), Denison (1963), and Moss (1968a) should be consulted for a flavor of these early debates.

Hall (1978, 1988) argued that the development of acellular bone from cellular tissues during the ontogeny of modern teleosts provides a necessary clue to understanding the evolutionary problem. Further studies in the late 1980s and early 1990s led Smith and Hall (1990) to conclude that cartilage was the primary tissue of the endoskeleton and that bone and dentine were the primary skeletal tissues of the exoskeleton. It may still be useful to know which came first, cartilage or bone. The question, however, no longer relates to which came first within the skeleton, for vertebrates have *two* skeletal systems. Cartilage came first in the endoskeleton, bone in the dermal (exo-) skeleton. Any cartilage in the dermal skeleton is secondary both evolutionarily and developmentally.

Remodeling of Skeletal Tissues

As with the question of the basic type of skeleton-forming cell, the question about the first (basic) type of skeleton-resorbing cell is intriguing. In extant vertebrates, both multinucleated giant cells and mononucleated cells are present, but there is evidence that a dominant mononucleated cell type is associated with acellular bone (osteocytes absent), whereas a dominant multinuclear cell type is found in bone with osteocytes (cellular bone) (Witten et al., 1999, 2000, 2001).

As we have seen, the vertebrate skeleton consists of an astoundingly plastic series of tissues. Because the skeleton is subject to constant reshaping (cartilage), remodeling (bone), and replacement (teeth), vertebrate hard tissue structures are not permanent; estimates of the percentage of bone that is resorbed and replaced by new bone in the human skeleton range between 4% and 10% per year (Delling and Vogel, 1992; Manolagas, 2000). Bone remodeling is triggered by the requirements of allometric growth, occurs as an adaptation to mechanical load, and is part of an animal's mineral homeostasis. Repair and maintenance of the skeleton are further causes for bone remodeling. Finally, after the reproduction period is over, bone can be remodeled to remove temporary skeletal elements that function as secondary sexual characters (Bertram and Swartz, 1991; Witten, 1997; Witten et al., 2001; Witten and Hall, 2002, 2003).

Whereas invertebrates replace and reshape their exoskeleton through molting, the vertebrate skeleton—endoskeleton and dermal skeleton, with the exception of teeth—remains inside the body. Molting is not an option. Therefore, and given the conserved structure of skeletal tissues, it is not surprising that signs of skeletal remodeling have been observed in Paleozoic (Devonian) vertebrates (Bystrow, 1959). Carroll (1988) reviewed different hypotheses regarding the function of the early dermal skeleton and cited Tarlo (1964), who argued that the observed variability of dermal skeletal plates in Paleozoic vertebrates could indicate the function of the dermal skeleton as a phosphorous reservoir, phosphorous limiting population size in many ecosystems. Mineral metabolism of extant teleosts supports these ideas; unlike calcium, which can be obtained from ambient water through the gills, phosphorous can only be absorbed by the gut (Simmons, 1971; Fenwick, 1974; Dacke, 1979). Given that ancestral aquatic vertebrates are assumed to have had a similar physiology, periods of starvation could have been periods of phosphorus shortage. Suggestions that the early skeleton functioned as a calcium reservoir are not substantiated by the physiology of modern bony fish. In teleosts, phosphorus deficiency can enhance bone resorption, whereas calcium deficiency has no effect on bone resorption (Witten, 1997; Roy et al., 2004) unless the most extreme laboratory (artificial) conditions are created (Weiss and Watabe, 1978a,b, 1979).

The cells that resorb skeletal tissues arise from the monocyte/macrophage cell lineage and are named according to the tissue they remove: *odontoclasts* remove dentine, *osteoclasts* remove bone, *chondroclasts* remove calcified cartilage. Despite different names, different environments, and different regulation mechanisms, there is much evidence that all skeleton-resorbing cells belong to the same cell type, suggesting that a basic type of skeleton-resorbing cells evolved together with a basic type of skeleton-forming cell. See Nordahl et al. (2000), Shibata and Yamashita (2001), Witten et al. (2001), Sasaki (2003), and Helfrich (2003) for skeleton-resorbing cells as modulations of a basic cell type.

On the basis of studies of resorbing cells for mammalian bone, teeth, and cartilage, key features of mammalian osteoclasts are widely believed to represent general key features of skeleton-resorbing cells. The most frequently cited morphological features are multinuclearity, a deeply folded apical cell membrane (ruffled border), and the formation of a resorption lacuna (Howship's lacuna; Plates 18 and 19; Chambers, 1985; Parfitt, 1988). However, skeleton-resorbing cells need not express any of these features. Studies on human bone have revealed that the majority (61%–81%) of active resorbing osteoclasts are mononucleated (Evans et al., 1979; Chambers, 1985; Ballanti et al., 1997), whereas Parfitt (1988) concluded that in human bone, multinucleated osteoclasts cause deep resorption lacunae, and mononucleated osteoclasts perform a smooth kind of bone resorption.

Findings on advanced osteichthyans with acellular bone also show that bone resorption can be accomplished by mononucleated cells that lack a ruffled border and do not create Howship's lacunae (Plate 18; Weiss and Watabe, 1978a,b, 1979; Ekanayake and Hall, 1987; Witten, 1997; Witten and Villwock, 1997). Most interestingly, similar features have recently been assigned by Kemp (2003) to the osteoclasts of an extant bony fish, the Australian lungfish, *Neoceratodus forsteri,* and we observed mononucleated osteoclasts in the gray bichir, *Polypterus senegalus* (Plate 18).

Despite the pronounced morphological differences between "typical" multinucleated mammalian osteoclasts and mononucleated osteoclasts, the molecular mechanisms by which bone resorption is accomplished appear to be identical in both cells types. A series of studies on mono- and multinucleated osteoclasts in different osteichthyans revealed that mononucleated bone-resorbing cells use the same gene products during bone degradation, as do multinucleated osteoclasts of mammals and fish, namely expression of α-napthyl acetate esterase in osteoclast precursor cells, and acid phosphatase, tartrate-resistant acid phosphatase, ATPase, and a vacuolar proton pump in resorbing cells (Witten, 1997; Witten and Villwock, 1997; Witten et al., 1999, 2000; Witten and Hall, 2002).

Although we assume that acellular bone is a derived and not a basal character (Witten et al., 2004a), a number of findings favor the idea that mononucleated cells represent the basic osteoclast type:

- During ontogeny, multinucleated cells arise through the fusion of mononucleated cells (Parfitt, 1988).
- The majority of bone-resorbing cells are mononucleated (Chambers, 1985; Ballanti et al., 1997).
- During teleostean development, resorbing mononucleated cells

precede the appearance of multinucleated cells (Sire et al., 1990; Witten et al., 2001).

- Mononucleated cells are found in basal extant osteichthyans (sarcopterygians and actinopterygians), in *Neoceratodus forsteri* (Kemp, 2003) and in *Polypterus senegalus*.
- Mononucleated osteoclasts use the same gene products (enzymes) for bone resorption as multinucleated osteoclasts (Witten et al., 1999, 2000).
- In zebra fish ontogeny, mononucleated osteoclasts occur together with the first traces of mineralized bone (Witten et al., 2001).

The lack of easily recognizable multinucleated bone-resorbing cells or their traces of activity (Howship's lacunae) can have important implications when it comes to the examination of fossilized bone structures (Witten et al., 2004a). In tetrapods and in bony fish with multinucleated bone-resorbing cells, resorption can be identified on the basis of the presence of typical resorption lacunae in the absence of cells. This might not be the case in species (such as many extant bony fish) that predominately rely on nonlacunar bone resorption by mononucleated cells. The lack of typical traces of bone resorption could explain the lack of evidence for bone resorption in some early gnathostomes (Donoghue and Sansom, 2002) and would further argue for mononucleated cells as the primitive type of bone-resorbing cell.

Reviewing the increasing evidence that all bone cells (osteoblasts, osteocytes, osteoclasts) act as a unit in regulating bone formation and bone resorption (Smit et al., 2002; Burger et al., 2003), and considering that skeletal resorption is an integral part of skeletal development, we can conclude that skeleton-resorbing cells evolved together with skeleton-forming cells and that the lack of traces of skeleton resorption in the dermal skeleton of early vertebrates relates to early skeletoclastic cells being mononucleated and performing nonlacunar resorption.

Mechanisms of Formation of Intermediate Skeletal Tissues

We have described both clearly categorized and intermediate skeletal tissue, and, having raised questions about the evolution of the intermediate skeletal tissues, we have described resorption of skeletal tissue and the replacement of resorbed tissue by new and often differently structured tissue as a necessary process in development, and as one way of changing the nature of a skeletal tissue (for example, cartilage changing into bone). Now we turn to the mechanism by which intermediate skeletal tissues and their cells develop.

Intermediate skeletal tissues and cells can differentiate early during ontogeny or can arise later in ontogeny by transformation (transdifferentiation, metaplasia) from one of the clearly categorized skeletal tissue types. Transformation is therefore another way to alter the nature of skeletal tissues and cells.

The Cellular Players

Normally, there is no difficulty in assigning vertebrate skeletal tissues and skeletogenic cells to one of the four categories of mineralized tissues. Box 1.2 provides a historical perspective. Recognizing the stability of differentiating or differentiated cell phenotypes and the tissues/organs they inhabit within the vertebrates, we readily identify four cell types and the four tissues/organs they create: (1) osteoblasts (bone), (2) chondroblasts (cartilage), (3) ameloblasts (enamel), and (4) odontoblasts (dentine).

Moss (1964a) grouped the cells that produce the four mineralized tissues and named them *scleroblasts* because they represent a family of related cells. Synthesis of the same type of collagen by cells producing two quite different mineralized tissues (bone and dentine) is one sign of affinity. Things become troublesome when we attempt to classify tissues on the basis of a specific molecule they produce: (1) Odontoblasts depositing predentine produce the same collagen (type I) as do osteoblasts depositing osteoid. (2) Cells

---

Box 1.2. Bone or cartilage, osteogenesis or chondrogenesis?

The distinction between cartilage and bone has been recognized since at least the time of Aristotle (384–322 BC), who recognized and separated the cartilaginous fishes, or Chondrichthyes, from the bony fishes, or Osteichthyes, by the presence of a cartilaginous or osseous skeleton.

Until the seventeenth century, cartilage was thought to transform into bone. Such a transformation *can* occur but is difficult to detect. The Belgian and Danish anatomists Adrian Spigelius (1631) and Thomas Bartholinus (1676) distinguished bone that forms from cartilage—what we now refer to as endochondral bone arising from endochondral ossification—from bone that forms directly—what we now refer to as membrane bone arising by intramembranous ossification.

Not all the bony material of an endochondral bone is the result of endochondral ossification; three modes of ossification can be involved in the generation of an endochondral bone. Primary ossification of the cartilaginous anlage occurs subperiosteally around the shaft by perichondral bone apposition. Endochondral replacement of cartilage of the shaft metaphysis is the second step followed by a third process, the extension of the skeletal element by appositional bone (*Zuwachsknochen*) in a process that reassembles intramembranous bone formation (Hall, 2005a; Witten and Huysseune, 2005).

So the word *endochondral* can be a misnomer. Furthermore, we know that *endochondral* is too limited a term; bone can replace marrow, tendon, or ligament without cartilage being involved. *Replacement bone* and *indirect ossification* may be better terms for endochondral bone and endochondral ossification, respectively. Membrane bone and intramembranous ossification reflect direct ossification (Hall, 1988, 2005a).

---

depositing notochord, cartilage, and the osteoid of membrane bone all synthesize type II collagen. (3) Lamprey cartilages do not produce or contain type II collagen but can be classified as cartilages on the bases of many other criteria.

Recognizing intermediate mineralized tissues takes Moss's scleroblasts to the tissue level. Although we do not fully understand the processes, sufficient examples are available to indicate that, at a variety of sites within embryonic and adult skeletons, and in vitro, skeletal cells can produce tissues whose characteristic features are intermediate between cartilage and bone, dentine and bone, or dentine and enamel. Before we discuss how intermediate skeletal tissues form, we need to distinguish between two different cellular processes, either of which could produce intermediate tissues such as cementum, enameloid, or chondroid bone. The two processes are *modulation of cellular activity*—for example, a cell switching from the synthesis of type I to the synthesis of type II; and *transformation of cell identity*—for example, a hypertrophic chondrocyte becoming an osteoprogenitor cell.

Modulation and Transformation

Modulation (transdetermination), which is a temporary change in cell behavior, structure, and/or the type of matrix product(s) produced, occurs in response to altered environmental conditions. Maintenance of the new state depends on the continued presence of the environmental stimulus. Remove the stimulus and the cell reverts to its original structure, function, or behavior. Modulation, therefore, is a physiological response. Transformation to a new cell identity is usually permanent; the new state remains after the stimulus is removed. A metaphor would be the difference between growth of a fetus during pregnancy (a permanent change) and the weight gained by the mother, a temporary modulation (although it may seem permanent to the mother). Intermediate tissues are permanent. They arise by transformation, not modulation.

Many studies document how altering the microenvironment in vivo or in vitro can modulate the synthetic activity of skeletal cells. Indeed, the same environmental change may have diverse effects on different cell populations, and different skeletogenic cells may respond differently to the same environmental change. We illustrate this by using alteration in oxygen tension as the environmental change.

Bone cell metabolism, especially glucose and glycine metabolism, is exquisitely sensitive to oxygen levels. This does not mean invariance of responses to altered levels, however; the neotropical teleost fish, the pacu (*Piaractus mesopotamicus*), develops extensions of dermal bone in its jaws when in oxygen-deficient waters. The extensions regress when the water is aerated (Saint-Paul and Bernardino, 1988). Several other responses to altered oxygen tension by different skeletogenic cell types illustrate these points.

Undifferentiated mesenchymal cells and prechondroblasts of prenatal murine mandibular condylar cartilages (Plate 7) are especially sensitive to increased oxygen tension; 28% O_2 is mitogenic, increasing the undifferentiated and prechondroblastic cells layers but decreasing glycosaminoglycan synthesis. Furthermore, different stages of chondroblast maturity (prechondroblast, chondroblast, chondrocyte, hypertrophic chondrocyte), chondrogenic cell populations from different regions of the condyle, and chondrocytes from different types of cartilage (primary, articular, secondary) respond differentially to changes of the oxygen regime (Plate 7; Vaes and Nichols, 1962; Kantomaa, 1986a,b; Hall, 2005a).

As many studies have shown, variation in oxygen tension also influences osteogenesis in vitro and in vivo. The optimal oxygen tension for osteogenesis is 35%; no osteogenesis occurs at 5%, while 95% initiates resorption. In one specific study, rabbit periostea were cultured for 6 weeks in agarose in oxygen tensions ranging from 1% to 90%. Aerobic conditions (12%–15% O_2) were optimal for osteogenesis. Although no bone formed in 5% O_2, 90% of the cultures that were maintained between 1% and 5% O_2 produced cartilage (but not bone), showing how variation in the same environmental signal can mediate the modulation of synthetic activity and differentiation (Shaw and Bassett, 1967; O'Driscoll et al., 1997).

The substrata on which skeletal tissues are maintained in vitro also can modulate differentiative pathways. For example, rabbit articular chondrocytes cultured on demineralized bone produce a *chondroid* matrix with features of cartilage and bone. When cultured on Gelfoam sponge, however, these same chondrocytes produce a chondromyxoid tissue (Green, 1971; Green and Ferguson, 1975).

Metaplasia: Cartilage from Fibrous Tissue

Modulation is a *physiological response*. Metaplasia (transdifferentiation) is an example of transformation in which one differentiated cell type or tissue transforms into another without any dedifferentiation or reinitiation of cell division. The transformation of differentiated osteocytes into osteoblasts (Roach, 1997), or of differentiated osteoblasts into chondroblasts (Li et al., 2004), are two examples.

An unusual fibrocartilage is produced by chondrogenic replacement of the patellar ligament during development of the epiphyseal tubercle of the tibiae in rats. The ligamentous tissue undergoes metaplasia to a fibrocartilage, which is then mineralized and replaced by bone (Badi, 1972). Mineralization occurs, however, *without* the chondrocytes undergoing hypertrophy or accumulating stores of glycogen, both of which are normally associated with the maturational changes in cartilage that precede mineralization.

Connective tissue around the tail vertebrae of 4-day-old rats undergoes metaplasia to a chondroid tissue after the vertebrae are transplanted. As growth pressure compresses the annulus fibrosus, the chondroid tissue is replaced by metaplastic bone, not unlike changes seen in late stages of ankylosing spondylitis in humans (Feik and Storey, 1982). A similar phenomenon is observed in Atlantic salmon (*Salmo salar*) that have shortened vertebral bodies; the notochordal tissue of the intervertebral space (in mammals represented by the intervertebral disk) is replaced by cartilage, which later begins to mineralize (Witten et al., 2005).

As indicated by the last example, metaplasia is common outside the mammals. Turtle femora contain substantial amounts of metaplastic bone, which forms as hypertrophic chondrocytes transform into osteoblasts. Lizards, too, have much metaplastic bone (Haines, 1969).

A further example comes from bullfrogs (*Rana catesbiana*), which possess an osteochondral ligament as a fibrous attachment between bone and articular cartilage. Osteoblasts are found on the inner area of the ligament near the bone. Mineralization of the growth cartilage and formation of bony trabeculae occur late in ontogeny and are dissociated. At 1 year of age, there is neither mineralization of cartilage nor replacement of cartilage by bone, although the chondrocytes are alkaline phosphatase positive, which usually is a sign of incipient mineralization or a marker for preosteoblasts. By 2 years, mineralization and replacement of cartilage are under way, but being so late in ontogeny, neither process contributes to long bone growth, giving the unusual situation of long bone growth without cartilage mineralization (Felisbino and Carvalho, 2000, 2001).

We now turn to a discussion of the formation of chondroid and chondroid bone as an instance of how such intermediate tissues form.

Formation of Chondroid

The category of tissues known as *chondroid* reflects our ability to recognize a permanent tissue type that has sufficient properties of cartilage and fibrous connective tissue that it cannot be classified as one or the other. We expect to find chondroid in what were once called *lower vertebrates,* a term that embraces groups known as fish, amphibians, and reptiles.

Chondroid is frequently found in the mandibular and maxillary bones of teleost fishes. Determining the cellular origin of such tissue is difficult; Benjamin (1986) could not decide whether these tissues were secondarily attached to the bones (like sesamoids) or secondary cartilages arising from the periosteal cells of the bones, as shown in Plates 9–11. Benjamin (1988) also identified an intermediate tissue, *mucocartilage,* defined as a mucous connective tissue, which is especially common beneath the skin and in the opercular valves. See Benjamin (1986, 1988, 1989a,b, 1990) and Benjamin and Ralphs (1991) for chondroid and other cartilage-like tissues in teleosts.

Chondroid often is a component of novel skeletal tissues that form outside the bounds of what we traditionally consider to be the "normal" skeleton. As one example, cartilage is present in the muscle at the base of the adipose fin in many teleost fish. In some species, the tissue is cartilage, in others chondroid (Matsuoka and

Iwai, 1983). A cartilaginous tissue in the cloacae of male caecilians has also been described as unmineralized chondroid (Wake, 1998).

Chondrogenic tissues, including chondroid, are frequently found in the sutures of mammalian skulls. Goret-Nicaise and co-workers published a series of studies identifying chondroid as a permanent tissue on the coronoid process in human infants where the temporal muscle inserts to allow migration of the muscle insertion, in the mandibular symphysis of newborns, at the sutural margins of the cranial vault, and in the midline suture between the frontal bones. In part, the identification of chondroid was based on the coexpression of collagen types I and II in the same tissue (Goret-Nicaise, 1981, 1982, 1986; Goret-Nicaise and Dhem, 1982; Goret-Nicaise et al., 1988; Manzanares et al., 1988).

Formation of Chondroid Bone

Chondroid bone is perhaps the most recognized class of a tissue with characters intermediate between cartilage and bone (Plates 11 and 12). Many tumors, sarcomas, and skeletal neoplasia fall into this category (Willis, 1962; Jaffe, 1968; Ashley, 1970; Wright and Cohen, 1983). Others are normal skeletal tissues in one taxon that resemble pathological situations in another taxon. The skeletal tissues in deer antlers (Hall, 2005a) or in the kype (a bony hook that develops on the distal tip of the lower jaw in the adult male) of Atlantic salmon, *Salmo salar* (Witten and Hall, 2002, 2003), have been compared to neoplastic bone.

According to Beresford (1981), chondroid bone arises commonly through one of two distinct ontogenetic pathways: transdifferentiation of skeletal cells (from osteoblast to chondroblast) within multipotential periostea (chondroid bone type I), or through the incomplete ossification of calcified cartilage (chondroid bone type II). The result of both processes is, however, cells of mostly chondrocytic form and dimensions in a mineralized collagenous matrix. Several examples follow:

- The cartilaginous articular surface of the temporal component of the human temporomandibular joint transforms into chondroid bone (Beresford, 1981). Transformation involves the formation of a predominantly collagenous extracellular matrix with the basophilic characteristic histology of bone but with nests of cells resembling chondrocytes in lacunae (Enlow, 1962).
- Hall (1972) presented evidence for a similar transformation of secondary cartilage to chondroid bone in paralyzed embryonic chicks, and analyzed the then available literature on such transformations. Matrix changes are the result of trapped chondroblasts and chondrocytes assuming osteoblastic activity and modifying the extracellular matrix toward an osseous type.
- Such alterations—true metaplasia—are also seen during enchondral ossification in chicks, during repair of fractured long bones in lizards and frogs, at the distal tip of penile bones, in human jaws in response to ill-fitting dentures, and in the formation of

bone and cartilage in mammary tumors (Ruth, 1934; Pritchard and Ruzicka, 1950; Smith and Taylor, 1969; Beresford, 1970, 1975; Cutright, 1972; Beresford and Clayton, 1977; Roach, 1997; Hall, 2005a).

Unraveling the mechanisms of such transformations will be a major step toward understanding interrelationships between skeletal cells and how they transform from one to another.

Mandibular condylar cartilage was introduced earlier (Plate 7). If mandibular condyles from 1- or 20-day-old fetuses are organ-cultured for 10 days, differentiated hypertrophic chondrocytes influence chondroprogenitor cells to transform to osteoblasts and deposit a mineralized, collagen type I–containing matrix, containing osteonectin and bone sialoprotein and interpreted as chondroid bone (Silbermann et al., 1983, 1987a,b). Chondroid bone identified in the rat glenoid fossa contains type I and type II and osteocalcin—that is, it also has features of hypertrophic chondrocytes and of osteoblasts/cytes (Tuominen et al., 1996; Mizoguchi et al., 1992, 1993, 1997a).

In addition to chondroid bone, various types of cartilage also have both type I and type II collagen in their extracellular matrices. Aspects of development that predispose cartilages to synthesize and deposit type I collagen include periosteal origin of the cartilage, function as an articular cartilage, mineralization, and/or hypertrophy and replacement by bone. A sample of studies is summarized in Box 1.3.

Parasites Can Induce Chondroid Bone

As early as 1915, reports appeared that yellow perch, *Perca flavescens*, infected with metacercaria of the trematode, *Apophallus brevis*, develop mineralized "bony" cysts in the muscle. *Apophallus brevis* produces the condition known as black spot in brook trout (*Salvelinus fontinalis*) and brown trout (*Salmo trutta*), with sand grain–sized black spots on the skin. Only yellow perch produce ossicles. These doughnut-shaped hollow cysts have pores plugged with connective tissue at two of the poles (Plate 13). Taylor et al. (1993, 1994) investigated these "bone cysts" to find that their structure is far more complex and interesting than originally thought; they contain collagen but little glycosaminoglycan, and they are organized into outer and inner layers of chondroid bone.

Chondroid Bone in the Visceral Arch Skeleton of Teleost Fish

As noted earlier, chondroid bone is a recognized feature on certain elements of the skull in many teleost fish species.

An adaptation found in many teleost fishes is the development of a second set of jaws known as pharyngeal jaws, which develop from the fifth branchial arch element (Plates 19 and 20). Pharyngeal jaws can be equipped with powerful muscles and very plastic teeth, which means that the morphology of the teeth can be modified from one tooth generation to the next in response to the kinds

Box 1.3. Some examples of cartilages that contain collagen types I and II

As part of their normal differentiation or as a result of mechanical loading, aging, or degenerative diseases, various different cartilages synthesize and deposit both type I and type II collagen into their extracellular matrices.[a] A number of examples follow.

Articular, fibro-, and secondary cartilages in rats contain both type I and type II collagen (Sasano et al., 1996).

Cartilages of the hagfish, *Eptatretus burgeri,* have two distinct types of type I collagen, neither of which is equivalent to type I in vertebrate cartilages (Kimura and Matsui, 1990; Matsui et al., 1990).

Tesserae of mineralized cartilage in sharks contain collagen types I and II as normal constituents (Kemp and Westrin, 1979).

Chondroid bone of the rat glenoid fossa has type I and type II and osteocalcin, i.e., this tissue has features of hypertrophic chondrocytes and of osteoblasts/cytes (Tuominen et al., 1996; Mizoguchi et al., 1997a).

Human thyroid cartilages contain substantial quantities of type I collagen (Claassen and Kirsch, 1994).

Mammalian hypertrophic chondrocytes contain type I collagen when endochondral osteoid is being deposited onto them, although some of the type I may not have been synthesized by the chondrocytes (von der Mark and von der Mark, 1977; Yasui et al., 1984).

Mandibular condylar cartilages of rodents, calves, and monkeys contain both type I and type II collagen (Hirschmann and Shuttleworth, 1976; Mizoguchi et al., 1990, 1992, 1997b; Silbermann and von der Mark, 1990; Silbermann et al., 1990; Ishii et al., 1998). Distribution of types I and II within condylar cartilage correlates with mechanical loading; type I is depressed and type II enhanced at sites of compression sites, and type I increased and type II depressed at sites of tension in tensile sites (Mizoguchi et al., 1996).

Osteoarthritic human articular cartilage synthesizes and deposits type I collagen and so is less able to bind glycosaminoglycans and therefore more liable to destruction from abrasion (Nimni and Deshmukh, 1973; Gay et al., 1976; Müller and Kühn, 1977). Here the presence of type I collagen is pathological.

[a] Absence of type I collagen from many avian cartilages reflects the presence of a cartilage-specific promoter within intron two of the chick *Cola2(I)* gene. As a consequence, cartilage has type I collagen mRNA but does not produce the protein product (Bennett and Adams, 1990).

of prey being eaten (Liem, 1973; Lauder, 1983; Huysseune, 1995; Huysseune et al., 1998). Pharyngeal jaws allow the regular jaws to specialize for prey capture, the pharyngeal jaws doing the breaking up of food (Plates 19 and 20). As one might expect, such a division of labor opens up evolutionary opportunities, which cichlid fishes and their marine relatives have exploited to an amazing degree. Speciation of cichlids of the African rift lakes is almost as rapid as has ever been measured. Other groups with pharyngeal jaws, such as cyprinids (with zebra fish as the most prominent extant member), can afford to reduce their teeth completely and rely on pharyngeal teeth only (Plate 21).

Liem (1973), who developed the thesis that pharyngeal jaws are an evolutionary innovation, studied them extensively, especially from the points of view of functional morphology and musculoskeletal adaptation. A shift in the insertion of the fourth levator externus muscle coupled with developmental flexibility of periostea to form cartilage when exposed to mechanical forces allows secondary fusion of the lower pharyngeal jaws with the basipharyngeal joint.

Development of pharyngeal jaws and their associated skeletal apparatus in the pharynx and branchial arches of the cichlid, *Haplochromis* (*Astatotilapia*) *elegans,* is known in detail, as is the microstructure of the bone of the pharyngeal jaws in two morphotypes (morphologically distinguishable stable variants) of the cichlid, *Astatoreochromis alluaudi* (Verraes et al., 1979; Huysseune and Verraes, 1990; Huysseune et al., 1994). Chondroid bone develops late in ontogeny at the articulation of the upper pharyngeal jaws with the base of the neurocranium. This chondroid bone resembles secondary cartilage but is formed by osteoblast-like cells rather than chondroblasts, has histological staining properties of bone rather than cartilage, forms on bone and on cartilage, and does not contain type II collagen.

Huysseune (1985) investigated the cartilage that develops on the opercular bone at its articulation with the hyosymplectic in the cichlid, *Haplochromis elegans*. Thus unusual cartilage does not hypertrophy, plays a limited role in growth of the opercular bone, and arises from the hyosymplectic, which is an endochondral bone. Huysseune identified the tissue as chondroid bone. Chondroid bone persists in adults on infrapharyngobranchials III and IV, and on the parasphenoid and basioccipital. In further studies of the contribution of chondroid bone to bone growth, incorporation of cells into chondroid bone on the parasphenoid was shown to correlate with local demand, chondroid bone being deposited whenever and wherever there was rapid growth. No cement lines—lines of arrested bone growth—develop between chondroid bone and the underlying acellular bone, a feature also suggestive of a switch in the type of matricial products deposited (Ismail et al., 1982; Huysseune, 1986, 1989, 2000; Huysseune et al., 1986, 1994; Huysseune and Verraes, 1986; Huysseune and Sire, 1990, 1992a,b).

Chondroid bone in the jewel cichlid, *Hemichromis bimaculatus,* has a mineralized collagenous extracellular matrix equivalent to bone, with osteocyte-like cells and matrix vesicles. This tissue is permanent, transforming neither into acellular bone nor into cellular bone. The carbohydrate histochemistry of this tissue supports its identification as chondroid bone.

Nigrelli and Gordon (1946) described spontaneous neoplasms (osteochondromas of the maxilla and opercular bones) in *H. bi-*

maculatus, consisting of hyaline cartilage and thought to arise from the periostea of these dermal bones, a potential that implies the ability to form secondary cartilage. Our studies on development of the lower jaw in Atlantic salmon (*Salmo salar*) show that the kype forms from the periosteum of the dentary bone. The kype skeleton grows rapidly and contains both bone and cartilage cells, the latter identified on the basis of expression of collagen type II, chondroitin sulfate, and hyaluronidase; continued proliferation; arrangement in isogenous groups; and the lack of cell processes. These features clearly distinguish the kype skeletal bone from regular dentary bone and qualify the cartilage component if the kype skeleton is secondarily arisen (Witten and Hall, 2002, 2003; Witten et al., 2004b). Our latest study show that a cartilage-to-bone transformation is already present in juvenile salmon and contributes to the elongation of the dentary (Plate 22; Gillis et al., 2005).

Conclusions

Vertebrate skeletons are composed of four long-recognized skeletal tissues—bone, cartilage, dentine, and enamel—and tissues that are intermediate between two skeletal tissues such as chondroid, chondroid bone, cementum, and various types of dentine. We describe each of the four well-categorized skeletal tissues and the intermediate tissues to emphasize their distinctive properties, distribution in extant and extinct taxa, and modes of development, and we conclude that they reflect the early evolution of highly plastic skeletogenic cells that could modulate their behavior in response to intrinsic and environmental signals. Once formed, skeletal tissues are not static but rather dynamic and subject to remodeling. We document the evolutionary origin of skeletal resorption as concomitant with the origin of the mineralized skeleton. The themes that emerge from the analysis are plasticity of skeletogenic cells and tissues, and skeletal tissues as part of a continuum with adjacent cells in the continuum capable of transformation to another skeletogenic tissue. We provide examples of the cellular responsiveness underlying plasticity and transition by considering the origin of intermediate tissues by two modes of transformation: modulation and metaplasia (transdifferentiation). Modulation, a temporary change in cell behavior, structure, and/or the type of matrix products produced, occurs in response to altered environmental conditions. Modulation is therefore the cellular equivalent of phenotypic plasticity. Metaplasia is the transformation of one differentiated cell type or tissue to another cell type or tissue. Differentiation, modulation, metaplasia, and resorption are four ancient processes that have accompanied the evolution of the vertebrate skeleton.

Acknowledgments

We thank Patricia Avendano for her careful checking of the manuscript. Research support from NSERC (Canada), the Killam Trust of the Canada Council for the Arts, and from funds associated with the George S. Campbell Chair and a University Research Professorship at Dalhousie (to BKH) is gratefully acknowledged, as is support from the Deutsche Forschungsgemeinschaft, from the Germany-Canada Science and Technology Cooperation Programme, and from the European COST Action B23 (to PEW). Some of the material and themes discussed in this chapter are included and expanded in greater detail and in different contexts in Hall (2005a).

References Cited

Andreucci, R. D., and G. Blumen. 1971. Radioautographic study of *Spheroides testudineus* denticles (checkered puffer). Acta Anatomica 79:76–83.

Armstrong, W. G., L. B. Halstead, F. B. Reed, and L. Wood. 1983. Fossil proteins in vertebrate calcified tissues. Philosophical Transactions of the Royal Society of London B 301:302–343.

Armstrong, W. G., and L. B. Halstead Tarlo. 1966. Amino-acid components in fossil calcified tissues. Nature 210:481–482.

Ashley, D. J. B. 1970. Bone metaplasia in trachea and bronchi. Journal of Pathology 102:186–188.

Ayad, S., R. P. Boot-Handford, M. J. Humphries, K. E. Kadler, and C. A. Shuttleworth. 1994. The Extracellular Matrix Facts Book. Academic Press, London and New York, 51 pp.

Bada, J. L., X. S. Wang, and H. Hamilton. 1999. Preservation of key biomolecules in the fossil record: current knowledge and future challenges. Philosophical Transactions of the Royal Society of London B 354:77–87.

Badi, M. H. 1972. Calcification and ossification of fibrocartilage in the attachment of the patellar ligament in the rat. Journal of Anatomy 112:415–422.

Ballanti, P., S. Minisola, M. T. Pacitti, L. Scarnecchia, R. Rosso, G. F. Mazzuoli, and E. Bonucci. 1997. Tartrate-resistant acid phosphate activity as osteoclastic marker: sensitivity of cytochemical assessment and serum assay in comparison with standardized osteoclast histomorphometry. Osteoporosis International 7:39–43.

Baratz, R. S. 1974. Fish tooth enameloid—an unusual mineralized system. Journal of Dental Research 53:77.

Bartholinus, C. 1676. Diaphragmatis structura nova. Acta Medica & Philosophica Hafniensia 4:14.

Benjamin, M. 1986. The oral sucker of *Gyrinocheilus aymonieri* (Teleostei: Cypriniformes). Journal of Zoology (London) B 1:211–254.

Benjamin, M. 1988. Mucochondroid (mucous connective) tissues in the heads of teleosts. Anatomy and Embryology 178:461–474.

Benjamin, M. 1989a. Hyaline-cell cartilage (chondroid) in the heads of teleosts. Anatomy and Embryology 179:385–303.

Benjamin, M. 1989b. The development of hyaline-cell cartilage in the head of the black molly, "*Poecilia sphenops.*" Evidence for secondary cartilage in a teleost. Journal of Anatomy 164:145–154.

Benjamin, M. 1990. The cranial cartilages of teleosts and their classification. Journal of Anatomy 169:153–172.

Benjamin, M., and J. R. Ralphs. 1991. Extracellular matrix of connective tissues in the heads of teleosts. Journal of Anatomy 179:137–148.

Bennett, V. D., and S. L. Adams. 1990. Identification of a cartilage-specific promoter within intron 2 of the chick α2(I) collagen gene. Journal of Biological Chemistry 265:2223–2230.

Beresford, W. A. 1970. Healing of the experimentally fractured os priapi of the rat. Acta Orthopedica Scandinavica 41:134–149.

Beresford, W. A. 1975. Growth cartilages of the mandibular condyle and penile bones: how alike? Journal of Dental Research 54:417.

Beresford, W. A. 1981. Chondroid Bone, Secondary Cartilage and Metaplasia. Urban und Schwarzenberg, Munich, 454 pp.

Beresford, W. A., and S. P. Clayton. 1977. Intracerebral transplantation of the genital tubercle in the rat: the fate of the penile bone and cartilages. Journal of Anatomy 123:297–312.

Bertram, J. E. A., and S. M. Swartz. 1991. The "law of bone transformation": a case of crying Wolff? Biological Reviews of the Cambridge Philosophical Society 66:245–273.

Bilezikian, J. P., L. G. Raisz, and G. A. Rodan (eds.). 1996. Principles of Bone Biology. Academic Press, San Diego, 1398 pp.

Biltz, R. M., and E. O. Pellegrino. 1969. The chemical anatomy of bone. I. A comparative study of bone composition in sixteen vertebrates. Journal of Bone and Mineral Research American Volume 51:456–466.

Bourne, G. H. (ed.) 1971–1976. The Biochemistry and Physiology of Bone. Second edition. Volumes 1–4. Academic Press, London and New York, 1949 pp.

Bosshardt, D. D. 2005. Are cementoblasts a subpopulation of osteoblasts or a unique phenotype? Journal of Dental Research 84:390–406.

Bromage, T. G. 1987. The scanning electron microscopy replica technique and recent applications to the study of fossil bone. Scanning Microscopy 1:607–614.

Brunn, A. 1891. Beiträge zur Kenntniss der Zahnentwicklung. Archiv für mikroskopische Anatomie 38:142–156.

Buffrènil, V. de, A. J. de Ricqlès, C. E. Ray, and D. P. Domning. 1990. Bone histology of the ribs of the archaeocetes (Mammalia: Cetacea). Journal of Vertebrate Paleontology 10:455–466.

Burger, E. H., J. Klein-Nulend, and T. H. Smit. 2003. Strain-derived canalicular fluid flow regulates osteoclast activity in a remodelling osteon—a proposal. Journal of Biomechanics 36:1453–1459.

Bystrow, A. P. 1959. The microstructure of skeleton elements in some vertebrates from lower Devonian deposits in the USSR. Acta Zoologica (Stockholm) 40:59–83.

Carroll, R. L. 1988. Vertebrate Paleontology and Evolution. W. H. Freeman and Company, New York, 698 pp.

Carter, J. G. (ed.) 1990. Skeletal Biomineralization: Patterns, Processes and Evolutionary Trends. Volumes 1 and 2. Van Nostrand Reinhold, New York, 832 pp. + atlas and index.

Chambers, T. J. 1985. The pathology of the osteoclast. Journal of Clinical Pathology 38:241–252.

Claassen, H., and T. Kirsch. 1994. Temporal and spatial localization of type I and II collagens in human thyroid cartilage. Anatomy and Embryology 189:237–242.

Cole, A. G., and B. K. Hall. 2004a. The nature and significance of invertebrate cartilages revisited: distribution and histology of cartilage and cartilage-like tissues within the Metazoa. Zoology 107:261–273.

Cole, A. G., and B. K. Hall. 2004b. Cartilage is a metazoan tissue: integrating data from non-vertebrate sources. Acta Zoologica 85:69–80.

Cutright, D. E. 1972. Osseous and chondromatous metaplasia caused by dentures. Oral Surgery, Oral Medicine, Oral Pathology 34:625–633.

Dacke, C. G. 1979. Calcium Regulation in Sub-Mammalian Vertebrates. Academic Press, London and New York.

Delling, G., and M. Vogel. 1992. Pathomorphologie der Osteoporose; pp. 7–26 in H. H. Schild and M. Heller (eds.), Osteoporose. Georg Thieme Verlag, Stuttgart.

Denison, R. H. 1963. The early history of the vertebrate calcified skeleton. Clinical Orthopedics and Related Research 31:141–152.

Dhem, A., M. Goret-Nicaise, R. Dambrain, C. Nyssen-Behets, B. Lengele, and M. C. Manzanares. 1989. Skeletal growth and chondroid tissue. Archives of Italian Anatomy and Embryology 94:237–241.

Diekwisch, T. G. H. 2001. The developmental biology of cementum. International Journal of Developmental Biology 45:695–706.

Diekwisch, T. G. H. (ed.). 2002. Development and histochemistry of vertebrate teeth. Microscopy Research and Technique 59:339–459.

Diekwisch, T. G. H., B. J. Berman, X. Anderton, B. Gurinsky, A. J. Ortega, P. G. Satchell, M. Williams, H. Arumugham, X. Luan, J. E. Mcintosh, A. Yamane, D. S. Carlson, J.-Y. Sire, and C. F. Shuler. 2002. Membranes, minerals, and proteins of developing vertebrate enamel. Microscopy Research and Technique 59:373–395.

Donoghue, P. C. J. 2002. Evolution of development of the vertebrate dermal and oral skeletons: unraveling concepts, regulatory theories, and homologies. Paleobiology 28:474–507.

Donoghue, P. C. J., and I. J. Sansom. 2002. Origin and early evolution of vertebrate skeletonization. Microscopical Research Techniques 59:352–372.

Donoghue, P. C. J., P. L. Forey, and R. J. Aldridge. 2000. Conodont affinity and chordate phylogeny. Biological Reviews of the Cambridge Philosophical Society 75:191–251.

Eastoe, J. E. 1970. The place of cartilage and bone among vertebrate mineralized tissues. Calcified Tissue Research 4 (Suppl.):24–27.

Ekanayake, S., and B. K. Hall. 1987. The development of acellularity of the vertebral bone of the Japanese medaka, Oryzias latipes (Teleostei, Cyprinidontidae). Journal of Morphology 193:253–261.

Ekanayake, S., and B. K. Hall. 1988. Ultrastructure of the osteogenesis of acellular vertebral bone in the Japanese medaka, Oryzias latipes (Teleostei, Cyprinidontidae). American Journal of Anatomy 182:241–249.

Enlow, D. H. 1962. A study of the postnatal growth and remodeling of bone. American Journal of Anatomy 110:79–102.

Enlow, D. H. 1969. The bone of reptiles; pp. 45–80 in C. Gans (ed.), Biology of the Reptilia. Volume 1, Morphology A. Academic Press, London and New York.

Enlow, D. H., and S. O. Brown. 1956. A comparative histological study of fossil and recent bone. Part I. Texas Journal of Science 8:405–443.

Enlow, D. H., and S. O. Brown. 1957. A comparative histological study of fossil and recent bone. Part II. Texas Journal of Science 9:186–214.

Enlow, D. H., and S. O. Brown. 1958. A comparative histological study of fossil and recent bone. Part III. Texas Journal of Science 10:187–230.

Evans, R. A., C. R. Dunstan, and D. J. Baylink. 1979. Histochemical identification of osteoclasts in undecalcified sections of human bone. Mineral and Electrolyte Metabolism 2:179–185.

Feik, S. A., and E. Storey. 1982. Joint changes in transplanted caudal vertebrae. Pathology 14:139–147.

Felisbino, S. L., and H. F. Carvalho. 2000. The osteochondral ligament: a fibrous attachment between bone and articular cartilage in *Rana catesbiana*. Tissue and Cell 32:527–536.

Felisbino, S. L., and H. F. Carvalho. 2001. Growth cartilage calcification and formation of bone trabeculae are late and dissociated events in the endochondral ossification of *Rana catesbiana*. Cell and Tissue Research 306:319–323.

Fenwick, J. C. 1974. The comparative in vitro release of calcium from dried and defatted eel and rat bone fragments and its possible significance to calcium homeostasis in teleosts. Canadian Journal of Zoology 52:755–764.

Fietzek, P. P., H. Allmann, J. Rauterberg, and E. Wachter. 1977. Ordering of cyanogen bromide peptides of type II collagen based on their homology to type I collagen: preservation of sites for cross linking formation during evolution. Proceedings of the National Academy of Sciences USA 74:84–86.

Franz-Odendaal, T., A. G. Cole, T. J. Fedak, M. Vickaryous, and B. K. Hall. 2003. Inside and outside skeletons. Palaeontological Association Newsletter 54:17–21.

Gay, S., P. K. Müller, C. Lemmen, K. Remberger, K. Matzen, and K. Kühn. 1976. Immunohistochemical study on collagen in cartilage-bone metamorphosis and degenerative osteoarthritis. Klinische Wochenschrift 54:969–976.

Gillis, J. A., P. E. Witten, and B. K. Hall. 2005. Chondroid bone and secondary cartilage contribute to apical dentary growth in juvenile Atlantic salmon, *Salmo salar* Linnaeus (1758). Journal of Fish Biology.

Glimcher, M. J., L. Cohen-Solal, D. Kossiva, and A. J. de Ricqlès. 1990. Biochemical analyses of fossil enamel and dentin. Paleobiology 16:219–232.

Glimcher, M. J., E. J. Daniel, D. F. Travis, and S. Kahmis. 1965. Electron optical and X-ray diffraction studies of the organization of the inorganic crystals in embryonic bovine enamel. Journal of Ultrastructural Research 7, Supplement:1–77.

Goret-Nicaise, M. 1981. Influence des insertions des muscles masticateurs sur la structure mandibulaire du nouveau-né. Bulletin l'Association d'Anatomie 65:287–296.

Goret-Nicaise, M. 1982. La symphyse mandibulaire du nouveau-né. Étude histologique et microradiographique. Revue Stomalologie Chirurgie Maxillo-faciale 83:266–272.

Goret-Nicaise, M. 1986. La croissance de la mandible humaine: conception actuelle. Thèse d'Agrégé de L'Enseignement Supérieur, Université Catholique de Louvain, Belgium, 147 pp.

Goret-Nicaise, M., and A. Dhem. 1982. Presence of chondroid tissue in the symphyseal region of the growing human mandible. Acta Anatomica 113:189–195.

Goret-Nicaise, M., and A. Dhem. 1985. Comparison of the calcium content of different tissues present in the human mandible. Acta Anatomica 124:167–172.

Goret-Nicaise, M., M. C. Manzanares, P. Bulpa, E. Nolmans, and A. Dhem. 1988. Calcified tissues involved in the ontogenesis of the human cranial vault. Anatomy and Embryology 178:399–406.

Graveson, A. C., M. M. Smith, and B. K. Hall. 1997. Neural crest potential for tooth development in a urodele amphibian: developmental and evolutionary significance. Developmental Biology 188:34–42.

Green, W. T., Jr. 1971. Behavior of articular chondrocytes in cell culture. Clinical Orthopedics and Related Research 75:248–260.

Green, W. T., Jr., and R. J. Ferguson. 1975. Histochemical and electron microscopic comparison of tissue produced by rabbit articular chondrocytes in vivo and in vitro. Arthritis and Rheumatism 18:273–280.

Haines, R. W. 1969. Epiphyses and sesamoids; pp. 81–116 in C. Gans (ed.), Biology of the Reptilia. Volume 1, Morphology A. Academic Press, London and New York.

Hall, B. K. 1972. Immobilization and cartilage transformation into bone in the embryonic chick. Anatomical Record 173:391–404.

Hall, B. K. 1975. Evolutionary consequences of skeletal differentiation. American Zoologist 15:329–350.

Hall, B. K. 1977. Chondrogenesis of the somitic mesoderm. Advances in Anatomy, Embryology and Cell Biology 53, IV:1–50.

Hall, B. K. 1978. Developmental and Cellular Skeletal Biology. Academic Press, London and New York, 304 pp.

Hall, B. K. 1988. The embryonic development of bone. American Scientist 76:174–181.

Hall, B. K. 1998a. Evolutionary Developmental Biology. Kluwer Academic Publishers, Dordrecht, 491 pp.

Hall, B. K. 1998b. The bone; pp. 289–307 in P. Ferretti and J. Géraudie (eds.), Cellular and Molecular Basis of Regeneration: From Invertebrates to Humans. John Wiley & Sons, London.

Hall, B. K. 1999. The Neural Crest in Development and Evolution. Springer-Verlag, New York, 313 pp.

Hall, B. K. 2005a. Bones and Cartilage: Developmental and Evolutionary Skeletal Biology. Elsevier/Academic Press, London, 792 pp.

Hall, B. K. 2005b. Skeletal biology in an Evo-Devo-Palaeo lab. Palaeontological Association Newsletter 59:26–34.

Hall, B. K. (ed.) 1990–1994. Bone. Volumes 1–9. CRC Press, Boca Raton.

Halstead, L. B. 1969a. The Pattern of Vertebrate Evolution. Oliver & Boyd, Edinburgh, 209 pp.

Halstead, L. B. 1969b. The origin and early evolution of calcified tissues in the vertebrates. Proceedings of the Malacological Society of London 38:552–553.

Halstead, L. B. 1969c. Calcified tissues in the earliest vertebrates. Calcified Tissue Research 3:107–124.

Halstead, L. B. 1973. The heterostracan fishes. Biological Reviews of the Cambridge Philosophical Society 48:279–332.

Halstead Tarlo, L. B., and J. R. Mercer. 1966. Decalcified fossil dentine. Journal of the Royal Microscopical Society 86:137–140.

Hanken, J., and B. K. Hall. 1983. Evolution of the vertebrate skeleton. Natural History 92:28–39.

Helfrich, M. H. 2003. Osteoclast diseases. Microscopical Research Techniques 61:514–532.

Hirschmann, P. N., and C. A. Shuttleworth. 1976. The collagen composition of the mandibular joint of the foetal calf. Archives of Oral Biology 21:771–774.

Ho, T.-Y. 1967a. Relationship between amino acid content of mammalian bone collagen and body temperature as a basis for estimation of body temperature of prehistoric mammals. Comparative Biochemistry and Physiology 22:113–120.

Ho, T.-Y. 1967b. Comparative biochemistry of amino acid composition of bone and dentine collagens in Pleistocene mammals. Biochimica Biophysica Acta 133:568–573.

Horner, J. R., A. de Ricqlès, and K. Padian. 1999. Variation in dinosaur skeletochronology indicators: implications for age assessment and physiology. Paleobiology 25:295–304.

Horner, J. R., A. de Ricqlès, and K. Padian. 2000. Long bone histology of the hadrosaurid dinosaur *Maiasaura peeblesorum*: growth dynamics and physiology based on an ontogenetic series of skeletal elements. Journal of Vertebrate Paleontology 20:115–129.

Horner, J. R., K. Padian, and A. de Ricqlès. 2001. Comparative osteohistology of some embryonic and perinatal archosaurs: developmental and behavioral implications for dinosaurs. Paleobiology 27:39–58.

Hughes, D. R., J. R. Bassett, and L. A. Moffat. 1994a. Structure and origin of the tooth pedicel (the so-called bone of attachment) and dental-ridge bone in the mandibles of the sea breams *Acanthopagrus australis*, *Pagrus auratus* and *Rhabdosargus sarba* (Sparidae, Perciformes, Teleostei). Anatomy and Embryology 189:51–69.

Hughes, D. R., J. R. Bassett, and L. A. Moffat. 1994b. Histological identification of osteocytes in the allegedly acellular bone of the sea breams *Acanthopagrus australis*, *Pagrus auratus* and *Rhabdosargus sarba* (Sparidae, Perciformes, Teleostei). Anatomy and Embryology 190: 163–179.

Hunt, A. M. 1959. A description of the molar teeth and investing tissues of normal guinea pigs. Journal of Dental Research 38:216–231.

Huysseune, A. 1985. The opercular cartilage in *Astatotilapia elegans*. Fortschritte der Zoologie 30:371–373.

Huysseune, A. 1986. Late skeletal development at the articulation between upper pharyngeal jaws and neurocranial base in the fish, *Astatotilapia elegans*, with the participation of a chondroid form of bone. American Journal of Anatomy 177:119–137.

Huysseune, A. 1989. Morphogenetic aspects of the pharyngeal jaws and neurocranial apophysis in postembryonic *Astatotilapia elegans* (Trewavas, 1933) (Teleostei, Cichlidae). Meddedelingen van de Koninklijke Academie voor Wetenschappen, Letteren en Schone Kunsten van Belgie 51:11–35.

Huysseune, A. 1995. Phenotypic plasticity in the lower pharyngeal jaw dentition of *Astatoreochromis alluaudi* (Teleostei, Cichlidae). Archives of Oral Biology 40:1005–1014.

Huysseune, A. 2000. Skeletal system; pp. 307–317 in G. K. Ostrander (ed.), The Laboratory Fish: Part 4. Microscopic Functional Anatomy. Academic Press, San Diego.

Huysseune, A., and J.-Y. Sire. 1990. Ultrastructural observations on chondroid bone in the teleost fish *Hemichromis bimaculatus*. Tissue and Cell 22:371–383.

Huysseune, A., and J.-Y. Sire. 1992a. Development of cartilage and bone tissues of the anterior part of the mandible in cichlid fishes: a light and TEM study. Anatomical Record 233:357–375.

Huysseune, A., and J.-Y. Sire. 1992b. Bone and cartilage resorption in relation to tooth development in the anterior part of the mandible in cichlid fishes: a light and TEM study. Anatomical Record 234:1–14.

Huysseune, A., and J.-Y. Sire. 1998. Evolution of patterns and processes in teeth and tooth-related tissues in nonmammalian vertebrates. European Journal of Oral Science 106(Suppl. 1):437–481.

Huysseune, A., J.-Y. Sire, and F. J. Meunier. 1994. Comparative study of lower pharyngeal jaw structure in two phenotypes of *Astatoreochromis alluaudi* (Teleostei: Cichlidae). Journal of Morphology 221:25–43.

Huysseune, A., C. Van der Heyden, and J.-Y. Sire. 1998. Early development of the zebrafish (*Danio rerio*) pharyngeal dentition (Teleostei, Cyprinidae). Anatomy and Embryology 198:289–305.

Huysseune, A., W. Van den Berghe, and W. Verraes. 1986. The contribution of chondroid bone in the growth of the parasphenoid bone of a cichlid fish as studied by oblique computer-aided reconstruction. Biologisch Jaarboek DODONAEA 54:131–141.

Huysseune, A., and W. Verraes. 1986. Chondroid bone on the upper pharyngeal jaws and neurocranial base in the adult fish *Astatotilapia elegans*. American Journal of Anatomy 177:527–535.

Huysseune, A., and W. Verraes. 1990. Carbohydrate histochemistry of mature chondroid bone in *Astatotilapia elegans* (Teleostei, Cichlidae) with a comparison to acellular bone and cartilage. Annales des Sciences Naturelles-Zoologie et Biologie Animale 11:29–43.

Ishii, M., N. Suda, T. Tengan, S. Susuki, and T. Kuroda. 1998. Immunohistochemical findings of type-I and type-II collagen in prenatal mouse mandibular condylar cartilage compared with the tibial anlage. Archives of Oral Biology 43:545–550.

Ismail, M. H., W. Verraes, and A. Huysseune. 1982. Developmental aspects of the pharyngeal jaws in *Astatotilapia elegans* (Trewavas, 1933) (Teleostei, Cichlidae). Netherlands Journal of Zoology 32:513–543.

Jaffe, H. L. 1968. Tumors and Tumorous Conditions of the Bones and Joints. Lea & Febiger, Philadelphia, 629 pp.

Janvier, P. 1996. Early Vertebrates. Oxford Monographs on Geology and Geophysics 33. Clarendon Press, Oxford, 393 pp.

Jarvik, E. 1959. Dermal fin-rays and Holmgren's principle of delamination. Kunglingar Svenska Vetenskapensakademiens Handlingar, ser. 4, 6(1):3–51.

Jundt, G., K.-H. Berghäuser, J. D. Termine, and A. Schulz. 1987. Osteonectin—A differentiation marker of bone cells. Cell and Tissue Research 248:409–415.

Kantomaa, T. 1986a. The effect of prenatally increased oxygen tension on the development of the mandibular condyle. Acta Odontologica Scandinavica 44:301–305.

Kantomaa, T. 1986b. The effect of increased oxygen tension on the growth of the mandibular condyle. Acta Odontologica Scandinavica 44:307–312.

Kemp, A. 2003. Ultrastructure of developing tooth plates in the Australian lungfish, *Neoceratodus forsteri* (Osteichthyes: Dipnoi). Tissue and Cell 35:401–426.

Kemp, N. E. 1985. Ameloblastic secretion and calcification of the enamel layer in shark teeth. Journal of Morphology 184:215–230.

Kemp, N. E., and S. K. Westrin. 1979. Ultrastructure of calcified cartilage in the endoskeletal tesserae of sharks. Calcification of the endoskeleton in elasmobranchs. Journal of Morphology 160:75–102.

Kimura, S., and R. Matsui. 1990. Characterization of two genetically distinct type I-like collagens from hagfish (*Eptatretus burgeri*). Comparative Biochemistry and Physiology B 95:137–143.

Kölliker, A. von. 1859. On the different types in the microstructure of the skeleton of osseous fishes. Proceedings of the Royal Society of London 9:656–668.

Langille, R. M., and B. K. Hall. 1993. In vitro calcification of cartilage from the lamprey, *Petromyzon marinus*. Acta Zoologica 74:31–41.

Lauder, G. V. 1983. Functional design and evolution of the pharyngeal jaw apparatus in euteleostean fishes. Zoological Journal of the Linnean Society 77:1–33.

Lecointre, G. 1994. Aspects historiques et heuristiques de l'ichtyologie systématique. Cybium 18:339–430.

Li, M., N. Amizuka, K. Oda, K. Tokunaga, T. Ito, K. Takeuchi, R. Takagi, and T. Maeda. 2004. Histochemical evidence of the initial chondrogenesis and osteogenesis in the periosteum of a rib fractured model: implications of osteocyte involvement in periosteal chondrogenesis. Microscopical Research Techniques 64:330–342.

Liem, K. F. 1973. Evolutionary strategies and morphological innovations: cichlid pharyngeal jaws. Systematic Zoology 22:425–441.

Linde, A. (ed.) 1984. Dentin and Dentinogenesis. Volume 1. CRC Press, Boca Raton, 165 pp.

Listgarten, M. A. 1973. Intracellular collagen fibrils in the periodontal ligament of the mouse, rat, hamster, guinea pig and rabbit. Journal of Periodontal Research 8:335–342.

Maisey, J. G. 1988. Phylogeny of early vertebrate skeletal induction and ossification patterns. Evolutionary Biology 22:1–36.

Manzanares, M. C., M. Goret-Nicaise, and A. Dhem. 1988. Metopic sutural closure in the human skull. Journal of Anatomy 161:203–215.

Manolagas, S. C. 2000. Birth and death of bone cells: basic regulatory mechanisms and implications for the pathogenesis and treatment of osteoporosis. Endocrinology Reviews 21:115–137.

Mathews, M. B. 1967. Macromolecular evolution of connective tissue. Biological Reviews of the Cambridge Philosophical Society 42:499–551.

Mathews, M. B. (ed.) 1975. Connective Tissue Macromolecular Structure and Evolution. Molecular Biology, Biochemistry and Biophysics, Volume 19. Springer-Verlag, Berlin, 318 pp.

Matsui, R., M. Ishida, and S. Kimura. 1990. Characterization of two genetically distinct type I-like collagens from lamprey (*Entosphenus japonicus*). Comparative Biochemistry and Physiology, B 95:669–675.

Matsuoka, M., and T. Iwai. 1983. Adipose fin cartilage found in some teleostean fishes. Japanese Journal of Ichthyology 30:37–46.

McBurney, K. M., and G. M. Wright. 1996. Chondrogenesis of a noncollagen-based cartilage in the sea lamprey, *Petromyzon marinus*. Canadian Journal of Zoology 74:2118–2130.

McKee, M. D., S. Zalzal, and A. Nanci. 1996. Extracellular matrix in tooth cementum and mantel dentin: localization of osteopontin and other noncollagenous proteins, plasma proteins, and glycoconjugates by electron microscopy. Anatomical Record 245:293–312.

Meunier, F.-J. 1989. The acellularisation process in osteichthyan bone. Fortschritte der Zoologie 35:443–446.

Meunier, F.-J., and G. Desse. 1986. Les hyperostoses chez les téléostéens: description, histologie et problèmes étiologiques. Ichthyophysiologia Acta 10:130–142.

Miles, A. E. W. (ed.) 1967. Structural and Chemical Organization of Teeth. Volumes 1 and 2. Academic Press, London and New York, 1014 pp.

Miles, A. E. W., and D. F. G. Poole. 1969. The history and general organization of dentitions; pp. 1:3–44 in A. E. W. Miles (ed.), Structural and Chemical Organization of Teeth. Academic Press, London and New York.

Miller, W. A. 1979. Observations on the structure of mineralized tissues of the coelacanth, including the scales and their associated odontodes. Occasional Papers of the California Academy of Sciences 134:68–78.

Mizoguchi, I., M. Nakamura, I. Takahashi, M. Kagayama, and H. Mitrani. 1990. An immunohistochemical study of localization of type I and type II collagens in mandibular condylar cartilage compared with tibial growth plate. Histochemistry 93:593–600.

Mizoguchi, I., M. Nakamura, I. Takahashi, M. Kagayama, and H. Mitrani. 1992. A comparison of the immunohistochemical localization of type I and type II collagens in craniofacial cartilages of the rat. Acta Anatomica 144:59–64.

Mizoguchi, I., M. Nakamura, I. Takahashi, Y. Sasano, M. Kagayama, and H. Mitrani. 1993. Presence of chondroid bone on rat mandibular condylar cartilage. An immunohistochemical study. Acta Anatomica 187:9–15.

Mizoguchi, I., I. Takahashi, M. Nakamura, Y. Sasano, S. Sato, M. Kagayama, M., and H. Mitrani. 1996. An immunohistochemical study of regional differences in the distribution of type I and type II collagens in rat mandibular condylar cartilage. Archives of Oral Biology 41:863–869.

Mizoguchi, I., I. Takahashi, Y. Sasano, M. Kagayama, Y. Kuboki, and H. Mitrani. 1997a. Localization of types I, II and X collagen and osteocalcin in intramembranous, endochondral and chondroid bone of rats. Anatomy and Embryology 196:291–297.

Mizoguchi, I., I. Takahashi, Y. Sasano, M. Kagayama, and H. Mitrani. 1997b. Localization of type I, type II and type III collagen and glycosaminoglycans in the mandibular condyle of growing monkeys—an immunohistochemical study. Anatomy and Embryology 195:127–135.

Moss, M. L. 1961a. Studies on the acellular bone of teleost fish. I. Morphological and systematic variation. Acta Anatomica 46:343–362.

Moss, M. L. 1961b. Osteogenesis of acellular teleost bone. American Journal of Anatomy 108:99–110.

Moss, M. L. 1964a. The phylogeny of mineralized tissues. International Reviews in General and Experimental Zoology 1:297–331.

Moss, M. L. 1964b. Development of cellular dentin and lepidosteal tubules in the bowfin, *Amia calva*. Acta Anatomica 58:333–354.

Moss, M. L. 1968a. Bone, dentin, and enamel and the evolution of vertebrates; pp. 37–65 in P. Person (ed.), Biology of the Mouth. American Association for the Advancement of Science, Publication Number 89, Washington, D.C.

Moss, M. L. 1968b. Comparative anatomy of vertebrate dermal bone and teeth. I. The epidermal co-participation hypothesis. Acta Anatomica 71:178–208.

Moss, M. L. 1977. Skeletal tissues in sharks. American Zoologist 17:335–342.

Moss-Salentijn, L., and M. L. Moss. 1975. Studies on dentin. 2. Transient vasodentin in the incisor teeth of a rodent (*Perognathus longimembris*). Acta Anatomica 91:386–404.

Müller, P. K., and K. Kühn. 1977. Kollagenbiosynthese als ein Beispiel für die Regulation der Genexpression in mesenchymalen Zellen. Arzneimittel-Forschung 27:199–202.

Nalla, R. K., J. H. Kinney, and R. O. Ritchie. 2003. Effect of orientation on the in vitro fracture toughness of dentin: the role of toughening mechanisms. Biomaterials 24:3955–3968.

Nanci, A. 2003. Enamel: composition, formation, and structure; pp. 145–191 in A. Nanci (ed.), Ten Cate's Oral Histology: Development, Structure and Function. C. V. Mosby, St. Louis.

Nigrelli, R. F., and M. Gordon. 1946. Spontaneous neoplasms in fishes. I. Osteochondroma in the jewelfish, *Hemichromis bimaculatus*. Zoologica 31:89–92.

Nimni, M. E., and K. Deshmukh. 1973. Differences in collagen metabolism between normal and osteoarthritic human articular cartilage. Science 181:751–752.

Nomura, S., A. J. Wills, D. R. Edwards, J. K. Heath, and B. L. M. Hogan. 1988. Developmental expression of 2ar (osteopontin) and SPARC (osteonectin) RNA as revealed by in situ hybridization. Journal of Cell Biology 106:441–450.

Nordahl, J., K. Hollberg, S. Mengarelli-Widholm, G. Andersson, and F. P. Reinholt. 2000. Morphological and functional features of clasts in low phosphate, Vitamin D–deficiency rickets. Calcified Tissue International 67:400–407.

O'Driscoll, S. W., J. S. Fitzsimmons, and C. N. Commisso. 1997. Role of oxygen tension during cartilage formation by periosteum. Journal of Orthopaedic Research 15:682–687.

Ørvig, T. 1951. Histologic studies of placoderms and fossil elasmobranchs. I. The endoskeleton with remarks on the hard tissues of lower vertebrates in general. Arkiv för Zoologi, ser. 2, 2:323–456.

Ørvig, T. 1957. Palaeohistological notes. 1. On the structure of the bone tissue in the scales of certain Palaeonisciformes. Arkiv för Zoologi, ser. 2, 10:481–490.

Ørvig, T. 1965. Palaeohistological notes. 2. Certain comments on the phyletic significance of acellular bone tissue in early vertebrates. Arkiv för Zoologi, ser. 2, 16:551–556.

Ørvig, T. 1967. Phylogeny of tooth tissues: evolution of some calcified tissues in early vertebrates; pp. 45–110 in A. E. W. Miles (ed.), Structural and Chemical Organization of Teeth. Volume 1. Academic Press, London and New York.

Ørvig, T. (ed.) 1968. Current Problems of Lower Vertebrate Phylogeny. Nobel Symposium 4. Almqvist & Wiksell, Stockholm.

Ørvig, T. 1989. Histologic studies of ostracoderms, placoderms and fossil elasmobranchs. 6. Hard tissue of Ordovician vertebrates. Zoologica Scripta 18:427–446.

Osborn, J. W. (ed.) 1981. Dental Anatomy and Embryology. Blackwell Scientific Publications, Oxford, 447 pp.

Owen, T. A., M. S. Aronow, L. M. Barone, D. Bettencourt, G. S. Stein, and J. B. Lian. 1991. Pleiotropic effects of vitamin D on osteoblast gene expression are related to the proliferative and differentiated state of the bone cell phenotype: dependency upon basal levels of gene expression, duration of exposure and bone matrix competency in normal rat osteoblast cultures. Endocrinology 128:1496–1504.

Parfitt, A. M. 1988. Bone remodelling: relationship to amount and structure of bone, and the pathogenesis and prevention of fractures; pp. 45–93 in B. L. Riggs and L. J. Melton (eds.), Osteoporosis: Etiology, Diagnosis, and Management. Raven Press, New York.

Patterson, C. 1977. Cartilage bones, dermal bones and membrane bones, or the exoskeleton versus the endoskeleton; pp. 77–122 in S. M. Andrews, R. S. Miles, and A. D. Walker (eds.), Problems in Vertebrate

Evolution. Linnean Society Symposium Series 4. Academic Press, London and New York.

Pawlicki, R. 1975a. Studies of the fossil dinosaur bone in the scanning electron microscope. Zeitschrift für mikroskopisch-anatomische Forschung (Leipzig) 89:393–398.

Pawlicki, R. 1975b. Bone canaliculus endings in the area of the osteocyte lacuna. Electron-microscopic studies. Acta Anatomica 91:292–304.

Pawlicki, R. 1977a. Topochemical localization of lipids in dinosaur bone by means of Sudan B black. Acta Histochemica 59:40–46.

Pawlicki, R. 1977b. Histochemical reactions for mucopolysaccharides in dinosaur bone: studies on epon-embedded and methacrylate-embedded semithin sections as well as on isolated osteocytes and ground sections of bone. Acta Histochemica 58:75–78.

Pawlicki, R. 1978. Morphological differentiation of the fossil dinosaur bone cells: light, transmission electron–, and scanning electron–microscopic studies. Acta Anatomica 100:411–418.

Pawlicki, R. 1983. Metabolic pathways of the fossil dinosaur bones. I. Vascular communication system. Folia Histochemica Cytochemica 21:253–262.

Pawlicki, R. 1984. Metabolic pathways of the fossil dinosaur bones. Part IV. Modes of linkage between osteocytes and a variety of nexuses of osteocytic processes. Intermediary and other osteocytes in the system of metabolic pathways of dinosaur bone. Folia Histochemica Cytochemica 22:99–104.

Pawlicki, R. 1985. Metabolic pathways of the fossil dinosaur bones. 4. Morphological differentiation of osteocyte lacunae and bone canaliculi and their significance in the system of extracellular communication. Folia Histochemica Cytochemica 23:165–174.

Pritchard, J. J., and A. J. Ruzicka. 1950. Comparison of fracture repair in the frog, lizard and rat. Journal of Anatomy 84:236–261.

Prostak, K., P. Seifert, and Z. Skobe. 1990. The effects of colchicine on the ultrastructure of odontogenic cells in the common skate, Raja erinanae. American Journal of Anatomy 189:77–91.

Reid, R. E. H. 1981. Lamellar-zonal bone with zones and annuli in the pelvis of a sauropod dinosaur. Nature 292:49–51.

Reid, R. E. H. 1984. The histology of dinosaurian bone, and its possible bearing on dinosaurian physiology; pp. 629–664 in M. W. J. Ferguson (ed.), The Structure, Development and Evolution of Reptiles. Symposium of the Zoological Society of London 52. Academic Press, London and New York.

Reif, W.-E. 1978. Types of morphogenesis of the dermal skeleton in fossil sharks. Paläontologische Zeitschrift 52:110–128.

Reif, W.-E. 1980a. Development of dentition and dermal skeleton in embryonic Scyliorhinus canicula. Journal of Morphology 166:275–288.

Reif, W.-E. 1980b. A model of morphogenetic processes in the dermal skeleton of elasmobranchs. Neues Jahrbuch für Geologie und Paläontologie, Abhandlungen 159:339–359.

Reif, W.-E., and M. Richter. 2001. Revisiting the lepidomorial and the odontode regulation theories of dermoskeletal morphogenesis. Neues Jahrbuch für Geologie und Paläontologie, Abhandlungen 219:285–304.

Ricqlès, A. J. de. 1973. Recherches paléohistologiques sur les os longs des Tétrapodes. Thèse de Doctorat d'Etat, Université de Paris VII, 321 pp.

Ricqlès, A. J. de. 1974. Evolution of endothermy: histological evidence. Evolutionary Theory 1:51–80.

Ricqlès, A. J. de. 1976. On bone histology of fossil and living reptiles, with comments on its functional and evolutionary significance; pp. 123–150 in A. d'A. Bellairs and C. B. Cox (eds.), Morphology and Biology of Reptiles. Linnean Society Symposium Series 3. Academic Press, London and New York.

Ricqlès, A. J. de. 1989. Les mécanismes hétèrochroniques dans le retour des tétrapodes au milieu aquatique. Géobios, Mémoire Spéciale 12:337–348.

Ricqlès, A. J. de. 2000. L'origine dinosaurienne des oiseaux et de l'endothermie avienne: les arguments histologiques. Année Biologie 39: 69–100.

Roach, H. I. 1997. New aspects of enchondral ossification in the chick: chondrocyte apoptosis, bone formation by former chondrocytes, and acid phosphatase activity in the enchondral bone matrix. Journal of Bone and Mineral Research 12:795–805.

Rodland, D. L., M. Kowalewski, D. L. Dettman, K. W. Flessa, V. Atudorei, and Z. D. Sharp. 2003. High-resolution analysis of delta[18]O in the biogenic phosphate of modern and fossil lingulid brachiopods. Journal of Geology 111:441–453.

Romer, A. S. 1963. The ancient history of bone. Annals of the New York Academy of Sciences 109:168–176.

Romer, A. S. 1970. The Vertebrate Body. W. B. Saunders Company, Philadelphia, 644 pp.

Roy, P. K., P. E. Witten, B. K. Hall, and S. P. Lall. 2004. Effects of dietary phosphorus on bone growth and mineralization of vertebrae in haddock (*Melanogrammus aeglefinus* L.). Fish Physiology and Biochemistry 27:35–48.

Ruth, E. B. 1934. The os priapi: a study in bone development. Anatomical Record 60:231–249.

Saint-Paul, U., and G. Bernardino. 1988. Behavioural and ecomorphological responses of the Neotropical pacu *Piaractus mesopotamicus* (Teleostei; Serrasalmidae) to oxygen-deficient waters. Experimental Biology (Berlin) 48:19–26.

Santone, P. 1935. Studien über den Aufbau, die Struktur und die Histogenese der Molaren der Säugetiere. Zeitschrift für mikroskopische und anatomische Forschung 34:49–100.

Sasagawa, I. 2002. Mineralization patterns in elasmobranch fish. Microscopical Research Techniques 59:396–407.

Sasaki, T. 2003. Differentiation and functions of osteoclasts and odontoclasts in mineralized tissue resorption. Microscopical Research Techniques 61:483–495.

Sasano, Y., M. Furusawa, H. Ohtani, I. Mizoguchi, I. Takahashi, and M. Kayogama. 1996. Chondrocytes synthesize type I collagen and accumulate the protein in the matrix during development of rat tibial articular cartilage. Anatomy and Embryology 194:247–252.

Satchell, P. G., X. Anderton, O. H. Ryu, X. Luan, A. J. Ortega, R. Opamen, B. J. Berman, D. E. Witherspoon, J. L. Gutmann, A. Yamane, M. Zeichner-David, J. P. Simmer, C. F. Shuler, and T. G. H. Diekwisch. 2002. Conservation and variation in enamel protein distribution during vertebrate tooth development. Journal of Experimental Zoology, Part B: Molecular Development and Evolution 284:91–106.

Schaedler, J. M., L. Krook, J. A. M. Wootton, B. Hover, B. Brodsky, M. D.

Naresh, D. D. Gillette, D. B. Madsen, R. H. Horne, and R. R. Minor. 1992. Studies of collagen in bone and dentin matrix of a Columbian Mammoth (late Pleistocene) of central Utah. Matrix 12:297–307.

Scott, J. E., and E. W. Hughes. 1981. Chondroitin sulphate from fossilized antlers. Nature 291:580–581.

Scott, J. H., and N. B. B. Symons. 1974. Introduction to Dental Anatomy. Seventh edition. Churchill, London, 460 pp.

Shaw, J. L., and C. A. L. Bassett. 1967. The effects of varying oxygen concentration on osteogenesis and embryonic cartilage in vitro. Journal of Bone and Mineral Research British Volume 49:73–80.

Shellis, R. P., and A. E. W. Miles. 1974. Autoradiographic study of the formation of enameloid and dentine matrices in teleost fishes using tritiated amino acids. Proceedings of the Royal Society of London B 185:51–72.

Shibata, S., and Y. Yamashita. 2001. An ultrastructural study of osteoclasts and chondroclasts in poorly calcified mandible induced by high doses of strontium diet to fetal mice. Annals of Anatomy 183:357–361.

Silbermann, M., and K. von der Mark. 1990. An immunohistochemical study of the distribution of matricial proteins in the mandibular condyle of neonatal mice. I. Collagens. Journal of Anatomy 170:11–22.

Silbermann, M., K. von der Mark, and D. Heinegard. 1990. An immunohistochemical study of the distribution of matricial proteins in the mandibular condyle of neonatal mice. II. Non-collagenous proteins. Journal of Anatomy 170:23–31.

Silbermann, M., D. Lewinson, H. Gonen, M. A. Lizarbe, and K. von der Mark. 1983. In vitro transformation of chondroprogenitor cells into osteoblasts and the formation of new membrane bone. Anatomical Record 206:373–383.

Silbermann, M., A. H. Reddi, A. R. Hand, R. Leapman, K. von der Mark, and A. Franzen. 1987a. Chondroid bone arises from mesenchymal stem cells in organ culture of mandibular condyles. Journal of Craniofacial Genetics and Developmental Biology 7:59–80.

Silbermann, M., H. Tenenbaum, E. Livne, E., R. Leapman, K. von der Mark, and A. H. Reddi. 1987b. The in vitro behavior of fetal condylar cartilage in serum-free hormone-supplemented medium. Bone 8:117–126.

Simmons, D. J. 1971. Calcium and skeletal tissue physiology in teleost fishes. Clinical Orthopedics and Related Research 76:244–280.

Simmons, N. B., and J. H. Marshall. 1970. The uptake of calcium[45] in the acellular-boned toadfish. Calcified Tissue Research 5:206–221.

Sire, J.-Y., and A. Huysseune. 2003. Formation of dermal skeletal and dental tissues in fish: a comparative and evolutionary approach. Biological Reviews of the Cambridge Philosophical Society 78:219–249.

Sire, J.-Y., A. Huysseune, and F. J. Meunier. 1990. Osteoclasts in teleost fish—light microscopical and electron microscopical observations. Cell and Tissue Research 260:85–94.

Smit, T. H., E. H. Burger, and J. M. Huyghe. 2002. A case for strain-induced fluid flow as a regulator of BMU-coupling and osteonal alignment. Journal of Bone and Mineral Research 17:2021–2029.

Smith, B. H., and H. B. Taylor. 1969. The occurrence of bone and cartilage in mammary tumours. American Journal of Clinical Pathology 51:610–618.

Smith, M. M. 1991. Putative skeletal neural crest cells in early Late Ordovician vertebrates from Colorado. Science 251:301–303.

Smith, M. M., and B. K. Hall. 1990. Developmental and evolutionary origins of vertebrate skeletogenic and odontogenic tissues. Biological Reviews of the Cambridge Philosophical Society 65:277–374.

Smith, M. M., and B. K. Hall. 1993. A developmental model for evolution of the vertebrate exoskeleton and teeth: the role of cranial and trunk neural crest. Evolutionary Biology 27:387–448.

Smith, M. M., and A. E. W. Miles. 1971. The ultrastructure of odontogenesis in larval and adult urodeles; differentiation of the dental epithelial cells. Zeitschrift für Zellforschung und mikroskopische Anatomie 121:470–498.

Smith, M. M., and I. J. Sansom. 1997. Exoskeletal micro-remains of an Ordovician fish from the Harding Sandstone of Colorado. Palaeontology 40:645–658.

Smith, M. M., and I. J. Sansom. 2000. Evolutionary origins of dentine in the fossil record of early vertebrates: diversity, development and function; pp. 65–81 in M. F. Teaford, M. M. Smith, and M. W. J. Ferguson (eds.), Development, Function and Evolution of Teeth. Cambridge University Press, Cambridge.

Smith, M. M., M. H. Hobdell, and W. A. Miller. 1972. The structure of the scales of *Latimeria chalumnae*. Journal of Zoology (London) 167: 501–509.

Smith-Vaniz, W. F., L. S. Kaufman, and J. Glowacki. 1995. Species-specific patterns of hyperostosis in marine teleost fishes. Marine Biology 121:573–580.

Sperber, G. H. 2001. Craniofacial Development. B. C. Decker, Hamilton, Ontario.

Spigelius, A. 1631. De Formato Foetu. Liber Singularis. Opera Posthuma Studio Liberalis Cremae Tarvisni. M. Merianus, Frankfurt.

Tarlo, L. B. H. 1964. The origin of bone; pp. 3–15 in H. J. J. Blackwood (ed.), Bone and Tooth. Proceedings of the First European Bone and Tooth Symposium. Pergamon Press, Oxford.

Taylor, L. H., B. K. Hall, and D. K. Cone. 1993. Experimental infections of yellow perch (*Perca flavescens*) with *Apophallus brevis* (Digenia, Heterophyidae): parasite invasion, encystment and bony ossicle development. Canadian Journal of Zoology 71:1886–1894.

Taylor, L. H., B. K. Hall, T. Miyake, and D. K. Cone. 1994. Ectopic ossicles associated with metacercaria of *Apophallus brevis* (Trematoda) in yellow perch, *Perca flavescens* (Teleostei): development and identification of bone and chondroid bone. Anatomy and Embryology 190:29–46.

Tuominen, M., T. Kantomaa, P. Pirttiniemi, and A. Poikela. 1996. Growth and type-II collagen expression in the glenoid fossa of the temporomandibular joint during altered loading—A study in the rat. European Journal of Orthodontics 18:3–9.

Tziafas, D. 2004. The future role of a molecular approach to pulp-dentinal regeneration. Caries Research 38:314–320.

Ulrich, M. M. W., W. R. K. Perizonius, C. F. Spoor, P. Sandberg, and C. Vermeer. 1987. Extraction of osteocalcin from fossil bones and teeth. Biochemical and Biophysical Research Communications 149:712–719.

Vaes, G. N., and G. Nichols. 1962. Oxygen tension and the control of bone cell metabolism. Nature 193:379–380.

Verraes, W., M. Ismail, and A. Huysseune. 1979. Developmental aspects of the pharyngeal jaws and the pharyngo-branchial neurocranial apophysis in *Haplochromis elegans* (Teleostei: Cichlidae). American Zoologist 19:1013.

Vickaryous, M., T. J. Fedak, T. Franz-Odendaal, B. K. Hall, and J. Stone. 2003. Dusting off bone ontologies: considering skeletons in the palaeo-closet. Palaeontological Association Newsletter 53:48–51.

Virchow, R. 1853. Das normale Knochenwachsthum und die rachitische Störung desselben. Archiv für Anatomie and Physiologie 5:409.

von der Mark, K., and H. von der Mark. 1977. The role of three genetically distinct collagen types in endochondral ossification and calcification of cartilage. Journal of Bone and Mineral Research British Volume 59:458–464.

Wake, M. H. 1998. Cartilage in the cloaca: phallodeal spicules in caecilians (Amphibia: Gymnophiona). Journal of Morphology 237:177–186.

Weidenreich, F. 1930. Das Knochengewebe; pp. 391–520 in W. von Möllendorff (ed.), Handbuch der mikroskopischen Anatomie des Menschen, Band 2. Die Gewebe, Teil 2. Springer, Berlin.

Weiss, R. E., and N. Watabe. 1978a. Studies on the biology of fish bone. I. Bone resorption after scale removal. Comparative Biochemistry and Physiology (A) 60:207–211.

Weiss, R. E., and N. Watabe. 1978b. Studies on the biology of fish bone. II. Bone matrix changes during resorption. Comparative Biochemistry and Physiology (A) 61:245–252.

Weiss, R. E., and N. Watabe. 1979. Studies on the biology of fish bone. III. Ultrastructure of osteogenesis and resorption in osteocytic (cellular) and anosteoctyic (acellular) bones. Calcified Tissue International 28:43–56.

Willis, R. A. 1962. The Borderland of Embryology and Pathology. Butterworth, London, 641 pp.

Wislocki, G. B., H. L. Weatherford, and M. Singer. 1947. Osteogenesis of antlers investigated by histological and histochemical methods. Anatomical Record 99:265–284.

Witten, P. E. 1997. Enzyme histochemical characteristics of osteoblasts and mononucleated osteoclasts in a teleost fish with acellular bone (*Oreochromis niloticus*, Cichlidae). Cell and Tissue Research 287:591–599.

Witten, P. E., and B. K. Hall. 2002. Differentiation and growth of the kype skeletal tissue in anadromous male Atlantic salmon (*Salmo salar*). International Journal of Developmental Biology 46:719–730.

Witten, P. E., and B. K. Hall. 2003. Seasonal change in the lower jaw skeleton in male Atlantic salmon (*Salmo salar* L.): remodelling and regression of the kype after spawning. Journal of Anatomy 203:435–450.

Witten, P. E., and A. Huysseune. 2006. Mechanisms of chondrogenesis and osteogenesis in fins; pp. 79–92 in B. K. Hall (ed.), Fins and Limbs: Evolution, Development and Transformation. University of Chicago Press, Chicago.

Witten, P. E., and W. Villwock. 1997. Growth requires bone resorption at particular skeletal elements in a teleost fish with acellular bone (*Oreochromis niloticus*, Teleostei: Cichlidae). Journal of Applied Ichthyology 13:149–158.

Witten, P. E., A. Hansen, and B. K. Hall. 2001. Features of mono- and multinucleated bone resorbing cells of the zebrafish *Danio rerio* and their contribution to skeletal development, remodeling and growth. Journal of Morphology 250:197–207.

Witten, P. E., H. Rosenthal, and B. K. Hall. 2004b. Mechanisms and consequences of the formation of a kype (hook) on the lower jaw of male Atlantic salmon (*Salmo salar* L.). Mitteilungen aus dem Hamburgischen Zoologischen Museum und Institut 101:149–156.

Witten, P. E., L. S. Holliday, G. Delling, and B. K. Hall. 1999. Immunohistochemical identification of a vacuolar proton pump (V-ATPase) in bone-resorbing cells of an advanced teleost species (*Oreochromis niloticus*, Teleostei, Cichlidae). Journal of Fish Biology 55:1258–1272.

Witten, P. E., W. Villwock, N. Peters, and B. K. Hall. 2000. Bone resorbing and bone remodeling cells in juvenile carp (*Cyprinus carpio*). Journal of Applied Ichthyology 16:254–261.

Witten, P. E., L. Gil-Martens, B. K. Hall, A. Huysseune, and A. Obach. 2005. Compressed vertebrae in Atlantic salmon (*Salmo salar*): evidence for metaplastic chondrogenesis as a skeletogenic response late in ontogeny. Diseases of Aquatic Organisms 64:237–264.

Witten, P. E., A. Huysseune, T. Franz-Odendaal, T. J. Fedak, M. Vickaryous, A. G. Cole, and B. K. Hall. 2004. Acellular teleost bone: dead or alive, primitive or derived? Palaeontological Association Newsletter 55:37–41.

Wright, B. A., and M. M. Cohen Jr. 1983. Tumors of cartilage, pp. 3:143–164 in B. K. Hall (ed.), Cartilage. Volume 3, Biomedical Aspects. Academic Press, London and New York.

Wright, D. M., and B. C. Moffett Jr. 1974. The post-natal development of the human temporomandibular joint. American Journal of Anatomy 141:235–250.

Wright, G. M., F. W. Keely, and P. Robson. 2001. The unusual cartilaginous tissues of jawless craniates, cephalochordates and invertebrates. Cell and Tissue Research 304:165–174.

Yagami, K., J.-Y. Suh, M. Enomoto-Iwamoto, E. Koyama, W. R. Abrams, I. M. Shapiro, M. Pacifici, and M. Iwamoto. 1999. Matrix GLA protein is a developmental regulator of chondrocyte mineralization and, when constitutively expressed, blocks endochondral and intramembranous ossification in the limb. Journal of Cell Biology 147:1097–1108.

Yamamoto, T., T. Domon, S. Takahashi, A. K. S. Arambawatta, and M. Wakita. 2004. Immunolocalization of proteoglycans and bone-related noncollagenous glycoproteins in developing acellular cementum of rat molars. Cell and Tissue Research 317:299–312.

Yasui, N., K. Ono, H. Konomi, and Y. Nagai. 1984. Transitions in collagen types during endochondral ossification in human growth cartilage. Clinical Orthopedics and Related Research 183:215–218.

2. Homologies and Evolutionary Transitions in Early Vertebrate History

Philippe Janvier

ABSTRACT

The major two morphological gaps in extant euchordate anatomy, the cephalochordate-vertebrate gap and the cyclostome-gnathostome gap, are reviewed in light of paleontological data to illustrate possible cases of evolutionary transitions in euchordate evolution. The Early Cambrian myllokunmingiids probably possessed neural crest– and epidermal placode–derived structures, but display character combinations that suggest they are stem vertebrates and, possibly alongside yunnanozoans, contribute to fill the cephalochordate-vertebrate gap. Although hagfishes and lampreys are known since the Carboniferous, their relationships to the other fossil jawless vertebrate taxa remain obscure. Euphaneropids and possibly anaspids display some lamprey-like features that may turn out to be general for the clade that includes lampreys and gnathostomes. Among the jawless stem gnathostomes ("ostracoderms"), osteostracans are currently regarded as the closest relatives to jawed gnathostomes, but share with lampreys some unique and strikingly similar characters, such as the structure of the nasohypophysial complex, that are currently regarded as homoplasies and therefore overlooked in narratives about early vertebrate evolution. The characters that support the current theories of vertebrate interrelationships are briefly reviewed and discussed. Despite some incongruent character distributions, "ostracoderms" provide a means for establishing the sequence in which gnathostome characters appeared. Finally, the question of the origin of jaws is discussed in light of recent data provided by developmental biology. Current vertebrate phylogenies, coupled with process-based narratives inferred from developmental data, may provide scenarios of evolutionary transitions. However, returning to problematical characters that display incongruent distribution patterns and searching for more data of good quality are probably more rewarding in the long term to generate new phylogenies.

Transitions and Character Transformations

Science has long turned to fossils as evidence for evolutionary transitions through time, particularly those transitions associated with the rise of the major vertebrate clades. *Vertebrate Paleontology and Evolution,* by Robert L. Carroll (1988), is a perfect example of the defense and justification of evolutionary paleontology, and is therefore widely used for teaching purposes. Like many of us, Carroll joined the cladistic revolution, but he did not argue much with the critics of paleontological data, even when fossils were under heavy fire, notably after the famous (or, to some, infamous) article by Rosen et al. (1981) about lungfish-tetrapod relationships. He pragmatically relies on characters—that is, the anatomical structures as we actually see them, and their homology—as evidence for relationships, but geological time remains his guide. Such paleontology-based scenarios of evolutionary transition, as the paired fin-limb transition or the rise of the mammalian middle ear, through a long series of "transitional" fossil forms are thus Carroll's favorite subjects. In the 1980s, Carroll and I were in different camps. He believed in the power of fossils as direct evidence for evolutionary transitions and evolution through time. In contrast, I considered that the antiquity of fossils was not a particular property that gave them any more weight: fossils were merely useful additional sources of characters, yet less informative than extant organisms in terms of actual observable data.

His reservations about certain issues in cladistics notwithstanding, Carroll published his first (and so far only) computer-assisted cladistic analysis of the choanates, with particular reference to early fossil tetrapods (Carroll, 1995). As a coeditor of the volume in which this article appeared, I remember Carroll sending me a note that alluded to his skepticism as to the trees he obtained. The conclusion of his article clearly showed that his preference goes to actual, chronologically ordered series of fossils, and ended with a quote from Scott and Janis (1993:300): "[C]omputer programs, . . . like the White Queen, can believe six impossible things before breakfast." Surprisingly, this echoed the criticism by Platnick et al. (1996:245) of standard parsimony and Farris's optimization, which considers all possible most-parsimonious cladograms, even though some are impossible things.

Now, I would perhaps write a somewhat similar conclusion as Carroll's (1995) regarding current cladistic computer programs. The weight of optimization, probabilities, and statistics in phylogeny reconstruction seems to eclipse the consideration of the characters and their definition, limitation, composition, or coding. There are more and more trees, some of which, although most parsimonious, make no sense in terms of plausible character distribution. However, the main problem lies in the fact that from these innumerable trees, optimizations generate ancestral character states at nodes. Programs therefore produce a series of virtual evolutionary transformations that are ultimately described and discussed as if they were actually observed from the fossil record or by any other material means. This strikingly recalls the evolutionary tradition of reconstructing hypothetical intermediates. The difference is that because of the mixture of homoplasy and homology (and missing data) in most data sets, the ancestral character states are now a consequence of optimizations from "data that are always significantly less than perfect" (Nelson, 1994:111). Such trees are said to be predictive, but are these virtual transformations mere predictions, amenable to a test by actual organisms, be they extant or fossil? Theoretically, yes, but their frequently poor or variable support at nodes and successive changes from one analysis to the other make them be quickly forgotten, before any test turns up. A strict consensus of most parsimonious trees that would display a majority of unresolved nodes would be dismissed as meaningless or poorly resolved, whatever the strength of the support to the few nodes it may display. Yet such poorly resolved trees are sometimes the best ones that can reasonably be obtained, and on the basis of which one may seriously investigate further. Kearney (2002) pointed out the contradiction between the search for fully resolved trees and the inevitable biases that result from optimization of missing data in fossil taxa. She offered the same solution: a consideration of the few, robust nodes, and a return to the characters that show conflicting distributions.

In common with many cladists of the 1980s, I was inclined toward considering evolutionary transitions as more or less plausible tales that could only be inferred from character distribution. Nevertheless, evolutionary transitions from ancestor to descendent could not be narrated without any appeal to paraphyly, and thus these transitions did not cope with phylogenetic systematics. In other words, the story that gnathostomes could be derived from "agnathans" was pointless, because "agnathans" do not exist as a taxon. Moreover, an evolutionary transition process requires an adaptationist scenario, generally constructed on the basis of an array of functional, environmental, or paleoenvironmental arguments, the core of which is the organism-environment interaction. Systematics is not supposed to use such causal explanations for justifying one particular theory of relationships, but once character transformation is invoked, they become necessary.

The notion of character transformation, from a plesiomorphic to one or more apomorphic states, has been regarded by Nelson (1994:127) as the last remnant of the evolutionary approach in cladistics. Saying that, for example, the jaws are derived from "nonjaws," be it the velar skeleton of lampreys or anything else, is just as uninformative as saying that the gnathostomes are derived from "agnathans." Although such statements still frequently appear in textbooks, they merely remain a veil thrown over ignorance, i.e., the ignorance of the precise homologue of jaws in any other taxon (Kuratani et al., 2002). Nelson's view is thus that only the presence of a character in different organisms or taxa (i.e., the same part of the organisms, sometimes with different names) has a meaning in term of systematics. One can tell whether a character is, or is not, but saying that it *was not* and then *is* is already a first step toward an imaginary process of evolutionary transition. Here begins the search for possible transformation processes based on such kinds of data, as ontogeny, biomechanics, or intermediate forms, generally fossils.

I might have had different views were I working on such remarkable evolutionary series as early synapsids. However, my experience of early jawless vertebrates was extremely frustrating in this respect. All the taxa I could consider, be they extant (hagfishes, lampreys, gnathostomes) or extinct (e.g., arandaspids, heterostracans, galeaspids, osteostracans), were clades (with the possible exception of thelodonts), long recognized as such, and any evolutionary transition or ancestor-descendent relationship between one group and others was ruled out. All that the fossil data and character distribution told me was that if gnathostomes evolved from jawless vertebrates, their closest relatives were some "ostracoderms"—that is, fossil jawless vertebrates possessing a calcified endo- and exoskeleton, but showing no evidence for incipient jaws, pre-jaws, or anything else, and currently referred to as *stem gnathostomes* (Donoghue et al., 2000). At any rate, there was no clear evidence that "agnathans" in general, and "ostracoderms" in particular, were clades. The anatomically informative fossil clades that are currently available and relevant to this question are barely more numerous than they were about a century ago, with the notable exception of galeaspids. All that we have are different tools for assessing character distributions, the most important of which is that absence means nothing.

The way I have been considering fossils during the past two decades or so was therefore necessarily different from the once classical, evolutionary approach that was based on assumptions about character transformations and ancestor-descendent relationships between higher taxa ordered through time. To me, it is an interesting exercise to write about evolutionary transitions in early vertebrates. In fact, these transitions, if any, only rest on an apparent hierarchy of certain homologues, less and less general in distribution, but there is no taxon among either jawless vertebrates or early gnathostomes that can be referred to as a fossil *transitional form* (by the virtue of their unique character combination)—reputations that have been bestowed on organisms such as *Panderichthys*, *Acanthostega*, or *Archaeopteryx*.

Major Morphological Gaps in Extant Euchordates

The amount of morphological data that can possibly throw some light on evolutionary transitions in the history of nongnathostome Euchordata (a name that predates Myomerozoa [Bjerring, 1984; see Carter, 1957]—i.e., cephalochordates and vertebrates) is desperately scarce. For example, all the characters that are currently used for reconstructing the progressive acquisition of the gnathostome body plan are known from such fossil jawless vertebrate taxa as heterostracans, galeaspids, and osteostracans, which are generally much derived in other respects (the so-called evolutionary dead ends). This situation compares to what would be the reconstruction of the evolutionary process that gave rise to the structure of eutherian mammals by only using such fossil taxa as edaphosaurs, dinocephalians, and dicynodonts. In addition, a long segment of early euchordate and vertebrate evolution took place before the rise of a mineralized skeleton and thus concerns organisms that are either rarely fossilized, or, if fossilized at all, provide little reliable information. Nevertheless, without these few fossil taxa, generally early Paleozoic in age, our conceptions of the relationships between the major present-day vertebrate taxa would perhaps be the same as those currently accepted, but our knowledge of the order in which the less and less general characters appeared throughout time would be virtually lacking (Donoghue and Sansom, 2002), unless inferred from the development of extant taxa, following von Baer's law.

The justification for the search for evolutionary transitions is the existence of gaps (generally morphological) in the overall resemblance of taxa. Without these gaps, evolutionary transitions would be as obvious as ontogeny, and thus their elucidation would be theoretically pointless. The major two morphological gaps among extant euchordates are between the cephalochordates and the vertebrates (in the classical sense; that is, including hagfishes, lampreys, and gnathostomes), and between the cyclostomes (hagfishes and lampreys, whatever their relationships) and the gnathostomes (Fig. 2.1).

How do we define such gaps and rate them as major? Here again, there is evidently the burden of anthropocentrism and subjectivity, as pointed out long ago throughout the cladist's criticisms of evolutionary systematics. In morphology, these gaps are regarded as the expression of disparity (however it may be measured), and in molecular systematics, they are assessed on the basis of large genetic distances (whatever this may actually mean). In fact, morphological gaps are assumed when no taxon can be used to demonstrate the imaginary transformations between two widely different taxa and when the latter lack readily recognizable homologies; that is, homologies that rest on "common sense" and can be used as a first theory of relationship with which to start an analysis (primary homologies, in the broad sense; de Pinna, 1991), such as the jaws of a cod and a cow, or pectoral fins and forelimbs. Thus, even though one may, on various grounds, assume homology between, say, the velar skeleton (or part of it) of lampreys and jaws of gnathostomes, there appears a broad morphological gap between these two taxa.

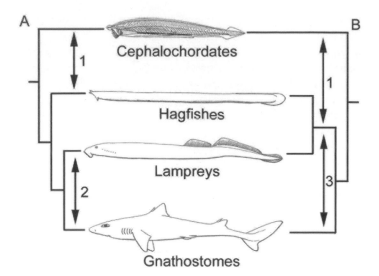

Figure 2.1. Two major morphological gaps in euchordate phylogeny. **A,** theory assuming cyclostome paraphyly; **B,** theory assuming cyclostome monophyly. **1,** cephalochordate-vertebrate gap; **2,** lamprey-gnathostome gap; **3,** cyclostome-gnathostome gap.

The relationships between the three chordate groups (i.e., tunicates, cephalochordates, and vertebrates) have been the subject of debate during the past three decades. Although there has long been a consensus on the sister-group relationship between cephalochordates and vertebrates, Jefferies (1986) advocated sister-group relationships between tunicates and vertebrates, and recent works on the developmental genetics of tunicates provided evidence for a vertebrate-like condition in the latter, as to certain structures, notably the placodes (Mazet et al., 2005). Assuming that the sister-group relationship between cephalochordates and vertebrates still is best supported by numerous anatomical, developmental, and molecular sequence data (Shimeld and Holland, 2000; Holland and Chen, 2001), there remains an impressive morphological gap between the two groups (1, Fig. 2.1). Therefore, attempts at reconstructing evolutionary transitions between the supposedly homologous characters of the two taxa remain largely imaginary (e.g., Butler, 2000), although ontogeny or developmental genetics may sometimes lend some support to one or the other of these inferred transformations, such as in the case of the median "eye" of cephalochordates and the paired eyes of vertebrates (Lacalli, 1996), or the placode precursors of tunicates and cephalochordates and the neurogenic placodes of vertebrates (Shimeld and Holland, 2000; Mazet et al., 2005). The cephalochordate-vertebrate gap begs at least two questions: first, how derived may cephalochordates be, relative to the common ancestor they shared with the vertebrates? And second, how much has the rise of the typical vertebrate neural crests and their adult derivatives contributed to give this gap an excessive importance?

The monophyly of extant cyclostomes is admittedly still debated (Delarbre et al., 2002), despite receiving increasingly strong molecular support (Hedges, 2001; Mallatt et al., 2001; Furlong and Holland, 2002; Fig. 2.1B). In contrast, the paraphyly of the cyclostomes, with lampreys being the sister-group of the gnathostomes among present-day vertebrates (Fig. 2.1A), is strongly supported by morphological and physiological characters (Løvtrup, 1977; Janvier, 1978, 1996a,b; Dingerkus, 1979; Janvier and Blieck, 1979; Forey, 1984; Jefferies, 1986; Donoghue et al., 2000; Donoghue and Smith, 2001). I shall not enter this debate again here. All I can say is that if cyclostomes form a clade, either hagfishes are the most extraordinary example of reversion among vertebrates, or lampreys and gnathostomes are the most extraordinary example of evolutionary convergence (Delarbre et al., 2002). There is admittedly a current trend toward considering that reversion is underestimated in morphology-based phylogenies, and that molecular sequence-based phylogenies can pinpoint where extensive morphological reversions have occurred (e.g., Jenner, 2004). This argument overlooks the numerous biases that occur when comparing morphology-based trees, which are generally generated by parsimony programs, with sequence-based trees, which are generated by a wide range of methods that involve increasingly more complex models of evolution. Character losses do occur, and the processes through which they occur are sometimes complex, but postulating them as soon as a new tree turns up on the basis of an inferred gene evolution probability is, I think, an abuse of inference.

Curiously, and despite the morphological and physiological differences between hagfishes on the one hand and lampreys and gnathostomes on the other (Hardisty, 1982), few authors have ever alluded to any major gap at the "cyclostome" level of vertebrate phylogeny, because hagfishes and lampreys share much the same overall structure (such as eel-shaped body, single median "nostril," more or less pouch-shaped gills, retractable lingual apparatus, and horny teeth), which is assessed as general relative to the gnathostome condition in the framework of cyclostome paraphyly (Løvtrup, 1977; Janvier, 1981b) and is unique in that of cyclostome monophyly (Schaeffer and Thomson, 1980; Yalden, 1985).

Whatever the status of the cyclostomes, the morphological gap between the gnathostomes and either hagfishes or lampreys remains impressive (Fig. 2.1). For example, a glimpse at the endoskeletal skulls of a hagfish, a lamprey, and a gnathostome, such as a shark (Fig. 2.2), shows that very few primary homologies that can be readily assumed in all three taxa. In fact, only the skeletal olfactory and otic capsules, although quite different in shape in the three taxa, can be regarded as homologous, essentially because they surround sensory organs that are, in turn, regarded as unlikely to be homoplastic. Some more likely homologies are only shared by two of the three taxa, such as the velar skeleton or the lingual apparatus in hagfishes and lampreys, or the arcualia of the axial skeleton in lampreys and gnathostomes (Fig. 2.2).

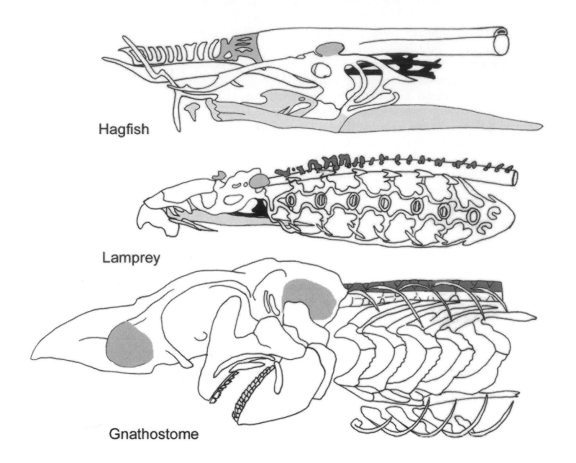

Hagfish

Lamprey

Gnathostome

Figure 2.2. Skull in the three major vertebrate taxa: hagfish, lampreys, and gnathostomes (exemplified here by a shark), showing some of the presumably homologous characters. The "lingual apparatus" (light gray) and the velar skeleton (black) are unique to hagfishes and lampreys and have no undisputed homologue in gnathostomes. The dorsal arcualia (dark gray) are unique to lampreys and gnathostomes. The olfactory and otic capsules (medium gray) are the only parts of the skull that are unanimously regarded as homologues in all three groups. Modified from Janvier (1996b).

Cambrian Euchordates and Presumptive Stem Vertebrates

Whether or not any fossil fills the cephalochordate-vertebrate gap remains unclear. Following the classical practice in paleontology, such fossils are generally looked for in the deep past of euchordate history, but on the vertebrate side of the gap, the earliest evidence for undisputed crown-group vertebrates are dermal armor fragments derived from arandaspids, from the Early Ordovician of Australia (Young, 1997). These, like other groups of "ostracoderms," are regarded as more closely related to the gnathostomes than to either lampreys or hagfish, essentially because they possess dermal bone (Fig. 2.3H). In current phylogenetic studies, arandaspids are grouped with heterostracans and astraspids into a very weakly supported clade Pteraspidomorphi (Fig. 2.3H–J), and, more recently, as the sister-group to astraspids (Sansom et al., 2005). They are thus not the most inclusive taxon of the dermal bone–bearing vertebrates (Janvier, 1996a; Donoghue et al., 2000; Donoghue and Smith, 2001; Donoghue and Sansom, 2002).

Deeper in time, a number of Cambrian fossils have been tentatively referred to the vertebrates on the basis of either their overall morphology or the histological structure of their mineralized parts, when present (Blieck, 1991; Janvier, 1997; Smith et al., 2001). Apart from the Myllokunmingiida (Fig. 2.3B), from the Early Cambrian of China (Shu et al., 1999, 2003a), none of them has yet been unanimously accepted as a vertebrate (Janvier, 2003). Among the presumed Cambrian vertebrates, two taxa have also received much attention during the last two decades: the euconodonts and *Anatolepis*.

Euconodonts

Euconodonts are known from the Late Cambrian to the Triassic (Fig. 2.3F), but the only two genera known from articulated, or partially articulated, specimens (i.e., showing both the assemblage of the denticles usually referred to as *conodonts,* and the imprint of the head and body soft tissues) are Early Ordovician and Early Carboniferous in age, respectively (Briggs et al., 1983; Gabbott et al., 1995). It is nevertheless assumed that euconodonts, as a whole, may have displayed much the same overall morphology. Although the few morphological features described from the head and body imprints of the articulated euconodonts are rather convincingly euchordate- and vertebrate-like (e.g., V-shaped myomeres; radials in caudal fin; large, paired optic capsules; Donoghue et al., 2000), most of the debates that arose in the 1990s about euconodont affinities were centered around the homology of the hard tissues that constitute their oral or pharyngeal denticles with those of the vertebrates. For reviews, see Aldridge and Purnell (1996), Schultze (1996), Donoghue and Aldridge (2001), and Kemp (2002). To date, there is no consensus on this question, but Donoghue et al. (2000) showed that hard tissue characters of euconodonts (i.e., whether they possess dentine, enamel, or bone) are not those that impose their position as crown-group vertebrates, and even as sister-group to "ostracoderms" plus gnathostomes in current vertebrate phylogenies. Euconodonts have also been regarded as the sister-group to either cephalochordates, crown-group vertebrates, hagfishes, or lampreys (see reviews in Aldridge and Donoghue, 1998; Donoghue et al., 2000), but an association with cephalochordates is less parsimonious because it would imply that the paired optic capsules either have appeared twice or are lost in cephalochordates. It has been pointed out that the myomeres of euconodonts seem to be V-shaped (as in cephalochordates), rather than W-shaped, as in crown-group vertebrates (and myllokunmingi ids; see below), but this is probably an artifact of preservation (Donoghue et al., 2000). Euconodonts share with crown-group vertebrates the presence of median fin radials and are best placed as stem gnathostomes, notably on the basis of their ability to develop mineralized skeletal elements made of apatite (Fig. 2.3F).

Anatolepis

Anatolepis is known from minute, tuberculate carapace fragments of Late Cambrian and Early Ordovician age; their histological structure recalls the classical vertebrate dentine (Smith and Sansom, 1995; Smith et al., 1996). However, none of the *Anatolepis* fragments recorded to date agrees with the structure of the individual dermal plates of the armor in classical "ostracoderm" taxa, nor

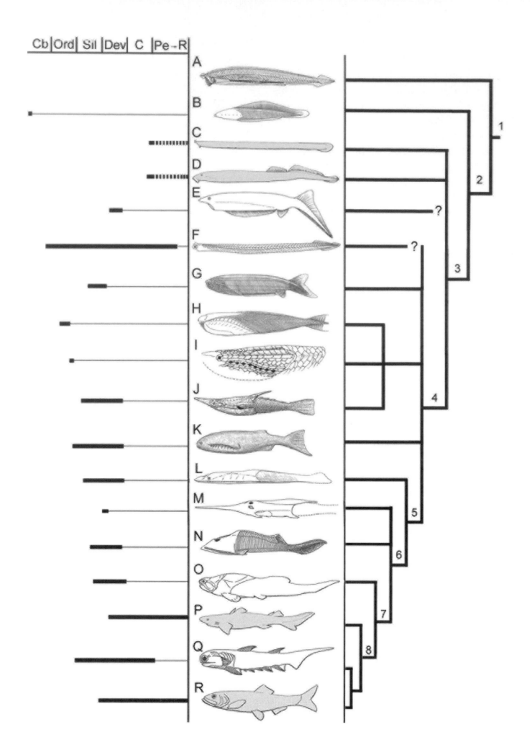

Figure 2.3. Synthesis (*not* a consensus tree generated by a computer program) of the major current theories of euchordate interrelationships (right), and stratigraphic distribution (thick bars, left) of the terminal taxa. All taxa are clades, except perhaps for thelodonts (**K**) and acanthodians (**Q**). Certain poorly known taxa (e.g., *Eriptychius*) are omitted. Extant taxa in gray. The conflict between the morphology- and molecular sequence–based theories regarding the hagfish-lamprey-gnathostome relationships is taken here into consideration, the relationships of the three groups being considered as unresolved. Taxa: **A**, cephalochordates;

do they show sensory-line canals or grooves that would be evidence for vertebrate epidermal placodes. In addition, some *Anatolepis* fragments are tubular in shape and have no equivalent in Paleozoic fishes. Yet Smith et al. (2001) tentatively compared them to the dermal bone sheath of the gill rakers in the extant whale shark. *Anatolepis* was once regarded as carapace fragments of an arthropod, but its histological structure apparently does not agree with this interpretation (Smith and Sansom, 1995). Currently, it is assumed that the name *Anatolepis* has been used for fragments of widely different origins, some of which may actually belong to arthropods, whereas others may belong to a vertebrate.

Admittedly, early evidence for dermal bone that can be readily referred to vertebrates known from articulated specimens may also display unusual histological structures, which differ from such classical vertebrate tissues as dentinous tissue (ortho-, meso-, and semi-dentine) or cellular bone. In anaspids and galeaspids, for example, the tubercles of the ornamentation are composed of a kind of acellular bone, and are thus entirely different from the dentinous tubercles of, for example, heterostracans, osteostracans, or gnathostomes. I even doubt that the galeaspid exoskeleton would have been referred to a vertebrate if complete head shields were not known. Thus, it cannot be ruled out that the early stages of the evolution of the vertebrate exoskeleton involved kinds of tissue structures outside the variation that we currently recognize in Paleozoic vertebrates. Judging from the histological structure of certain Ordovician vertebrate exoskeleton remains from Australia (Young, 1997), the United States (Sansom et al., 2001; Donoghue and Sansom, 2002), and Siberia (Karatajute-Talimaa and Smith, 2004), we can predict that entirely new types of histological structures may soon turn up in the earliest skeletonized vertebrates, and may make sense in systematics only

Figure 2.3. *(continued)*
B, Myllokunmingiida; C, hagfishes; D, lampreys; E, Euphaneropidae (also possibly included in the Anaspida); F, Euconodonta (relationships to or within the vertebrates still debated); G, Anaspida; H, Arandaspida; I, Astraspida; J, Heterostraci; K, Thelodonti (possibly nonmonophyletic); L, Galeaspida; M, Pituriaspida; N, Osteostraci; O, Placodermi; P, Chondrichthyes; Q, Acanthodii (possibly nonmonophyletic); R, Osteichthyes. Cb, Cambrian; Ord, Ordovician; Sil, Silurian; Dev, Devonian; C, Carboniferous; Pe–R, Permian to Recent. (A, C, D, J–R, after Janvier, 1996b; B, after Mallatt and Chen (2003a); E, based on pers. obs.; F, after Aldridge and Purnell, 1996; G, after Ritchie, 1964; H, after Gagnier, 1993a; I, based on Sansom et al., 2001). Higher taxa and selected morphological characters inferred at nodes: 1, Euchordata (=Myomerozoa): chevron-shaped myomeres, endostyle, sinus venosus in blood vascular system, segmentally arranged spinal nerves; 2, Vertebrata: neural crest and epidermal placodes in development, olfactory, optic and otic capsules, gill filaments, W-shaped myomeres; 3, crown group Vertebrata: cartilaginous radials in fins, nonsegmental gonads; 4, Gnathostomata (in stem-based classifications): extensive dermal skeleton on head and body, well-developed cerebellum, well-developed vertical semicircular canals forming distinct loops; 5, unnamed taxon: perichondrally calcified endoskeleton, externally open endolymphatic ducts; 6, unnamed taxon: distinct pectoral fins and endoskeletal shoulder girdle, cellular perichondral and dermal bone, sclerotic ring and scleral ossification (unless present in arandaspids), slit-shaped external gill openings, epicercal tail (the latter four characters evidenced only in osteostracans among "ostracoderms"); 7, Gnathostomata (in apomorphy-based classifications): jaws, pelvic fins and girdles, nasal cavities opening to the exterior by separate nostrils and disconnected from Rathke's pouch; 8, crown group Gnathostomata: superior oblique muscle attached in anterior part of the orbit, adductor jaw muscles lateral to palatoquadrate.

Figure 2.4. *(opposite page)* **A–C**, attempted reconstruction of the myllokunmingiid *Haikouichthys ercaicunensis* Shu et al. (1999), from the Lower Cambrian of Chengjiang, China. **A**, reconstruction in lateral view (size and shape of external gill openings hypothetical); **B**, presumed internal structures in lateral view; **C**, stains of the head in dorsal view (from Janvier, 2003, modified and completed on the basis of new data on the fin and tail structure provided by Zhang and Hou, 2004). **D**, possible phylogenetic position of the myllokunmingiids. Characters at nodes: **1**, neural crest and epidermal placodes in development, olfactory, optic and otic capsules, gill filaments, W-shaped myomeres; **2**, cartilaginous radials in fins, nonsegmental gonads. **Abbreviations used in figures: a**, anus; **af**, anal fin; **afac**, canal for the facial artery; **ama**, anterior mesenteric artery; **anc**, annular cartilage; **arc?**, possible arcualia; **asc**, anterior semicircular canal; **atr**, atrium; **bra?**, branchial arches; **bra1**, first gill arch; **bro**, external branchial openings; **brpl**, branchial plates; **cart**, hypothetical cartilage plate bearing the dermal oral plates; **cer**, cerebellum; **cf**, caudal fin; **cmvs**, canal for the marginal vein sinus; **cnhyd**, circum-nasohypophysial depression; **crhab**, recess for the right habenular ganglion; **cvcl**, canal for the lateral head vein; **da**, dorsal aorta; **da + oesc**, canal for the dorsal aorta and the esophagus; **dar**, dorsal arcualia; **dC**, duct of Cuvier; **dend**, open endolymphatic duct; **df**, dorsal fin; **df2**, second dorsal fin; **dt**, digestive tract; **el**, epichordal lobe; **exbrc**, extrabranchial cartilages; **gaimp?**, possible gill-arch impressions; **gf**, gill filaments; **gon?**, possible gonads; **gp**, gill pouches; **gs**, gill slits; **h**, pericardic cartilage or heart; **hl**, ventral horizontal lobe of caudal fin; **hm**, anterior head margin (oral hood?); **ht**, impressions of horny teeth; **hv**, hepatic vein; **hy**, hyoid arch; **hyhbrp**, posterior hyoid hemibranch; **hypc**, hypophysial cavity; **hyt**, hypophysial tube; **ic**, canal for the internal carotid artery; **lvsk**, lateral velar skeleton; **m**, mouth; **ma**, mandibular arch; **mdo**, median dorsal opening; **mins**, pits for muscle insertion; **ml**, hypothetical levator muscle; **mm**, ectomesenchyme surrounding the mandibular mesoderm; **mvp**, medial ventral processes for attachment of gill arches; **mvs**, marginal vein sinus; **mvsk**, medial velar skeleton; **my**, myomeres; **nac**, nasal cavity; **nc**, neural canal; **nch**, notochord; **nchc**, canal for the notochord; **nho**, nasohypophysial opening; **nhplac**, nasohypophysial placode; **nhv**, nasohypophysial valve; **es**, esophagus; **ol**, optic lobes; **olf**, olfactory capsule/organ; **olfplac**, olfactory placode; **oltr**, olfactory tract; **opl**, oral plates; **opt**, optic capsule; **optf**, optic fenestra; **orb**, orbit; **oro**, oral opening; **orv**, oral valve; **ot**, otic capsule; **paf**, preanal fin-fold; **pc**, piston cartilage; **pcard**, pericardial cavity; **pef**, pectoral fin; **pefe**, pectoral fin endoskeleton; **pelf**, pelvic fin; **pelfrad**, pelvic fin radials; **pf**, paired fin; **pfrad**, paired fin radials; **pif**, pineal foramen; **pm**, ectomesenchyme surrounding the premandibular mesoderm; **psc**, posterior semicircular canal; **rhab**, right habenular ganglion; **Rp**, Rathke's pouch; **sc**, single semicircular canal; **sk**, imprint of the skin; **spi**, spiracular canal; **st**, suspensory connective tissues for the esophagus; **tc**, tectal cartilages; **tcl**, tentacles; **tel**, telencephalic division of the brain cavity; **var**, ventral arcualia; **vcl**, lateral head vein or dorsal jugular vein (vena capitis lateralis); **vel**, velum; **ven**, ventricle; **vfl**, velar flap; **vj?**, possible ventral jugular vein; **V2–3**, maxillary and mandibular branches of the trigeminal nerve; **VII**, facial nerve.

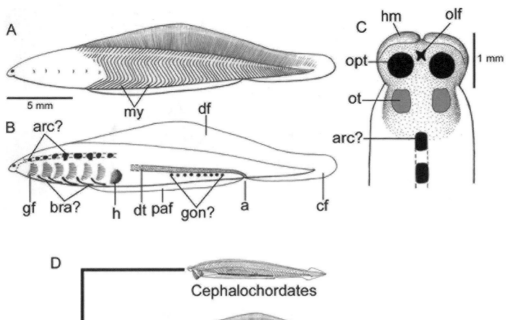

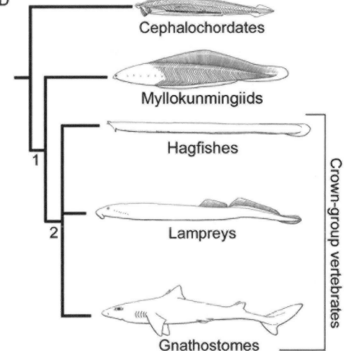

when these isolated remains are assigned to articulated specimens. However, whatever vertebrate *Anatolepis* may turn out to be, it is probable that it will fall, according to current phylogenies, somewhere among "ostracoderms"—that is, stem gnathostomes.

Myllokunmingiids

The Myllokunmingiida (*Haikouichthys, Myllokunmingia,* and *Zhongjianichthys*; Shu et al., 1999; Shu, 2003), from the Lower Cambrian of Chengjiang, China, are probably the most convincing Cambrian vertebrates known to date, although lacking mineralized tissues (Figs. 2.3B, 2.4). The differences among these three genera remain difficult to assess and may well be due to differences in size and preservation (Hou et al., 2002). However, their basic structure is now better known as a result of the discovery of about 500 specimens, referred to *Haikouichthys ercaicunensis,* which are preserved in different aspects and allow a more accurate reconstruction (Janvier, 2003; Shu et al., 2003a). To date, the identification of the myllokunmingiids as vertebrates has not been called into question. They provide quite convincing evidence for paired olfactory, optic and possibly otic capsules (olf, opt, ot, Fig. 2.4C), and six gills borne by skeletal arches (bra?, gf, Fig. 2.4B). This suggests the presence of epidermal placodes and neural crest–derived tissues, which are currently regarded as the most reliably unique vertebrate characters (Baker and Bronner-Fraser, 1997; Le Douarin and Kalcheim, 1999). Holland and Chen (2001) erected the taxon Cristozoa, i.e., organisms with neural crest, to accommodate the discovery of possible stem vertebrates, such as myllokunmingiids. In addition, their myomeres seem to display the characteristic W shape of the vertebrate myomeres, rather that the V shape in cephalochordates (Fig. 2.4A).

The first description of *Haikouichthys* and *Myllokunmingia* (Shu et al., 1999) raised questions about their having ventrolateral paired fins or fin folds, and the first phylogenetic analysis performed on the basis of only two specimens (referred to these two genera, respectively) suggested that *Myllokunmingia* was the sister-group of lampreys plus all other vertebrates except for hagfishes, whereas *Haikouichthys* appeared as the sister-group to lampreys (Shu et al., 1999; Donoghue et al., 2003). Further discoveries failed to provide evidence for paired fins in myllokumningiids, and it is now clear that they all possessed a single median dorsal fin or fin fold, extending from the head to the tip of the tail (df, cf, Fig. 4A; Zhang and Hou, 2004). Ventrally, a similar median fin extends from the tip of the tail to the posterior limit of the branchial apparatus (paf, Fig. 2.4B), with an interruption at the level of the anus (a, Fig. 4B). There is thus a preanal skin fold (or fin fold), as in adult hagfishes, larval lampreys, and certain larval gnathostomes. One of the puzzling features in the first descriptions of the myllokunmingiids was the numerous, closely set, and forward-tilted radials in the dorsal fin (Shu et al., 1999; Janvier, 2003). This is at odds with the condition in vertebrates, where the dorsal fin radials are generally tilted backward. However, a recently discovered specimen shows a perfectly preserved body and tail, and demonstrates that these presumed radials become vertical at the midlength of the body, and then increasingly tilted backward toward the tip of the tail (Fig. 2.4A; Zhang and Hou, 2004). Moreover, this specimen suggests that the radials may in fact not be true radials (i.e., endoskeletal structures), but either epidermal folds or collagenous structures comparable to the ceratotrichia of the gnathostomes (Zhang and Hou, 2004). The lack of endoskeletal radials in the unpaired fins could thus suggest that myllokunmingiids are not crown-group vertebrates, which all have radials (Figs. 3, 4D;

Zhang and Hou, 2004). The position of the notochordal lobe, relative to the dorsal and ventral fin webs, is still unclear in myllokunmingiids, although the specimen described by Zhang and Hou (2004: fig. 1) seems to show a slightly larger dorsal web that could foreshadow the hypocercal condition of most present-day and fossil "agnathans" (Figs. 2.4, 2.12).

Another odd feature of myllokunmingiids is the structure of the gills (cf. Fig. 2.4B). These display relatively long, posteriorly directed, filamentous structures, which recall those also found in yunnanozoans. Whether these structures are gill rays supporting the gill filaments, or whether they armed the interbranchial septa and gill covers remains undecided. Zhang and Hou (2004) even suggested that the gill ray–like structures extended externally through the gill openings, the shape and size of which remain unclear. These structures are suggestive of the gnathostome gill rays, which are lacking in extant jawless vertebrates. Yet the recent discovery of numerous gill ray–like endoskeletal elements in the Late Devonian anaspid-like vertebrate *Euphanerops* (P. Janvier, pers. obs., 2004) suggests that this character is possibly more general than previously believed.

Haikouichthys displays, in the anterior part of the body, a number of more or less elongated, sinuous stains, which have been interpreted as imprints of cartilages, some being gill arches, others possibly arcualia of the vertebral column (arc? bra?, Fig. 2.4B; Shu et al., 2003b). This, however, should be considered with great reservation. Notably, the presumed arcualia would be anomalously large, relative to the size of the animal, and do not extend beyond the posterior limit of the branchial region. I assume that all these imprints are more likely derived from the branchial skeleton. The presence of a pericardial cartilage (or the heart proper), in the form of the well-marked stain at the rear of the branchial apparatus, is probable (h, Fig. 2.4B), although it is not ruled out that this stain was left by the liver, as is frequently observed in soft-bodied fishes preserved as imprints.

Myllokunmingiids show no clear evidence for a braincase, but specimens preserved in dorsal aspect display an unusually smooth area surrounding the sensory capsules, which suggests the presence of some kind of either cartilaginous or fibrous braincase (Fig. 2.4C), possibly comparable to the partly fibrous braincase of hagfishes. The peculiar bilobate anterior margin of the myllokunmingiid head (hm, Fig. 2.4C) somewhat resembles the cartilage of the oral hood of larval lampreys, but this is the only feature that could possibly suggest a closer relationship to lampreys than to any other chordate. Yet it also recalls the somewhat bilobate anterior head margin of yunnanozoans (Mallatt and Chen, 2003a: fig. 6A; see below). Although the olfactory organ of myllokunmingiids is small and apparently composed of two closely set nasal capsules (olf, Fig. 2.4C), there is no evidence that it opened dorsally and formed a nasohypophysial complex of the lamprey type.

The phylogenetic position of the myllokunmingiids remains ambiguous. It is still unclear whether they are crown-group or stem-group vertebrates. However, two characters could possibly support the latter theory (2, Fig. 2.4D): the apparent lack of cartilaginous radials in the median fins, and the possible presence of serially arranged gonads. As a matter of fact, some specimens of *Haikouichthys* display, ventral to the imprint of the digestive tract (dt, Fig. 4B), a series of rounded stains, which have been interpreted as possible evidence for serially arranged gonads (gon?, Fig. 4B; Shu et al., 1999, 2003b; Hou et al., 2002). Such a structure of the gonads, also inferred in yunnanozoans (Mallatt and Chen, 2003a,b; see below), would recall the condition in the cephalochordates (also controversially recorded in the branchial region of certain enteropneust species) and is unknown in extant vertebrates. If such a condition of the gonads can ever be confirmed, then myllokunmingiids would show the first evidence of a combination of a reputedly unique cephalochordate character (serially arranged gonads) and reputedly unique vertebrate characters (head with paired sensory capsules, gill arches, W-shaped myomeres, etc.). It should nevertheless be pointed out here that these presumed gonad impressions in myllokunmingiids are remarkably well marked (Shu et al., 2003b: fig. 1J), as are also those of yunnanozoans (Mallatt and Chen, 2003a: figs. 10, 12), and strikingly recall the series of impressions left on the surface of the body musculature by the ventrolateral series of large slime glands of hagfish (Marinelli and Strenger, 1956: figs. 78, 82, 108). This is perhaps mere coincidence, but it is worth keeping in mind.

In sum, there is thus some support, albeit tenuous, to the theory that myllokunmingiids are the sister-group of the crown-group vertebrates, somehow filling the morphological gap between the cephalochordates and vertebrates.

Yunnanozoans

The Yunnanozoa are yet another taxon from the Early Cambrian of China, which has received much attention in connection with the question of vertebrate ancestry. Yunnanozoans (*Yunnanozoon* and *Haikouella*) are known from hundreds of exquisitely preserved specimens but remain the subject of heated controversies. They were first regarded as the closest fossil relatives to the vertebrates (Chen et al., 1995, 1999; Holland and Chen, 2001; Mallatt and Chen, 2003a,b), then to chordates or hemichordates (Shu et al., 1996), cephalochordates (Gould, 1995), and finally as stem deuterostomes, possibly related to the Cambrian vetulicolians (Shu et al., 2001, 2003b; Shu and Conway Morris, 2003). The detailed study of *Haikouella* by Mallatt and Chen (2003a), however, suggests some resemblance to larval lampreys in the organization of the head imprints. Yet these authors may have overinterpreted certain rather vague imprints, notably their claim that paired eyes, nostrils, and a brain are visible on the specimens (Mallatt and Chen, 2003a: fig. 7). Moreover, the very deep body and vertically straight myomeres of yunnanozoans are at odds with the V- or W-shaped myomeres of euchordates. In contrast, their elongated "gill rays" borne by the six gill arches resemble the gills of myllokunmingiids. To some extent, the deep body, straight myomeres, and shallow head region of yunnanozoans also recall the aspect of the Middle Cambrian Burgess Shale fossil *Pikaia gracilens*, often referred to as a chordate or a vertebrate relative (Gould, 1995; Conway Morris, 1998). At any rate, yunnanozoans lack a number of characters uniquely shared by myllokunmingiids and vertebrates, such as the well-defined optic and olfactory capsules, but, like myllokunmingiids, they show some possible indication of serially arranged gonads (unless, as suggested above for the myllokunmingiids, these are impressions of ventral slime glands).

Considering some evidence for a relatively complex organization of the branchial apparatus, and despite their lack of chevron-shaped myomeres, yunnanozoans could thus possibly be euchordates, and more closely related to the myllokunmingiids and vertebrates than to the cephalochordates, as proposed by Mallatt and Chen (2003a,b). In such a case, they would also contribute in filling the morphological gap between the latter and the vertebrates, and provide a plausible example of an organism in which gill arch formation involved neural crest cell migration, before the appearance of epidermal placodes and thus vertebrate-like sensory capsules. However, the recent suggestion by Shu et al. (2003b) and Shu and Conway Morris (2003) that yunnanozoans represent stem deuterostomes is also worth considering as an alternative. It would imply that some of the characters regarded as unique to chordates or euchordates are, in fact, general for deuterostomes.

Are There Fossil Cephalochordates?

Strangely, it seems now that the main problem in euchordate evolutionary history is less with the early stages in vertebrate evolution than in those of their sister-group, the cephalochordates. Despite some dubious fossils either expressly referred to the cephalochordates (the Permian *Palaeobranchiostoma*; Oelofsen and Loock, 1981), or questionably so (e.g., *Emmonsaspis*; Resser and Howell, 1938; discarded by Conway Morris, 1993), there is no unambiguous evidence for fossil cephalochordates (Blieck, 1991). Molecular clock data, however reliable they may be (see criticisms by Donoghue et al., 2003), suggest divergence times between cephalochordates and vertebrates that can be as early as 750 Ma ago (e.g., Hedges, 2001). Thus, theoretically, cephalochordates (or stem cephalochordates) are likely to occur as early as the Cambrian, but we may be unable to recognize them, unless they are some of the mitrates, as claimed by Jefferies (1986). Between the mid-nineteenth and mid-twentieth centuries, the lack of an organized head in cephalochordates has been variously regarded as either primitive or the result of degeneracy. For example, Holmgren and Stensiö (1936: fig. 206) reconstructed a hypothetical ancestral cephalochordate that superficially looks like a naked osteostracan lacking paired eyes, and more recently, Bjerring (1984) figured a branching diagram where cephalochordates are the sister-group to a clade Agnatha (thereby implying an unlikely number of either losses in cephalochordates, or convergences between "agnathans" and gnathostomes). Current molecular phylogenies and developmental data demonstrate quite clearly that cephalochordates are not nested within the vertebrates (see Holland and Chen, 2001). Nevertheless, as pointed out above, it remains undecided how much their anatomy has been modified since their divergence from the lineage that led to the vertebrates. Despite their possibly having neural crest and placode precursors (Holland et al., 1996; Shimeld and Holland, 2000; Holland and Holland, 2001; Holland and Chen, 2001), they show no evidence for migrating neural crest cells, and thus they are unlikely to have developed a skull with gill arches (in the vertebrate sense) in their evolutionary history. To date, cephalochordates are thus the only major deuterostome group for which there are no undoubted fossil data.

Evolutionary Transition in Stem Vertebrates

There is thus a slight possibility that myllokunmingiids and, less likely, yunnanozoans are stem-group vertebrates that have diverged before the common ancestor to all extant vertebrates and after the cephalochordate-vertebrate divergence (Figs. 2.3, 2.4D). Do these taxa illustrate an evolutionary transition? Not really, because myllokunmingiids are morphologically very close to crown-group vertebrates. Whether yunnanozoans branch off deeper in euchordate phylogeny or are stem deuterostomes remains debated. As a whole, the gap between the cephalochordates and crown-group vertebrates remains void, or almost so. However, because we are in the realm of speculation, it is also possible that myllokunmingiids (assuming that they actually are "cristozoans") tell us that the gap in question may not have been that large, and that once the neural crest and epidermal placodes gained their role in head morphogenesis, the so-called vertebrate *Bauplan* turned up in perhaps no more time than the rise of paired limbs from fins. Such a simplistic scenario is probably doomed to be refuted by recent research on the presumed placode homologues of tunicates, which throws a new light on the diversity and early role of these developmental characters (Mazet et al., 2005).

It is also possible that we pay excessive attention to myllokunmingiids because they have a very early age. Would we do the same if they were merely Carboniferous in age? After all, some Carboniferous Konservat-Lagerstätten have yielded peculiar soft-bodied vertebrate-like fossils (such as *Pipiscius, Gilpichthys,* and *Conopiscius;* Bardack and Richardson, 1977; Briggs and Clarkson, 1987) that may also turn out to be stem-group vertebrates.

Whence the Cyclostomes?

Whether the cyclostomes are monophyletic or not, the relationships of hagfishes and lampreys to the fossil jawless vertebrate taxa remain largely an enigma. In the mid-nineteenth century, the armored jawless vertebrates, now informally referred to as "ostracoderms," were first regarded as bony fishes (e.g., Lankester, 1868–1870; Fig. 2.5A), until it was proven that they are jawless and that some of their anatomical features (e.g., the lack of horizontal semi-circular canal) also occur in extant cyclostomes, notably lampreys. However, apart from their being jawless, "ostracoderms" could not easily be related to either hagfishes or lampreys. Stensiö (1927) assumed that hagfish were the closest relatives of heterostracans, whereas lampreys were that of anaspids, and that the latter two groups were in turn the sister-group of osteostracans. Contrary to the classical definition of the cyclostomes, Stensiö's one was thus stem based (Fig. 2.5B). This theory, referred to as the "diphyletism of the cyclostomes" (i.e., independent appearance of the characters that are regarded unique to hagfishes and lampreys, within a clade Agnatha), has been the subject of debates until the 1970s, mainly regarding Stensiö's assumption that hagfishes were related to heterostracans (Stensiö, 1927, 1932, 1958, 1964, 1968; see review in Janvier and Blieck, 1993). Yet nobody rejected his theory, outlined by Kiaer (1924), that lampreys were the closest relatives of anaspids and osteostracans, all three groups forming the clade Cephalaspidomorphi, characterized by a dorsal nasohypophysial opening. It was thus assumed, except by Stensiö, that hagfishes were more closely related to lampreys than to any extinct vertebrate taxon, and thus that hagfish morphology was derived from a cephalaspidomorph, and most probably a lamprey-like one. Some "ostracoderms" were thus stem cyclostomes, whereas others (namely heterostracans) were possibly stem gnathostomes because of their paired olfactory capsules (Fig. 2.5C). Only Moy-Thomas and Miles (1971) removed hagfishes from the cephalaspidomorphs (in Stensiö's sense), but they nevertheless regarded them as the sister-group of the latter because of their monorhinal condition (i.e., single median nostril). Then, after the rise of the theory of cyclostome paraphyly (Løvtrup, 1977), only lampreys and gnathostomes were thought to be derived from "ostracoderms" through loss of a calcified skeleton, but cephalaspidomorph monophyly remained the received wisdom (Janvier, 1978; Fig. 2.5D). Cephalaspidomorph monophyly was questioned much later (Forey, 1984; Janvier, 1984), and more clearly so after the first computer-assisted parsimony analysis of vertebrate interrelationships by Gagnier (1993b), as well as by subsequent analyses (Forey and Janvier, 1993; Janvier, 1996b; Donoghue et al., 2000, 2003; Donoghue and Smith, 2001), all of which showed hagfishes and lampreys as more inclusive than all vertebrates with a mineralized skeleton (i.e., euconodonts, "ostracoderms," and gnathostomes; Fig. 2.5E,F).

Hagfishes and lampreys are almost unknown as fossils, and there is no hint for knowing how much derived or generalized they may be relative to the early Paleozoic "ostracoderms." All the fossils referred to hagfishes and lampreys are Carboniferous in age and preserved as imprints. The two fossils referred to lampreys, *Hardistiella* and *Mayomyzon* (Fig. 2.6A,B), both look like a small present-day lamprey, except for their lack of separate anterior and posterior dorsal fins and their shorter branchial apparatus (gp, Fig. 2.6A2; Bardack and Zangerl, 1968, 1971; Janvier and Lund, 1983; Janvier et al., 2004b). The head of *Mayomyzon* displays imprints

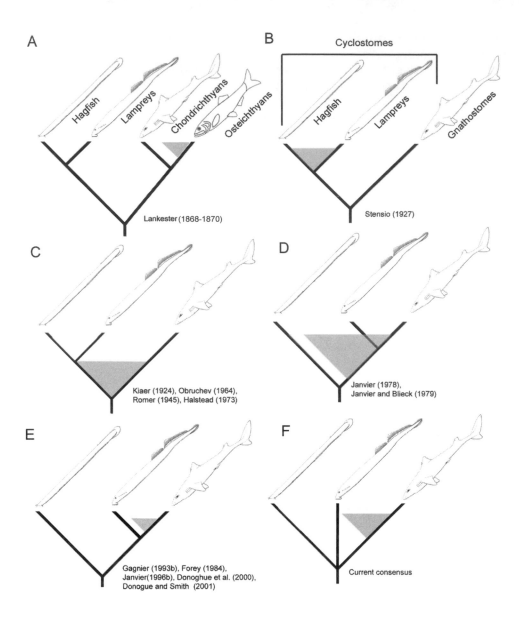

A   Hagfish Lampreys Chondrichthyans Osteichthyans

Lankester (1868-1870)

B   Cyclostomes

Hagfish Lampreys Gnathostomes

Stensio (1927)

C

Kiaer (1924), Obruchev (1964),
Romer (1945), Halstead (1973)

D

Janvier (1978),
Janvier and Blieck (1979)

E

Gagnier (1993b), Forey (1984),
Janvier(1996b), Donoghue et al. (2000),
Donogue and Smith (2001)

F

Current consensus

of cartilages that somewhat match the shape and position of the characteristic piston, tectal and annular cartilages, and the small olfactory capsule of adult lampreys (pc, tc, anc, olf, Fig. 2.6A2). There is thus little doubt that these two Carboniferous forms actually represent lampreys and are adults, despite their small size.

The fossil record of hagfishes is less clear. Of the two presumed fossil hagfish, *Myxinikela* and *Myxineides,* the former displays the characteristic tentacles at the tip of its snout (tcl, Fig. 2.6C2), but its body is much stouter than that of extant hagfishes (Bardack, 1991; Fig. 2.6C). The few imprints of internal structures in *Myxinikela* only provide some indication of a hagfish-like cartilaginous nasal basket (corbiculum nasale) and prenasal sinus (olf, Fig. 2.6C2). In contrast, *Myxineides* shows no clear evidence for tentacles nor for a nasal basket, but it has the characteristic eel-like body shape of modern hagfishes and shows indications of two V-shaped rows of horny teeth (ht, Fig. 2.6D) in the form of impressions in the natural cast of the oral cavity (dt, Fig. 2.6D3; Poplin et al., 2001). None of these specimens, provided that they are correctly identified as lampreys and hagfish, respectively, shows character combinations such as, for example, an annular cartilage in a hagfish or large tentacles in a lamprey, which would suggest that one taxon is nested in the other. Only *Myxinikela* suggests that the branchial apparatus was primitively closer to the braincase than in modern hagfishes, as was also inferred earlier from hagfish development (e.g., Holmgren, 1946). In addition, these fossils provide no indication of any character loss, at least since the Carboniferous, in hagfishes and lampreys, such as the long-presumed loss of exoskeleton or paired fins implied by the traditional "degeneracy" theory of cyclostome evolution. Only *Mayomyzon,* which displays no separate dorsal fins, suggests that the two dorsal fins of present-day lampreys and gnathostomes could be homoplastic.

The Question of the Cephalaspidomorphs

Hagfishes and lampreys display very few characters that are uniquely shared with any particular group of the fossil jawless vertebrates referred to as "ostracoderms." The presence of a common external branchial opening in some hagfishes, the Myxinidae, and heterostracans, long invoked by Stensiö (1964, 1968) in support to his theory of hagfish-heterostracan relationships, is regarded as homoplastic (Janvier, 1996a,b). Only lampreys and osteostracans (and possibly anaspids) share some uniquely derived characters, which have been highlighted by Stensiö (1964) and others (e.g., Jarvik, 1980; Janvier, 1981a, 1984) and long recognized as robust support for the monophyly of the clade Cephalaspidomorphi. Most of these characters (e.g., enlarged right habenular ganglion; shift of the dorsal aorta to the right side, dorsally to the heart; ventricle and atrium of the heart closely set and lying side by side; lateral and dorsal expansions of the membranous labyrinth) have often been dismissed as either highly variable in vertebrates, ambiguous, or shared also with either some hagfishes or some gnathostomes (Janvier, 1984, 1996b; Janvier et al., 1991; see below). However, the presence of a dorsal nasohypophysial opening in osteostracans, lampreys, and possibly anaspids remains an intriguing character, which, in current vertebrate phylogenies, is assumed to have evolved separately in lampreys and osteostracans.

In lampreys, the olfactory organ is much reduced and housed in a small cartilaginous capsule that is independent from the braincase proper (olf, Fig. 2.7B1). In ontogeny, it is derived from an anterior median olfactory placode, which is situated immediately rostral to the hypophysial tube, long presumed to be the homologue of the Rathke's pouch in gnathostomes. For recent reviews of the question, see Kuratani et al. (2001) and Ushida (2003). During early embryonic development, the two structures remain closely united as the na-

Figure 2.5. *(opposite page)* Simplified history of the conceptions about the relationships of "ostracoderms." The areas in gray indicate the extent of all possible relationships for taxa referred to as "ostracoderms." A, Lankester's (1868–1870) theory that "ostracoderms" were primitive osteichthyans; B, Stensiö's (1927) theory of "cyclostome diphyletism," where "ostracoderms" are paraphyletic and subgroups ancestral to hagfishes and lampreys, respectively; C, theory defended by various opponents to Stensiö, assuming that some "ostracoderms" are either ancestral to or the closest relatives of the cyclostomes, whereas others (generally heterostracans) are either ancestral to or the closest relatives of the gnathostomes; D, Janvier's (1978) theory, incorporating Løvtrup's (1977) theory of cyclostome paraphyly, and assuming "ostracoderm" paraphyly; E, current morphology-based theory, assuming that "ostracoderms" are paraphyletic stem gnathostomes; F, current consensus, incorporating the conflicting results from morphology- and molecular sequence–based analyses of crown-group vertebrate relationships.

Figure 2.6. Presumed fossil hagfish and fossil lampreys. **A,** *Mayomyzon pieckoensis* Bardack and Zangerl (1968), Upper Carboniferous of Illinois; **A1,** overall shape; **A2,** details of the cartilage imprints in the head (based on Bardack and Zangerl, 1968, and Janvier, 1993). **B,** *Hardistiella montanensis* Janvier and Lund (1983), Upper Carboniferous of Montana (from Janvier and Lund, 1983). **C,** *Myxinikela siroka* Bardack, 1991, Upper Carboniferous of Illinois; **C1,** overall shape; **C2,** details of the cartilage imprints in the head (from Bardack, 1991). **D,** *Myxineides gononorum* Poplin, Sotty and Janvier (2001), Upper Carboniferous of Allier, France; **D1, D2,** head and anterior part of the body (**D1**) and detail of the internal cast of the oral cavity of the same specimen (**D2,** framed in **D1**), showing impressions of the horny teeth; **D3,** internal cast of the oral cavity of the holotype, showing impressions of two pairs of rows of horny teeth (from Poplin et al., 2001). Abbreviations as in Figure 2.4.

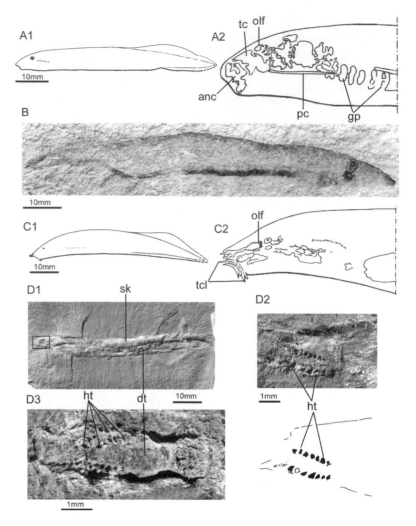

sohypophysial placode, but their common external opening, the nasohypophysial opening, migrates dorsally, as a result of the development of the "upper lip," or oral hood (Gorbman and Tamarin, 1985). In the larval lamprey, the nasohypophysial opening is slit-shaped (nho, Fig. 2.7A), and in the adult lamprey, it is rounded (nho, Fig. 2.7B2) but continued internally by a short duct, which ends ventrally with a valve, the nasohypophysial valve (nhv, Fig. 2.7B), that surrounds a keyhole-shaped opening and marks the entrance to both the olfactory organ and the posteriorly closed hypophysial tube (olf, hyp, Fig. 2.7; Janvier, 1975b). The ontogeny of osteostracans is, of course, unknown, but adult osteostracans display, anteriorly to the orbits, a keyhole-shaped opening (nho, Fig. 2.7C3), which strikingly resembles the nasohypophysial valve of lampreys (Fig. 2.7B3). Because the endoskeleton is perichondrally ossified, it can be shown that the posterior division of this opening is prolonged internally by a small pear-shaped cavity (nac, Fig. 2.7C1), which is confluent with the telencephalic division of the brain cavity and supposed to have

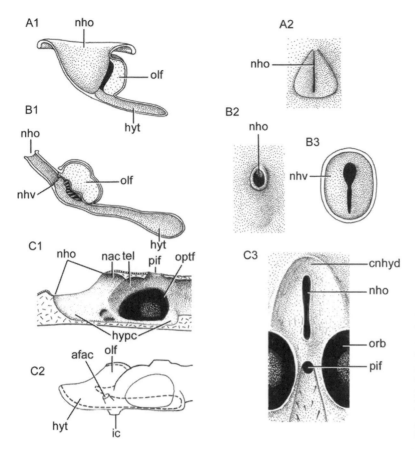

Figure 2.7. Nasohypophysial complex in lampreys and osteostracans (not to scale). **A**, nasohypophysial complex of a larval lamprey; **A1**, sagittal section; **A2**, nasohypophysial opening in dorsal view. **B**, nasohypophysial complex of an adult lamprey; **B1**, sagittal section; **B2**, nasohypophysial opening in dorsal view; **B3**, nasohypophysial valve in dorsal view. **C**, nasohypophysial complex in the Early Devonian osteostracan *Belonaspis*; **C1**, sagittal section through the ethmoid and anterior brain cavity of the head shield; **C2**, attempted reconstruction of the olfactory organ and hypophysial tube (dashed line); **C3**, dorsal view of the nasohypophysial opening. (**A**, **B**, after Janvier, 1975b; **C**, after Janvier, 1985a.) Abbreviations as in Figure 2.4.

housed the olfactory organ (olf, Fig. 2.7C1). The anterior division of the nasohypophysial opening is prolonged internally by an elongated cavity (hypc, Fig. 2.7C1), which is confluent dorsally with the ventral part of the diencephalic division of the brain cavity and the cavity for the olfactory organ, and is considered to have housed the hypophysial tube (hyt, Fig. 2.7C2). As already shown by Stensiö (1927), the structure of the nasohypophysial complex (Janvier, 1974) in lampreys and osteostracans is strikingly similar, and this is further emphasized by the quite similar distribution of the surrounding arteries derived from the internal carotids in this area (Janvier, 1975b: fig. 4). Nevertheless, the late Colin Patterson (London) once asked me, "How do you know that osteostracans have a nasohypophysial opening? Can you see the hypophysis?" Admittedly, we cannot see the hypophysis of osteostracans, nor can we see any recess of the brain cavity of the braincase where it could possibly be housed, because of the coalescence between the brain cavity and the lodge for the hypophysial tube. Thus, in common with many fossil data, this interpretation of osteostracan anatomy remains only a strong probability that is based on striking resemblance.

Kiaer (1924) and Stensiö (1927) argued that the same condition of the nasohypophysial complex also occurred in anaspids on

Evolutionary Transitions in Early Vertebrates • 79

account of the median, keyhole-shaped opening, which is situated between two dermal plates, anterior to the orbits. However, anaspids lack a calcified endoskeleton, and the actual shape of the nasohypophysial complex is unknown in this taxon.

The question of the significance of this remarkable resemblance in the structure of the nasohypophysial complex between lampreys and osteostracans is no longer at the core of current debates about vertebrate phylogeny, but it has long been the cornerstone of cephalaspidomorph monophyly and the theory of lamprey "degeneracy" (relative to "ostracoderms"). In the 1960s, the reliability of this character to paleontologists, as a signature (i.e., an indisputable synapomorphy), was comparable to that of, for example, the choana of *Eusthenopteron* and tetrapods.

Although Gross (1964) already hinted at the possibility of the dorsal nasohypophysial opening of lampreys and osteostracans being independently derived, I feel somewhat responsible for the way this character is currently regarded as being inconsequential, and even overlooked, as an obvious homoplasy. In fact, early cladistic analyses of fossil and extant vertebrates did not clearly break through the "cephalaspidomorph barrier," although they supported both cyclostome and "ostracoderm" paraphyly (Fig. 2.5D; Janvier, 1978, 1981a; Janvier and Blieck, 1979). The idea that cephalaspidomorphs (at any rate lampreys and osteostracans) might not be a group has developed progressively, in the wake of computerized cladistic analyses, and as characters were given equal weight. The unique structure of the nasohypophysial complex shared by lampreys and osteostracans was thus outnumbered by characters that are uniquely shared by the latter and gnathostomes (e.g., Forey, 1984, 1995; Janvier, 1984, 1986, 1996a; Gagnier, 1993b; Fig. 2.5E). Osteostracans admittedly do share a larger number of characters with gnathostomes (e.g., true pectoral fins, well-developed, perichondrally ossified endoskeleton, cellular bone, and epicercal tail) that Stensiö (1927) considered as general vertebrate characters, lost in lampreys, whereas other paleontologists regarded them as convergences between osteostracans and gnathostomes.

The discovery of the galeaspids and pituriaspids (L, M, Fig. 2.3), two groups that, like osteostracans, possess an exo- and endoskeletal head shield, and the elucidation of part of their internal anatomy have also cast doubts on cephalaspidomorph monophyly (Wang, 1991; Young, 1991). Galeaspids lack paired fins and possess a median, dorsally opening "nostril," as in lampreys and osteostracans, but their "nostril" is in fact the external opening of a duct, which communicates ventrally with the oralobranchial (or orobranchial) cavity (more or less like the nasopharyngeal duct of hagfishes), and into which open distinctly paired olfactory organs (Janvier, 1996b). Remarkably, this dorsal opening is slit-shaped in some galeaspids, thereby mimicking the nasohypophysial opening of osteostracans, but this condition is currently regarded as derived within galeaspids (Janvier, 1984, 1996b). Pituriaspids possess pectoral fins, like osteostracans, but seem to lack any dorsal nasohypophysial or nasopharyngeal opening (the olfactory organs may have opened ventrally, anterior to the mouth; Young, 1991). Galeaspids suggested that a dorsally opening nasohypophysial complex that retains an incurrent function could coexist with a massive, osteostracan-like, endoskeletal head shield, and thus that a convergence between lampreys and osteostracans was plausible. Moreover, pituriaspids showed that an osteostracan-like head shield structure could possibly coexist without any dorsal nasal or nasohypophysial opening, and thus that a head shield–shaped skull was not necessarily linked to a dorsal "nostril," be it incurrent or not. In sum, the nasohypophysial complex of fossil "agnathans" seemed to have had a greater morphological plasticity than was believed 50 years ago.

All computerized cladistic analyses of extant and extinct vertebrate taxa published during the past decade show all "ostracoderms" (i.e., pteraspidomorphs, galeaspids, pituraspids, thelodonts, osteostracans, and anaspids) and even euconodonts as more closely related to the gnathostomes than to either lampreys and hagfishes (Figs. 2.3, 2.5E,F; Gagnier, 1993b; Forey, 1995; Janvier, 1996a; Donoghue et al., 2000; Donoghue and Smith, 2001; Donoghue and Sansom, 2002). In all these analyses, cephalaspidomorphs are nonmonophyletic, with osteostracans being almost always the sister-group to the gnathostomes. Only one of the analyses performed by Donoghue et al. (2000: fig. 17, "preferred phylogeny") shows osteostracans, galeaspids, and pituriaspids (i.e., all taxa with an endoskeletal head shield) as a clade that, in turn, is the sister-group to gnathostomes and the poorly known Ordovician genus *Eriptychius* (not considered here), and this would lend further support to the theory that the lamprey-osteostracan type of nasohypophysial complex is homoplastic. Yet this clade including galeaspids, pituriaspids, and osteostracans is no longer considered as supported by Donoghue and Smith (2001), Donoghue and Sansom (2002), and Donoghue et al. (2003). As a whole, the current phylogenies yielded by standard parsimony, in which "ostracoderms" are stem-group gnathostomes, display a character distribution from the more particular to the more general, but, frustratingly, the most diverse character complex in these taxa—that is, the nasohypophysial complex—displays incongruent distributions. Some scenarios have been proposed to account for the convergence in the structure of the nasohypophysial complex of lampreys and osteostracans (Schaeffer and Thomson, 1980; Janvier, 2001), but none of them is fully satisfactory.

Lampreys, Euphaneropids, and Anaspids

Assuming that the monophyly of cephalaspidomorphs is refuted, this leaves us with very few hints as to the relationships of lampreys with any particular fossil vertebrate taxon, and even less so as to the relationships of hagfishes. Nevertheless, recent investigations on the Late Devonian naked anaspid, *Euphanerops,* possibly throw new light on the old hypothesis of a sister-group relationship between lampreys and anaspids. The anaspids, like osteostracans, have long

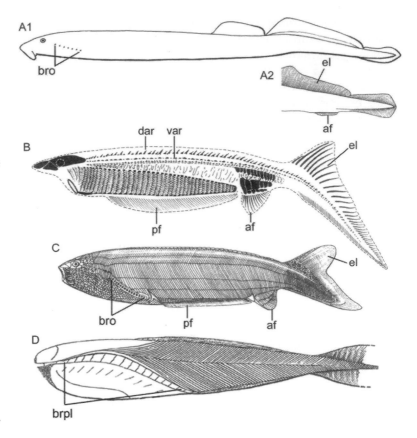

Figure 2.8. Lampreys, euphaneropids, anaspids, and arandaspids (not to scale). **A**, extant lamprey *Petromyzon marinus*; **A1**, lateral aspect; **A2**, tail of an abnormal female individual showing fin radials in the position of an anal fin (drawn after Vladikov, 1973). **B**, *Euphanerops longaevus*, Upper Devonian of Canada, reconstruction of the calcified endoskeleton (black) in lateral view (based on personal observations by P. Janvier and M. Arsenault). **C**, *Pharyngolepis oblongus*, Lower Silurian of Norway, reconstruction in lateral view (after Ritchie, 1964). **D**, *Sacabambaspis janvieri*, Ordovician of Bolivia, reconstruction in lateral view (after Gagnier, 1993a). Abbreviations as in Figure 2.4.

been regarded as cephalaspidomorphs, but are currently regarded as stem gnathostomes alongside other "ostracoderms" (Fig. 2.3G).

The overall structure of lampreys compares best with that of anaspids (slender body shape, small gill openings arranged in a slanting line behind the eyes, hypocercal tail, body musculature—inferred from the squamation—extending dorsally and ventrally to the branchial apparatus, and possibly a similar dorsal nasohypophysial opening; Fig. 2.8A,C). However, contrary to anaspids, lampreys lack paired and anal fins, although an anal fin–like structure supported by radials sometimes occurs as an abnormality in extant lampreys (af, Fig. 2.8A2; Vladikov, 1973). Conversely, anaspids lack an anterior dorsal fin, but their epichordal lobe is regarded as homologous to the posterior dorsal fin of lampreys. Unfortunately, no anaspid displays any trace of endoskeletal structures, but an extensively calcified endoskeleton can be observed in large individuals of the anaspid-like form *Euphanerops longaevus* (Janvier and Arsenault, 2002). Although it is uncertain whether this calcification is biogenic or the result of early diagenesis, it does provide information about the morphology of some endoskeletal structures (Arsenault and Janvier, 1991; P. Janvier and M. Arsenault, pers. obs.). *Euphanerops* shares with anaspids a markedly hypocercal tail, an anal fin, and elongated ventrolateral paired fins, but seems to lack a dermal skeleton (Fig. 2.8B), and thus cannot reliably be proven to belong to the clade Anaspida, which is diagnosed on the basis of dermoskeletal characters.

In contrast, its endoskeleton is somewhat suggestive of that of lampreys, with a large branchial basket, although the latter is composed of more than 30 sinuous branchial arches (Figs. 2.2B, 2.8B). Some other endoskeletal elements recall certain lamprey characters, such as a ring-shaped structure and a median ventral rod, which vaguely recall the annular and piston cartilages, respectively. In addition, the calcification of its endoskeleton forms a thin lining around very large, bubble-like spaces, which were presumably occupied by chondrocytes and are grouped into cell nests, quite comparable in size and arrangement to the chondrocytes of lamprey cartilage (Langille and Hall, 1993; Janvier and Arsenault, 2002). Yet this still unexplained structure is at odds with the spherulitic calcified cartilage of "ostracoderms" (such as *Eriptychius,* galeaspids, and osteostracans) and gnathostomes. As a whole, the head endoskeleton of *Euphanerops,* whatever the nature of its calcification, is more similar to that of lampreys than to the massive endoskeletal head shield of galeaspids and osteostracans, but we know virtually nothing of the endoskeleton of the pteraspidomorphs (Fig. 2.3H–J; i.e., astraspids, arandaspids, and heterostracans), and it cannot be ruled out that they possessed a lightly built, lamprey- or *Euphanerops*-like endoskeleton, housed in an elongated dermal armor.

Other early Paleozoic "naked agnathans," namely *Achanarella* and *Cornovichthys,* closely resemble *Euphanerops* and could readily be lumped in the family Euphaneropidae Woodward, 1900, which would thus range in time at least from the early Middle to the Late Devonian (Newman, 2002; Newman and Trewin, 2001). It is also possible that the Early Silurian "naked agnathan," *Jamoytius* (Ritchie, 1968), also belongs to this clade.

Euphanerops thus displays an anaspid-like overall structure and some lamprey-like features (Fig. 2.8). These shared characters could be regarded as supporting the once classical theory that lampreys are derived from anaspids, and that *Euphanerops* would illustrate an intermediate form, which has lost the dermal skeleton (assuming that the number of gill arches in lampreys have been reduced and the paired fins lost). Other possible relationships can be inferred either among these three taxa, or among each of them and the other vertebrate taxa, but these are strongly dependent on how much one can rely on their interpretation based on such material, with radically different character quality (i.e., an extant taxon and a fossil taxon essentially preserved as imprints, both lacking a dermal skeleton, and a fossil taxon known exclusively from its dermal skeleton). Moreover, the homology of most of the characters observed in *Euphanerops* remains putative and depends on how one interprets such flattened structures. Very few of these characters can be readily homologized with a structure found in either lampreys or other vertebrates. Therefore, attempts at including *Euphanerops* in a data matrix of all major extant and extinct vertebrate taxa show that it groups with lampreys and anaspids, generally as the sister-group to anaspids, because of its paired fin morphology and more strongly hypocercal tail.

I shall not discuss further here the possible interpretations of this topology, but it may be worth raising the question of the overall resemblance of these three taxa, which contrasts to that of most other fossil and extant vertebrates, notably "ostracoderms" and gnathostomes. In previously published phylogenies, anaspids and *Euphanerops* appear more closely related to the gnathostomes than to lampreys, with various positions within the "ostracoderms." Anaspids appear as the sister-group to all "ostracoderms" and gnathostomes, except for either euconodonts (Donoghue and Smith, 2001; Donoghue and Sansom, 2002), or euconodonts and pteraspidomorphs (Donoghue et al., 2000). It is nevertheless possible that the lamprey-like aspect of anaspids and *Euphanerops* (i.e., slender body shape, hypocercal tail, branchial basket extending far behind the eyes, and possibly dorsal nasohypophysial opening) is, in fact, general for the clade that includes lampreys, "ostracoderms," and gnathostomes. Interestingly (or coincidentally), this structure somewhat recalls that of the Ordovician arandaspids, the earliest completely known vertebrates (Fig. 2.8D). Notably, anaspids share with arandaspids numerous external branchial openings (or compartments) arranged in slanting line (bro, brpl, Fig. 2.8), a strongly oblique head/trunk limit, rod-shaped body scales arranged in chevrons, and open sensory-line grooves.

The presence of a lamprey-like branchial basket in *Euphanerops*, as well as some other endoskeletal structures that vaguely resemble lamprey characters, are tenuous hints at a close relationship between the two taxa because they could well be more general characters, also present in anaspids and even arandaspids, astraspids, and heterostracans, for which we still lack direct information about the endoskeleton. The presence of ventral arcualia in the axial skeleton of *Euphanerops* and gnathostomes (var, Fig. 2.8B), but not in lampreys, may also suggest that *Euphanerops* is a stem gnathostome, more inclusive than anapids and still lacking an extensive dermal skeleton. The intuitive theory that lampreys are derived from anaspids through a loss of the dermal skeleton and paired fins is thus barely better supported by the new data on *Euphanerops*, however lamprey-like they may be, as long as endoskeletal data remain virtually unknown in euconodonts, anaspids, and pteraspidomorphs.

Nevertheless, apart perhaps from euphaneropids, no fossil provides information about any evolutionary transition between non-lampreys and lampreys, non-hagfishes and hagfishes, and even non-cyclostomes and cyclostomes, if the latter are a clade.

"Ostracoderms" and the "Agnathan"-Gnathostome Transition

Despite the uncertainty over the phylogenetic position of the lampreys and hagfishes, or the cyclostomes as a whole, relative to the fossil vertebrate taxa, there is currently a consensus on the paraphyly of the "ostracoderms," with certain taxa, such as the osteostracans or galeaspids, being more closely related to the gnathostomes than to the two cyclostome taxa and other "ostracoderm" taxa (Figs. 2.3L–N, 2.5E,F; Forey and Janvier, 1993; Gagnier, 1993b; Forey, 1995; Janvier, 1996a; Donoghue et al., 2000, 2003; Donoghue and Smith, 2001; Donoghue and Sansom, 2002). This hypothesis of phylogenetic relationships provides interesting insights as to the degree of generality and the relative dates of appearance for certain gnathostome characters among extant vertebrates, such as dermal bone, perichondral bone, paired fins, epicercal tail, cerebellum, open endolymphatic ducts, or cellular bone (Forey, 1995; Janvier, 1996a, 2001; Donoghue and Sansom, 2002). However, one must not forget that the distribution of these characters is essentially that which gives structure to the current trees. These trees tell us that such primary homologies as the pectoral fins of osteostracans and crown-group gnathostomes are not overturned by optimizations.

The theory that "ostracoderms" are paraphyletic is not new, and many opponents to Stensiö's theory of "agnathan" monophyly considered that, among "ostracoderms," heterostracans were possibly the closest relatives of the gnathostomes (Fig. 2.5C), essentially because of their paired olfactory organ (unknown in galeaspids until the mid-1980s; Fig. 2.9D; Kiaer, 1924; Romer, 1945; Obruchev, 1964; Halstead, 1973; Novitskaya, 1983). However, after 1927, no author ever thought of osteostracans as gnathostome relatives, this taxon being then well established as a cephalaspidomorph, alongside lampreys and anaspids. The current vertebrate phylogenies that place the paraphyletic "ostracoderms" as stem gnathostomes may be regarded as somehow illustrating an evolutionary transition between jawless and jawed vertebrates (Fig. 2.3G–R).

One of the best supported of these nodes is generally the sistergroup relationship between osteostracans (and possibly pituriaspids) and gnathostomes (6, Fig. 2.3). It is based on homology relationships, which are briefly outlined below, with particular references to problematic interpretations of some of them. Illustrations of the characters currently regarded as synapomorphies shared by the gnathostomes, and a variable number of "ostracoderms" have been repeatedly published or cited during the last decade (e.g., Janvier, 1984, 1996a,b, 2001; Forey and Janvier, 1993; Donoghue et al., 2000, 2003; Donoghue and Smith, 2001; Donoghue and Sansom, 2002). This time, I shall play the devil's advocate by emphasizing the characters whose distribution is incongruent with the current trees, however these may be interpreted. Three of these characters have been discussed earlier (Janvier, 1984; and see above), but others are considered here.

Brain

The brain cavity of osteostracans is supposed to provide a relatively accurate cast of the brain (Fig. 2.9F). Its overall structure is rather similar to that of the brain cavity in galeaspids (Fig. 2.9E),

Figure 2.9. Brain, olfactory, and otic capsules (dorsal view) in extant hagfishes (A), lampreys (B), and crown-group gnathostomes (C, elasmobranch). D, impression of the brain, olfactory and otic capsules in the internal surface of the dorsal head armor of a heterostracan (not to scale). E, F, internal cast of the cavity for the brain, olfactory capsules, and labyrinth (dorsal view) in a galeaspid (E) and an osteostracan (F). G, habenulopineal region of the brain of a lamprey in dorsal view (G1) and vertical section (G2), showing the enlarged habenular ganglion of the right side. H, roof of the habenulopinal region of the brain cavity in ventral view (H1) and internal cast of the same region (H2) of an Early Devonian osteostracan (*Belonaspis*), showing the enlarged recess for the habenular ganglion of the right side. (A–D, F, after Janvier, 1996b; E, after Wang, 1991; G, H, after Janvier, 1984.) Abbreviations as in Figure 2.4.

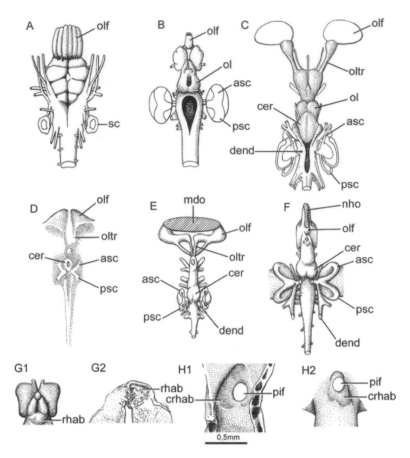

and its dorsal surface also matches the brain impression in the internal surface of the dorsal exoskeleton of heterostracans (Fig. 2.9D). In many respects, it also resembles the shape of the brain of lampreys (Fig. 2.9B), but it lacks the strong dorsal flexure of the adult lamprey brain. One of the most conspicuous features of the osteostracan brain cavity is the paired dorsal recess, regarded by Stensiö (1927) as housing a paired cerebellum (cer, Fig. 2.9F). The cerebellum is lacking in lampreys, but the optic lobes form a pair of prominent dorsal swellings (ol, Fig. 2.9B). Therefore, it has been suggested that the recess reconstructed by Stensiö (1927) as having housed the cerebellum perhaps in fact housed the optic lobes (Halstead Tarlo and Whiting, 1965; Janvier, 1975c). Nevertheless, this interpretation is now regarded as unlikely, because such a posterior position of the optic lobes would not agree with the position of the canals for the cranial nerve roots, notably that of the trochlear nerve (Janvier, 1985a; Janvier and Blieck, 1993). The elongated and straight brain of larval lampreys, galeaspids, osteostracans, and possibly heterostracans seems to be a general feature for vertebrates (except perhaps for hagfishes; Fig. 2.9A), but there is no evidence that the cerebellum is secondarily reduced in hagfishes and lampreys. The large paired cerebellum of osteostracans, galeaspids, gnathostomes, and possibly heterostracans is thus likely to be a homology (cer, Fig. 2.9C–F).

The brain cavity of osteostracans displays a single character that was regarded by Stensiö (1927) as uniquely shared with the brain of lampreys: the enlarged habenular ganglion of the right side (rhab, crhab, Fig. 2.9G,H). This character was later dismissed because of its variability, notably in extant gnathostomes (Janvier, 1984), where the size of the habenular ganglia depends largely on the relative development of the pineal and parapineal organ. This asymmetry has been observed in a number of osteostracans belonging to widely different taxa (Stensiö, 1927, Janvier, 1985a), which admittedly all have a pineal foramen. It would be interesting to check its presence in the few osteostracans that lack a pineal foramen, such as the Acrotomaspidinae. Moreover, this character should also be looked for in galeaspids. To date, the limited available information on the galeaspid brain cavity does not show evidence for such an asymmetrical habenular recess (Fig. 2.9E; Wang, 1991: fig. 4).

Cranial Nerves

Osteostracans are the only fossil "agnathans" in which the path of the cranial nerves within the endoskeletal head shield has been described in detail. Their pattern is, however, somewhat difficult to compare with that of either lampreys or gnathostomes, because of the unique forward shifting of the entire branchial apparatus in this group, which makes the postotic nerves (IX, X) bend rostrally and pass through the cavity for the otic capsule. Nevertheless their arrangement is globally similar to that of the cranial nerves in certain placoderms (e.g., acanthothoracids, petalichthyids), in which the branchial apparatus lies well beneath the braincase (Janvier, 1996a). What is known of the cranial nerves in galeaspids generally agrees with the condition in osteostracans, apart from the lack of forward shifting of the branchial nerves. In contrast to osteostracans, lampreys, and hagfishes, galeaspids resemble gnathostomes in having a pair of long olfactory tracts (which are probably also present in heterostracans; oltr, Fig. 2.9C–E). The condition of the visceromotor nerves in lampreys (and hagfishes) is relatively different from that in osteostracans, notably as to the path of the trigeminal, glossopharyngeal, and vagal nerves (although this is probably linked to the unique development of the lingual apparatus in hagfishes and lampreys). In addition, osteostracans and galeaspids resemble gnathostomes in possessing an occipital region to the endoskeleton that encloses the glossopharyngeal and vagal nerve roots.

Olfactory Capsule and Nasohypophysial Complex

As mentioned above, the olfactory capsule of osteostracans was enclosed in a small median cavity, and thus bears no resemblance to the paired olfactory capsules of the gnathostomes, but it is strikingly similar to the condition in lampreys (olf, Figs. 2.7, 2.10B,F). In addition, the anatomical relations between the olfactory organ and the proximally closed hypophysial tube are likely

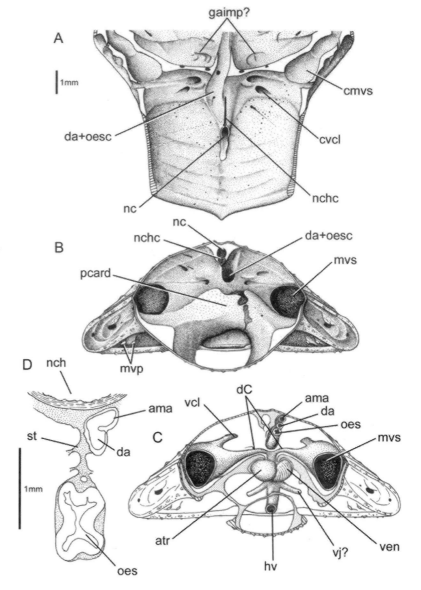

Figure 2.10. Heart and associated blood vessels in the lampreys and osteostracans. A–C, abdominal division and postbranchial wall of the head shield of the Early Devonian osteostracan *Norselaspis* in ventral (**A**) and posterior (**B**) views; **C**, attempted reconstruction of the heart and major blood vessels, showing the shift of the dorsal aorta, the esophagus, and probably the anterior mesenteric artery to the right. **D**, transverse section of the notochord, esophagus, dorsal aorta, and anterior mesenteric artery of a larval lamprey at the level of the heart, showing the shift of the dorsal aorta and anterior mesenteric artery to the right side. (**A–C**, after Janvier, 1985a, and Janvier et al., 1991, modified; **D**, drawn after Percy and Potter, 1991.) Abbreviations as in Figure 2.4.

to be the same as in lampreys (hyt, Fig. 2.7C). Remarkably, the position of the endocranial division for the hypophysial tube in osteostracans relative to the internal carotid arteries and the facial arteries is virtually the same as in lampreys (afac, ci, Fig. 2.7C2; Janvier, 1975c). In contrast, the paired olfactory capsules of galeaspids (olf, Fig.2.10E) strikingly resemble the gnathostome condition (olf, Fig. 2.9C), differing from the latter only by the fact that they open into a large median, supposedly incurrent duct (mdo, Fig. 2.9E), instead of opening separately to the exterior by means of external nostrils. The little we know of the paired olfactory organ in heterostracans would agree with the condition in galeaspids (olf, Fig. 2.9D), except that there is no material evidence for a separate incurrent duct (Purnell, 2002). The osteostracan-lamprey homology relationship of the nasohypophysial complex thus remains, after nearly a century, a major problem of vertebrate morphology, currently solved by an appeal to homoplasy.

Optic Capsule

Osteostracans are the only "agnathans" for which we have reliable information about the anatomy of the optic capsule and the extrinsic eye muscles. Galeaspids only provide a possible indication of a posterior ventral myodome for the insertion of the recti muscles.

Osteostracans and gnathostomes (notably placoderms; Goujet, 1984; Goujet and Young, 2004) share a perichondrally ossified sclera, which is attached to the sclerotic ring. Nevertheless, there remains the question of the possible presence of a sclerotic ring and calcified sclera in arandaspids (Gagnier, 1993a), which are only observed in three distorted specimens. Recent reexamination of the material confirms the presence of a sclerotic ring (P. Janvier, pers. obs.). To date, there is no way to know whether osteostracans possessed an intrinsic eye musculature, like gnathostomes. However, no paraxial musculature could extend anteriorly to the occipital region, and this precludes the presence of a corneal muscle of lamprey type, supposedly derived in development from the foremost myomeres (Janvier, 1975a; yet the postotic derivation of the corneal muscle remains debatable; Nicol, 1989). The accommodation of the eye lens in osteostracans, if any, was thus most likely effected by an intrinsic eye musculature, as in extant gnathostomes. Some indications about the arrangement of the extrinsic eye muscles in osteostracans are provided by the myodomes in the wall of the orbital cavity. At least two large myodomes are clearly visible: a posterodorsal one, into which opens the canal for the trochlear nerve and thus is inferred to have housed the superior oblique muscle; and a large, ventral one, for either all or part of the recti muscles (Stensiö, 1964). The posterodorsal insertion of the trochlear-innervated superior oblique muscle is comparable to the condition in lampreys (Stensiö, 1964), but a similar condition is also inferred from the myodomes of placoderms (Young, 1986; Goujet and Young, 2004). It is thus likely to be a general condition, relative to the anterodorsal insertion of the superior oblique muscle in chondrichthyans and osteichthyans (Goujet and Young, 2004).

Labyrinth

Among fossil "agnathans," only galeaspids and osteostracans can be proven to have only two vertical semicircular canals, with respective ampullae, like lampreys (asc, psc, Fig. 2.9E,F). Heterostracans only display imprints of two vertical canals (asc, psc, Fig. 2.9E,F), but the medial extension of a branchial compartment between them renders unlikely the presence of a horizontal canal, as once advocated by Halstead (1973). The semicircular canals of osteostracans and galeaspids (and probably heterostracans) differ from those of lampreys in forming distinct loops, widely separated from the saccular division of the labyrinth, whereas those of lampreys are closely appressed against the utricular wall (asc, psc, Fig. 2.9B; Janvier, 1996a; Mazan et al., 2000). Again, it is unclear whether this condition in lampreys is general or unique because hagfishes only have a single semicircular canal with two ampullae (sc, Fig. 2.9A), the homology of which is still debated.

Jarvik's (1965) suggestion of a homology between the large ciliated recesses in the utricular wall of lampreys and the "sel" canals that connect the labyrinth cavity to the lateral and dorsal cephalic fields of osteostracans has been discarded on the basis that the only resemblance between the two structures was in their number (Janvier, 1985a). It is nevertheless not clearly refuted because we still ignore what passed through the "sel" canals (nerves or endolymphatic expansions of the labyrinth). Jarvik's interpretation was aimed at providing yet another character supporting the sister-group relationship between lampreys and osteostracans. We know that galeaspids, which lack cephalic fields, have a small utricular division of the labyrinth, which shows neither "sel" canals nor utricular recesses that could suggest the presence of ciliated recesses (Fig. 2.9E).

Like placoderms, some chondrichthyans, and possibly acanthodians among gnathostomes, all osteostracans have an externally open endolymphatic duct (dend, Fig. 2.9C,F). This is unknown in all other fossil "agnathans," except in the primitive galeaspid *Xiushuiaspis* (dend, Fig. 2.9E; Wang, 1991). However, it is possible that the presumed paired pineal foramina of arandaspids (Gagnier, 1993a) are, in fact, external endolymphatic openings, as suggested by their funnel-like shape, and this character could be either more general or more variable than currently believed. Lampreys and hagfishes have a possible homologue of the endolymphatic duct, but the latter never opens to the exterior, and this has long been regarded as a secondary condition. Sahney and Wilson (2001) showed that exogenous sand grains could penetrate through the endolymphatic duct into the labyrinth of osteostracans and acanthodians, as in some extant chondrichthyans, and serve as statoliths. This is an additional shared trait shared by osteostracans and gnathostomes, hitherto unknown in other vertebrate taxa.

Blood Vessels

The blood vascular system of galeaspids and osteostracans seems to have been quite similar, with the exception of the large marginal arteries and veins (a possible homologue of the anterior cardinal veins; Janvier et al., 1991), which are lacking in the former. The most striking resemblance between galeaspids and osteostracans in the anatomy of the blood vascular system is the very large lateral head vein that drains the blood from the anterior part of the head (and possibly part the branchial region in galeaspids). It was regarded by Stensiö (1927) as the homologue of the lateral head vein of larval lampreys, although in the latter it is far thinner and much reduced in the adult. In contrast, it compares fairly well with the large but more laterally situated dorsal jugular vein of placoderms and other gnathostomes (Forey and Janvier, 1993: fig. 3).

The blood vascular system of osteostracans displays a character that can be regarded as uniquely shared with lampreys: the canal for the dorsal aorta and the groove that prolongs it posteriorly are displaced to the right side (da + oesc, Fig. 2.10A,B). Stensiö (1927) considered that the anterior cardinal veins flanked the dorsal aorta in osteostracans, and that this asymmetry was evidence for the lack of the left duct of Cuvier, as in adult lampreys, with both anterior cardinal veins emptying into the right duct of Cuvier. A reinterpretation of this peculiar asymmetry has been suggested in the light of larval lamprey anatomy (Janvier et al., 1991). In larval lampreys, the dorsal aorta and anterior mesenteric artery (ama, da, Fig. 2.10D) have to pass to the right side above the heart because of the median wall of connective tissues that suspend the esophagus from the notochord at this level (st, Fig. 2.10D). Although this could also apply to osteostracans, there remains a difference in the position of the esophagus that, in osteostracans, had to pass through the postbranchial wall through the same asymmetrical canal as the dorsal aorta (oes, Fig. 2.10C). Consequently, the median connective tissues that in lampreys connect the esophagus dorsally to the heart were probably lacking in osteostracans—or at any rate, they played no role in shifting the dorsal aorta to the right side. This asymmetry, to date unique to lampreys and osteostracans (we ignore the condition in other "ostracoderms"), may thus be homoplastic, the condition in osteostracans being possibly a consequence of the dorsoventral flattening of the head shield, which leaves little space for the esophagus for passing dorsally to the heart (Fig. 2.10C). Moreover, the symmetry of the marginal vein sinuses and posterior end of the lateral head vein canals in osteostracans (mvs, Fig. 2.10B,C), as well as the large dorsal opening of the pericardial cavity (pcard, Fig. 2.10B), rather suggest that both ducts of Cuvier were retained, leaving even less space for the esophagus (dC, Fig. 2.10C; Janvier et al., 1991).

The organization of the heart in osteostracans is known from the internal shape of the pericardial cavity, which shows that the atrium and ventricle were lying side by side and closely set, exactly as in lampreys (atr, ven, Fig. 2.10C; Janvier et al., 1991). Although the condition in other "ostracoderms" and in placoderms remains unknown, current phylogenies with a paraphyletic group, Cephalaspidomorpha, imply that this condition is intermediate between that of hagfishes (atrium and ventricle side by side, but well apart) and that of extant gnathostomes (atrium and ventricle closely set, but with the atrium expanding dorsally to the ventricle). Lamprey heart structure displays a unique character, with the venous network lining the wall of the pericardium and involved in the lubrication of the heart (Percy and Potter, 1991). The grooves on the internal surface of the pericardial cavity, described by Stensiö (1927: fig. 35) from horizontal grinding sections of the osteostracan *Mimetaspis* (then interpreted as impressions of the pronephric tubules), could either be impressions of this venous network or a mere artifact of preservation. Vertical grinding sections through *Norselaspis* and direct observation of the pericardial wall in *Axinaspis* (Wängsjö, 1952: pl. 99:2) failed to yield similar grooves (Janvier, 1985a).

Gills

Much has been written about the organization of the gills and gill arches in fossil jawless vertebrates, sometimes in connection with the question of the origin of jaws (e.g., Stensiö, 1927, 1964; Wänsgjö, 1952; Damas, 1954; Watson, 1954; Halstead, 1973; Whiting, 1977; Novitskaya, 1983; Janvier, 1985a, 1996a,b). However, there is very little evidence of detailed gill structure, apart from the path of branchial nerves and some vague imprints of the gill filaments. Strangely, even in the extensively ossified head shields of osteostracans and galeaspids, the gill arches seem to vanish beyond the medial ventral processes of the oralobranchial cavity (mvp, Fig. 2.10B) on which they are supposed to be attached. Janvier's (1985a, 1996b) default option was to consider that the gills in osteostracans were organized in much the same way as in larval lampreys, the cartilaginous branchial skeleton as a whole being housed in the oralobranchial cavity, the roof of which did not incorporate dorsal gill-arch components (contra Stensiö, 1927), and that the same condition could be generalized to galeaspids. There are some hints suggesting that the gills of "ostracoderms" were supported by sinuous cartilaginous arches of lamprey type, notably some sinuous imprints in the roof of the oralobranchial cavity of osteostracans (gaimp?, Fig. 2.10A; Janvier, 1981b: fig. 4).

The external gill openings of most osteostracans (except for some tremataspidids) somewhat resemble those of elasmobranchs in being slit-shaped (gs, Fig. 2.11B), in contrast to those of the extant and most fossil "agnathans" (e.g., anaspids, astrapidids, galeaspids), which are smaller and rounded in shape.

The anatomy of osteostracans implies that the posteriormost gill compartments (or gill pouches) were connected to the pharynx by means of a pharyngobranchial duct, as in adult lampreys (Janvier, 1984: fig. 2.5C,D). It is probable that the same condition also prevailed in galeaspids, anaspids, and euphaneropids (Janvier, 2004).

Shoulder Girdle and Paired Fins

The shoulder girdle and paired fins of osteostracans bear a striking resemblance to those of the gnathostomes, in particular placoderms (Janvier, 1978, 1984, 1985a; Goujet, 1984, 2001). In both taxa, the girdle is a massive endoskeletal structure, on which the fin endoskeleton articulates by means of a small articular facet. The foramina for the brachial artery and brachial veins occupy exactly the same position in both groups, relative to the articular surface. Additionally, the surface of the fin insertion area displays a number of depressions for muscle insertions, which indicate a relatively complex musculature. The paired fin endoskeleton of osteostracans

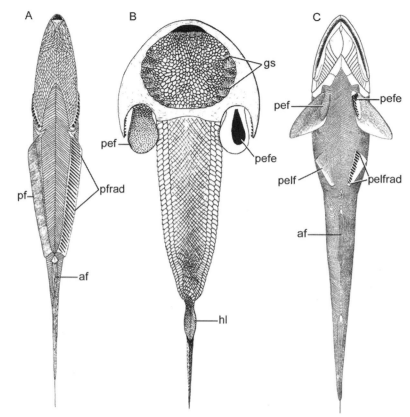

Figure 2.11. Comparative organization of the paired fins in an anaspid (**A**, *Pharyngolepis,* Early Silurian), an osteostracan (**B**, *Zenaspis,* Early Devonian), and a gnathostome (**C**, *Cheirolepis,* Late Devonian) in ventral view (not to scale). Endoskeleton (in black) of the paired fins reconstructed on the right side (inferred from the conditions in *Euphanerops* in **A**, *Escuminaspis* in **B**, and *Mimia* for the pelvic fin in **C**). (**A**, after Ritchie, 1964; **B**, based on Heintz, 1967, Janvier, 1996b, and Janvier et al., 2004a; **C**, after Pearson and Westoll, 1979.) Abbreviations as in Figure 2.4.

has be reconstructed on the basis of a single specimen of *Escuminaspis laticeps,* where it is partially preserved in the form of patches of calcified cartilage (Janvier et al., 2004a), and the recent discovery of a second specimen of the same species showing similar structures suggests that it is unlikely to be a fabric due to diagenesis. The paired fin endoskeleton was probably a leaf-shaped cartilage plate (pcfe, Fig. 2.11B) that recalls the one or two radials of most placoderms, and more so the cartilaginous disk of the embryonic pectoral fins of most extant fishes (Grandel and Schulte-Merkel, 1998; Cohn et al., 2002; Coates, 2003). The main difference between osteostracans and gnathostomes is that the endoskeletal shoulder girdle is continuous with the braincase (i.e., it is part of the endoskeletal head shield) in osteostracans, whereas it is always separate from the latter in the gnathostomes. It is assumed that pituriaspids possessed much the same kind of paired fins as osteostracans, yet this is merely a guess based on the similar aspect of the fin insertion area.

The presence of endoskeletal radials in the ribbon-shaped paired fins of *Euphanerops* suggests that similar radials were also present in the similarly shaped paired fins of anaspids (pfrad, Fig. 2.11A; P. Janvier and M. Arsenault, pers. obs., 2004). Thus, if true paired fins (i.e., with endoskeletal supports), as a whole, are regarded as homologous, their absence in all pteraspidomorphs, galeaspids, and presumably thelodonts (at least *Turinia;* Donoghue and Smith, 2001; but see Wilson et al., Chapter 3) poses a problem in the framework of most current phylogenetic hypotheses because this would imply that paired fins were lost independently in several "ostracoderm" groups (pteraspidomorphs, galeaspids, and probably most thelodonts). In fact, there are important differences between the paired fins of *Euphanerops* and anaspids on the one hand, and the osteostracan and early gnathostome pectorals on the other. The former are ribbon-shaped webs (pf, Fig. 2.11A) extending from the anus to the branchial region and supported by numerous thin, parallel radials, lacking any fin support or girdle (pfrad, Fig. 2.12A), whereas the latter are stout, paddle-shaped structures concentrated behind the branchial apparatus and supported by a few large endoskeletal elements that articulate with a massive girdle (pef, Fig. 2.11B,C).

The paired fins of *Euphanerops* and anaspids, with their more ventral position and their numerous parallel radials, rather resemble the pelvic fins of such early gnathostomes, as acanthodians and generalized actinopterygians (pelf, Fig. 2.11C; yet the pelvics of the latter do not extend as far anteriorly, and their radials articulate on a pelvic girdle; see Wilson et al., Chapter 3), whereas the paired fins of osteostracans and the pectorals of the gnathostomes are basically similar. This also relates to the difference between the pelvic and pectoral fins of the gnathostomes as to their early embryonic development: the radials of the pelvics arise from the beginning as separate cartilage rods, whereas those of the pectorals appear later, by subdivision of a single cartilage disk (Grandel and Schulte-Merkel, 1998; Cohn et al., 2002; Coates, 2003; Davis et al., 2004). Apart from some indications of possible pelvic fins in a thelodont in the form of scaled skin flaps in addition to pectoral fin folds (Märss and Ritchie, 1998; Wilson et al., Chapter 3), no fossil "agnathan" displays both pectoral and pelvic fins with indisputable endoskeletal radials. Maisey (1986, 1988) invoked this resemblance between the anaspid paired fins and the gnathostome pelvics as support for a sister-group relationship between anaspids and gnathostomes. However, this would pose the problem of the characters shared by osteostracans and gnathostomes, but lacking in anaspids. Thus, apart from the pectoral fins of gnathostomes and osteostracans (and possibly pituriaspids), for which a hypothesis of primary homology is acceptable and which is congruent with other uniquely shared characters, the distribution of the paired fins in extant and fossil vertebrates raises problems of either homoplasy or loss, as in the case of the nasohypophysial complex.

Axial Skeleton

Curiously, and contrary to what is observed in decay experiments (Briggs and Kear, 1994), the notochord, which is made up by relatively tough tissues, rarely leaves any imprint in fossil vertebrates, even when preserved under conditions that allow soft-tissue

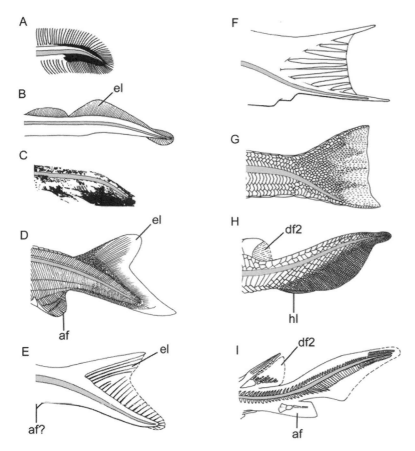

Figure 2.12. Tail in crown-group vertebrates, in lateral view (not to scale). The position of the notochord (**gray**) is hypothetical, except in **A**–**C** and **I**. **A**, hagfish (endoskeleton in black); **B**, lamprey; **C**, euconodont; **D**, anaspid; **E**, thelodont (*Loganellia*); **F**, thelodont (furcacaudiform); **G**, heterostracan; **H**, osteostracan; **I**, crown-group gnathostome (hybodontiform chondrichthyan). (**A**, after Janvier, 1997; **B**, after Janvier, 1996b; **C**, after Aldridge et al., 1986; **D**, after Ritchie, 1964; **E**, based on specimen GBP 367, Muséum National d'Histoire Naturelle, Paris; **F**, after Wilson and Caldwell, 1993; **G**, after Gross, 1963; **H**, after Heintz, 1967; **I**, after Maisey, 1989.) Abbreviations as in Figure 2.4.

preservation. The position of the notochord in the various fossil "agnathans" reconstructed in Figure 2.13 (gray) is entirely hypothetical and is only inferred from the overall shape of the tail, except possibly in euconodonts. However, the path of the notochord can be reconstructed with some confidence only when the arcualia bordering it are visible (Figs. 2.8B, 2.13I).

Lampreys and gnathostomes are the only extant vertebrates that possess vertebral components, in addition to the notochord. They both have dorsal arcualia (basidorsals and interdorsals), and gnathostomes have, in addition, ventral arcualia (basiventrals and interventrals; Fig. 2.2B,C). *Euphanerops* is the only fossil "agnathan" in which the arcualia can actually be observed, forming complete series of dorsal and ventral arcualia, as in gnathostomes (dar, var, Fig. 2.8B; P. Janvier and M. Arsenault, pers. obs.), suggesting that euphaneropids (and possibly anaspids) are more closely related to gnathostomes than to lampreys. The very elongated occipital region of galeaspids and certain osteostracans suggests that it incorporates vertebral elements, at any rate dorsal arcualia, but provides no information as to their posterior extension. Only an enigmatic series of five calcified elements in the trunk of the osteostracan *Ateleaspis* (Ritchie, 1967: pl. 4, fig. 1) is suggestive of an axial skele-

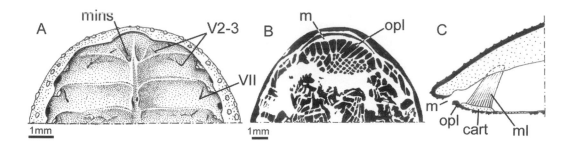

Figure 2.13. Organization of the oral region in osteostracans. **A,** *Norselaspis* (Early Devonian), anterior part of the roof of the oralobranchial cavity in ventral view, showing a pair of probable muscle insertions in the oral roof; **B,** *Hirella* (Early Silurian), dermal covering of the oralobranchial membrane in ventral view, showing the oral plates bordering the mouth; **C,** reconstructed sagittal section of the oral region of an osteostracan head, showing the possible position of the hypothetical cartilage plate bearing the oral plates and its associated musculature. Abbreviations as in Figure 2.4.

ton. Nevertheless, these elements do not resemble the arcualia of *Euphanerops*.

Median Fins and Tail

Osteostracans and gnathostomes are the only vertebrates in which the tail is unambiguously epicercal (Fig. 2.12H,I). In fact, the only difference between the tail of an osteostracan and that of a gnathostome is the small horizontal ventral lobe that underlies the hypochordal web of the osteostracan tail (hl, Figs. 2.11B, 2.12H), and which is an autapomorphy of this group, although possibly representing a modified anal fin. The evolutionary transformations that can be inferred from the various tail structures in extant and fossil vertebrates (Fig. 2.12) seem relatively simple in the framework of current phylogenies, assuming that the apparently homocercal tail in heterostracans and furcacaudiform thelodonts (Fig. 2.12F,G) is merely a particular case of the hypocercal condition (Fig. 2.12A–E), much in the same way as the homocercal tail of teleosts is a particular case of the epicercal condition of gnathostomes. Nevertheless, the transition from a generalized hypocercal tail (as exemplified by the thelodont *Loganellia* or anaspids; Fig. 2.12D,E) to the epicercal tail of osteostracans and gnathostomes (Fig. 2.12H,I) remains undocumented by transitional states.

The primitive number of dorsal fins in gnathostomes remains debatable. It is generally assumed that two dorsal fins is the general condition for gnathostomes. This may be true for crown-group gnathostomes, but placoderms seem to have a single dorsal fin, except perhaps for ptyctodonts. The most generalized osteostracans have two dorsal fins, but the anterior one lacks evidence for a fin web and appears as a mere hump that becomes reduced to a large median scute in cornuates, and finally disappears. Wängsjö (1952:110, 234) showed that the vascular canals in the endoskeletal dorsal spine (or hump) of osteostracans are arranged segmentally and suggested that it was originally a real fin supported by radials. In contrast, the second dorsal fin of osteostracans clearly displays a fin web with lepidotrich-like series of scales (df2, Fig. 2.12H), and may well be the homologue of the anterior part of the epichordal lobe of anaspids, thelodonts, and lampreys (el, Fig. 2.12B,D,E). In "agnathans" that display an anal fin (euphaneropids, anaspids, and probably some thelodonts; af, Fig. 2.12D,E), the latter is situated well in front of the level of the epichordal lobe. Assuming that the anterior part of the epichordal lobe is the homologue of the second dorsal fin of osteostracans and gnathostomes, this suggests that the modular development of anal and posterior dorsal fin invoked by Mabee et al. (2002) for the actinopterygian unpaired fins did not apply to "ostracoderms."

Organization of the Exoskeleton

Apart from the pineal plate and sclerotic ring, none of the dermal bones of the osteostracan head shield displays any plausible homology with the dermal head and shoulder girdle skeleton of the macro- or mesomeric gnathostomes. The long-presumed homology between the cornual processes of osteostracans and the pectoral fin spines or spinal plates of gnathostome does not hold because the most plesiomorphous osteostracans are all noncornuates. Whether the structure and organization of the body scales have any significance remains debated. Admittedly, the macrosquamose, diamond-shaped body scales of osteostracans bear some superficial resemblance to those of placoderms and osteichthyans, but the most striking resemblance is perhaps between their closely set dermal rays of their median fins and the osteichthyan lepidotrichs. Unfortunately, there is to date no consensus as to the general condition for the gnathostome exoskeleton; yet the primitiveness of the microsquamose or micromeric elasmobranch condition remains a widespread assumption—essentially for historical reasons—and because thelodonts, among "agnathans," also have a microsquamose exoskeleton (Janvier, 1981a; for an extensive review see Donoghue, 2002). The macro- or mesomeric condition of placoderms and generalized osteichthyans, as to both scales and dermal bones, would rather support the theory that this condition is general for the gnathostomes, and therefore somewhat similar to that in osteostracans.

Histology

Histological characters are generally difficult to assess because of their considerable variability in certain taxa and, in some cases, uncertainties as to their definition and characterization. Certain hard tissues, however, can be regarded as unique to particular taxa, or are empirically known to display little homoplasy, such as the semidentine of placoderms. Osteostracans and gnathostomes are the only vertebrates in which the exoskeleton and endoskeleton show bone cell lacunae. Admittedly, some acanthodians and chondrichthyans have an acellular exoskeleton, and some acanthomorph teleosts have an entirely acellular skeleton, but this is currently regarded as a particular condition relative to cellular bone. Similarly, the Late Devonian cornuate osteostracans *Escuminaspis* shows no evidence for bone cell lacunae, and this is also assumed to be a derived condition because the exo- and endoskeleton of all other osteostracans, including the generalized noncornuate taxa, show the same cellular structure as those of the placoderms and generalized osteichthyans.

The presence of perichondral bone seems restricted to osteostracans and gnathostomes, but perichondral calcification, lacking cell lacunae, occurs in galeaspids. The nature of the possible perichondral calcification described in arandaspids (Gagnier, 1993a), notably in the sclera, remains debated. It is far from being as extensive as in galeaspids, osteostracans, and gnathostomes.

Spherulitic (globular) calcified cartilage has long been regarded as a widespread and structurally monotonous type of hard tissue in "ostracoderms" (in fact, known only in *Eriptychius,* galeaspids, and osteostracans) and gnathostomes. The unusual type of calcified cartilage of *Euphanerops,* with very large chondrocyte spaces lined with a thin calcified shell (Janvier and Arsenault, 2002) and a spherulitic matrix, is still a matter of debate; it may either be of diagenetic origin or may represent a mode of calcification that occasionally occurs in vivo in lampreys (Bardack and Zangerl, 1971) but is still undescribed in detail.

Homology Relationships of the Characters in Crown-Group Vertebrates

As a whole, and except for the vexing question of the nasophypophysial complex, most of the characters that can be observed in osteostracans have their sister character in the gnathostomes, rather than in either lampreys, hagfishes, or other "ostracoderms." However, the problem lies in the fact that most of the latter characters are less clear-cut than those of the nasohypophysial complex, and according to a paleontological tradition, regarded as adaptive or subject to variability. The nasohypophysial complex of osteostracans is always present with the same structure. Admittedly, the presence of cellular bone, epicercal tail, and complex pectoral fin musculature and endoskeleton seem generally unique to osteostracans and gnathostomes, but there are some exceptions (e.g., the acellular bone of *Escuminaspis* and certain gnathostome taxa, the lack of paired fins in tremataspidid osteostracans, and the variable state of the caudal fin in gnathostomes), which may raise questions about the distribution of these characters. Tail structure is often regarded as a typically adaptive character because of its diversity in extant gnathostomes, although neither the latter nor osteostracans display any instance of a hypocercal tail. Moreover, if one assumes a loss of the paired fins in tremataspidids or a loss of bone cells in *Escuminaspis,* then why not assume the same for the acellular dermal bone and lack of paired fins of more inclusive "ostracoderms," such as arandaspids or heterostracans? The only reason for not doing this is perhaps that in contrast to osteostracans, all members of these clades share this lack of cellular bone and pectoral fins.

When considering the distribution of the morphological and histological characters that can be readily observed in fossil and extant vertebrates (e.g., the character-taxon matrices provided by Janvier, 1996a; Donoghue et al., 2000; Donoghue and Smith, 2001), one can notice that they fall into three classes: (1) those that have a broadly congruent distribution and largely define the shape of the current trees, in which either some, or all, "ostracoderms" are more closely related to the gnathostomes than to either lampreys or hagfishes (e.g., large paired cerebellum, dermal bone, perichondral bone, pectoral fins); (2) those that have a poorly informative, or disjunct, distribution, but can nevertheless be accommodated by the preceding character distribution (e.g., orthodentine), by an appeal to plausible processes of loss (or autapomorphic modification) rather than to homoplasy; and (3) those whose distribution shows no congruence with the majority of other homologies and are thus currently regarded as homoplastic (e.g., nasohypophysial complex of the lamprey and osteostracan type). These differences in data congruence tell us that the current view of "agnathan"-gnathostome transition is far from being as remarkably illustrated as other evolutionary transitions in vertebrate history, and much imagination is still required to make a narrative from the current trees. For example, current trees generally imply that the shield-shaped head endoskeleton of galeaspids, osteostracans, and pituriaspids is a general condition, relative to the narrower braincase and independent shoulder girdle of the gnathostomes. The "agnathan"-gnathostome transition would thus involve a general reduction of the head and pectoral endoskeleton. No character state clearly documents such a process of reduction, except perhaps for the broad expansion of the placoderm braincase over the branchial region, which could represent a transitional condition between that of galeaspids or osteostracans and that of crown-group gnathostomes (Janvier, 2001).

If osteostracans, and probably the other "ostracoderms," best fill the morphological gap between extant "agnathans" and gnathostomes and provide some examples of evolutionary transitions at the level of particular characters or character complexes, the transition from a jawless to a jawed mouth remains virtually undocumented by fossils. The situation is somewhat comparable to that of the question of the fin-limb transition before the discovery of the Devonian tetrapods and elpistostegalians. The scenarios about the origin of jaws thus must rest essentially on developmental data. To some extent, this could also apply to the origin of paired fins, although thelodonts possess structures referred to as paired fin folds, which cannot be clearly proven to be actual fins with endoskeletal radials, but are at least potential precursors to fins (see Wilson et al., Chapter 3).

Question of the Origin of Jaws

What we know of the organization of the oral region in fossil "agnathans" (apart perhaps from fossil hagfishes and lampreys, and euconodonts) suggests that none of them possessed the highly specialized lingual apparatus of extant cyclostomes. They may nevertheless have possessed a protraction and retraction device, the basic function of which was somehow similar, and possibly general to all vertebrates, but lost in the gnathostomes (Janvier, 1981a). The mouth of arandaspids, heterostracans, galeaspids, and osteostracans is generally bordered ventrally by a series of dermal oral plates (opl, Fig. 2.13B; a single plate in galeaspids), which, one may imagine, could be slightly expanded as the mouth opened (Kiaer, 1928; Janvier, 1996b; Purnell, 2002; Elliott et al., 2004). The presence of large muscle insertion pits in the supraoral field of osteostracans (mins, Fig. 2.13A) and the fact that the oral plates of osteostracans sometimes remain articulated when displaced (e.g., in *Hirella*; Heintz, 1939: pl. 24, fig. 3, pl. 25, fig. 2) suggest the presence of some still unknown endoskeletal structure involved in the movements of the oral region (cart, Fig. 2.13C; Janvier, 1985a:49), which was moved up and down by means of muscles (ml, Fig. 2.13C). Possibly, this hypothetical element was articulated laterally to the foremost endoskeletal medial ventral process of the oralobranchial cavity, which is sometimes much developed (e.g., in tremataspidids; Janvier, 1985b: figs. 11, 26a). The median part of the oral roof is covered with dermal denticles in some osteostracans, which, together with certain thelodonts (such as *Loganellia*) and gnathostomes, are the only vertebrates in which oral or pharyngeal denticles can be seen. The question of the presence of a velum in fossil "agnathans" remains debated, although the prebranchial fossae of osteostracans are best interpreted as having housed a velum of larval lamprey type (Janvier, 1985a, 1996b). In contrast, Mallatt (1996) followed Stensiö's (1927, 1964) interpretation and considered that osteostracans possessed a respiratory spiracular pouch, with a posterior mandibular hemibranch and an anterior hyoidean hemibranch, but no velum. In sum, none of the fossil "agnathans" displays any structure that would possibly foreshadow a mandibular arch of gnathostome type—that is, with a palatoquadrate and a Meckelian cartilage.

Theories about the origin of jaws are thus entirely based on either an imaginary transformation series based on presumed homologies in the oral apparatus of extant vertebrates (e.g., Romer, 1945; Jollie, 1962; Mallatt, 1996; Janvier, 1996b), or, more recently, inferences based on data from developmental genetics (see reviews in Kuratani et al., 2001; Kuratani, 2004; Shigetani et al., 2005). In a key article on this question, Kimmel et al. (2001:116) refer to the origin of jaws as a "wonderful problem, long debated in the literature." It is indeed, and studies on this subject have become a small industry among developmental biologists. Since the nineteenth century, jaws have been regarded as serial homologues of the more posterior visceral arches ("repeated parts" of Schmitt, 2004), but the debates about their origin began with the question of the homology between the visceral arches (as a whole) of the extant "agnathans" and gnathostomes. To make a long story short, Goette (1901) discovered that lampreys differ from gnathostomes by the fact that their gill filaments are of endodermal origin, and, alongside the branchial nerves and blood vessels, are medial to the skeletal gill arches (bra1, Fig. 2.14A). In the gnathostomes, they are lateral to the skeletal gill arches (bra1, Fig. 2.14B) and can be shown to be of ectodermal origin, at any rate in osteichthyans. This later became a key argument for Stensiö (1927) and others for arguing that vertebrates fell into two sister clades, the Agnatha

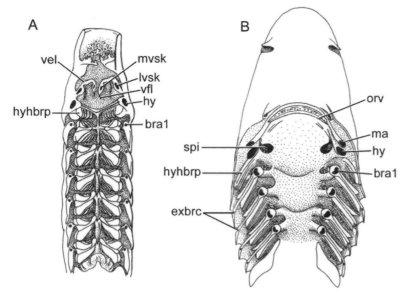

A

vel
mvsk
lvsk
vfl
hy
hyhbrp
bra1

B

orv
ma
hy
bra1

spi
hyhbrp
exbrc

Figure 2.14. Horizontally sectioned heads of a larval lamprey (**A**) and a crown-group gnathostome (**B**, elasmobranch), in ventral view, showing the relative position of the endoskeletal branchial arches and gill filaments. Not to scale. (After Mallatt, 1984, modified.) Abbreviations as in Figure 2.4.

and Gnathostomata (or Entobranchiata and Ectobranchiata), precluding any evolutionary transition from one group to the other. After a century-long debate about which of these two gill conditions could possibly be primitive relative to the other, Jarvik (1980, 1981) finally concluded that the gill arches of lampreys and gnathostomes were homologous, but not the gill filaments and their associated musculature and vessels, whereas others (e.g., Mallatt, 1984; Janvier, 1981a, 1996a,b) considered that the gills, musculature, and blood vessels were homologous, but the arches were not.

Because the jaws of the gnathostomes were unanimously regarded as serially homologous to the more posterior gill arches, they had to be organized according to the same pattern; that is, they should be medial relative to their nerves and blood vessels, which they essentially are in extant gnathostomes. Only placoderms possess mandibular arch elements that are reconstructed as being situated laterally to the adductor jaw musculature, but it is still undecided whether this represents a general gnathostome condition, or an autapomorphy of this particular extinct group. Because "agnathans" have no jaws but share gill arches with the gnathostomes, the received view used to be that gill arches have preceded jaws in evolutionary history, and it was thus necessary to have fossil evidence for a gill-bearing mandibular arch and, of course, to assume homology between the "agnathan" and gnathostome visceral arches. This was basically the aim of Stensiö's (1927) interpretation of osteostracans. This evidence is now discarded, and the homology of the lamprey and gnathostome visceral arches is still debated, despite much detailed investigations by developmental biologists. Nevertheless, most textbooks, and even more so popular books, about vertebrate evolution still propagate the classical and naive idea that the first (i.e., mandibular) or first two (i.e., mandibular and premandibular) respiratory branchial arch or arches of some unknown extinct jawless vertebrate once turned into jaws.

The foremost visceral arch (or arch 1, in current terminology) of embryonic lampreys is now almost unanimously regarded as the homologue of the mandibular arch of the gnathostomes, and therefore also termed the *mandibular arch*. In neither of these two groups does it develop into a respiratory gill arch (leaving aside the complex question of the pseudobranch of the gnathostomes; Mallatt, 1996), and none of the anatomically well-known extant and extinct "agnathans," notably osteostracans, can be proven to have possessed a mandibular gill (for details, see Janvier, 1981b, 1985b, 1996b). Instead, the mandibular arch of lampreys is connected to a complex skeleton, which supports a special water-pumping and antireflux organ, referred to as the *velum* (hagfish development remains almost undocumented, and it is difficult to clearly know which parts of the adult hagfish skull are mandibular arch derivatives).

Two morphology-based scenarios about the origin of jaws have been proposed during the past decade. Mallatt (1996) considered that the common ancestor to all vertebrates possessed both the lateral (lamprey-like) and medial (gnathostome-like) visceral arches, and that the former progressively disappeared in the gnathostomes, whereas the latter disappeared in lampreys (and hagfishes), except perhaps at the level of the velum. In larval lampreys, the velar skeleton actually displays two components: a lateral one, level with the more posterior arches, and a medial one, closer to the pharynx (lvsk, mvsk, Fig. 2.14A). Janvier (1993, 1996b) argued that the medial visceral arches of the gnathostomes were a neomorph, resulting from the invasion of the medial part of the interbranchial septa by neural crest–derived skeletogenous tissues, which started in the mandibular arch, as suggested by the medial component of the lamprey velar skeleton. Then this process was continued by the development of the medial, respiratory, hyoid, and gill arches. The joint between the dorsal and ventral portions of the visceral arches in general, which is absent in lampreys, arose first in the mandibular arch and, for functional reasons, progressively extended to the hyoid and gill arches. Subsequently, the lateral gill arches disappeared, unless they remain in elasmobranchs in the form of the extrabranchial cartilages (exbrc, Fig. 2.14B), which, however, are apparently not neural crest–derived (this remains to be checked by means of current marking techniques). Both Mallatt (1996) and Janvier (1996a,b) considered that at least part of the velar skeleton and the jaws could be homologous, but Mallatt rejected Janvier's (1993) hypothesis that jaws could be derived from any velum-like structure.

Kimmel et al. (2001) showed that the difference in the embryonic branchiomeres (the precursors of the mandibular, hyoid, and gill arches) of lampreys and gnathostomes can be traced as early as the neural crest cells have migrated ventrally into them. These authors considered that in lampreys, the neural crest cells, which will give rise to the visceral arch cartilages, lie lateral to the mesoderm bars of the branchiomeres, whereas in gnathostomes, they surround them, forming a cylinder. Subsequently, the gnathostome arches form from the cells that are in the medial part of this cylinder. Kimmel et al. (2001) therefore concluded that the lamprey and gnathostome arches are homogenic, forming from the same embryonic tissue, but that the chondrogenic cells have migrated medially in the gnathostomes to form medially placed arches, a process that these authors depict as the outside-in theory. Thus, visceral arch homology in lampreys and gnathostomes remains undecided, depending on whether homogeny is or is not evidence for homology ("causes of homologues are neither causes for homology nor homology itself" [Nelson, 1994:123]; see also discussion in Patterson, 1982). Kimmel et al. (2001) pointed out the fact that at an early stage, the lateral component of the velar cartilage of lampreys was described by Damas (1944) as showing a cylinder of neural crest–derived cells, surrounding a core of mesodermal cells, as in a gnathostome arch. However, subsequently, Meulemans and Bronner-Fraser (2002) and McCauley and Bronner-Fraser (2003) showed that even in lampreys, the neural crest cells surround the mesodermal components of all arches, after having migrated simultaneously along both lateral (subjacent to the ectoderm) and medial (between the mesoderm and endoderm) routes. This seemed to refute the promising outside-in theory by Kimmel et al. (2001) as an explanation for the origin of the gnathostomes, but Cerny et al. (2004b:266) did not regard the question as settled. They concluded that "still more precise cell tracing experiments are needed in both vertebrate groups to understand how developmental pathways were changed during the agnathan transition." In particular, they pointed out that in gnathostomes (at any rate, in the axolotl), only the neural crest cells that lie medial to the mesoderm and close to the endodermal wall of the pharynx condense into cartilage. Thus the outside-in theory would only concern the chondrogenic potentiality of the neural crest cells. McCauley and Bronner-Fraser (2003:2321) tentatively concluded that the difference in the position of the branchial arches in the lampreys and gnathostomes may "depend on a mechanism that is independent of neural crest migration." This is also favored by Kimmel et al. (2003) in the wake of the discovery by Couly et al. (2002) of the role of endoderm-ectomesenchyme induction in the patterning of the visceral arches. In sum, the position (lateral or medial) of the endoskeletal visceral arches would not so much be determined by the position of the neural crest cells, but rather by the proximity of the endoderm to the latter.

To sum up what we currently know of the early development of the gill arches in lampreys and gnathostomes, it seems that: (1) in lampreys, early neural crest cell migration simultaneously follows both lateral and medial routes in all arches, but only the lateral neural crest populations become chondrogenic and form arches, except perhaps in the medial component of the velar skeleton; (2) in gnathostomes, neural crest cells fail to migrate along the medial route from the beginning, but progressively surround (outside-in) the mesodermal bars of the arches; (3) in gnathostomes, only the medially placed neural crest cells of the cylinder surrounding the mesodermal bars become chondrogenic and form visceral arches, the position and structure of which is induced by the endoderm. As to the lateral arches of lampreys, it cannot be ruled out that a similar induction, but from the ectoderm, may also occur (G. Crump, pers. comm., 2004). Cerny et al. (2004b:266) thus considered that the "the failure of cranial neural crest cells to migrate along the medial migratory pathway in gnathostomes was the first step in the modification of the gnathostome body plan away from their agnathan forerunners."

This seems quite a different scenario than those proposed by morphologists (e.g., Janvier, 1996b; Mallatt, 1996). Yet despite the complexity of the arguments now involved in this debate, it seems that we are coming closer and closer to a solution, and that the signaling from the endoderm or ectoderm to the neural crest cells at an early stage of their migration is a crucial factor in determining the lateral or medial position of the pharyngeal arches. The only contribution of paleontology to this debate has been to provide some evidence that the transition from the "agnathan" to the gnathostome condition of the mandibular arch has occurred later than the acquisition of other features, such as the exoskeleton, pectoral fins, and epicercal tail, and was a comparatively sudden event in vertebrate evolution. In a sense, the rise of jaws appears as sudden as (and coeval with) that of the horizontal semicircular canal (Mazan et al., 2000).

In a somewhat different field than the mere lamprey-gnathostome visceral arch homology, fundamental data have been recently published by S. Kuratani and his collaborators on the developmental genetics of the lamprey head, with particular reference to the origin of jaws (Kuratani, 2003; Kuratani et al., 2004, Shigetani et al., 2002; see also reviews in Kuratani, 2004; Shigetani et al., 2005). Although these authors agreed with others on the fact that the early development of the pharyngeal arches of lampreys and gnathostomes is basically similar, with homologous mandibular-arch primordia in both taxa, they point out a number of fundamental differences as to the extension of the ectomesenchyme in the oral and rostral regions of the vertebrate head. Notably, these authors demonstrated that the neural crest–derived tissues of the lamprey "upper lip" (i.e., the oral hood) are entirely premandibular in origin (pm, Fig. 2.15A1). Here, of course, they did not mean that they are derived from a once-existing premandibular arch, but only from the ectomesenchyme that extends anteriorly to the mandibular arch proper and gives rise to the trabeculae cranii in the gnathostomes (pm, Fig. 2.15A2). They thus inferred that the so-called trabeculae cranii of lampreys are not homologous to those of the gnathostomes, but rather are a mere prolongation of the mesoderm-derived

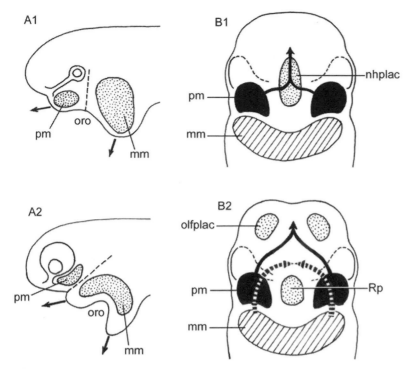

Figure 2.15. **A**, distribution of the ectomesenchyme surrounding the premandibular and mandibular mesoderm (dotted) in the embryonic head of a lamprey (**A1**) and a gnathostome (**A2**); the arrows indicate the direction of the growth of the upper and lower lips. **B**, schematic distribution of the premandibular (black) and mandibular (hatched) ectomesenchyme in a lamprey (**B1**) and a gnathostome (**B2**), as well as the nasohypophysial placode, the olfactory placodes, and the Rathke's pouch (dotted). The arrows indicate the direction of the extension of the premandibular (black arrow) and mandibular (dashed arrow) ectomesenchyme. (**A**, after Kuratani et al., 2004; **B**, after Kuratani et al., 2001.) Abbreviations as in Figure 2.4.

parachordals. Consequently, the oral apparatus (which, as Kuratani [2004] correctly noted, has much more diverse derivations than the mandibular arch alone) of lampreys and gnathostomes would be radically different in this respect because the ectomesenchyme extends much farther anteriorly in lampreys (Fig. 2.15A1).

The heterotopy theory of jaw evolution proposed by Shigetani et al. (2002) and Kuratani (2004) suggests that the premandibular ectomesenchyme, which in gnathostomes forms the prechordal plate and anterior braincase floor, underwent a heterotopic shift in ectoderm-ectomesenchyme interactions and differentiated into a large upper lip (pm, Fig. 2.15A1) that somehow parallels the anterodorsal expansion of the more posterior mandibular arch of gnathostomes (mm, Fig. 2.15A2). Kuratani et al. (2004) and Kuratani (2004) thus concluded that the oral hood of lampreys is in no way homologous to the dorsal part of the gnathostome jaw. However, it should be pointed out that most early anatomists and embryologists never envisaged such a homology because all of them regarded that the lamprey oral hood as being derived from the posthypophysial region of the head. Kuratani et al. (2004:459) stated that heterotopy leads to a "loss of homology" between the mandibular arches of lampreys and gnathostomes. In other words, heterotopy implies that changes in the position of gene expressions and tissue inductions make two anatomical structures be no longer homologous. Again, such homology concepts based on causes and aimed at justifying one particular theory of character relationship

or, more frequently, of character transformation, have been reviewed and criticized by Nelson (1994). There was once a belief
that the knowledge of the processes (the causes or pathways) involved in developmental genetics would solve the homology problem (Gilbert and Bolker, 2001), but it only provides yet another
subset of homology relationships (assuming that gene-expression
patterns can be regarded as homologues), sometimes congruent
with the previously proposed ones, and sometimes not.

What this evidence tells us is perhaps not that the oral apparatus of lampreys and gnathostomes displays no homology, but
rather that the oral opening is not exactly in the same position in
the two groups, relative to the distribution of the mandibular and
premandibular ectomesenchyme (oro, Fig. 2.15A). Which of these
two conditions can be regarded as general for the vertebrates or as
ancestral to the other remains undecided, for lack of a third set of
characters, which could perhaps be provided by hagfish development. It thus seems that heterotopy is not just displacing gene expressions and tissue interactions; it also results in displacing the homology problem to a different level of organization. These debates
about heterotopy are, I think, the result of the long-lived confusion
of homology relationships and ancestor-descendant character relationships (or character transformations). This is reflected in such
sentences as this: "the velum would be homologous with the jaws
only as a derivative of the mandibular arch, but neither of them
would represent an ancestral condition to the other" (Kuratani et
al., 2001:1629). Again, the question is not whether there has been
a transformation, but whether two structures are related characters
(or character states) relative to a third one.

Now, this question of the lamprey-gnathostome homologies as
to the oral apparatus may turn out to be considered in a different
way, following the recent suggestion by Cerny et al. (2004a) that
the palatoquadrate of the gnathostomes is in fact a composite
structure formed by an anterior portion, which is derived from the
same "premandibular" ectomesenchyme as the trabeculae cranii,
and a posterior (pterygoquadrate) portion, which is derived, like
the Meckelian cartilage, from the mandibular ectomesenchyme
that accumulates in the ventral (mandibular) prominence of the
mandibular arch. Surprisingly, this new interpretation of the palatoquadrate is almost identical to Jarvik's (1981: fig. 48) fossil-based
theory of a composite palatoquadrate derived from both the premandibular and mandibular arches, a theory long dismissed as
mere "anatomical philosophy" (see Janvier, 1996b). In this way,
the mandibular arch proper of the gnathostomes, lacking an anterior expansion, would be more similar to that of lampreys. However, the interpretation offered by Cerny et al. (2004a) is based on
a tetrapod (axolotl), and one may wonder whether this reduction
of the mandibular ectomesenchyme-derived portion of the palatoquadrate is not a consequence of the reduction of the palatoquadrate as a whole in tetrapods. The same reduction process of
the palatoquadrate in lungfishes had also led Bjerring (1977:164,

fig. 31D) to a similar conclusion concerning the nature of the dipnoan upper jaw.

Another important result provided by Kuratani's team is the evidence for a motor innervation of the lamprey upper lip musculature (derived from premandibular ectomesenchyme) by fibers from the maxillary branch (V2) of the trigeminal nerve (Kuratani et al., 2004). Back in the 1920s, this was a hotly debated question in connection with the premandibular arch theory, and the finding of Kuratani et al. (2004) may lead us to revisit Stensiö's (1927) old interpretation of osteostracan cranial nerves (with a visceromotor maxillary branch). The foremost nerve canal opening in the oralobranchial cavity of osteostracans (V2–3, Fig. 13A) was first regarded by Stensiö (1927) as having transmitted a visceromotor profundus nerve, then a visceromotor maxillary branch of the trigeminus (Wängsjö, 1952; Damas, 1954; Janvier, 1981a), and, finally, both the maxillary and mandibular branches of the trigeminal nerve (Whiting, 1977; Janvier, 1985b, 1996a,b), the former being exclusively viscerosensory. Now, it is no longer ruled out that the maxillary branch actually was also visceromotor and involved in the innervation of a premandibular ectomesenchyme-derived oral musculature, as in lampreys.

Janvier (1996a, 2001) pointed out that a major event in the "agnathan"-gnathostome transition was the disconnection of the olfactory placode (or placodes) from the hypophysial tube, resulting in the separate opening of the two nasal capsules to the exterior. Kuratani et al. (2001) and Ushida et al. (2003) also regarded this disconnection of the nasal placodes from the hypophysial anlage as having had a decisive impact on the reorganization of the premandibular and mandibular ectomesenchyme in the gnathostome head, notably in allowing the premandibular ectomesenchyme (pm, Fig. 15B2) to form the internasal septum, and the mandibular ectomesenchyme (mm, Fig. 15B2) to extend laterally to the Rathke's pouch (Rp, Fig. 15B2). Instead, the premandibular ectomesenchyme (pm, Fig. 15B1) of lampreys has to expand ventrally to the nasohypophysial placode (nhplac, Fig. 15B1), forming the oral hood.

Finally, Cohn (2002) suggested a most promising process for the rise of jaws. The neural crest cells that migrate into the mandibular arch of the gnathostomes are *Hox*-free. In contrast, those in the same arch (giving rise to the velar skeleton) of the lamprey *Lampetra fluviatilis* were shown by Cohn (2002) to express a *Hox* gene, *HoxL6*, whose ortholog in the cephalochordates (*AmphiHox6*) also extends far anteriorly, beyond the expression domains of *AmphiHox2–4*. It seems thus that there has occurred at some time after the divergence of lampreys and in the lineage that ultimately led to the gnathostomes a posterior retraction of the expression domain of *Hox6*, which freed the mandibular arch from any *Hox* gene. Because *Hox6* is known for having a role in homogenizing the structure of the serially homologous structures and was thought to be involved in the similar shaping of the hyoid and gill arches, Cohn (2002) assumed that by becoming *Hox*-free, the mandibular arch could expand ventrally and develop a Meckelian cartilage. This discovery would have lent some support to the theory that jaws have appeared first and preceded in time the "gnathostomization" of the more posterior arches. However, further investigations failed to corroborate Cohn's finding (Takio et al., 2004), as neither *HoxL6* nor any other *Hox* gene expression occurs in the mandibular arch of the closely related lamprey *Lethenteron japonicum* ("*Lampetra japonica*"), which is thus similar to the gnathostome condition in this respect. Because it is unlikely that such a major difference could occur in such closely related taxa, one may conclude that Cohn's theory remains controversial.

One may try and imagine the kind of fossil data that could provide a hint for determining at which node of the current vertebrate phylogeny could occur conditions that would somehow foreshadow the gnathostome jaws. The presence of an extensive denticle covering in the pharyngeal cavity of certain thelodonts (Van der Brugghen and Janvier, 1993; Smith and Coates, 2001) could be an indication for an early skeletogenic neural crest–pharyngeal endoderm induction process, and thus the presence of precursors of the medial arches, but this awaits precise knowledge of the manner in which these denticles are organized in the thelodont pharynx. Mallatt's (1996) suggestion of a homology between the oral valve of gnathostomes (orv, Fig. 2.14B) and the medial flap of the "agnathan" velum (vfl, Fig. 2.14A) seems an interesting hint (see, however, reservations expressed by Kuratani et al., 2001:1629). This antireflux valve attached to inner side of the mandibular arch of the gnathostomes has virtually the same anatomical relationships and function as the velum of larval lampreys and, when denticle covered, could be potentially preserved in stem gnathostomes. How much such a discovery would help in explaining the origin of jaws or provide evidence for evolutionary transition is uncertain, but it could provide an interesting character that is functionally related to the presumed early role of jaws—that is, increased ventilation and ultimately suction feeding (Mallatt, 1996).

Conclusions

Certain fossil vertebrates, be they stem- or crown-group vertebrates, provide examples of character combinations that may support scenarios of evolutionary transition in early vertebrate history, notably as to the rise of the gnathostomes. Possible stem vertebrates, such as the Early Cambrian myllokunmingiids, represent a proxy for the kind of organism from which crown-group vertebrates may have derived. However, they already display most of the characters of the latter, and they thus barely fill the reputedly major morphological gap that still separates cephalochordates from vertebrates. The interpretation of the Early Cambrian yunnanozoans remains controversial, but they do bear some resemblance to the myllokunmingiids and crown-group vertebrates, notably in the organization of their gills. Whether this resemblance is a particular vertebrate or a general deuterostome character is still debated. The absence to date of any fossil cephalochordates precludes any assessment of the derivedness of the extant representatives of this group relative to the common ancestor they are supposed to have shared with vertebrates.

However, the structure of the myllokunmingiids and yunnanozoans, which combines general euchordate characters, hitherto reputedly unique cephalochordate characters, and some unique vertebrate characters, does not suggest a very long evolutionary history from the cephalochordate-vertebrate divergence to the earliest fossil evidence for vertebrates. A cascade of developmental innovations, such as the neural crest and its derivatives, could have rapidly led to profound modifications of stem-vertebrate anatomy to such an extent that they would differ widely from cephalochordates. The history of vertebrate paleontology and phylogeny shows that many major gaps (e.g., "agnathan"-gnathostome, fish-tetrapod, reptile-mammal, reptile-bird, ape-human gaps) were once regarded as evidence for deep divergence times and slow transitional process, but new or previously unnoticed character combinations showed that these evolutionary transitions were in fact much more simple and rapid than previously thought. The same perhaps also applies to the gap between cephalochordates and vertebrates.

Whatever the relationships between hagfishes and lampreys may be, the various clades of fossil armored "agnathans," colloquially referred to as "ostracoderms" or stem gnathostomes (although lacking jaws), provide some examples of evolutionary transitions, at any rate at the level of certain characters or character complexes. In fact, these assumptions about evolutionary transitions are merely the consequence of the co-occurrence of the lack of jaws on the one hand, and some characters that are only known in gnathostomes among extant vertebrates on the other. This may indeed give the impression that the stem of the gnathostomes is documented by a more or less acceptable series of fossil transitional forms, leading from the naked hagfishes or lampreys to the bony gnathostomes, but there is a danger that this scenario becomes doctrine. Therefore, instead of considering relationships between taxa and inferring transformation scenarios from character optimizations at nodes, as most current computer programs for phylogenetic reconstruction do, it may be sometimes useful to return to a consideration of the homology relationships at the level of individual characters in the form of a three-item statement. By doing so, there would be admittedly a temptation to return to the key characters established as such by an authority, as in the time of Stensiö's "cephalaspidomorph" characters. However, this would also make incongruent patterns of homology relationships less readily overlooked or dismissed forever as homoplasies by virtue of optimization. The nasohypophysial complex of lampreys and osteostracans is such an example.

Admittedly, in classical adaptationist scenarios, evolutionary transitions are perhaps more readily illustrated by taxa than by characters, but this exclusive consideration of the end product of character analysis, i.e., a taxon tree supported by optimized character transformations at nodes, may be abused to generate narratives, in particular in the mind of researchers to whom the unending quest of phylogenetics is not familiar. The current vertebrate phylogeny is merely one hypothesis, more parsimonious by a few steps than Stensiö's (1927) assumption of "agnathan" monophyly, and most narrative evolutionary scenarios that can be inferred from it are mere incarnations of an optimization procedure. Perhaps this is entirely wrong; perhaps "agnathans" actually do form a clade and are far more derived than the gnathostomes, relative to the common ancestor of the vertebrates, as suggested by Jarvik (1980). However, I doubt it. At present, the evolutionary transition from "agnathans" to gnathostomes remains best illustrated by some characters or character complexes, most of which gain significance in the light of fossil data. However, whether or not these homology relationships (and the character transformations that may be inferred from them) can be integrated in an adaptationist scenario based on taxa remains a matter of personal choice and imagination.

The future of this field of research, as to its paleontological aspect, lies essentially in the methods used for analyzing the data (Nelson, 1994:111), but also in the good and equal quality of the latter. The source of data on the structure and diversity of extinct jawless vertebrates has not dried up yet. The anatomy of galeaspids could technically be as well known as that of osteostracans. A better knowledge of pituriaspids only depends on further fieldwork (in admittedly difficult areas). Exceptionally well-preserved thelodonts are being found in Canada, Estonia, and Scotland, and a hitherto unsuspected diversity of Ordovician vertebrates is turning up in Australia, Canada, and the United States. Moreover, certain early Paleozoic Konservat-Lagerstätten, such as Miguasha (Canada), the Soom Shale (South Africa), or Chengjiang (China), provide more and more data on soft-bodied jawless vertebrates. Early fossil vertebrates may not be able to overturn theories of relationships based on morphological data of extant taxa, but they may call into question some character homology relationships, and thus radically change the way we imagine evolutionary transitions at the level of certain characters.

Acknowledgments

This article is dedicated to Robert L. Carroll, as an acknowledgment of his defense of the role of paleontology in comparative biology. I am grateful to Georges Kimmel, Gage Crump (Oregon State University), Shigeru Kuratani (Okayama University), and John Maisey (American Museum of Natural History) for their extremely helpful information and comments, notably as to the section on developmental data. I also thank all the other developmental geneticists I have been pestering all these years regarding various questions of jaw development. I am grateful to the reviewers of this article, Peter Forey and Mark Wilson, for their most helpful comments.

References Cited

Aldridge, R. J., and P. C. J. Donoghue. 1998. Conodonts: a sister-group to hagfish?; pp. 16–31 in J. M. Jørgensen, J. P. Lomholt, R. E. Weber, and H. Malte (eds.), The Biology of Hagfishes. Chapman and Hall, London.

Aldridge, R. J., and M. Purnell. 1996. The conodont controversies. Trends in Ecology and Evolution 11:463–467.

Aldridge, R. J., D. E. G. Briggs, and E. N. K. Clarkson. 1986. The affinities of conodonts—new evidence from the Carboniferous of Edinburgh, Scotland. Lethaia 19:279–291.

Arsenault, M., and P. Janvier. 1991. The anaspid-like craniates of the Escuminac Formation (Upper Devonian) from Miguasha (Quebec, Canada) with remarks on anaspid-petromyzontid relationships; pp. 19–44 in M. M. Chang, Y.-H. Liu, and G.-R. Zhang (eds.), Early Vertebrates and Related Problems of Evolutionary Biology. Science Press, Beijing.

Baker, C., and M. Bronner-Fraser. 1997. The origin of the neural crest. Part II: an evolutionary perspective. Mechanisms of Development 69:13–29.

Bardack, D. 1991. First fossil hagfish (Myxinoidea): a record from the Pennsylvanian of Illinois. Science 254:701–703.

Bardack, D., and E. S. Richardson. 1977. New agnathous fishes from the Pennsylvanian of Illinois. Fieldiana (Geology) 33:289–510.

Bardack, D., and R. Zangerl. 1968. First fossil lamprey: a record from the Pennsylvanian of Illinois. Science 1962:1265–1267.

Bardack, D., and R. Zangerl. 1971. Lampreys in the fossil record; pp. 67–84 in M. W. Hardisty and I. C. Potter (eds.), The Biology of Lampreys. Volume 1. Academic Press, London and New York.

Bjerring, H. C. 1977. A contribution of structural analysis of the head of craniate animals. Zoological Scripta 6:127–183.

Bjerring, H. C. 1984. Major anatomical steps toward craniotedness: a heterodox view based largely on embryological data. Journal of Vertebrate Paleontology 4:17–29.

Blieck, A. 1991. At the origin of chordates. Géobios 25:101–113.

Briggs, D. E. G., and E. N. K. Clarkson. 1987. An enigmatic chordate from the Lower Carboniferous Granton "shrimp-bed" of the Edinburgh district, Scotland. Lethaia 20:107–115.

Briggs, D. E. G., and A. J. Kear. 1994. Decay of the lancelet *Branchiostoma lanceolatum* (Cephalochordata): implications for the interpretation of soft-tissue preservation in conodonts and other primitive chordates. Lethaia 26:275–287.

Briggs, D. E. G., E. N. K. Clarkson, and R. J. Aldridge. 1983. The conodont animal. Lethaia 20:1–14.

Butler, A. B. 2000. Chordate evolution and the origin of craniates: an old brain in a new head. Anatomical Record 261:111–125.

Carroll, R. L. 1988. Vertebrate Paleontology and Evolution. W. H. Freeman and Company, New York, 598 pp.

Carroll, R. L. 1995. Problems of the phylogenetic analysis of Paleozoic choanates; pp. 389–445 in M. Arsenault, H. Lelièvre, and P. Janvier (eds.), Études sur les Vertébrés inférieurs. Bulletin du Muséum National d'Histoire Naturelle, Paris, sér. C, 17:389–445.

Carter, G. S. 1957. Review of "Chordate Phylogeny." Systematic Zoology 6:187–193.

Cerny, R., P. Lwigale, D. Ericsson, D. Meulemans, H.-H. Epperlein, and M. Bronner-Fraser. 2004a. Developmental origins and evolution of jaws: new interpretation of "maxillary" and "mandibular." Developmental Biology 266:225–236.

Cerny, R., D. Meulemans, J. Berger, M. Wilsch-Bräuninger, T. Kuth, M. Bronner-Fraser, and H.-H. Epperlin. 2004b. Combined intrinsic and extrinsic influences pattern cranial neural crest migration and pharyngeal arch morphogenesis in axolotl. Developmental Biology 266:252–269.

Chen, J.-Y., J. Dzik, G. Edgecombe, L. Ramsköld, and G.-Q. Zhou. 1995. A possible Early Cambrian chordate. Nature 377:720–722.

Chen, J.-Y., D.-Y. Huang, and C. W. Li. 1999. An Early Cambrian craniate-like chordate. Nature 402:518–522.

Coates, M. I. 2003. The evolution of paired fins. Theory in Biosciences 122:266–287.

Cohn, M. J. 2002. Lamprey Hox genes and the origin of jaws. Nature 416:386–387.

Cohn, M. J., C. O. Lovejoy, L. Wolpert, and M. I. Coates. 2002. Branching, segmentation and the metapterygial axis: pattern versus process in the vertebrate limb. BioEssays 24:460–465.

Conway Morris, S. 1993. Ediacaran-like fossils from the Cambrian Burgess Shale type faunas of North America. Palaeontology 36:593–635.

Conway Morris, S. 1998. The Crucible of Creation. Oxford University Press, Oxford, 276 pp.

Couly, G., S. Creuzet, S. Bennaceur, C. Vincent, and N. M. Le Dourarin. 2002. Interactions between Hox-negative cephalic neural crest cells and the foregut endoderm in patterning the facial skeleton in the vertebrate head. Development 129:1061–1073.

Damas, H. 1944. Recherches sur le développement de Lampetra fluviatilis. I. Contribution à l'étude de la céphalogénèse des Vertébrés. Archives de Biologie 55:1–285.

Damas, H. 1954. La branchie préspiraculaire des Céphalaspides. Annales de la Société Royale zoologique de Belgique 85:89–102.

Davis, M. C., N. H. Shubin, and A. Force. 2004. Pectoral fin and girdle development in the basal actinopterygians Polyodon spathula and Acipenser transmontanus. Journal of Morphology 262:608–628.

Delarbre, C., C. Gallut, V. Barriel, P. Janvier, and G. Gachelin. 2002. Complete mitochondrial DNA of the hagfish, Eptatretus burgeri: the comparative analysis of mitochondrial DNA sequences strongly supports the cyclostome monophyly. Molecular Phylogenetics and Evolution 22:184–192.

de Pinna, M. C. C. 1991. Concepts and tests of homology in the cladistic paradigm. Cladistics 7:367–394.

Dingerkus, G. 1979. Chordate cytogenetic studies: an analysis of their phylogenetic implications with particular reference to fishes and the living coelacanth. California Academy of Sciences, Occasional Papers 134:111–127.

Donoghue, P. C. J. 2002. Evolution and development of the vertebrate dermal and oral skeletons: unravelling concepts, regulatory theories, and homologies. Paleobiology 28:474–507.

Donoghue, P. C. J., and R. J. Aldridge. 2001. Origin of a mineralized skeleton; pp. 85–105 in P. E. Ahlberg (ed.), Major Events in Early Vertebrate Evolution. Palaeontology, Phylogeny, Genetics and Development. Taylor & Francis, London and New York.

Donoghue, P. C. J., and I. J. Sansom. 2002. Origin and evolution of vertebrate skeletonization. Microscopical Research Techniques 59:352–
372.

Donoghue, P. C. J., and M. P. Smith. 2001. The anatomy of *Turinia pagei*
(Powrie) and the phylogenetic status of the Thelodonti. Transactions
of the Royal Society of Edinburgh: Earth Sciences 92:15–37.

Donoghue, P. C. J., P. L. Forey, and R. J. Aldridge. 2000. Conodont affinity and chordate phylogeny. Biological Reviews of the Cambridge
Philosophical Society 75:191–351.

Donoghue, P. C. J., M. P. Smith, and I. J. Sansom. 2003. The origin and
early evolution of chordates: molecular clocks and the fossil record;
pp. 190–223 in P. C. J. Donoghue and M. P. Smith (eds.), Telling the
Evolutionary Time: Molecular Clocks and the Fossil Record. Taylor
& Francis, London and New York.

Elliott, D. K., R. C. Reed, and E. J. Loeffler. 2004. A new species of *Al-
locryptaspis* (Heterostraci) from the Early Devonian, with comments
on the structure of the oral area in cyathaspidids; pp. 455–472 in G.
Arratia, M. V. H. Wilson, and R. Cloutier (eds.), Recent Advances
in the Origin and Early Radiation of Vertebrates. Verlag Dr. Friedrich
Pfeil, Munich.

Forey, P. L. 1984. Yet more reflections on agnathan-gnathostome relationships. Journal of Vertebrate Paleontology 4:330–343.

Forey, P. L. 1995. Agnathans recent and fossil, and the origin of jawed vertebrates. Reviews in Fish Biology and Fisheries 5:267–303.

Forey, P. L., and P. Janvier. 1993. Agnathans and the origin of jawed vertebrates. Nature 361:129–134.

Furlong, R. F., and P. W. H. Holland. 2002. Bayesian phylogenetic analysis
supports monophyly of Ambulacria and of cyclostomes. Zoological
Science 19:593–599.

Gabbott, S., R. J. Aldridge, and J. Theron. 1995. A giant conodont with
preserved muscle tissue from the Upper Ordovician of South Africa.
Nature 374:800–803.

Gagnier, P.-Y. 1993a. *Sacabambaspis janvieri*, Vertébré ordovicien de Bo-
livie. 1. Analyse morphologique. Annales de Paléontologie (Vertébrés)
79:19–69.

Gagnier, P.-Y. 1993b. *Sacabambaspis janvieri*, Vertébré ordovicien de Bo-
livie. 2. Analyse phylogénétique. Annales de Paléontologie (Vertébrés)
79:119–166.

Gilbert, S. F., and J. A. Bolker. 2001. Homologies of process and modular el-
ements in embryonic construction; pp. 345–354 in G. P. Wagner (ed.),
The Character Concept in Evolutionary Biology. Academic Press, San
Diego.

Goette, A. 1901. Über die Kiemen der Fische. Zeitschrift für wissenschaftliche Zoologie 69:533–577.

Gorbman, A., and A. Tamarin. 1985. Early development of oral, olfactory
and adenohypophyseal structures of agnathans and its evolutionary
implications; pp. 165–185 in R. E. Foreman, A. Gorbman, J. M.
Dodd, and R. Olsson (eds.), Evolutionary Biology of Primitive Fishes.
Plenum Press, New York and London.

Goujet, D. 1984. Les Poissons Placodermes du Spitsberg. Cahiers de
Paléontologie. Éditions du Centre National de la Recherche Scientifique, Paris, 284 pp.

Goujet, D. 2001. Placoderms and basal gnathostome apomorphies; pp.
209–222 in P. E. Ahlberg (ed.), Major Events in Early Vertebrate Evo-

lution. Palaeontology, Phylogeny, Genetics and Development. Taylor & Francis, London and New York.

Goujet, D., and G. C. Young. 2004. Placoderm anatomy and phylogeny: new insights; pp. 109–126 in G. Arratia, R. Cloutier, and M. V. H. Wilson (eds.), Recent Advances in the Origin and Early Radiation of Vertebrates. Verlag Dr. Friedrich Pfeil, Munich.

Gould, S. J. 1995. Of it, not above it. Nature 377:681–682.

Grandel, H., and S. Schulte-Merkel. 1998. The development of the paired fins in the zebrafish (*Danio rerio*). Mechanisms of Development 79:99–120.

Gross, W. 1963. *Drepanaspis gemuendenensis* Schlüter. Neuuntersuchung. Palaeontographica, A 121:135–155.

Gross, W. 1964. Polyphyletische Stämme im System der Wirbeltiere? Zoologischer Anzeiger 173:1–22.

Halstead, L. B. 1973. The heterostracan fishes. Biological Reviews of the Cambridge Philosophical Society 48:279–332.

Halstead Tarlo, L. B., and H. P. Whiting. 1965. A new interpretation of the internal anatomy of the Heterostraci (Agnatha). Nature 206:148–150.

Hardisty, M. W. 1982. Lampreys and hagfishes: analysis of cyclostome relationships; pp. 165–260 in M. W. Hardisty and I. C. Potter (eds.), The Biology of Lampreys, Volume 4a. Academic Press, London and New York.

Hedges, S. B. 2001. Molecular evidence for the early history of living vertebrates; pp. 119–134 in P. E. Ahlberg (ed.), Major Events in Early Vertebrate Evolution. Palaeontology, Phylogeny, Genetics and Development. Taylor & Francis, London and New York.

Heintz, A. 1939. Cephalaspida from the Downtonian of Norway. Skrifter utgitt av det Norske Videnskaps Akademin (Matematisk-naturvidenskapslige Klasse) 1939(5):1–119.

Heintz, A. 1967. Some remarks about the structure of the tail in cephalaspids; pp. 21–36 in J.-P. Lehman (ed.), Évolution des Vertébrés. Colloques internationaux du Centre National de la Recherche Scientifique 163. Éditions du Centre National de la Recherche Scientifique, Paris.

Holland, L. Z., and Holland N. D. 2001. Evolution of neural crest and placodes: *Amphioxus* as a model for ancestral vertebrate? Journal of Anatomy 199:85–98.

Holland, N. D., and J. Chen. 2001. Origin and early evolution of the vertebrates: new insights from advances in molecular biology, anatomy, and palaeontology. BioEssays 23:142–151.

Holland, N. D., G. Panganiban, E. Henley, and L. Holland. 1996. Sequence and developmental expression of *AmphiDII*, an amphioxus *Distal-less* gene transcribed in the ectoderm, epidermis and nervous system: insight into the evolution of craniate forebrain and neural crest. Development 122:2911–2920.

Holmgren, N. 1946. On two embryos of *Myxine glutinosa*. Acta Zoologica (Stockholm) 27:1–90.

Holmgren, N., and E. Stensiö. 1936. Kranium und Visceralskelett der Akranier, Cyclostomen und Fische; pp. 233–500 in L. Bolk, E. Göppert, E. Kallius, and W. Lubosch (eds.), Handbuch der vergleichenden Anatomie der Wirbeltiere, Volume 4. Urban and Schwarzenberg, Berlin and Vienna.

Hou, H.-G., R. Aldridge, D. Siveter, D. Siveter, and X.-H. Feng. 2002. New evidence on the anatomy and phylogeny of the earliest vertebrates. Proceedings of the Royal Society of London B 269:1865–1869.

Janvier, P. 1974. The structure of the naso-hypophysial complex, and the mouth in fossil and extant cyclostomes, with remarks on amphiaspiforms. Zoologica Scripta 3:193–200.

Janvier, P. 1975a. Les yeux des Cyclostomes et le problème de l'origine des Myxinoïdes. Acta Zoologica (Stockholm) 56:1–9.

Janvier, P. 1975b. Remarques sur l'orifice naso-hypophysaire des Céphalaspidomorphes. Annales de Paléontologie (Vertébrés) 61:3–16.

Janvier, P. 1975c. Spécialisations précoces et caractères primitifs du système circulatoire des Ostéostracés; pp. 15–30 in J.-P. Lehman (ed.), Problèmes Actuels de Paléontologie. Évolution des Vertébrés. Colloque internationaux du Centre National de la Recherche Scientifique 218. Éditions du Centre National de la Recherche Scientifique, Paris.

Janvier, P. 1978. Les nageoires paires des Ostéostracés et la position systématique des Céphalaspidomorphes. Annales de Paléontologie (Vertébrés) 64:113–142.

Janvier, P. 1981a. The phylogeny of the Craniata, with particular reference to the significance of fossil "agnathans." Journal of Vertebrate Paleontology 1:121–159.

Janvier, P. 1981b. *Norselaspis glacialis* n.g., n.sp. et les relations phylogénétiques entre les Kiaeraspidiens (Osteostraci) du Dévonien inférieur du Spitsberg. Palaeovertebrata 11:19–131.

Janvier, P. 1984. The relationships of the Osteostraci and Galeaspida. Journal of Vertebrate Paleontology 4:344–358.

Janvier, P. 1985a. Les Céphalaspides du Spitsberg. Anatomie, Phylogénie et Systématique des Ostéostracés Siluro-Dévoniens. Révision des Ostéostracés de la Formation de Wood Bay (Dévonien inférieur du Spitsberg). Cahiers de Paléontologie. Centre National de la Recherche Scientifique, Paris, 244 p.

Janvier, P. 1985b. Les Thyestidiens (Osteostraci) du Silurien de Saaremaa (Estonie). Première partie: morphologie et anatomie. Annales de Paléontologie 71:83–147.

Janvier, P. 1986. Les nouvelles conceptions de la phylogénie et de la classification des "Agnatha" et des Sarcoptérygiens. Oceanis 12:123–138.

Janvier, P. 1993. Patterns of diversity in the skull of jawless fishes; pp. 131–188 in J. M. Hanken and B. K. Hall (eds.), The Skull. Volume 2, Patterns of Structural and Systematic Diversity. University of Chicago Press, Chicago.

Janvier, P. 1996a. The dawn of the vertebrates: characters versus common ascent in the rise of current vertebrate phylogenies. Palaeontology 39:259–287.

Janvier, P. 1996b. Early Vertebrates. Oxford Monographs on Geology and Geophysics 33. Clarendon Press, Oxford, 393 pp.

Janvier, P. 1997. Les Vertébrés avant le Silurien. Géobios 30:931–951.

Janvier, P. 2001. Ostracoderms and the shaping of the gnathostome characters; pp. 172–186 in P. E. Ahlberg (ed.), Major Events in Early Vertebrate Evolution: Palaeontology, Phylogeny, Genetics and Development. Taylor & Francis, London and New York.

Janvier, P. 2003. Vertebrate characters and Cambrian vertebrates. Compte Rendus Palevol 2:523–531.

Janvier, P. 2004. Early specializations of the branchial apparatus in jawless vertebrates: a consideration of gill number and size; pp. 29–52 in G. Arratia, R. Cloutier, and M. V. H. Wilson (eds.), Recent Advances in the Origin and Early Radiation of Vertebrates. Verlag Dr. Friedrich Pfeil, Munich.

Janvier, P., and M. Arsenault. 2002. Calcification of early vertebrate cartilage. Nature 417:609.

Janvier, P., and A. Blieck. 1979. New data on the internal anatomy of the
Heterostraci (Agnatha), with general remarks on the phylogeny of the
Craniota. Zoologica Scripta 8:287–296.

Janvier, P., and A. Blieck. 1993. L. B. Halstead and the heterostracan controversy. Modern Geology 18:89–105.

Janvier, P., and R. Lund. 1983. *Hardistiella montanensis* n. gen. et sp.
(Petromyzontida) from the Lower Carboniferous of Montana, with
remarks on the affinities of the lampreys. Journal of Vertebrate Paleontology 2:407–413.

Janvier, P., M. Arsenault, and S. Desbiens. 2004a. Calcified cartilage in the
paired fins of the osteostracan *Escuminaspis laticeps* (Traquair 1880),
from the Late Devonian of Miguasha (Québec, Canada), with a consideration of the early evolution of the pectoral fin endoskeleton in
vertebrates. Journal of Vertebrate Paleontology 24:773–779.

Janvier, P., E. Grogan, and R. Lund. 2004b. Further consideration of the
earliest known lamprey, *Hardistiella montanensis* Janvier and Lund,
1983, from the Carboniferous of Bear Gulch, Montana, USA. Journal
of Vertebrate Paleontology 24:742–743.

Janvier, P., R. Percy, and I. C. Potter. 1991. The arrangement of the heart
chambers and associated blood vessels in the Devonian osteostracan
Norselaspis glacialis. A reinterpretation based on recent studies of the
circulatory system in lampreys. Journal of Zoology 223:567–576.

Jarvik, E. 1965. Die Raspelzunge der Cyclostomen und die pentadactyle
Extremität der Tetrapoden als Beweise für monophyletische Herkunft.
Zoologischer Anzeiger 175:101–143.

Jarvik, E. 1980. Basic Structure and Evolution of Vertebrates. Volume 1.
Academic Press, London and New York, 575 pp.

Jarvik, E. 1981. Basic Structure and Evolution of Vertebrates. Volume 2.
Academic Press, London and New York, 337 pp.

Jefferies, R. P. S. 1986. The ancestry of the vertebrates. British Museum
(Natural History), London, 376 pp.

Jenner, R. A. 2004. When molecules and morphology clash: reconciling
conflicting phylogenies of the Metazoa by considering secondary loss.
Evolution and Development 6:372–378.

Jollie, M. 1962. Chordate Morphology. Reinhold, New York, 478 pp.

Karatajute-Talimaa, V., and M. M. Smith. 2004. *Tessakoviaspis concen-
trica:* microskeletal remains of a new order of vertebrate from the
Upper Ordovician and Lower Silurian of Siberia; pp. 53–64 in G. Ar-
ratia, R. Cloutier, and M. V. H. Wilson (eds.), Recent Advances in the
Origin and Early Radiation of Vertebrates. Verlag Dr. Friedrich Pfeil,
Munich.

Kearney, M. 2002. Fragmentary taxa, missing data, and ambiguity: mistaken assumptions and conclusions. Systematic Biology 51:369–381.

Kemp, A. 2002. Amino acid residues in conodont elements. Journal of Paleontology 76:518–528.

Kiaer, J. 1924. The Downtonian fauna of Norway. I. Anaspida with a geological introduction. Videnskapsselskapets Skrifter, I. Matematisk-
naturvidenskapslige Klasse 6:1–139.

Kiaer, J. 1928. The structure of the mouth of the oldest known vertebrates,
pteraspids and cephalaspids. Palaeobiologica 1:117–134.

Kimmel, C. B., C. T. Miller, and R. J. Keynes. 2001. Neural crest patterning and the evolution of jaws. Journal of Anatomy 199:105–119.

Kimmel, C. B., B. Ullmann, M. Walker, C. T. Miller, and J. Crump. 2003. Endothelin 1-mediated regulation of pharyngeal bone development in zebrafish. Development 130:1339–1351.

Kuratani, S. 2003. Evolution of the vertebrate jaw—homology and developmental constraints. Paleontological Research 7:89–102.

Kuratani, S. 2004. Evolution of the vertebrate jaw: comparative embryology and molecular developmental biology reveals the factors behind evolutionary novelty. Journal of Anatomy 205:335–347

Kuratani, S., S. Kuraku, and Y. Mukami. 2002. Lamprey as an evo-devo model: lessons from comparative embryology and molecular phylogenetics. Genesis 34:175–183.

Kuratani, S., Y. Nobusada, N. Horigome, and Y. Shigetani. 2001. Embryology of the lamprey and evolution of the vertebrate jaw: insights from molecular and developmental perspectives. Philosophical Transactions of the Royal Society of London B 356:15–32.

Kuratani, S., Y. Murakami, Y. Nobusada, R. Kusakabe, and S. Hirano. 2004. Development fate of the mandibular mesoderm in the lamprey, *Lethenteron japonicum:* comparative morphology and development of the gnathostome jaw with special reference to the nature of the trabecula cranii. Journal of Experimental Zoology, Part B: Molecular and Developmental Evolution 302:458–468.

Lacalli, T. 1996. Frontal eye circuitry, rostral sensory pathways and brain organisation in amphioxus larvae: evidence from 3D reconstruction. Philosophical Transactions of the Royal Society of London B 351:243–263.

Langille, R. M., and B. K. Hall. 1993. Calcification of cartilage from the lamprey *Petromyzon marinus* in vitro. Acta Zoologica (Stockholm) 74:31–41.

Lankester, E. R. 1868–1870. The Cephalaspididae; pp. 1–62 in J. Powrie and E. R. Lankester (eds.), A Monograph of the Old Red Sandstone of Britain. Palaeontographical Society, London.

Le Douarin, N., and C. Kalcheim. 1999. The Neural Crest. Cambridge University Press, Cambridge, 445 pp.

Løvtrup, S. 1977. The Phylogeny of the Vertebrata. Wiley, New York, 330 pp.

Mabee, P. M., P. L. Crotwell, N. C. Bird, and A. C. Burke. 2002. Evolution of median fin modules in the axial skeleton of fishes. Journal of Experimental Zoology, Part B: Molecular and Developmental Evolution 294:77–90.

Maisey, J. G. 1986. Heads and tails: a chordate phylogeny. Cladistics 2:201–256.

Maisey, J. G. 1988. Phylogeny of early vertebrate skeletal induction and ossification pattern. Evolutionary Biology 22:1–36.

Maisey, J. G. 1989. *Hamiltonichthys mapesi* g. et sp. nov. (Chondrichthyes; Elasmobranchii) from the Upper Pennsylvanian of Kansas. American Museum Novitates 2931:1–42.

Mallatt, J. 1984. Early vertebrate evolution: pharyngeal structure and the origin of gnathostomes. Journal of Zoology (London) 204:169–183.

Mallatt, J. 1996. Ventilation and the origin of jawed vertebrates: a new mouth. Zoological Journal of the Linnean Society 117:329–404.

Mallatt, J., and J. Chen. 2003a. Fossil sister group of craniates: predicted and found. Journal of Morphology 258:1–31.

Mallatt, J., and J. Chen. 2003b. Comment on "A new species of yunnanozoan with implications for deuterostome evolution." Science 300:1372c.

Mallatt, J., J. Sullivan, and C. J. Winchell. 2001. The relationships of lampreys to hagfishes: a spectral analysis of ribosomal DNA sequences; pp. 106–118 in P. E. Ahlberg (ed.), Major Events in Early Vertebrate Evolution. Palaeontology, Phylogeny, Genetics and Development. Taylor & Francis, London and New York.

Marinelli, W. von, and A. Strenger. 1956. *Myxine glutinosa* L. Vergleichende Anatomie und Morphologie der Wirbeltiere. Franz Deuticke, Vienna, 80 pp.

Märss, T., and A. Ritchie. 1998. Silurian thelodonts (Agnatha) of Scotland. Transaction of the Royal Society of Edinburgh: Earth Sciences 88:143–195.

Mazan, S., D. Jaillard, B. Baratte, and P. Janvier. 2000. *Otx1* gene-controlled morphogenesis of the horizontal semicircular canal and the origin of the gnathostome characteristics. Evolution and Development 2:186–193.

Mazet, F., J. A. Hutt, J. Milloz, J. Millard, A. Graham, and S. M. Shimeld. 2005. Molecular evidence from *Ciona intestinalis* for the evolutionary origin of vertebrate sensory placodes. Developmental Biology 282:494–508.

McCauley, D. W., and M. Bronner-Fraser. 2003. Neural crest contributions to the lamprey head. Development 130:2317–2327.

Meulemans, D., and M. Bronner-Fraser. 2002. Amphioxus and lamprey AP-2 genes: implications for neural crest evolution and migration patterns. Development 129:4953–4962.

Moy-Thomas, J. A., and R. S. Miles. 1971. Palaeozoic Fishes. Chapman and Hall, London, 259 pp.

Nelson, G. 1994. Homology and systematics; pp. 101–149 in B. K. Hall (ed.), Homology: The Hierarchical Basis of Comparative Biology. Academic Press, San Diego.

Newman, M. 2002. A new naked jawless vertebrate from the Middle Devonian of Scotland. Palaeontology 45:933–941.

Newman, M., and N. Trewin. 2001. A new jawless vertebrate from the Middle Devonian of Scotland. Palaeontology 44:43–51.

Nicol, J. A. C. 1989. The Eye of Fishes. Clarendon Press, Oxford, 308 pp.

Novitskaya, L. I. 1983. [Morphology of ancient agnathans. Heterostracans and the problem of relationships of agnathans and gnathostome vertebrates]. Trudy Paleontologicheskogo Instituta Akademiya Nauk SSSR 196:1–182. [In Russian.]

Obruchev, D. V. 1964. [Agnathans and fishes]. Volume 11 of I. A. Orlov (ed.), [Fundamentals of Palaeontology]. Nauka, Moscow, 522 pp. [In Russian, translated by the Israel Program for Scientific Translations in 1967.]

Oelofsen, B. W., and J. C. Loock. 1981. A fossil cephalochordate from the Early Permian Whitehill Formation of South Africa. South African Journal of Sciences 77:178–180.

Patterson, C. 1982. Morphological characters and homology; pp. 21–74 in K. A. Joysey and A. E. Friday (eds.), Problems of Phylogenetic Reconstruction. Systematics Association Special Volume 25. Academic Press, London and New York.

Pearson, M. D., and T. S. Westoll. 1979. The Devonian actinopterygian *Cheirolepis* Agassiz. Transactions of the Royal Society of Edinburgh 70:337–399.

Percy, R., and I. C. Potter. 1991. Aspects of the development and functional morphology of the pericardia, heart and associated blood vessels of lampreys. Journal of the Zoological Society of London 223:49–66.

Poplin, C., D. Sotty, and P. Janvier. 2001. Un Myxinoïde (Craniata, Hyperotreti) dans le Konservat-Lagerstätte Carbonifère supérieur de Montceau-les-Mines (Allier, France). Comptes Rendus de l'Académie des Sciences, Paris, sér. II, 332:345–350.

Platnick, N. I., C. J. Humphries, G. Nelson, and D. Williams. 1996. Is Farris optimization perfect? Three-taxon statements and multiple branching. Cladistics 12:243–252.

Purnell, M. A. 2002. Feeding in extinct jawless heterostracan fishes and testing scenarios of early vertebrate evolution. Proceedings of the Royal Society of London B 269:83–88.

Resser, P. E., and B. F. Howell. 1938. Lower Cambrian *Olenellus* zone of the Appalachians. Bulletin of the Geological Society of America 49:195–248.

Ritchie, A. 1964. New light on the morphology of the Norwegian Anaspida. Skrifter uitgitt av det Norske Videnskaps-Akademi i Oslo, I. Matematiske-naturvidenskapslige Klasse, Ny Serie, 14:1–35.

Ritchie, A. 1967. *Ateleaspis tessellata* Traquair, a non-cornuate cephalaspid from the Upper Silurian of Scotland. Zoological Journal of the Linnean Society of London 47:69–81.

Ritchie, A. 1968. New evidence on *Jamoytius kerwoodi* White, an important ostracoderm from the Silurian of Lanarkshire, Scotland. Palaeontology 11:21–39.

Romer, A. S. 1945. Vertebrate Paleontology. Second Edition. University of Chicago Press, Chicago, 687 pp.

Rosen, D. E., P. L. Forey, B. G. Gardiner, and C. Patterson. 1981. Lung-fishes, tetrapods, paleontology, and plesiomorphy. Bulletin of the American Museum of Natural History 167:157–276.

Sahney, S., and M. V. H. Wilson. 2001. Extrinsic labyrinth infillings imply open endolymphatic ducts in Lower Devonian osteostracans, acanthodians, and putative chondrichthyans. Journal of Paleontology 21:660–669.

Sansom, I. J., P. C. J. Donoghue, and G. Albanesi. 2005. Histology and affinity of the earliest armoured vertebrate. Biology Letters 1: 446–449.

Sansom, I. J., M. M. Smith, and M. P. Smith. 2001. The Ordovician radiation of vertebrates; pp. 156–171 in P. E. Ahlberg (ed.), Major Events in Early Vertebrate Evolution. Palaeontology, Phylogeny, Genetics and Development. Taylor & Francis, London and New York.

Schaeffer, B., and K. S. Thomson. 1980. Reflections on agnathan-gnathostome relationships; pp. 19–33 in L. L. Jacobs (ed.), Aspects of Vertebrate History. Museum of Northern Arizona Press, Flagstaff.

Schmitt, S. 2004. Histoire d'une question anatomique: la répétition des parties. Collection Archives. Muséum National d'Histoire Naturelle, Paris, 704 pp.

Schultze, H.-P. 1996. Conodont histology: an indicator of vertebrate relationship? Modern Geology 20:275–285.

Scott, K. M., and C. M. Janis. 1993. Relationships of the Ruminantia (Artiodactyla) and an analysis of the characters used in ruminant taxonomy; pp. 282–302 in F. S. Szalay, M. J. Novacek, and M. C. McKenna (eds.), Mammal Phylogeny. Placentals. Springer-Verlag, New York.

Shigetani, Y., F. Sugahara, and S. Kuratani. 2005. A new evolutionary scenario for the vertebrate jaw. BioEssays 27:331–338.

Shigetani, Y., F. Sugahara, Y. Kawakami, Y. Murakami, S. Hirano, and S. Kuratani. 2002. Heterotopic shift of epithelial-mesenchymal interactions for vertebrate jaw evolution. Science 296:1319–1321.

Shimeld, S. M., and P. W. H. Holland. 2000. Vertebrate innovations. Proceedings of the National Academy of Sciences of USA 97:4449–4452.

Shu, D. 2003. A paleontological perspective of vertebrate origins. Chinese Science Bulletin 48:725–735.

Shu, D., and S. Conway Morris. 2003. Response to comments on "A new species of yunnanozoan with implications for deuterostome evolution." Science 300:1372.

Shu, D., X. Zhang, and L. Chen. 1996. Reinterpretation of *Yunnanozoon* as the earliest known hemichordate. Nature 380:428–430.

Shu, D., S. Conway Morris, J. Han, L. Chen, X.-L. Zhang, Z.-F. Zhang, H.-Q. Liu, Y. Li, and J.-N. Liu. 2001. Primitive deuterostomes from the Chengjiang Lagerstätte (Lower Cambrian, China). Nature 414: 419–424.

Shu, D., S. Conway Morris, Z.-F. Zhang, J.-N. Liu, J. Han, L. Cheng, X.-L. Zhang, K. Yasui, and L. Yong. 2003a. A new species of yunnanozoan with implications for deuterostome phylogeny. Science 299:1380–1384.

Shu, D., H.-L. Luo, S. Conway Morris, X.-L. Zhang, S.-X. Chen, J. Han, M. Zhu, Y. Li, and L.-Z. Chen. 1999. Lower Cambrian vertebrates from South China. Nature 402:42–46.

Shu, D., S. Conway Morris, J. Han, Z. F. Zhang, K.Yasui, P. Janvier, L. Chen, X. L. Zhang, J. N. Liu, Y. Li, and H. K. Liu. 2003b. Head and backbone of the early Cambrian vertebrate *Haikouichthys*. Nature 421:526–529.

Smith, M. M., and M. I. Coates. 2001. The evolution of vertebrate dentitions: phylogenetic patterns and developmental models; pp. 223–240 in P. E. Ahlberg (ed.), Major Events in Early Vertebrate Evolution. Palaeontology, Phylogeny, Genetics and Development. Taylor & Francis, London and New York.

Smith, M. P., and I. J. Sansom. 1995. The affinity of *Anatolepis* Bockelie and Fortey; pp. 61–63 in H. Lelièvre, S. Wenz, A. Blieck, and R. Cloutier (eds.), Premiers Vertébrés et Vertébrés Inférieurs. Géobios, Mémoire Spécial 19.

Smith, M. P., I. J. Sansom, and K. Cochrane. 2001. The Cambrian origin of vertebrates; pp. 67–84 in P. E. Ahlberg (ed.), Major Events in Early Vertebrate Evolution. Palaeontology, Phylogeny, Genetics and Development. Taylor & Francis, London and New York.

Smith, M. P., I. J. Sansom, and J. Repetski. 1996. Histology of the first fish. Nature 380:702–704.

Stensiö, E. A. 1927. The Devonian and Downtonian vertebrates of Spitsbergen. Part 1. Family Cephalaspidae. Skrifter om Svalbard og Nordishavet 12:1–391.

Stensiö, E. A. 1932. The Cephalaspids of Great Britain. Trustees of the British Museum, London, 220 p.

Stensiö, E. A. 1958. Les Cyclostomes fossiles ou Ostracodermes; pp. 173–425 in P. P. Grassé (ed.), Traité de Zoologie, Volume 13(1). Masson et Cie., Paris.

Stensiö, E. A. 1964. Les Cyclostomes fossiles ou Ostracodermes; pp. 96–383 in J. Piveteau (ed.), Traité de Paléontologie, Volume 4(1). Masson et Cie, Paris.

Stensiö, E. A. 1968. The cyclostomes with special reference to the diphyletic origin of the Petromyzontida and the Myxinoidea; pp. 13–71 in T. Ørvig (ed.), Current Problems in Lower Vertebrate Phylogeny. Almqvist and Wiksell, Stockholm.

Takio, Y., M. Pasqualetti, S. Kuraku, S. Hirano, F. M. Rijli, and S. Kuratani. 2004. Lamprey *Hox* genes and the evolution of jaws. Nature 409:2 pp. following p. 262.

Ushida, K., Y. Murakami, S. Kuraku, S. Hirano, and S. Kuratani. 2003. Development of the adenohypophysis in the lamprey: evolution of epigenetic patterning programs in organogenesis. Journal of Experimental Zoology, Part B: Molecular and Developmental Evolution 300:32–47.

Van der Brugghen, W., and P. Janvier. 1993. Denticles in thelodonts. Nature 364:107.

Vladikov, V. D. 1973. A female sea lamprey (*Petromyzon marinus*) with a true anal fin, and the question of the presence of an anal fin in the Petromyzontidae. Canadian Journal of Zoology 51:221–224.

Wang, N. Z. 1991. Two new Silurian galeaspids (jawless craniates) from Zhejian Province, China, with a discussion of galeaspid-gnathostome relationships; pp. 41–65 in M. M. Chang, G. R. Zhang, and Y. H. Liu (eds.), Early Vertebrates and Related Problems of Evolutionary Biology. Science Press, Beijing, China.

Wängsjö, G. 1952. The Downtonian and Devonian vertebrates of Spitsbergen. 9. Morphologic and systematic studies of the Spitsbergen cephalaspids. Results of Th. Vogt's Expedition 1928 and the English-Norwegian-Swedish Expedition in 1939. Norsk Polarinstitutt Skrifter 97:1–611.

Watson, D. M. S. 1954. A consideration of ostracoderms. Philosophical Transactions of the Royal Society of London B 238:1–25.

Whiting, H. P. 1977. Cranial nerves in lampreys and cephalaspids; pp. 1–23 in S. M. Andrews, R. S. Miles, and A. D. Walker (eds.), Problems in Vertebrate Evolution. Linnean Society Symposium Series 4. Academic Press, London and New York.

Wilson, M. V. H., and M. W. Caldwell. 1993. New Silurian and Devonian fork-tailed "thelodonts" and jawless vertebrates with stomachs and deep bodies. Nature 361:442–444.

Woodward, A. S. 1900. On a new ostracoderm fish (*Euphanerops longaevus*) from the Upper Devonian of Scaumenac Bay, Quebec, Canada. Annals and Magazine of Natural History (7)5:416–419.

Yalden, D. W. 1985. Feeding mechanisms as evidence of cyclostome monophyly. Zoological Journal of the Linnean Society 84:291–300.

Young, G. C. 1986. The relationships of placoderm fishes. Zoological Journal of the Linnean Society 88:1–57.

Young, G. C. 1991. The first armoured agnathan vertebrates from the Devonian of Australia; pp. 67–86 in M. M. Chang, G. R. Zhang, and Y. H. Liu (eds.), Early Vertebrates and Related Problems of Evolutionary Biology. Science Press, Beijing, China.

Young, G. C. 1997. Ordovician microvertebrate remains from the Amadeus Basin, central Australia. Journal of Vertebrate Paleontology 17:1–25.

Zhang, Y.-G., and X.-G. Hou. 2004. Evidence for a single median fin-fold in tail in the Lower Cambrian vertebrate, *Haikouichthys ercaicunensis*. Journal of Evolutionary Biology 17:1162–1166.

3. Paired Fins of Jawless Vertebrates and Their Homologies across the "Agnathan"-Gnathostome Transition

Mark V. H. Wilson, Gavin F. Hanke, and Tiiu Märss

ABSTRACT

The origin of jawed vertebrates (Gnathostomata) is one of the greatest events in vertebrate evolution, and one of the most poorly understood to this day. Among the many features of gnathostomes shared with possible precursors in jawless ("agnathan") vertebrates are paired fins. "Agnathan" paired fin-like structures occurred in many species of anaspids, thelodonts, and osteostracans. These early paired fins are not all the same: some taxa have pectoral precursors, and others have pelvic precursors. At least one thelodont probably had both, the only "agnathan" known to share this feature with gnathostomes. Some "agnathan" lineages likely lost either pectoral fins (furcacaudiforms) or pelvic fins (osteostracans, perhaps some thelodonts) that were present in their ancestors. Pectoral and pelvic fins or their precursors differed fundamentally in position and structure even before the origin of jaws, and within most of the major groups of early jawed vertebrates.

Introduction

One of the greatest unsolved events in vertebrate evolution is the origin of Gnathostomata (vertebrates with jaws). The origins of

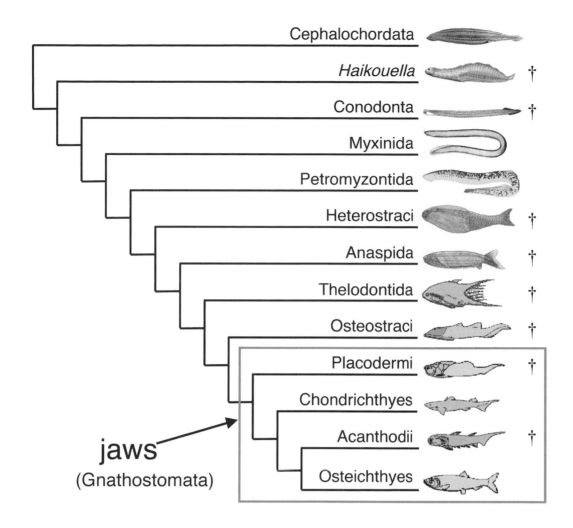

Cephalochordata

Haikouella †

Conodonta †

Myxinida

Petromyzontida

Heterostraci †

Anaspida †

Thelodontida †

Osteostraci †

Placodermi †

Chondrichthyes

Acanthodii †

Osteichthyes

jaws
(Gnathostomata)

Figure 3.1. Suggested phylogeny of craniates giving a framework for the origin of paired fins in vertebrates and illustrating the importance of fossil taxa. Of the four main groups of gnathostomes (jawed vertebrates), two are exclusively fossil groups. In addition, most of the jawless vertebrates that are closely related to gnathostomes are known only from fossils. Neither of the two extant groups of jawless vertebrates (Myxinida and Petromyzontida) has paired fins. Only selected taxa from among the many early fossil craniates that lack paired fins are shown.

many other major vertebrate clades (e.g., teleosts, tetrapods, amniotes, birds, and mammals) are now reasonably well outlined, with numerous fossilized taxa that illustrate the acquisition of the major features of the clade. However, when undoubted gnathostomes (Fig. 3.1) appear in the fossil record, they are already very distinct from what are believed to be their closest relatives among jawless vertebrates.

Biologists studying the extant members of the jawless vertebrates (the "agnathans," sometimes called cyclostomes: the hagfishes and the lampreys) and the present-day jawed vertebrates (the chondrichthyans or cartilaginous fishes, and the osteichthyans or bony fishes) can list many differences between them. Examples include movably articulated jaws formed from the skeleton of an anterior gill arch, marginal jaw teeth forming in an epithelial invagination with a regular replacement pattern, large eyes, a third semicircular canal in the inner ear, gill arches and their respective openings that are slit-like, a branchial skeleton internal to the gills rather than external to them, an epicercal caudal fin, a gut containing a stomach as a storage and digestive chamber anterior to the intestine, lateral abdominal veins draining the left and right body walls, renal portal veins, myelinated nerves, and paired pectoral and pelvic fins with their suite of muscles and their internal skeleton of radials and basals attached to internal skeletal girdles. The origin of gnathostomes was clearly one of the most significant events in vertebrate evolution.

Many of these gnathostome features, including large eyes, jaws, teeth, stomach, and paired fins, have been conventionally regarded (for example, in textbooks of vertebrate biology) as correlated with a change in diet and lifestyle associated with the origin of predatory, macrophagous feeding habits. In this traditional view, large eyes were an adaptation for more acute vision, paired fins were an adaptation for maneuverability for catching prey and/or for avoiding capture, myelinated nerves were an adaptation for rapid reactions and vigorous swimming, the third semicircular canal was an adaptation for superior balance, jaws and teeth were an adaptation for capturing and eating prey, and the stomach was an adaptation for storing and beginning the digestion of the larger prey items.

Recent discoveries, however, have challenged the tight linkage of these adaptations with the origin of jaws. For example, there is evidence that placoderms, regarded as the most basal group of gnathostomes, lacked teeth in their primitive members, although some derived members evolved tooth-like structures (Johanson and Smith, 2003; Smith, 2003; Smith and Johanson, 2003). Similarly, new research shows that stomach-like storage chambers predated the origin of jaws, occurring in many thelodonts that probably fed on organic-rich mud (Wilson and Caldwell, 1993, 1998). The same thelodonts had exceptionally large eyes. Osteostracans, the "agnathans" usually regarded as the closest fossil relatives of gnathostomes, share with gnathostomes an epicercal caudal fin (Janvier, 1996). Osteostracans also share with some primitive gnathostomes and with modern sharks the ability to incorporate sand grains into the ballast of the inner ear, likely improving their sense of balance or hearing (Sahney and Wilson, 2001). Neither the thelodonts nor the osteostracans were predaceous in the usual sense of the word (i.e., macrophagous), and no specimens exist with ingested macroscopic animal remains preserved. Thus, recent studies show that many of the features of extant gnathostomes could not have been adaptations for predation, having evolved in nonpredaceous jawless vertebrates and only later becoming useful to animals with a predaceous lifestyle, while others (e.g., teeth) evolved long after the origin of jaws, and possibly independently in several early gnathostome lineages.

The question of the origin of the paired fins of gnathostomes has interested paleontologists, anatomists, and embryologists for more than a century, but definitive answers remain elusive. Among the most influential theories on the origin of paired fins is the fin-fold theory, first articulated independently by Balfour (1876) and Thacher (1876). In the fin-fold theory, fins were thought to have appeared first in the form of a continuous median fin fold extending from the dorsal midline, around the tail, to the region of the anus. Just posterior to the anus, the fold was thought to divide, extending forward on either side of the ventrolateral body wall to a point behind the branchial chambers. Existence of a median fold finds support in the embryological development of primitive vertebrates such as paddlefishes (Bemis and Grande, 1999). Median fins were then thought to have evolved as subdivisions of the median portions of the fin fold, much as they develop through subdivision of the median fold in embryos. Paired fins were thought to have evolved similarly, from subdivision of the hypothetical paired, ventral fin fold.

The theory initially became popular (e.g., Mivart, 1879; Goodrich, 1906), received support from experimental studies of supernumerary limbs in tetrapods (e.g., Balinsky, 1933) and developmental studies of fishes (Ekman, 1941), and made its way into textbooks of vertebrate biology as the standard explanation for the origin of paired fins (e.g., Neal and Rand, 1939; Parker and Haswell, 1943; Jarvik, 1980–1981). During the period when the fin-fold theory was most popular, certain "agnathans" with paired fin-like structures were thought to provide support for the theory. This was especially true for the Anaspida, some of which have a pair of elongate, ventrolateral fins reminiscent of the hypothetical, paired fin fold. More recently, the fin-fold theory has been mentioned with approval by molecular developmental biologists because of its apparent correspondence with the anteroposterior sequence of expression of *Hox* genes (Kessel, 1993; Tabin and Laufer, 1993; Shubin et al., 1997; Tanaka et al., 2002).

On the contrary, others (e.g., Borkhvardt, 1986; Coates, 1993, 1994, 2003; Janvier, 1996) expressed reservations about the theory, and Bemis and Grande (1999) recently summarized numerous lines of evidence suggesting that paired fins do not develop from paired fin-fold-like embryological precursors in primitive bony fishes. Stahl (1974) suggested that the different examples of paired fins had evolved in parallel in different "agnathans" and gnathostomes, but other workers (e.g., Maisey, 1986, 1988) implied homology by citing paired fins as a synapomorphy linking osteostracans, anaspids, and gnathostomes. Some authors have argued that true pectoral fins, exemplified by the paired fins of osteostracans, evolved before pelvic fins (e.g., Coates, 1993, 1994; Forey and Janvier, 1993; Shubin et al., 1997; Coates and Cohn, 1998).

What, then, are we to make of the diversity of paired fin-like structures in extinct vertebrates? Are they simply examples of convergent evolution? We consider "true" paired fins to be those of crown-group gnathostomes, attached to an endochondral (in part at least) girdle, to which is attached a system of internal skeletal elements usually including basals and radials, and fringed by a fin web supported by either dermal rays (lepidotrichia) or keratinous, connective-tissue rays (ceratotrichia). In this chapter, we review the diversity of structure in other early vertebrates and draw attention to two distinct categories of possible paired-fin precursors among "agnathans," suggesting a revised interpretation of homologies relative to the pectoral and pelvic fins of jawed vertebrates.

Paired Fin-like Structures in Jawless Vertebrates

"Agnathans" without Paired Fins

Protochordates, extant "agnathans," and the earliest fossil vertebrates all lack paired fins. The list of basal groups lacking paired fins, based on present evidence, includes conodonts, hagfishes, lampreys, arandaspidiforms, astraspidiforms, and heterostracans (Janvier, 1996; Figs. 3.1, 3.2C). The Cambrian craniates *Myllokunmingia* and *Haikouichthys* (the latter synonymized with *Myllokunmingia* by Hou et al., 2002) were initially thought to possess "paired fins" (Shu et al., 1999), but specimens showing their undoubted paired nature and fin-like structure have not been found. Shu et al. (2003) expressed doubt that the structure in question was paired, Janvier (2003) suggested that it more likely was a preanal median fin fold as seen in hagfishes and larval lampreys, and Zhang and Hou (2004) supported this interpretation.

Anaspida

In his initial description of three genera of anaspids based on well-preserved fossils from Ringerike, Norway, Kiaer (1924) did not mention paired fins, although his identification of a postbranchial spine led him to conclude that the spine represented a remnant of a continuous lateral fin. A row of spines in a similar position, numbering about seven to nine and termed *postcephalic rods,* occurs in the anaspid *Lasanius* (Parrington, 1958). The homology of the anaspid postbranchial spine or spines, which appear to have been immobile, has yet to be demonstrated conclusively. However, one possibility is that it is a precursor of the cornual spine of osteostracans. Indeed, Stensiö (1932:181) argued that the "pectoral spine apparatus [of anaspids] obviously corresponds to the pectoral fin [of osteostracans]." Against this view, the most plesiomorphic osteostracans, such as *Ateleaspis* and *Aceraspis,* lack cornual spines while retaining such primitive features as two dorsal fins (Janvier, 1996). It is thus probable that cornual spines evolved independently within the Osteostraci (P. Janvier, pers. comm., 2005) and are not homologous to the postbranchial spines of anaspids.

Undoubted paired fin-like structures were described for the Devonian anaspid-like vertebrate *Endeiolepis* by Stensiö (1939, 1964), who suggested that the long, fold-like fins of this fossil from Miguasha corresponded to both pectoral and pelvic fins. Another Devonian anaspid-like genus, *Euphanerops,* also from Miguasha, has ribbon-shaped fins with parallel and well-spaced radials according to Janvier et al. (2004). The series of paired-fin radials overlaps anteroposteriorly with the greatly extended branchial structures (P. Janvier, pers. comm., 2005). The Silurian naked,

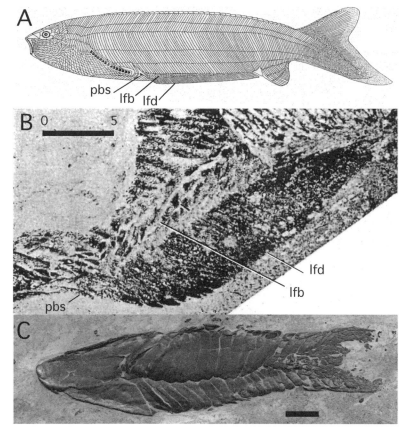

A

pbs lfb lfd

B 0 5

lfd

lfb

pbs

C

Figure 3.2. Anaspids and heterostracans. **A, B,** the Silurian anaspid *Pharyngolepis heintzae.* **A,** reconstruction of the whole animal showing location and structure of the paired fins, from Ritchie (1964: fig. 1); **B,** detail of a specimen showing the anterior end of a paired fin, after Ritchie (1964: pl. 1, fig. 3). **C,** the Early Devonian heterostracan *Dinaspidella elisabethae,* illustrating the absence of paired fins in this group. **lfb,** basal scale-covered region of the ventrolateral paired fin; **lfd,** distal parallel-rayed region of the paired fin; **pbs,** postbranchial (postcephalic) spine. **B,** scale bar = 5 mm; **C,** scale bar = 1 cm.

anaspid-like form *Jamoytius* White, 1946, redescribed by Ritchie (1960, 1968a, 1984), is thought also to have had long, ribbon-like paired fins containing stiffening elements, which have loosely been termed *rays.*

Paired fins in typical, scaled birkeniid anaspids were first identified by Ritchie (1964) in two species of *Pharyngolepis* (Figs. 3.2A,B, 3.3A). The fins extend from just behind the single pair of post-branchial (postcephalic) spines, posteriorly along the body margin, reaching almost to the anus in *P. heintzi* but not as far in *P. oblongus.* Each fin has a long basal region covered by larger, roughly rectangular scales and a long peripheral region covered by smaller, parallel, ray-like scales (Fig. 3.2A,B). Ritchie later (1980) identified paired fins in *Rhyncholepis,* the fins of which are also posterior to the postbranchial (postcephalic) spines, but may not have been scale covered and are shorter than those in *Pharyngolepis.* There is no evidence of paired fins in *Pterygolepis* according to Ritchie (1980). The scale-covered basal region of the paired fins in *Pharyngolepis* implies muscular control (Ritchie, 1980). Rather than being immobile, stabilizing structures, they may have been capable of undulatory movement for slow forward or backward motion (Janvier, 1987), as in extant notopterid and gymnotid fishes.

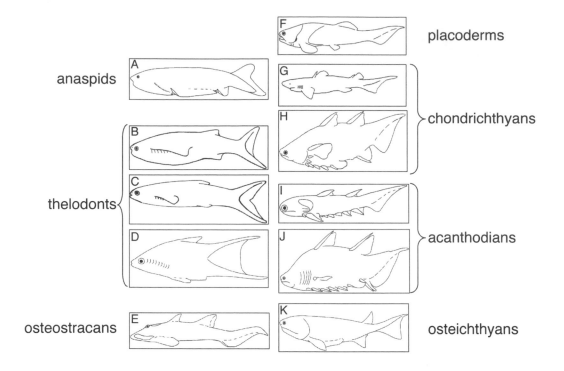

Figure 3.3. Outline drawings of selected early vertebrates showing the shape and position of paired fins. Drawings based on: A, *Pharyngolepis*; B, *Phlebolepis* (the condition in *Turinia*, *Loganellia*, and *Lanarkia* is similar); C, *Shielia*; D, *Furcacauda*; E, *Ateleaspis*; F, *Dunkleosteus*; G, *Squalus*; H, *Kathemacanthus*; I, *Climatius*; J, *Brochoadmones*; K, *Cheirolepis*.

Researchers have disagreed about whether anaspid paired fins are homologous with both pairs of fins in gnathostomes, or only with pectorals, or with neither. Ritchie (1964) and Coates (2003) did not believe that anaspid paired fins were homologous with the paired fins of gnathostomes, but rather were independent developments. Janvier (1987) agreed, saying that they were a derived feature compared with other craniates, although primitive within anaspids. Borkhvardt (1986) suggested that anaspid fins should be considered as long *anterior* fin folds, apparently assuming that they corresponded to pectorals.

Moy-Thomas and Miles (1971) suggested that lateral fins of anaspids were not differentiated into pectoral and pelvic regions, although they said that *Pharyngolepis heintzi* had short fins almost in the position of pectorals. They (Moy-Thomas and Miles, 1971:26) further stated that "[l]ateral fins of anaspids are homologous with the pelvic fins [*sic*] plus the ventrolateral ridges of osteostracans" and drew a connection with the "restated Lateral Fin-fold Theory" of Westoll (1958). Their statement of homology with pelvic fins was likely an error, because at that time, osteostracans were known to have anterior, presumably pectoral paired fins, and the ventrolateral body ridges of osteostracans were assumed to represent the posterior expression of the hypothetical fin fold. The possible homology of the postbranchial spine or spines with the cornual spine of osteostracans (considered unlikely because of the absence of cornual spines in primitive osteostracans), and the homology of anaspid paired fins with osteostracan pectorals (an idea considered worthy of further study by one of us [GFH]) would be considered more likely under a phylogenetic scheme in which anaspids and osteostracans are united as a clade, as was commonly believed when the book by Moy-Thomas and Miles (1971) was written.

Maisey (1986, 1988) implied homology with both pairs of gnathostome paired fins by postulating a synapomorphy uniting the paired fins of anaspids, osteostracans, and gnathostomes, while he also joined anaspids with gnathostomes (to the exclusion of osteostracans) on their joint possession of endoskeletal fin radials and paired fins extending posteriorly to the cloaca.

In summary, anaspid paired fins are ventrolateral, postbranchial structures (extending anteriorly beneath the branchial region in *Euphanerops*), probably under muscular control, and extending posteriorly almost as far as the anus/cloaca in some taxa (Figs. 3.2A,B, 3.3). Anaspid paired fins have variously been said to represent the undivided fin fold, the pectoral fins, or neither.

Thelodonti

Thelodonts possess a variety of paired fin-like structures (Figs. 3.3–3.5). Although there has been some controversy about monophyly or paraphyly of thelodonts (e.g., Turner, 1991; Turner and van der Brugghen, 1992; Wilson and Caldwell, 1998; Donoghue and Smith, 2001), the most recent phylogenetic study of thelodonts (Wilson and Märss, 2004) suggests that they were monophyletic. Turner and Miller (2005) have proposed a radical view that thelodonts are more closely related to chondrichthyans and teleostomes than they are to placoderms (i.e., effectively that they are gnathostomes, a sister-taxon to crown-group gnathostomes), an idea that we consider very unlikely. Additional views of thelodont relationships include their possible paraphyletic status relative to osteostracans and gnathostomes (e.g., Donoghue and Smith, 2001), and their possible sister-group relationship with gnathostomes (e.g., Gagnier, 1995; Wilson and Caldwell, 1998). Evaluation of these alternatives is beyond the scope of the present work, although they were not supported by the most comprehensive analysis of thelodont phylogeny to date (Wilson and Märss, 2004). We here follow Wilson and Märss (2004) in treating thelodonts as "agnathans" that are plausibly sister-group to osteostracans (and perhaps galeaspids and pituriaspids) plus gnathostomes.

Studies of relationships within the Thelodonti are still in their infancy. However, it appears (Wilson and Märss, 2004) that there are several distinct clades whose structure can be examined in one or more articulated skeletons (squamations). The paired fin-like structures of representative genera of these clades are discussed here.

Phlebolepis

Articulated specimens of *Phlebolepis elegans,* originally described under the name *Coelolepis luhai* Kiaer and Heintz, 1932, have been known from the Himmiste locality, Saaremaa Island, Estonia, since the early twentieth century (Ritchie, 1968b; Märss, 1986). Thou-

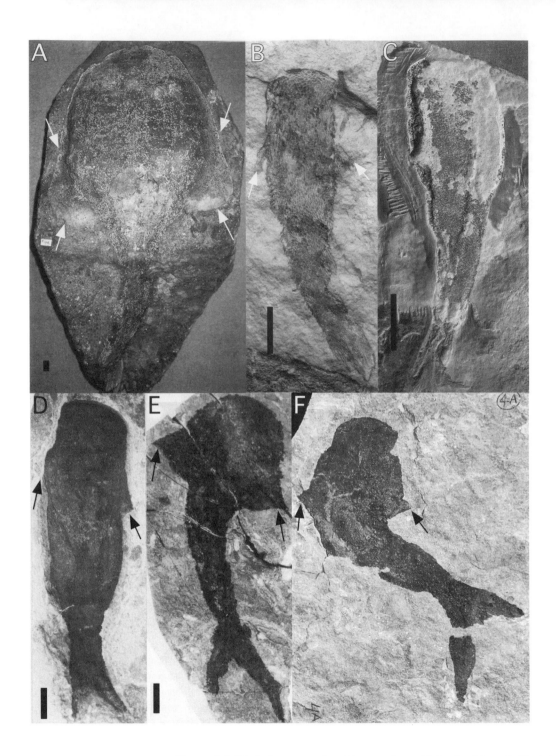

Figure 3.4. Thelodonts mostly with suprabranchial paired fins. **A,** *Turinia pagei;* **B,** *Phlebolepis elegans;* **C,** *Archipelepis turbinata* (paired fins unknown); **D,** *Loganellia scotica;* **E,** *Lanarkia horrida;* **F,** *Shielia taiti* (photograph courtesy of W. van der Brugghen). Arrows indicate location of paired suprabranchial fins. Scale bars = 1 cm.

sands of articulated specimens have been recovered from the same locality, yet ventral abdominal paired fins have never been seen in any individual. However, more than a few specimens demonstrate the presence of a pair of anterior fins (Figs. 3.3B, 3.4B). Recent restudy by two of us (MW, TM) of hundreds of specimens in the collection of Tartu University confirmed the interpretation of Ritchie (1968b), who based his conclusions on a single specimen with the fin preserved. These fins are triangular; they originate high on the side of the animal not far behind the orbit, and their posterior extremity corresponds approximately with the expected posterior end of the branchial region (although undoubted branchial openings have never been described). We refer to this pair of fins by the term *suprabranchial* in light of its position, to avoid making initial assumptions about homology, although authors have often called these fins pectorals. The suprabranchial fins of *Phlebolepis* differ from those of other thelodonts in being composed of rows of narrow scales somewhat resembling the fin rays of bony fishes (Ritchie, 1968b; MW and TM, pers. obs.; Fig. 3.4B), the anteriormost/outermost row being composed of slightly larger scales than those within the fin. Ritchie (1968b) contrasted the anatomical position of the paired fins of *Phlebolepis* with that of the paired fins of anaspids.

Loganellia *and* Lanarkia

Articulated specimens of both *Loganellia* and *Lanarkia* spp. were first recovered from the Silurian of Scotland in the nineteenth century (Powrie, 1870; Märss and Ritchie, 1998). Specimens are usually preserved as dorsoventral compressions such that the paired fins can often be seen extending from either side of the cephalobranchial region (Figs. 3.4D,E). These fins are also suprabranchial in position. They are fleshy, scale-covered outgrowths from the sides of the head. The scales of the trailing portions of the fin are generally smaller than those covering the adjoining head and body, and in *Loganellia* the scales along their outer/leading edge have smoother crowns and are larger than the scales of the trailing parts of the fin. According to Märss and Ritchie (1998), leading-edge scales are not distinct in young individuals of *Lanarkia horrida* but are larger and more flattened in *Lanarkia spinulosa* and in adult specimens of *L. horrida* (=*L. spinosa*), in which trailing-edge scales are very fine (Märss and Ritchie, 1998). Immediately behind the suprabranchial fins, the body narrows abruptly (Fig. 3.4D,E). Turner (1991) stated that these fins did not differ from true fins in function, and Märss and Ritchie (1998) concluded that they were homologous with pectoral fins.

Ventral abdominal fins are not certainly known for either genus, although there is some suggestion that they occur in *Lanarkia lanceolata* (Märss and Ritchie, 1998:189).

Turinia

There is only a single nearly complete specimen (Fig. 3.4A) of the Devonian thelodont *Turinia* from Scotland; its anatomy was reviewed recently by Donoghue and Smith (2001). Like *Loganellia* and *Lanarkia*, *Turinia* has paired, scale-covered extensions from the posterolateral edges of the cephalothorax. It can be seen clearly in the type specimen of *Turinia pagei* (Fig. 3.4A; Donoghue and Smith, 2001: fig. 4a; MW, pers. obs.) that these flaps originate close to the anterior end of the presumed branchial impressions. The flaps broaden posteriorly, and behind their posterior extremities the body narrows abruptly. Because they span much the same anteroposterior extent as do the branchial structures, these fins most likely were either suprabranchial or subbranchial in position. By comparison with other thelodonts, we suggest that they were suprabranchial.

Donoghue and Smith (2001) did not report any evidence of internal skeletal supports within these "pectoral flaps," and both they and Janvier (this volume) suggest, therefore, that these are not true fins. However, the former authors apparently did not consider that they originated as far anteriorly as we suggest here (restricting their extent to what we would call the posterior part of the fin web), because they figure a strut within the area that we would interpret as the anterior base of the fin. If our interpretation of their greater extent is correct, these pectoral flaps are similar in arrangement and shape to the paired suprabranchial fins of *Loganellia* and *Lanarkia*. Whether they are considered to be true fins, fin flaps, or fin precursors depends on one's hypothesis of homology and on one's definition of the characteristics of a true fin. It is likely that they lack some of the features of the true fins of gnathostomes, but we suggest that they are early homologues of them.

Pezopallichthys *and* Archipelepis

No paired fins are known in any specimen of the Canadian Silurian furcacaudiform thelodont *Pezopallichthys* (Wilson and Caldwell, 1998). The same is true for *Archipelepis*, a nonfurcacaudiform thelodont with depressed body from the same deposit and also known as a different species from the Canadian Arctic Archipelago (Soehn et al., 2001; Märss et al., 2002, 2006). The two known species of *Archipelepis* are represented by articulated squamations (e.g., Fig. 3.4C), but preservation is poor, and any paired fins might have been delicate. For *Pezopallichthys*, specimens are laterally compressed (Fig. 3.5A), but again, preservation is comparatively poor. By comparison with other furcacaudiforms, suprabranchial paired fins might not be expected in *Pezopallichthys*, but ventrolateral fins might be predicted; only exceptional preservation and preparation will rule out their presence. In the phylogeny of Wilson and Märss (2004), these two genera are not closely related, yet each appears to be descended from ancestors that had some sort of paired-fin structures.

Furcacaudidae

Furcacaudids (Fig. 3.5B,C) are laterally compressed thelodonts (Fig. 3.3D), which are considered a relatively derived clade by Wilson and Märss (2004). Unlike most thelodonts, furcacaudids lack paired

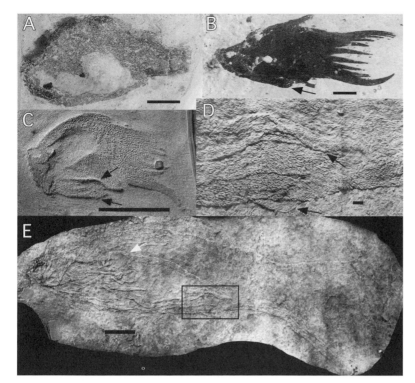

Figure 3.5. Thelodonts mostly with paired, postbranchial, ventrolateral fins. A–C, furcacaudiforms: **A**, *Pezopallichthys ritchiei*, a basal furcacaudiform apparently lacking paired fins; **B**, *Furcacauda* sp. (photograph courtesy of the Museum of Natural History, Berlin); **C**, *Drepanolepis maerssae*. **D, E,** *Shielia taiti*, which has both paired suprabranchial fins and paired ventrolateral fins: **D**, enlarged area showing the paired ventrolateral fins; **E**, entire specimen preserved as a lateral compression, with enlarged area of **D** outlined in black. Black arrows indicate ventrolateral paired fins; white arrow indicates margin of suprabranchial paired fin. **A–C, E**, scale bar = 1 cm; **D**, scale bar = 1 mm.

suprabranchial fins. However, several species have paired, ventrolateral fin flaps that originate posterior to the rearmost branchial openings, a position similar to that of anaspid paired fins (Fig. 3.5B,C). These fin-like structures have larger scales more proximally, finer scales more distally, and smooth-crowned scales along their outer leading edge. The ventrolateral fin flaps widen somewhat posteriorly, and end abruptly about two-thirds the distance to the anal opening.

Shielia

Märss and Ritchie (1998) were the first to recognize the distinct body form of a species that was originally classified in *Thelodus* by its describer Stetson (1931), in *"Logania"* (a preoccupied name) by Gross (1967), and in *Loganellia* by Turner (1991). *Shielia taiti* is more fusiform than are other Scottish thelodonts (Fig. 3.3C). Some well-preserved specimens are preserved dorsoventrally (Fig. 3.4F), and others are preserved laterally (Fig. 3.5E), suggesting a roughly circular cross section of the body. Both types of preservation reveal paired suprabranchial fins as scale-covered extensions from the side of the head, but apparently more flexible than the paired fins of *Loganellia* and *Lanarkia* (Märss and Ritchie, 1998: fig. 23b). Trailing portions of the fin are covered with fine, tripartite, narrow, imbricated scales, with smoother, more compact scales along the outer/leading edge. In at least two specimens, the basal part of the fin, where it attaches to the body, has short indentations that could represent impressions of cartilaginous skeletal elements (see Märss and Ritchie, 1998: figs. 23b, 25).

One well-preserved specimen (Fig. 3.5D,E) has, in addition to its suprabranchial fins, small, paired fin flaps in an abdominal position, anterior and lateral to the estimated position of the anus, which is indicated by a patch of fine, outwardly oriented scales. Both left and right ventrolateral fin flaps bear finer scales on the fin web and larger, smoother scales along the outer/leading edge (Fig. 3.5D). The paired nature of these fins is determined by relationship to the row of ventral midline scales between them (Märss and Ritchie, 1998). Another specimen reveals one of the presumed two ventral paired fins (Märss and Ritchie, 1998: fig. 23d). Märss and Ritchie (1997) called these fins *pelvics,* but in their 1998 paper, they referred to them by the more neutral term *ventral fins.*

Galeaspida

This group of "agnathans" is known mainly from China and is usually considered closely related to osteostracans (Liu, 1965; Tarlo, 1967; Janvier, 1996), or to be slightly more basal than osteostracans, yet paired fins are lacking in all well-preserved specimens. Galeaspids do have paired ventrolateral ridges, along which each scale consists of a series of four or five minute scales somewhat resembling dermal fin rays (P. Janvier, pers. comm., 2005). Janvier (1984) pointed out that a close relationship of galeaspids with osteostracans would imply loss of paired fins in galeaspids.

Pituriaspida

Another group possibly related to osteostracans is the Pituriaspida, from the Devonian of Australia (Young, 1991a). Although the fossils are natural molds that do not preserve bone, anterior paired fins something like those of osteostracans were likely present because there is a pair of openings in the shield in an appropriate position (Young, 1991a; Janvier, 1996). There is no solid evidence as to the presence or absence of ventrolateral paired fins, although ventrolateral projections from the rear of the shield, similar to those on the PVL plates of some placoderms, might hint at the existence of pelvic-fin structures farther posteriorly (P. Janvier, pers. comm., 2005).

Osteostraci

Osteostracans are usually considered the closest relatives of gnathostomes (Forey and Janvier, 1993, 1996; Goujet, 2001; Janvier, 2004, this volume; Fig. 3.1). Osteostracans possess a pair of lobate, paddle-shaped fins (Figs. 3.3E, 3.6) that originate in the sinus located medial to the cornua (where a cornua is present; Fig. 3.6B,C). Cornua are immovable spine-like extensions from the posterolateral corners of the cephalic shield. In well-preserved examples (Fig. 3.6B), it can be seen that the proximal and central parts of each fin are covered with larger, flatter scales, surrounded in the illustrated example by a zone of smaller, presumably more flexible

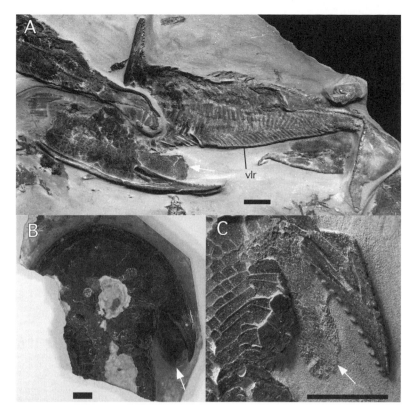

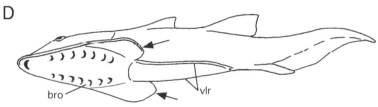

Figure 3.6. Paired fins in osteostracans. **A**, *Superciliaspis gabrielsei* preserved in lateral view, showing the suprabranchial (pectoral) paired fin in the sinus behind the cornual spine, and the row of ventrolateral ridge scales (**vlr**). **B**, *Waengsjoeaspis nahanniensis* preserved in dorsal view. **C**, *Superciliaspis gabrielsei* preserved in ventral view; **D**, sketch of *Ateleaspis* showing the suprabranchial location of the paired fins relative to the inwardly curving row of branchial openings (**bro**) on the ventral surface, and the ventrolateral ridge scales (**vlr**). Arrows indicate paired fins. Scale bars = 1 cm.

scales; the most peripheral parts of the fin (especially the medial edge) are covered in very small scales, sometimes arranged in diagonal rows, suggesting greater flexibility peripherally than centrally (Adrain and Wilson, 1994: fig. 7b, reproduced in Coates, 2003: fig. 2c; Janvier et al., 2004). Belles-Isles (1989) was the first to draw attention to the possible existence of an internal cartilaginous skeleton in the paired fin; Janvier et al. (2004) have recently confirmed the presence in *Escuminaspis laticeps* of a calcified cartilaginous disk something like that seen in embryonic *Polypterus*, acipenseriforms, and teleosts (e.g., Davis et al., 2004). The paired fins of osteostracans appear to have had a narrow attachment to the scapulocoracoid, indicated by concentrated articular surfaces, muscle origins, and foramina for nerves and blood vessels. This narrow attachment is considered similar to the monobasal attachment of the pectoral fins of placoderms (Goujet, 2001; Johanson, 2002).

Osteostracans do not have paired ventrolateral fins, although some authors have suggested that the ventrolateral scale ridges in the abdominal region of osteostracans (Fig. 3.6A,D) correspond to the hypothetical ventrolateral fin fold (Stensiö, 1964) or to the pelvic fins in other vertebrates (Moy-Thomas and Miles, 1971). The ventrally flattened body form of osteostracans might suggest that paired fin-like structures in the ventrolateral body wall could have been lost if they had been present earlier; however, there are many extant benthic fishes that retain large pelvic fins. Coates (1993, 1994) used the presence of putative fin-fold structures (i.e., the ventrolateral scale ridges) together with pectoral fins in osteostracans to argue that one (paired fin folds) could not have evolved into the other (pectoral fins).

Janvier (1987) suggested that paired fins of some osteostracans, such as *Ateleaspis*, might have been capable of undulatory movement useful in slow swimming. We suggest that the paired fins of osteostracans might also have created currents to carry away the outflow from the external branchial openings when in contact with or buried in bottom sediments, or to assist with the process of becoming partially buried in sediment by expelling sediment-laden water from under the shield. Morrissey et al. (2004) described specimens from the Upper Silurian in Wales, referable to the trace fossil genus *Undichna*, that appear to show independent action of the paired fins of an osteostracan against the substrate.

Homology between osteostracan suprabranchial paired fins and gnathostome pectoral fins seems very likely, and indeed they have many of the features of the true paired fins of gnathostomes, but what should we conclude about their possible homology with paired fins of other "agnathans"? We draw attention here to the similarity of anatomical position between osteostracan paired fins and the suprabranchial paired fins of thelodonts. Indeed, in most osteostracans the paired fins are attached some distance lateral to the ventral midline, and also lateral to the inwardly curving paired rows of branchial openings (Fig. 3.6D). Osteostracan ancestors were undoubtedly not so ventrally flattened. In a deeper-bodied ancestor, the row of branchial openings would have been slanted diagonally as in anaspids and furcacaudiform thelodonts, and the point of attachment of the paired fins would have been dorsal to the openings, much like that of thelodonts such as *Loganellia, Turinia,* and *Shielia*. Thus, the paired fins of osteostracans are also suprabranchial in position.

Paired Fins in Basal Gnathostomes

Placodermi

Placoderms possess both well-developed pectoral fins and, where the condition can be assessed, pelvic fins with the exception of antiarchs (Janvier, 1996; Fig. 3.3F). Pelvic fins are attached to a girdle in some taxa. The pelvic fins of Phyllolepida and Ptyctodontida contain enlarged metapterygia (sexually dimorphic in ptyctodontids, with hook-shaped, clasper-like structures resembling those of chondrichthyans; e.g., Long, 1997; Ritchie, 2005). Concerning the pectorals, Goujet (2001) and Johanson (2002) have presented evidence that these fins were monobasal, a single basal fin element articulating at a small articular surface on the pectoral girdle. Antiarchs have peculiar, externally armored pectoral appendages. Johanson (2002) argued that antiarch paired fins resemble those of osteostracans in that blood vessels and nerves pass through a postbranchial lamina before reaching the scapulocoracoid, but this resemblance might be a convergence. In other placoderms and in higher gnathostomes, the vessels and nerves are posterior to the lamina (Johanson, 2002).

Chondrichthyes

Crown-group chondrichthyans have well-developed pectoral and pelvic fins (Fig. 3.3G), the pectorals attached to a cartilaginous scapulocoracoid and the pelvics to a cartilaginous pelvic girdle. The fin is supported proximally by cartilaginous radials and more distally by connective tissue fibers called *ceratotrichia*. Most fossil chondrichthyans for which paired fins are preserved have long, stiff radials supporting large parts of the fin and supported in turn by three large basal elements or a single basal element (Zangerl, 1981; Stahl, 1999). In some Carboniferous holocephalans, the pectoral fins have large spines and are inserted high on the flank of the fish (Zangerl, 1973; Lund, 1986). There is some doubt about the primitive mode of attachment of chondrichthyan pectoral fins to the girdle, whether through a single basal or by several (Janvier, 1996). In any case, most pectoral fins in chondrichthyans have a narrow base from which internal supports radiate. Pelvic fins, in contrast, are constructed with their internal radials nearly parallel. Males of most chondrichthyans have portions of their pelvic-fin skeleton modified to form claspers, used for internal fertilization of the female.

Many Ordovician through Early Devonian taxa have been allied with chondrichthyans on the basis of the structure of their scales, but the condition of their paired fins is generally not known. An exception is the Early Devonian genus *Kathemacanthus* Gagnier and Wilson, 1996 (Figs. 3.3H, 3.7C). In light of new material and restudy of original specimens (Hanke, 2001; Hanke and Wilson, unpubl. data), and assuming that features of scales are a reliable guide to higher-level phylogenetic relationships, *Kathemacanthus* is arguably a stem-group chondrichthyan. It possesses chondrichthyan features such as globular calcified cartilage and areally growing scales similar to those of the presumed chondrichthyan *Seretolepis* Karatajute-Talimaa, 1968, yet it shares other features with primitive acanthodians. *Kathemacanthus* has a large, lobate pectoral fin and a long-based pelvic fin (Fig. 3.7C; Gagnier and Wilson, 1996: pl. 1, fig. 1, reproduced in Coates, 2003: fig. 3c). The pectoral fin is inserted high on the flank and has a large, curved spine at its leading edge. A series of prepectoral paired spines is distributed between the fin and the ventral midline of the body; these spines decrease in size

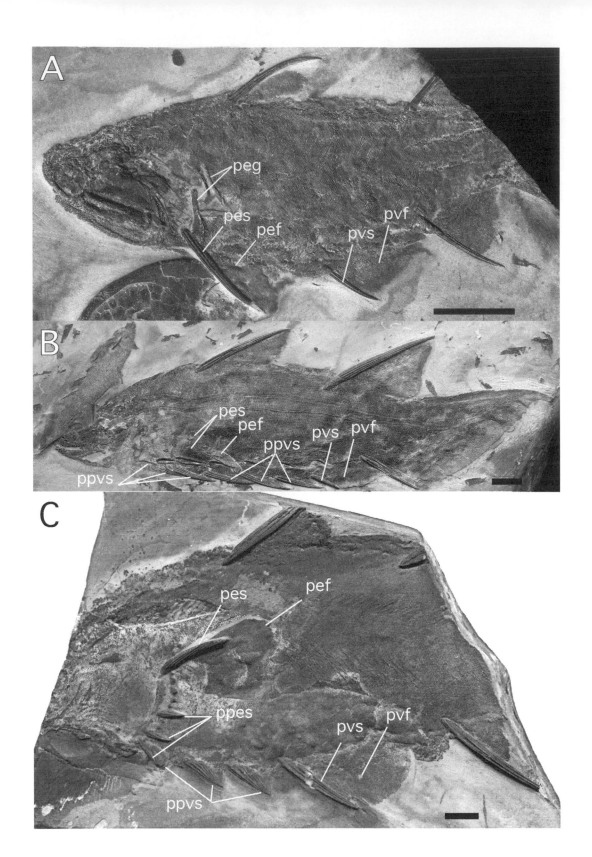

ventrally. The central part of the fin is covered by large, flat scales, implying a muscular and probably skeletal lobe within the fin; the more distal parts of the fin are covered by very fine scales arranged in closely spaced, radiating rows (Gagnier and Wilson, 1996). The pelvic fins appear not to have had a muscular/skeletal lobe extending from the body margin. Instead, the pattern of scales on the fin suggests a long-based attachment with parallel internal supports (Fig. 3.7C). The leading edge is reinforced by a large spine, and pairs of prepelvic spines are arranged anterior to the pelvic ones along the belly, decreasing in size anteriorly.

Other early chondrichthyans are now also believed to have had paired fin spines, including the Early Devonian *Doliodus problematicus*, which has pectoral spines (Miller et al., 2003; Turner and Miller, 2005), and two acanthodian-like putative chondrichthyans, *Lupopsyroides* and *Obtusacanthus*, from the same Lochkovian-age locality that has yielded *Kathemacanthus*. Both *Lupopsyroides* and *Obtusacanthus* have pectoral, pelvic, and a variable number of prepectoral and prepelvic spines (Hanke and Wilson, 2004).

The type specimen of *Antarctilamna prisca* has a fin spine lying immediately behind the branchial chamber (Young, 1982), and this spine lacks the expanded insertion areas characteristic of elasmobranch dorsal-fin spines. Spines with a prominent insertion, appropriately identified as median fin spines, have also been assigned to *Antarctilamna* (Young, 1982: pl. 87, fig. 3). The position relative to the branchial chamber of the questionable spine on the type specimen and its lack of an insertion area suggest that original interpretations that it is a median spine (see reconstructions in Janvier, 1996; Young, 1989, 1991b) were incorrect. We suggest instead that this chondrichthyan also had pectoral spines.

Acanthodii

Acanthodians are usually united with Osteichthyes in the group Teleostomi, which is conventionally considered the sister-group of Chondrichthyes, although some recent studies suggest that acanthodians may be paraphyletic with respect to osteichthyans (e.g., Hanke and Wilson, 2004). Acanthodians typically have well-developed pectoral and pelvic fins (Fig. 3.3I). In some acanthodians, the pectoral fin is attached to an internal cartilaginous or ossified pectoral girdle (Fig. 3.7A). In most species, each paired fin is reinforced by a leading-edge spine (Fig. 3.7A). Prepectoral spines as seen in the putative chondrichthyan *Kathemacanthus* are represented in many acanthodians by ornamented, often spine-like plates anterior to and medial to the pectoral fins (Gagnier and Wilson, 1996). The pelvic fin of acanthodians is long-based (Fig. 3.7A), and the pelvic spines in many taxa are preceded by one to many pairs of prepelvic spines (formerly called intermediate spines) in many taxa (Fig. 3.3I). These prepelvic spines are reduced in number in more derived groups of acanthodians such as diplacanthids, ischnacanthids, and acanthodids (Hanke and Wilson, 2004). Some acanthodians (*Paucicanthus*

Figure 3.7. *(opposite page)* Paired fins in selected gnathostomes. **A,** *Ischnacanthus* sp.; **B,** *Brochoadmones milesi*; **C,** *Kathemacanthus rosulentus.* **pef,** pectoral fin; **peg,** pectoral girdle; **pes,** pectoral-fin spine; **ppes,** prepectoral paired spines; **ppvs,** prepelvic paired spines; **pvf,** pelvic fin; **pvs,** pelvic-fin spine. Scale bars = 1 cm.

vanelsti Hanke, 2002; *Yealepis douglasi* Burrow and Young, 1999) lack paired fin spines entirely.

The Early Devonian acanthodian *Brochoadmones milesi* (Fig. 3.3J, 3.7B) illustrates both the difference between pectoral and pelvic fin morphology and the extreme development of the pelvic series. It has a greatly reduced and delicate, but probably still lobate, pectoral fin inserted higher on the side of the fish than the pelvics. The pectoral spines are also reduced to small, flattened structures that resemble little bivalve shells. There are no prepectoral spines (Fig. 3.7B). The pelvic fins, in contrast, are large and long-based, and covered by small scales in parallel rows, the scales decreasing in size toward the fin margin. The large pelvic spines are preceded by a long series of paired prepelvic spines (Gagnier and Wilson, 1996; Hanke and Wilson, 2006), now known to number six pairs (Fig. 3.7B). The pectoral and pelvic fins of *Brochoadmones* are thus distinct in form and position from each other, and the degree of their development was under separate control.

A somewhat similar situation occurs in other acanthodians such as the diplacanthid *Gladiobranchus*, in which the anteriormost prepelvic spine (known as an *admedian* in earlier literature) in most specimens is hidden by the pectoral fin and its spine (Bernacsek and Dineley, 1977; Hanke, 2001). The prepelvic spine series of *Gladiobranchus* is thus ventral to the pectoral spine and fin as it is in *Brochoadmones*. Species of *Diplacanthus* have a similar condition.

Osteichthyes

Bony fishes, usually considered the sister-group of the Acanthodii, are divided into two main groups, the Actinopterygii (ray-finned fishes) and the Sarcopterygii (fleshy-finned fishes). In both groups, pectoral and pelvic fins are well developed (Fig. 3.3K), and supported basally by cartilaginous or bony radials and distally by segmented, bony fin rays. Especially in early actinopterygians but also in some early sarcopterygians, the pectoral and pelvic fins differ in structure.

The basalmost actinopterygian preserved as articulated skeletons, on current evidence, is *Dialipina*, recently redescribed by Schultze and Cumbaa (2001) on the basis of abundant, well-preserved specimens from the Lower Devonian of northern Canada. *Dialipina* possesses a number of features that resemble characteristics of the Sarcopterygii, yet it is classified as a basal actinopterygian. Its pectoral fins are attached high on the flank, and the pelvic fins, in contrast, are abdominal and long-based.

One of the best known of early actinopterygians is *Cheirolepis canadensis* (Fig. 3.3K) from the Upper Devonian of eastern Canada (Arratia and Cloutier, 1996). It is often considered (e.g., Cloutier and Arratia, 2004; Gardiner et al., 2005) as the most basal actinopterygian apart from *Dialipina*. Reinterpretation of its structure by Arratia and Cloutier (1996) shows that the pectoral fins are lobate, with the rays attaching in a radiating pattern. The pelvic fins, in contrast, are not lobed; they have parallel rays and a very long base (attachment to the body wall). The Late Devonian actinopterygian *Howqualepis*, sometimes itself considered the most basal actinopterygian, also has long-based pelvic fins (Long, 1988).

The basalmost crown-group actinopterygian clade consists of *Polypterus* and its relatives in the Polypteriformes. Polypterids have lobate, narrow-based pectoral fins supported by a cartilaginous plate composed of three elements, and a radiating arrangement of fin rays. The pelvic fin of *Polypterus* (a pelvic fin is absent in *Erpetoichthys*) is unlike that of early fossil actinopterygians in having, like the pectoral fin, a concentrated base with a single basal element resulting from a fusion of embryonic cartilages (Goodrich, 1930; Cloutier and Arratia, 2004).

Among sarcopterygians, lobate pectoral and pelvic fins are the rule. Onychodontiformes have fins attached high on the side of the fish, the pectoral fin being monobasal. Onychodonts are considered basal with respect to most osteichthyans, with the possible exception of actinopterygians.

Coelacanths are often considered basal sarcopterygians, and one of the best known of the basal coelacanths is *Miguashaia*, which has lobed pectoral and pelvic fins (Cloutier, 1996). The similarity between pectoral and pelvic fins as seen in *Miguashaia* is probably a derived feature of most sarcopterygians (and of the homologous pectoral and pelvic limbs of tetrapods).

Revised Homology of "Agnathan" Fins

The above survey of paired fin characteristics of early vertebrates shows that rudimentary paired fin-like structures are present in several groups of fossil "agnathans" (Fig. 3.3) and that the paired fins of many basal gnathostomes have distinctive characteristics that argue against their origin from a common ancestral fin fold. Paired fins of osteostracans are usually considered homologous to pectoral fins of gnathostomes, but the homologies of other "agnathan" paired fins are less certain. Here we suggest that precursors (homologues possessing some but not all of the features of true pectoral and pelvic fins of gnathostomes) occur in several fossil "agnathan" taxa (Plate 23). Some "agnathans" possess homologues of pectoral fins, some of pelvic fins, and perhaps a few possess homologues of both.

Pectoral-Fin Precursors

We propose that precursors of pectoral fins in "agnathans" were situated high on the side of the cephalothorax, dorsal to the branchial chambers and branchial openings. In thelodonts they originate close behind the orbits and extend as far as the posterior end of the branchial region. Suprabranchial fins occur in osteostracans and in thelodonts such as *Phlebolepis*, *Loganellia*, *Lanarkia*, and *Shielia* (Plate 23B,C,E). In osteostracans and less certainly in some thelodonts, there is evidence of internal skeletal support, as might be expected for a functional control structure. In gnatho-stomes, the pectoral fins of many primitive forms (Plate 23H,J,K) differ from pelvic fins in being lobate rather than having parallel structure and in originating higher on the side of the fish than the pelvics.

Pelvic-Fin Precursors

Pelvic-fin precursors in "agnathans" are long-based, post-branchial, and ventrolateral in position (Plate 23A,C,D), extending almost as far posteriorly as the anus in some. In thelodonts, well-developed pelvic-fin precursors occur in most species of furcacau-dids. In *Shielia,* small pelvic-fin precursors occur together with pec-torals but are more ventral. The pelvic precursors in *Shielia* are much smaller than those in furcacaudids.

The paired fins of anaspids have a long basal region covered by larger scales and presumably containing muscles and parallel radi-als, and a more flexible, distal region covered in finer, parallel scales. A case might be made that they are homologues of pectorals, pri-marily involving an inferred homology of the postbranchial spines with the cornual spines of osteostracans (but see discussion above). Alternatively, as argued here, anaspid fins can be regarded as homo-logues of pelvics (Plate 23A). Evidence in favor of this view includes their postbranchial, abdominal position, and their long-based, parallel-rayed construction.

In many primitive gnathostomes, the pelvic fins are more ven-tral than the pectorals (Plate 23I–K). They are usually long-based, with parallel internal radials and parallel bony or connective tissue rays. In acanthodians and some putative chondrichthyans, the pelvics are end points of a series of paired fin spines (Plate 23H–J). In at least two acanthodian taxa, the pelvic series of paired struc-tures extends even anterior to the pectoral fins and ventral to the branchial region (Plate 23J).

Phylogenetic Distribution of Paired-fin Precursors

If both pectoral and pelvic precursors occur in jawless verte-brates long before the origin of Gnathostomata, we may ask whether pectorals appeared before pelvics, as some authors have suggested (e.g., Coates, 1993, 1994), whether pelvics appeared first, or whether both types of fins appeared together. When we plot the dis-tribution of paired fins on the suggested phylogeny of early verte-brates (Plate 24), and assuming for the sake of this discussion that this phylogeny and our proposed homologies are correct, it is appar-ent that there is no support for the idea that pectorals predate pelvics, and only weak support for the possibility that pelvics predate pectorals. The most basal taxon to possess undoubted paired fins is the Anaspida, the fins of which are, according to some authors, pre-cursors of pectoral fins, or, as suggested here, precursors of pelvic fins (Plate 24). However, the next most basal taxon, the Thelodonti, has well-developed pectoral precursors in many species and pelvic pre-cursors in others. It is possible that furcacaudids are descended from ancestors that had pectoral-fin precursors, and that osteostracans and many thelodonts are descended from ancestors that had pelvic-fin precursors.

Discussion

Although long-based paired fins of anaspids might appear to give support to the lateral fin-fold theory, it is more likely that these fins are just early examples of either pectoral or pelvic fins. For the fin-fold theory to be correct, we would expect pectoral fins to originate low on the flank in line with the pelvic fins. Instead, the earliest examples of pectoral fins are suprabranchial, anterior outgrowths from the cephalothorax of thelodonts, distinct from the ventrolateral, abdominal position of the pelvic precursors seen in *Shielia*, furcacaudids, and perhaps anaspids.

Shubin et al. (1997) suggested a sequence of paired-fin evolution involving elongate, fold-like fins first (in anaspid-like vertebrates), then pectoral fins only (in osteostracans), and finally two pairs of paired fins (in gnathostomes). We suggest instead that both pelvic precursors and pectoral precursors had appeared by the origin of thelodonts, and if the consensus phylogeny is correct, that osteostracans must have lost the (rudimentary) pelvic fins of their ancestors. Although there is no strong evidence to support a homology between the ventrolateral scale ridge of osteostracans and galeaspids and a ventrolateral fin fold, as suggested by early workers (Stensiö, 1964; Moy-Thomas and Miles, 1971), it is possible that the scale ridge evolved where pelvic fins formerly existed. In this way, the scale ridge might be said to have replaced the pelvic fins.

Coates and Cohn (1998) argued for a causal link between the appearance of pectoral fins together with a stomach-like gut chamber in furcacaudid thelodonts, both being examples of anteroposterior regionalization of the vertebrate body. They assumed that pelvic fins were a later innovation of gnathostomes. However, our proposed homology of furcacaudid paired fins with the pelvic fins of gnathostomes, together with the presence of pelvic precursors in *Shielia* and perhaps in anaspids, questions both the association between pectoral fins and stomach-like gut chambers and the later evolution of pelvic fins.

In summary, "agnathan" paired fin-like structures occurred in many taxa before the origin of gnathostomes (Plate 24). These early paired fins should not all be assumed to be the same: we suggest that some taxa have pectoral precursors, and others have pelvic precursors. At least one thelodont probably had both. Some "agnathan" lineages likely lost either pectoral fins (e.g., furcacaudiforms) or pelvic fins (e.g., osteostracans, perhaps some thelodonts) that were present in their ancestors. Homologues of pectorals and pelvics differed in position and structure even before the origin of jaws, and within most of the major groups of early jawed vertebrates.

Acknowledgments

This research was supported by Natural Sciences and Engineering Research Council of Canada Discovery Grant A9180 to MVHW and Estonian Science Foundation grant 5726 and the target financed project 0331760s01 to TM. We thank P. Gagnier, P. Janvier, and K. Soehn for helpful discussion, A. Lindoe for specimen preparation, J. Bruner for technical assistance, and John Long and Philippe Janvier for helpful comments on the manuscript.

References Cited

Adrain, J. M., and M. V. H. Wilson. 1994. Early Devonian cephalaspids (Vertebrata: Osteostraci: Cornuata) from the southern Mackenzie Mountains, N.W.T., Canada. Journal of Vertebrate Paleontology 14:301–319.

Arratia, G., and R. Cloutier. 1996. Reassessment of the morphology of *Cheirolepis canadensis* (Actinopterygii); pp. 165–197 in H.-P. Schultze and R. Cloutier (eds.), Devonian Fishes and Plants of Miguasha, Quebec, Canada. Verlag Dr. Friedrich Pfeil, Munich.

Balfour, F. M. 1876. On the development of elasmobranch fishes. Journal of Anatomy and Physiology (London) 11:128–172.

Balinsky, B. I. 1933. Das Extremitätenseitenfeld, seine Ausdehnung und Beschaffenheit. Wilhelm Roux's Archiv für Entwicklungsmechanik der Organismen 130:704–736.

Belles-Isles, M. 1989. Mise en évidence de l'endosquelette postcrânien chez un Ostéostracé, *Alaspis macrotuberculata* ørvig, du Dévonien supérieur de Miguasha (Québec). Comptes Rendus de l'Académie des Sciences (Paris), Sér. II, 308:1497–1501.

Bemis, W. E., and L. Grande. 1999. Development of the median fins of the North American paddlefish (*Polydon spathula*), and a reevaluation of the lateral fin-fold hypothesis; pp. 41–68 in G. Arratia and H.-P. Schultze (eds.), Mesozoic Fishes 2: Systematics and the Fossil Record. Verlag Dr. Friedrich Pfeil, Munich.

Bernacsek, G. M., and D. L. Dineley. 1977. New acanthodians from the Delorme Formation (Lower Devonian) of N.W.T., Canada. Palaeontographica A 159:1–25.

Borkhvardt, V. G. 1986. Proiskhozhdeniye parnykh plavnikov: sostoyaniye problemy. Voprosy Ikhtiologii 26(5):707–714.

Burrow, C. J., and G. C. Young. 1999. An articulated teleostome fish from the Late Silurian (Ludlow) of Victoria, Australia. Records of the Western Australian Museum, Supplement 57:1–14.

Cloutier, R. 1996. The primitive actinistian *Miguashaia bureaui* (Sarcopterygii); pp. 227–247 in H.-P. Schultze and R. Cloutier (eds.), Devonian Fishes and Plants of Miguasha, Quebec, Canada. Verlag Dr. Friedrich Pfeil, Munich.

Cloutier, R., and G. Arratia. 2004. Early diversification of actinopterygians; pp. 217–270 in G. Arratia, M. V. H. Wilson, and R. Cloutier (eds.), Recent Advances in the Origin and Early Radiation of Vertebrates. Verlag Dr. Friedrich Pfeil, Munich.

Coates, M. I. 1993. *Hox* genes, fin folds and symmetry. Nature 364:195–196.

Coates, M. I. 1994. The origin of vertebrate limbs. Development 1994(Suppl.):169–180.

Coates, M. I. 2003. The evolution of paired fins. Theory in Biosciences 122:266–287.

Coates, M. I., and M. J. Cohn. 1998. Fins, limbs, and tails: outgrowths and axial patterning in vertebrate evolution. BioEssays 20:371–381.

Davis, M., N. H. Shubin, and A. Force. 2004. Pectoral fin and girdle development in the basal actinopterygians *Polyodon spathula* and *Acipenser transmontanus*. Journal of Morphology 262:608–628.

Donoghue, P. C. J., and M. P. Smith. 2001. The anatomy of *Turinia pagei* (Powrie), and the phylogenetic status of Thelodonti. Transactions of the Royal Society of Edinburgh: Earth Sciences 92:15–37.

Ekman, S. 1941. Ein laterales Flossensaumrudiment bei Haiembryonen. Nova Acta Regiae Societatis Scientiarum Upsaliensis, Series IV 12(7):1–43.

Forey, P., and P. Janvier. 1993. Agnathans and the origin of jawed vertebrates. Nature 361:129–134.

Forey, P., and P. Janvier. 1994. Evolution of the early vertebrates. American Scientist 82:554–565.

Gagnier, P.-Y. 1995. Ordovician vertebrates and vertebrate phylogeny. Bulletin du Muséum National d'Histoire Naturelle, Section C, Sciences de la Terre, Série 4, 17:1–37.

Gagnier, P.-Y., and M. V. H. Wilson. 1996. Early Devonian acanthodians from northern Canada. Palaeontology 39:241–258.

Gardiner, B. G., B. Schaeffer, and J. A. Masserie. 2005. A review of the lower actinopterygian phylogeny. Zoological Journal of the Linnean Society 144:511–525.

Goodrich, E. S. 1906. Notes on the development, structure and origin of the median and paired fins of fish. Quarterly Journal of Microscopical Science 50:333–376.

Goodrich, E. S. 1930. Studies on the Structure and Development of Vertebrates. Macmillan, London, 837 pp.

Goujet, D. 2001. Placoderms and basal gnathostome apomorphies; pp. 209–222 in P. E. Ahlberg (ed.), Major Events in Early Vertebrate Evolution. Palaeontology, Phylogeny, Genetics and Development. Systematics Association Special Volume 61. Taylor & Francis, London and New York.

Gross, W. 1967. Über Thelodontier-Schuppen. Palaeontographica A 127: 1–67.

Hanke, G. F. 2001. Comparison of an Early Devonian acanthodian and putative chondrichthyan assemblage using both isolated and articulated remains from the Mackenzie Mountains, with a cladistic analysis of early gnathostomes. Ph.D. dissertation, University of Alberta, Edmonton, Canada, 566 pp.

Hanke, G. F. 2002. *Paucicanthus vanelsti* gen. et sp. nov., an Early Devonian (Lochkovian) acanthodian that lacks paired fin-spines. Canadian Journal of Earth Sciences 39:1071–1083.

Hanke, G. F., and M. V. H. Wilson. 2004. New teleostome fishes and acanthodian systematics; pp. 189–216 in G. Arratia, M. V. H. Wilson, and R. Cloutier (eds.), Recent Advances in the Origin and Early Radiation of Vertebrates. Verlag Dr. Friedrich Pfeil, Munich.

Hanke, G. F., and M. V. H. Wilson. 2006. Anatomy of the Early Devonian acanthodian *Brochoadmones milesi* based on nearly complete body fossils, with comments on the evolution and development of paired fins. Journal of Vertebrate Paleontology 26:526–537.

Hou, X.-G., R. J. Aldridge, D. J. Siveter, and X.-H. Feng. 2002. New evidence on the anatomy and phylogeny of the earliest vertebrates. Proceedings of the Royal Society of London B 269:1865–1869.

Janvier, P. 1984. The relationships of the Osteostraci and Galeaspida. Journal of Vertebrate Paleontology 4:344–358.

Janvier, P. 1987. The paired fins of anaspids; one more hypothesis about their function. Journal of Paleontology 61:850–853.

Janvier, P. 1996. Early Vertebrates. Oxford Monographs in Geology and Geophysics 33. Clarendon Press, Oxford, 393 pp.

Janvier, P. 2003. Vertebrate characters and the Cambrian vertebrates. Comptes Rendus Palevol 2:523–531.

Janvier, P. 2004. Early specializations in the branchial apparatus of jawless vertebrates: a consideration of gill number and size; pp. 29–52 in G. Arratia, M. V. H. Wilson, and R. Cloutier (eds.), Recent Advances in the Origin and Early Radiation of Vertebrates. Verlag Dr. Friedrich Pfeil, Munich.

Janvier, P., M. Arsenault, and S. Desbiens. 2004. Calcified cartilage in the paired fins of the osteostracan *Escuminaspis laticeps* (Traquair 1880) from the Late Devonian of Miguasha (Québec, Canada), with a consideration of the early evolution of the pectoral fin endoskeleton in vertebrates. Journal of Vertebrate Paleontology 24:773–779.

Jarvik, E. 1980–1981. Basic Structure and Evolution of Vertebrates. Volumes 1 and 2. Academic Press, London and New York, 575 pp. + 337 pp.

Johanson, Z. 2002. Vascularization of the osteostracan and antiarch (Placodermi) pectoral fin: similarities, and implications for placoderm relationships. Lethaia 35:169–186.

Johanson, Z., and M. M. Smith. 2003. Placoderm fishes, pharyngeal denticles, and the vertebrate dentition. Journal of Morphology 257:289–307.

Karatajute-Talimaa, V. N. 1968. Noviye Telodonti, Heterostraki i Artrodiri iz Chortkovskogo Horizonta Podolii; pp. 33–42 in D. V. Obruchev (ed.), Ocherki po Foligenii i Sistematike Iskopaemikh Rib i Bescheliustnikh. Nauka, Moscow.

Kessel, M. 1993. *Hox* genes, fin folds and symmetry. Nature 364:197.

Kiaer, J. 1924. The Downtonian fauna of Norway. 1. Anaspida with a geological introduction. Videnskapsselskapets Skrifter, I. Matematiske-naturvidenskapslige Klasse, 6:1–139.

Kiaer, J., and A. Heintz. 1932. New coelolepids from the Upper Silurian on Oesel (Esthonia). Eesti Loodusteaduse Arhiiv, Seeria 1, Köide 10, Vihk 3:1–8.

Liu, Y.-H. 1965. New Devonian agnathans of Yunnan. Vertebrata PalAsiatica 9:125–134.

Long, J. A. 1988. New palaeoniscoid fishes from the Late Devonian and Early Caroniferous of Victoria; pp. 1–64 in P. A. Jell (ed.), Devonian and Carboniferous Fish Studies. Association of Australasian Palaeontologists, Sydney, Australia. Memoir 7.

Long, J. A. 1997. Ptyctodontid fishes (Vertebrata, Placodermi) from the Late Devonian Gogo Formation, Western Australia, with a revision of the European genus *Ctenurella* Ørvig, 1960. Geodiversitas 19:515–555.

Lund, R. 1986. The diversity and relationships of the Holocephali; pp. 97–106 in T. Uyeno, R. Arai, T. Taniuchi, and K. Matsuura (eds.), Indo-Pacific Fish Biology. Ichthyological Society of Japan, Tokyo.

Märss, T. 1986. Squamation of the thelodont agnathan *Phlebolepis*. Journal of Vertebrate Paleontology 6:1–11.

Märss, T., and A. Ritchie. 1997. Articulated thelodonts of Scotland. Journal of Morphology 232:293.

Märss, T., and A. Ritchie. 1998. Silurian thelodonts (Agnatha) of Scotland. Transactions of the Royal Society of Edinburgh: Earth Sciences 88: 143–195.

Märss, T., M. V. H. Wilson, and R. Thorsteinsson. 2002. New thelodont (Agnatha) and possible chondrichthyan (Gnathostomata) taxa established in the Silurian and Lower Devonian of the Canadian Arctic Archipelago. Proceedings of the Estonian Academy of Sciences, Geology 51:8–120.

Märss, T., M. V. H. Wilson, and R. Thorsteinsson. 2006. Silurian and Lower Devonian thelodonts and putative chondrichthyans from the Canadian Arctic Archipelago. Special Papers in Palaeontology 75:1–144.

Maisey, J. G. 1986. Heads and tails: a chordate phylogeny. Cladistics 2:201–256.

Maisey, J. G. 1988. The phylogeny of early vertebrate skeletal induction and ossification patterns. Evolutionary Biology 22:1–36.

Miller, R. F., R. Cloutier, and S. Turner. 2003. The oldest articulated chondrichthyan from the Early Devonian period. Nature 425:501–504.

Mivart, St. G. 1879. Notes on the fins of elasmobranchs, with considerations on the nature and homologues of vertebrate limbs. Transactions of the Zoological Society of London 10(12):439–484.

Morrissey, L. B., S. J. Braddy, J. P. Bennertt, S. B. Marriott, and P. R. Tarrant. 2004. Fish trails from the Lower Old Red Sandstone of Tredomen Quarry, Powys, southeast Wales. Geological Journal 39:337–35.

Moy-Thomas, J. A., and R. S. Miles. 1971. Palaeozoic Fishes. Chapman and Hall, London, 259 pp.

Neal, H. V., and H. W. Rand. 1939. Chordate Anatomy. P. Blakiston's Son & Co., Philadelphia, 467 pp.

Parker, T. J., and W. A. Haswell. 1943. A Text-Book of Zoology. Volume 2. Revised sixth edition. Macmillan, London, 758 pp.

Parrington, F. R. 1958. On the nature of the Anaspida; pp. 108–128 in T. S. Westoll (ed.), Studies on Fossil Vertebrates. University of London, Athlone Press, London.

Powrie, J. 1870. On the earliest known vestiges of vertebrate life; being a description of the fish remains of the Old Red Sandstone of Forfarshire. Transactions of the Geological Society of Edinburgh 1:284–301.

Ritchie, A. 1960. A new interpretation of *Jamoytius kerwoodi* White. Nature 188:648–649.

Ritchie, A. 1964. New light on the morphology of the Norwegian Anaspida. Skrifter uitgitt av det Norske Videnskaps-Akademi i Oslo, I. Matematiske-naturvidenskapslige Klasse, Ny Serie, 14:1–35.

Ritchie, A. 1968a. New evidence on *Jamoytius kerwoodi* White, an important ostracoderm from the Silurian of Lanarkshire, Scotland. Palaeontology 11:21–39.

Ritchie, A. 1968b. *Phlebolepis elegans* Pander, an Upper Silurian thelodont from Oesel, with remarks on the morphology of thelodonts; pp. 81–88 in T. Ørvig (ed.), Current Problems of Lower Vertebrate Phylogeny. Almqvist & Wiksell, Stockholm.

Ritchie, A. 1980. The Late Silurian anaspid genus *Rhyncholepis* from Oesel, Estonia, and Ringerike, Norway. American Museum Novitates 2699:1–18.

Ritchie, A. 1984. Conflicting interpretations of the Silurian agnathan, *Jamoytius*. Scottish Journal of Geology 20:249–256.

Ritchie, A. 2005. *Cowralepis*, a new genus of phyllolepid fish (Pisces, Palcodermi) from the late Middle Devonian of New South Wales, Australia. Proceedings of the Linnean Society of New South Wales 126:215–259.

Sahney, S., and M. V. H. Wilson. 2001. Extrinsic labyrinth infillings imply open endolymphatic ducts in Lower Devonian osteostracans, acanthodians, and putative chondrichthyans. Journal of Vertebrate Paleontology 21:660–669.

Schultze, H.-P., and S. L. Cumbaa. 2001. *Dialipina* and the characters of basal actinopterygians; pp. 315–332 in P. E. Ahlberg (ed.), Major Events in Early Vertebrate Evolution. Palaeontology, Phylogeny, Genetics and Development. Systematics Association Special Volume 61. Taylor & Francis, London and New York.

Shu, D.-G., H.-L. Luo, S. Conway Morris, X.-L. Zhang, S.-X. Hu, L. Chen, J. Han, M. Zhu, Y. Li, and L.-Z. Chen. 1999. Lower Cambrian vertebrates from south China. Nature 402:42–46.

Shu, D.-G., S. Conway Morris, J. Han, Z.-F. Zhang, K. Yasui, P. Janvier, L. Chen, X.-L. Zhang, J.-N. Liu, Y. Li, and H.-Q. Liu. 2003. Head and backbone of the Early Cambrian vertebrate *Haikouichthys*. Nature 421:526–529.

Shubin, N., C. Tabin, and S. Carroll. 1997. Fossils, genes and the evolution of animal limbs. Nature 388:639–648.

Smith, M. M. 2003. Vertebrate dentitions at the origin of jaws: when and how pattern evolved. Evolution and Development 5:394–413.

Smith, M. M., and Z. Johanson. 2003. Separate evolutionary origins of teeth from evidence in fossil jawed vertebrates. Science 299:1235–1236.

Soehn, K. L., T. Märss, M. W. Caldwell, and M. V. H. Wilson. 2001. New and biostratigraphically useful thelodonts from the Silurian of the Mackenzie Mountains, Northwest Territories, Canada. Journal of Vertebrate Paleontology 21:651–659.

Stahl, B. J. 1974. Vertebrate History: Problems in Evolution. McGraw-Hill, New York, 489 pp.

Stahl, B. J. 1999. Chondrichthyes III: Holocephali. Handbook of Paleoichthyology, Part 4. Verlag Dr. Friedrich Pfeil, Munich, 164 pp.

Stensiö, E. A. 1932. The Cephalaspids of Great Britain. Trustees of the British Museum, London, 220 pp.

Stensiö, E. A. 1939. A new anaspid from the Upper Devonian of Scaumenac Bay in Canada, with remarks on other anaspids. Kungliga Svenska Vetenskapsakademiens Handlingar (3)18:1–25.

Stensiö, E. A. 1964. Les Cyclostomes fossiles ou Ostracodermes; pp. 96–382 in J. Piveteau (ed.), Traité de Paléontologie. Volume 4(1). Masson et Cie, Paris.

Stetson, H. C. 1931. Studies on the morphology of the Heterostraci. Journal of Geology 39:141–154.

Tabin, C., and E. Laufer. 1993. *Hox* genes and serial homology. Nature 361:692–693.

Tanaka, M., A. Münsterberg, W. G. Anderson, A. R. Prescott, N. Hazon, and C. Tickle. 2002. Fin development in a cartilaginous fish and the origin of vertebrate limbs. Nature 416:527–531.

Tarlo, L. B. H. 1967. Agnatha; pp. 629–636 in W. B. Harland, C. H. Holland, M. R. House, N. F. Hughes, A. B. Reynolds, M. J. S. Rudwick, G. E. Satherthwaite, L. B. H. Tarlo, and E. C. Tilley (eds.), The Fossil Record. Geological Society of London.

Thacher, J. K. 1876. Medial and paired fins, a contribution to the history of vertebrate limbs. Transactions of the Connecticut Academy of Arts and Science 3:281–310.

Turner, S. 1991. Monophyly and interrelationships of the Thelodonti; pp. 87–120 in M.-M. Chang, Y.-H. Liu, and G.-R. Zhang (eds.), Early Vertebrates and Related Problems of Evolutionary Biology. Science Press, Beijing, China.

Turner, S., and R. F. Miller. 2005. New ideas about old sharks. New Scientist 93:244–252.

Turner, S., and W. van der Brugghen. 1992. The Thelodonti, an important but enigmatic group of Palaeozoic fishes. Modern Geology 18:1–29.

Westoll, T. S. 1958. The lateral fin-fold theory and the pectoral fins of ostracoderms and early fishes; pp. 180–211 in T. S. Westoll (ed.), Studies on Fossil Vertebrates. University of London, Athlone Press, London.

White, E. I. 1946. *Jaymoytius kerwoodi*, a new chordate from the Silurian of Lanarkshire. Geological Magazine 83:89–97.

Wilson, M. V. H., and M. W. Caldwell. 1993. New Silurian and Devonian fork-tailed "thelodonts" are jawless vertebrates with stomachs and deep bodies. Nature 361:442–444.

Wilson, M. V. H., and M. W. Caldwell. 1998. The Furcacaudiformes: a new order of jawless vertebrates with thelodont scales, based on articulated Silurian and Devonian fossils from northern Canada. Journal of Vertebrate Paleontology 18:10–29.

Wilson, M. V. H., and T. Märss. 2004. Toward a phylogeny of the thelodonts; pp. 95–108 in G. Arratia, M. V. H. Wilson, and R. Cloutier (eds.), Recent Advances in the Origin and Early Radiation of Vertebrates. Verlag Dr. Friedrich Pfeil, Munich.

Young, G. C. 1982. Devonian sharks from south-eastern Australia and Antarctica. Palaeontology 25:817–843.

Young, G. C. 1989. The Aztec fish fauna (Devonian) of southern Victoria Land: evolutionary and biogeographic significance; pp. 43–62 in J. A. Crame (ed.), Origins and Evolution of the Antarctic Biota. Geological Society Special Publication 47.

Young, G. C. 1991a. The first armoured agnathan vertebrates from the Devonian of Australia; pp. 67–86 in M.-M. Chang, Y.-H. Liu, and G.-R. Zhang (eds.), Early Vertebrates and Related Problems of Evolutionary Biology. Science Press, Beijing.

Young, G. C. 1991b. Fossil fishes from Antarctica; pp. 538–567 in R. J. Tingey (ed.), The Geology of Antarctica. Oxford Monographs on Geology and Geophysics 17. Clarendon Press, Oxford.

Zangerl, R. 1973. Interrelationships of early chondrichthyans; pp. 1–14 in P. H. Greenwood, R. S. Miles, and C. Patterson (eds.), Interrelationships of Fishes. Academic Press, London and New York.

Zangerl, R. 1981. Chondrichthyes I: Paleozoic Elasmobranchii. H.-P. Schultze (ed.), Handbook of Paleoichthyology, Part 3A. Gustav Fischer Verlag, Stuttgart, 115 pp.

Zhang, X.-G., and X.-G. Hou. 2004. Evidence for a single median fin-fold and tail in the Lower Cambrian vertebrate, *Haikouichthys ercaicunensis*. Journal of Evolutionary Biology 17:1162–1166.

4. MODEs of Developmental Evolution: An Example with the Origin and Definition of the Autopodium

Hans C. E. Larsson

ABSTRACT

Evolutionary novelty and modularity are important factors in macroevolution. These terms are discussed and shown to lack properties to be adequately applied to the dynamic process of evolution. A novel comparative system is formalized to combine definitions of evolutionary novelty and modularity with data from evolutionary and developmental biology. This system introduces modules of developmental evolution (MODEs) to explain how macroevolutionary transformations and developmental changes can better explain the origin and evolution of structures. The transition from the fin of fish-like sarcopterygians to the tetrapod limb is discussed in detail as an example. The skeletal anatomy of the series of known extinct and extant sarcopterygian fish fins is reviewed with current information of the development of paired appendages in fish and tetrapods. A comparative framework is established to equate the evolution of an autopodial field at the dipnoan-tetrapodomorph split, evolution of digits within the autopodial field at the node Tetrapodomorpha, and evolution of digit identity at the node Tetrapoda. A novel definition for the autopodium and digits is derived that suggests the metapterygial axis contributes to a single digital anlage within the autopodium, but that all other digits are repetitions of either the same axis or represent a novel axis within the autopodium. This hypothesis incorporates aspects of earlier models concerning the formation of the digital arch and neomorphic structures.

Introduction

Evolutionary novelties abound. We can all agree on their existence and importance in evolution. The presence of hair in mammals or feathers in birds is easy enough to identify today. Each structure is well characterized by its anatomy, histology, chemical composition, and development. The issue of describing these structures in the context of an evolutionary novelty is more difficult. Mayr (1960:351) stated, "The exact definition of an 'evolutionary novelty' faces the same insuperable difficulty as the definition of the species." Although this may seem a nearly impossible task, attempts have been made to address the definition of evolutionary novelty. Some workers suggest evolutionary novelties are so pervasive that all apomorphies (derived characters) should be considered evolutionary novelties (Arthur, 2000; Hall, 2005). Others would prefer to restrict the definition of evolutionary novelties to a subset of apomorphies. Müller and Wagner (1991:243) suggested that a "morphological novelty is a structure that is neither homologous to any structure in the ancestral species nor homonomous to any other structure of the same organism." They went on to suggest evolutionary novelties are novel characters that "open the door to new opportunities." This definition could be expanded to other realms of evolution, such as genes, proteins, and behavior. The principle behind their definition is that the evolutionary novelties have overcome ancestral constraints and in doing so are open to new *variational opportunities* (Müller and Wagner, 1991; Wagner and Larsson, 2007). Classic examples of evolutionary novelties that use this definition are hair in mammals, feathers in birds, and limbs in tetrapods. Each is a persistent structure within the respective lineage and has surely contributed to the radiation of that clade.

But with sufficient examination, all novel characters in the evolution of a lineage could be considered evolutionary novelties. Each, like novel phalangeal numbers in the therapsid ancestors of mammals (Hopson, 1995), the semilunate carpal of neotetanuran theropods (Sereno, 1999a), or the novel incorporation of *engrailed* into the regulatory complex of *yellow* in species of *Drosophila* with wing eyespots (Gompel et al., 2005), surely opens up new opportunities to lineages. On the other hand, all evolved characters may be considered as mere modifications of existing structures. Novel morphologies develop (and evolve) from existing structures, be they from the morphological or genetic level of exaptation. Thus, restricting novelties to nonhomologous or nonhomonomous structures, as suggested by Müller and Wagner (1991), may be founded on bases that are currently not rigorously defined.

I suggest that one difficulty in understanding many evolutionary novelties is that they have been based on a lack of phylogenetic resolution, either on the order of speciation events, or sequence of character transformations, or both. These deficiencies may lead to artificial inflation of a particular evolutionary event. On a different level, the current use of evolutionary novelties is restricted to historical usage. Only after the novelty has achieved the recognition of facilitating new variational opportunities to a lineage can it be identified as an evolutionary novelty.

Explanations of evolutionary transformations are traditionally within the realm of microevolutionary studies concerning adaptations. Hypotheses to explain the origin and maintenance of traits can be made only after a set of selective differentials are qualified. Local changes in allele frequencies among and within populations form the principal tool to explain the adaptation of traits within individuals under particular selective regimes. The explanation that allows the use of this tool to describe the change is natural selection. The mathematics of population genetics is subject to intensive investigation and allows biologists to measure allele frequency, life histories, and local features in order to derive models of evolutionary adaptations within those populations under study. In rare cases, the underlying genetic changes have been related to developmental modifications that, in turn, generate the transformed phenotype. These include the evolution of wing eyespots in butterflies (Nijhout, 2001) and *Drosophila* (Gompel et al., 2005), and abdominal bristle numbers in *Drosophila* (Long et al., 1998). Quantitative trait loci data are also available for adaptations in stickleback fish (Shapiro et al., 2004). All of these studies investigate relatively recent evolutionary transformations that may not be considered major phenotypic modifications when compared with the vast range of phenotypic changes in the fossil record.

Evolutionary novelties have traditionally fallen into a different category of observation. The transformations associated with evolutionary novelties are relatively large phenotypic changes that have occurred in deep time. Because these transformations are historical and undoubtedly involved complex genotypic changes, the proximate explanation for evolutionary novelties cannot be allele frequency changes. Alleles were certainly involved in the change; however, allele frequencies of evolutionary novelties within populations in the past are entirely lost to science. The proximate explanation for evolutionary novelties must lie somewhere else.

To address some of these issues of evolutionary novelties, I will begin with a discussion of some recent advances in understanding the evolution of feathers while highlighting elements I consider important to investigating evolutionary novelties. These elements are characterized, resulting in the derivation of a new term to aid in explaining the origin of novel features. This term and its associated methodology are subsequently applied to the origin of the autopodium.

Origin of Feathers

The presence of hair in mammals may seem a trivial matter, but the exact timing of the origin of hair in mammals is unknown because the fossil record of the evolution of hair is exceedingly sparse. The developmental origin of hair is equally mysterious; to date, no well-tested (or testable) scenario has been put forth to explain what developmental transformations may account for the evolution of hair. The origin of feathers in birds is a more interesting case. The most basal bird is, by definition, *Archaeopteryx,* and it already has a full plumage (Ostrom, 1976; Sereno, 1999b). The resulting variational opportunities are evident in the nearly 10,000 species of birds alive today; a large number of passerines are identified on the basis of subtle feather colorations.

The sequence of the evolutionary transformation of protofeathers to fully formed asymmetrical feathers has been reviewed by numerous authors. Because clear reviews are presented by Prum (1999), Prum and Williamson (2001b), and Prum and Brush (2002), I will only provide a brief summary for the sake of completeness. The earliest feather-like structures within the closest ancestry of birds appear to be elongate tube-like body coverings (Fig. 4.1). These elements were coincidentally the first described integumentary structures of nonavian dinosaurs. Long nonbranching integumentary structures were first reported for Early Jurassic coelophysoid theropod footprints from Massachusetts by Hitchcock (1836:323), who intuitively identified them as "wiry feathers." These structures were reviewed by Kundrát (2004) and found to compare well with the natal down of paleognathous birds. The quality of preservation required for these structures is rare in the fossil record. However, these elements are now known from the tail of psittacosaurs (Mayr et al., 2002), the feet of coelophysoids (Hitchcock, 1836; Kundrát, 2004), and the entire body of some basal coelurosaurs (Chen et al., 1998). This distribution suggests the origin of these structures may date back to at least the origin of Dinosauria. Some of these integumentary structures are branched in *Sinosauropteryx* (Chen et al., 1998; Ji et al., 1998; Currie and Chen, 2001) and may represent the next step in the transformation of these early feathers (Prum, 1999; Prum and Brush, 2002). Dinosaurian grades more closely related to birds exhibit symmetrical feathers with a well-formed rachis, as seen in *Caudipteryx* (Ji et al., 1998). A further modification of these feathers is an asymmetrical arrangement of barbs about the rachis to stabilize the feather during flight, found in taxa such as the dromaeosaurid *Microraptor, Archaeopteryx,* and all flying birds (Prum, 2003; Xu et al., 2003). Although a secondary function, coloration patterns that feathers exhibit in present-day birds are equally elaborate in fossils, with different species highlighting bars, stripes, spots, and flecks of patterns.

The question of feathers as an evolutionary novelty now be-

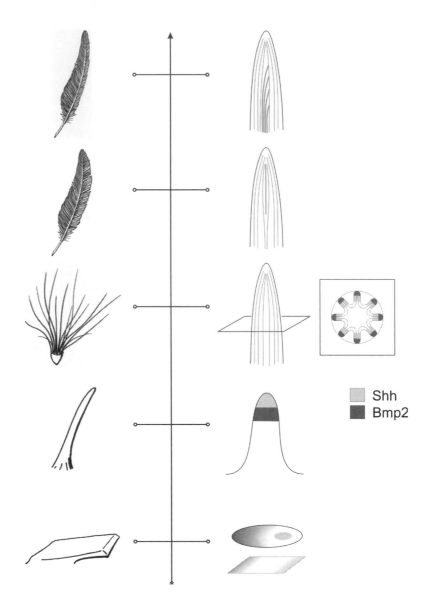

Figure 4.1. Evolutionary and developmental transformations of feathers.

comes blurred. Feather-like structures appear to have a long history with a number of discrete evolutionary transformations. Something else must be done to retain focus on the issue of feathers as an evolutionary novelty. Rather than simply attempting to record the origin of an evolutionary novelty, an attempt to explain the novelty seems more productive. Explanations of novelties seem to be the problem. The lack of a fully functional and rigorous definition of *evolutionary novelty* does not help the matter.

Proximate explanations of evolutionary novelties may help. As discussed above, ultimate explanations surely lie within the population genetics of evolutionary change; however, the majority of evolutionary novelties are found at times and in clades not accessible to this level of research. We can assume that population genetic changes are the foundations of the phenotypic changes, in response to environmental cues in modern systems (West-Eberhard, 2003). Although individual phenotypes within these polyphenic reaction norms are not specifically allied to particular genotypes, the set of possible polyphenisms is certainly associated with a particular genotype. Phenotypic changes that are dissociated from genotypic changes have been hypothesized for phenotypic evolution (Budd, 1999; Newman and Müller, 2000). Newman and Müller termed this form of phenotypic evolution *epigenetic evolution*. Proximate explanations for these changes may be more difficult to examine because they would require knowledge of the environmental conditions and changes, as well as the particular genetic architectures present in the taxa at the time, to assess what reaction norms may be experienced by the phenotypes. A simplified situation was presented by Müller and Streicher (1989) to hypothesize how a skeletal reaction to mechanical stress might have become later canalized to maintain the fibular crest of the tibia to support the reduced fibula of extant birds.

But regardless of the mediation of the genotypic change, such change must still result in a developmental change to account for the phenotypic transformation. Developmental change may offer the insight needed to explain phenotypic change at the most proximate level we can hope for. It is with this notion that I will attempt to explore explanations of evolutionary novelties.

Prum (1999) has presented an elegant example of examining feathers as an evolution of a complex evolutionary novelty. He compared the early morphogenesis of feathers in present-day birds to that of their hypothesized evolution. Feather development was divided into five discrete stages. Each subsequent stage elaborates on its predecessor and represents discrete phenotypic outcomes. Prum ordered the stages such that the developmental stages in the series are conditionally dependent on the stage preceding it. This arrangement presents a biologically grounded and ordered developmental series for examining evolutionary changes. The elegant step Prum made was to associate specific developmental stages with evolutionary stages of feathers (Fig. 4.1). The morphogenetic stages have also been corroborated with gene expression stages of *Shh* and *Bmp2* that appear to support the early stages of feather development (Harris et al., 2002). A further discussion of the development of feather shapes and coloration patterns was presented by Prum and Williamson (2001a,b). Their model was able to encompass most of the range of feather patterns by applying relatively simple changes to parameters of feather development.

I must add here that many earlier workers have also suggested the use of comparable developmental and taxonomic patterns. A great history of comparisons between developmental and taxonomic patterns has extended across the centuries (see reviews in Gould, 1977, and Rieppel, 1988). Von Baer (1828) devised a method to compare embryological events to a taxonomic hierarchy to support that particular taxonomic arrangement, although this was before the advent of Darwinian evolution (Rieppel, 1985). Similarly, Haeckel (1874) used developmental data to defend his scenarios of animal phylogeny (Gould, 1977). Although these approaches add support and justification to particular preevolutionary taxonomic arrangements or phylogenies, they do nothing to examine the basis for the phenotypic transformations common to the developmental and taxonomic patterns. A leap forward on this front was the concept of the *epigenetic trap* (Wagner, 1989) to provide an explanation for the origin, maintenance, and evolution of homology. This explanation was based on the interaction between morphogenetic rules and the hierarchical nature of ontogenetic networks: once a particular phenotype has been established by these two factors, the very presence of these factors acts to constrain possible variations of the phenotype. Moreover, an epigenetic trap subsequently constrains the evolution of that phenotype to maintain its original developmental information and thereby creates a type of module.

The comparison of hierarchical data presented above allows congruencies to emerge. Prum's (1999) comparison of stages of feather development with stages of feather evolution was based on shared similarities between the two series. His approach was founded on the insight that each stage of feather development is contingent on preceding stages of feather development. This forced Prum's series of developmental stages into an ordered arrangement that he compared with a similar ordered arrangement of feather evolution. I will term congruencies between developmental and evolutionary data *symmetries* of data. Symmetrical data must have equivalent representative structures. For example, the redeployment of *Shh* and *Bmp* expressions during the second stage of feather development to form an elongate hollow structure has symmetry with the elongate hollow epidermal protofeathers in some dinosaurs because they share similar phenotypes, and presumably similar underlying genetic and morphogenetic mechanisms. The symmetries of these two sets of pattern hierarchies offer an explanation for the underlying developmental (mechanism) changes that may be attributed to the evolution of feathers seen in the phylogenetic record (pattern). In fact, the symmetries offer the initial pieces of information to justify the evolutionary novelty, and they may offer insight to a special class of novelties called *modules*.

Biological Modules

Biological modules have led workers to similar extremes as evolutionary novelties. Modules have been coined for single *cis*-regulatory elements to give a modular status for genetic processes (Arnone and Davidson, 1997). Indeed, models and data of gene duplication and subfunctionalization reveal a mechanism for creating *cis*-regulatory modularity within genomes (Force et al., 1999). Modules have also been used for discrete developmental fields in embryology (Raff, 1996). A slight modification to these definitions is presented by Atchley and Hall (1991), who used the term *funda-*

mental developmental unit to describe the mechanisms required to develop a complex morphological feature. Essentially, developmental modules are defined by a mechanism level of integration. In turn, evolutionary biologists have used modules to describe morphological regions that appear to evolve relatively independently from other regions (Raff, 1996). Definitions of evolutionary modules may vary between workers and specific cases because morphological subunits can be identified in a number of ways to reflect the level of integration being used. Integration at an evolutionary level may focus on functional, topological, or phylogenetic aspects, or on combinations of these. The essential feature of modules in all cases is that they are more integrated internally than externally (Raff, 1996; Bolker, 2000).

As could be expected, the limits to modules are not well defined. At the morphological and developmental level, the entire organism may be considered a module if one considers the development of each multicellular organism from a single zygote. At the genetic level, the entire genome may also be considered a module as a result of its inherent integration. Likewise, modules may be broken down to individual *cis*-regulatory regions or to single functional, morphological, physiological, and developmental components. There appears to be no limit without introducing special definitions of modularity in individual cases.

How to Explain Evolutionary Novelties and Modules

Evolutionary biology is often satisfied with narrative hypotheses (Gould and Lewontin, 1978). Although these stories may seem ingenious, they are generally untestable. An interesting feature of many classically identified evolutionary novelties is their complex and modular nature. Take feathers in birds, for instance. Feather evolution took place over multiple stages, and the development of feathers has many levels of developmental control and morphogenesis. This complexity lends itself to a special form of justification for the evolution and development of the feather.

The justification of the evolution and development of a module must lie within the interactions between evolution and development. The symmetries that may be found between them may be considered structural similarities. The shared structures between the two patterns are hypothesized to be equivalent representations of the same event. The developmental representations may be considered the causal equivalents of the evolutionary representations that, in turn, may be considered the effects. At this point, I would like to bring into the argument the symmetry principle, which states that the cause must have at least as much structure as the effect (Rosen, 1983). Thus, if at least as many symmetries can be found in the developmental data to represent the evolutionary data, a justification of symmetry may be applied. A similar use of the symmetry principle has been used by Larsson and Wagner (in preparation) to justify and construct a test of single developmental transformations with single historical evolutionary transformations.

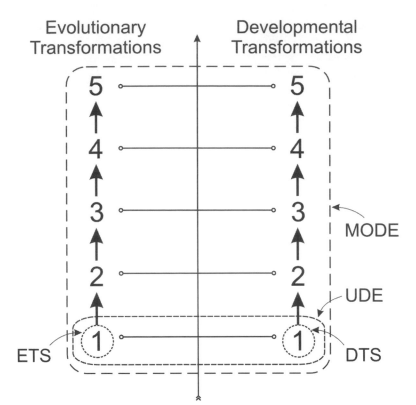

Figure 4.2. Schematic representation of comparing evolutionary and developmental transformations that share symmetrical information. A module of developmental evolution (**MODE**) is composed of a series of units of developmental evolution (**UDEs**) that are constructed by shared information (symmetry) between an evolutionary transformational stage (**ETS**) and a developmental transformational stage (**DTS**).

The symmetries lend support for the interactions of evolutionary changes with developmental mechanisms. Although this sort of correlation does not test the relationship in terms of falsification, the data do offer consilience to the hypothesis. Symmetries in evolutionary and developmental data offer only consilience to the hypotheses of evolutionary change, but the interactions between the symmetries may, in some cases, be tested experimentally. Harris et al. (2002) did just that by experimentally altering the expressions of *Shh* and *Bmp2*. Their results indicated that feather-barb formation is dependent on the local expression repatterning of these genes and suggests a causal relation between these stages of feather development and form.

I will now formalize the symmetries between evolutionary and developmental patterns to help construct a methodology for examining evolutionary transformations. Schematically, the method can be outlined as a series of evolutionary transformations, with equivalent developmental transformations (Fig. 4.2). The associated structures are hypothesized to have causal connections; that is to say, a series of cause-and-effect relationships can be hypothesized. Each paired set of relationships is dependent on previous sets. The more sets of cause-and-effect relationships, the more power the hypothesis has to provide for the proximate explanation of the evolutionary transformation.

Modules of Developmental Evolution (MODEs)

Rather than further confuse the issues of modularity, evolutionary novelty, and evolutionary transformation, I will present a new term, *Modules of Developmental Evolution* (MODEs), to describe instances where a sequence of evolutionary transformations can be causally explained by a sequence of developmental transformations with equivalent structural information (mechanism and pattern). MODEs are not meant to be rigid demarcations of evolutionary and developmental trajectories, but rather a sliding window along evolutionary and developmental transformations that looks at the causal interactions between the two trajectories. MODEs are composed of an ordered set of units of developmental evolution (UDEs). Each UDE is justified by symmetry between an evolutionary transformational stage (ETS) and a developmental transformational stage (DTS). The series of ETSs are a set of evolutionary transformations that are temporally contingent on each other. Likewise, the series of DTSs are developmental transformations that are also contingent on preceding transformations.

MODEs cannot be discrete. There are certainly genotypic and phenotypic traits that are related proximally and distally to events in the evolutionary and developmental trajectories. All novel phenotypes or genotypes evolved from some predecessor. In fact, the origin of many genes may be grounded in the duplication of previously existing genes (Force et al., 1999; Zhang, 2003) and others through molecular mechanisms that act on an existing genome (Long et al., 2003). Likewise, morphological structures are generally derived from preexisting ones—for example, feathers from scales. Developmental trajectories are similar in that early developmental events set the basis for later events. For instance, feathers are dependent on the development of scale-like epidermal placodes, and feather colorations are dependent on the presence of feathers. In fact, the suite of possible coloration patterns is itself a derivative of the morphogenesis of feathers (Prum and Williamson, 2001a). One can go beyond the probable demarcations of modules and extend interactions between morphological regions. The truncated snout of hominid primates may depend on elongate hind limbs for long-distance running, which in turn depend on physiological adaptations for endurance (Bramble and Lieberman, 2004). Because these traits cannot be completely separated from each other within the same organism or the same lineage across time, MODEs should not be defined in isolation as *natural kinds* might. Natural kinds exist in any conceivable condition and are time-independent. For example, an atom of gold remains an atom of gold in the presence or absence of atoms of other elements and will remain an atom of gold across vast periods of time. Because MODEs represent a dynamic exchange between evolution and development over time, MODEs must be defined within a sliding window along the evolutionary and developmental axes of the lineage in discussion. This usage of MODEs may better reflect the dynamic nature of evolutionary and developmental transformations. In this treatment, specific dynamic components of the evolutionary and developmental data require explicit arguments to make the hypothesis of overall evolutionary and developmental interaction justifiable.

The degree to which the set of evolutionary and developmental transformations can be associated results in the degree of explanatory power for the evolutionary transformation. As Haeckel noted, the order of matches between evolutionary and developmental events need not be a simple progressive order (Richardson and Keuck, 2002). Mismatches in the symmetries may occur from heterochronic events, truncations or unrecognizable modifications to the developmental transformations, or missing phylogenetic data from poor taxon-sampling or poor phylogenetic reconstruction. In some cases, mismatches in the sequences may indicate hierarchical divisions of a set of MODEs along axes of evolutionary and developmental modules.

With these new tools to examine evolutionary and developmental transformations, I will discuss the origin and definition of one of the major features in the transition from fins to limbs: the autopodium.

Transition from Fin to Limb

The example of Prum (1999) clearly demonstrates the utility of matching developmental and evolutionary trajectories. A major feature of its elegance is the division of the evolutionary and developmental trajectories into stages that incorporate a minimal number of transformational events. That is to say, each stage requires a minimum number of changes to explain the differences between it and its preceding and succeeding stages. By means of this approach, I shall explore the origin of the tetrapod limb.

The origin of the tetrapod limb has long been considered an evolutionary novelty. In fact, it probably aided many workers to formulate definitions of evolutionary novelties. The tetrapod limb is clearly derived from the paired appendages of fish-like sarcopterygians (Fig. 4.3); for recent reviews, see Coates et al. (2002), Coates (2003), Coates and Ruta (2007), Carroll and Holmes (2007), and Wagner and Larsson (2007). I will not reproduce these and other works on sarcopterygian fins and limbs in full; however, I will present a brief summary of the current phylogenetic and paleontological knowledge.

The paired appendages of basal Actinopterygii (such as *Polypterus*), and probably the osteichthyan stem, are composed of three skeletal axes: an anterior propterygium, a posterior metapterygium, and an intermediate mesopterygium. Each of these skeletal axes appears to be relatively separate from each other in early fin chondrification. For example, the early chondrification of the pectoral fins of *Polyodon* suggests that the three axes develop separately (Mabee and Noordsy, 2004), or at least that the pro- and mesopterygium develop separate from the metapterygium within an endochon-

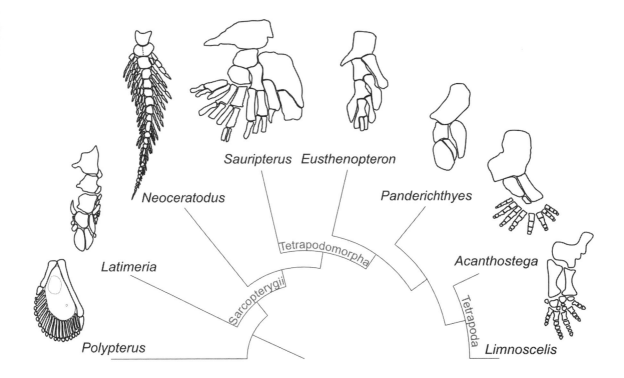

Neoceratodus

Sauripterus Eusthenopteron

Panderichthyes

Tetrapodomorpha

Latimeria

Acanthostega

Sarcopterygii

Tetrapoda

Polypterus

Limnoscelis

dral disk (Davis et al., 2004). Teleosts lose the metapterygium, and the propterygium is either reduced or absent (Mabee, 2000; Wagner and Larsson, 2007).

The paired appendages of Sarcopterygii are composed of a single skeletal axis. No indication of a pro- or mesopterygium has been documented in the fossil record or in the development of extant forms. The remaining metapterygium comprises the entire endochondral fin skeleton. The metapterygium is divided proximodistally into a series of mesomeres. The mesomeres of basal clades such as Actinistia (the coelacanths including *Latimeria*) comprise a string of repeating elements that often have accessory endochondral elements juxtaposed on the preaxial side of each segmental joint. After four mesomeres, the metapterygium terminates in a fan of variously sized plates supporting the lepidotrichia.

The paired appendages of the more derived dipnoans and porolepiforms are similar to those of Actinistia. The metapterygium is elaborated into a longer string of more than 20 undivided mesomeres with endochondral elements (termed *radials*) juxtaposed on the pre- and postaxial sides of the mesomeres. Two of the present-day lungfish, *Lepidosiren* and *Protopterus,* have lost these radials to achieve an even more elongate fin. The second mesomere of at least *Neoceratodus,* the third extant lungfish, has some features that suggest an elaboration closer to that of tetrapods than that of more basal clades. Early in development, the second mesomere is composed of two cartilaginous elements that later fuse (Joss and Longhurst, 2001). This structure is reminiscent of the paired elements in sarcopterygians

Figure 4.3. Phylogeny of Sarcopterygii. Right forefins or forelimbs in dorsal aspect are illustrated above each taxon. Figures are not to scale.

more closely related to tetrapods, called the zeugopodium. It lies distal to the stylopodium (humerus/femur) and is composed of the ulna/radius and fibula/tibia.

The next most inclusive clade of sarcopterygians has been named the Tetrapodomorpha (Ahlberg, 1991). This clade contains all sarcopterygians more closely related to tetrapods than dipnoans, such as *Sauripterus* and the rhizodontids. The paired fins of these taxa represent a novel repatterning of the mesomeres distal to the zeugopodium. The known fin diversity of rhizodontids suggests that this group had extensive elaboration of distal endochondral elements to form a fan of elongate radial structures (Jeffery, 2001). Tetrapodomorphs more closely related to tetrapods include the tristichopterids, such as *Eusthenopteron* and *Panderichthys*. These taxa retain the single stylopodium and paired zeugopodial elements, but reduce the number of distal elements compared with rhizodontids. The reduction is most extreme in *Panderichthys,* which retains only two endochondral elements distal to the ulna.

The oldest well-documented tetrapods are *Acanthostega* and *Ichthyostega* (Clack, 2002, and references therein). These taxa have well-defined digits distal to the zeugopodial elements. There are up to seven digits in the hind limb and eight in the forelimb of *Acanthostega* (Coates, 1996). They also exhibit some degree of individual identity through differential sizes and phalangeal numbers. The digits lie within a segment called the autopodium. Soon after the origin of tetrapods, the autopodium was reduced to five digits and a stable pattern of wristbones and anklebones. The wrist (carpal) and ankle (tarsal) bones are called the mesopodium, and the digits are called acropodium.

Fin-limb Development and Hypothesized Fin-limb UDEs

One of the fundamental features of the tetrapod limb is the presence of digits. A vast body of data is available for the development of paired appendages in vertebrates. Rather than attempt to summarize all of it here, I will instead suggest candidate UDEs that appear to indicate symmetries in the evolution and development of limbs.

The principal question at hand is, to what degree can the origin of the tetrapod limb be parceled into discrete evolutionary and developmental stages that have complementary information? Wagner and Chiu (2001) made an important step in distilling the origin of the limb into the origin of the autopodial developmental field. The autopodial field may be considered a UDE. Developmental gene expressions and gene knockouts have lent good mechanistic support for an autopodial field.

Autopodial Field

The autopodial field is initiated by the expression boundaries of *Hoxa-11* and *Hoxa-13* during development (Haack and Gruss, 1993). *Hoxa-11* is expressed in the early field of the zeugopodium, and the *Hoxa-13* is expressed in the early field of the autopodium. The field appears to be established through the nonoverlapping expressions of *Hoxa-11* and *Hoxa-13* (Wagner and Chiu, 2001). These genes have overlapping expressions in the developing paired fins of zebra fish (Sordino et al., 1995; Neumann et al., 1999) and paddlefish (Metscher et al., 2005), but these genes do not overlap in the frog, *Xenopus* (Blanco et al., 1998), chickens (Nelson et al., 1996), and mice (Haack and Gruss, 1993). Hypodactylous (Hd) mice have a deletion of part of the coding region of *Hoxa-13* (Mortlock et al., 1996). Heterozygous Hd mutants lack digit I, while homozygous mutants lack all but one digit (Robertson et al., 1997; Post and Innis, 1999). However, although the phenotypic changes of knockout mutations of *Hoxa-13* in mice appear localized to the autopodium, they only lack digit I (Fromental-Ramain et al., 1996). This suggests some redundancy in autopodial formation; double knockouts of *Hoxa-13* and *Hoxd-13* produce limbs that lack even a vestige of an autopodium in mice and chickens (Fromental-Ramain et al., 1996). Together, these data suggest a partial functional overlap between these two genes during autopodial development.

Additional support for the initiation of the autopodial field by the nonoverlapping expression boundaries of *Hoxa-11* and *-13* comes from misexpressions of these genes. Ectopic expression of *Hoxa-13* in proximal regions of chick wings results in nodular cartilaginous elements similar to those of the wrist (Yokouchi et al., 1995). Shifting the *Hoxa-11/13* expression boundary distally results in a malformation of the carpals toward a long-bone morphology (Mercanter et al., 1999). Together, these data suggest that establishing the autopodial field is one of the first events in autopodial development, and this pattern may also be one of the first stages in the evolution of the autopodium.

Digits and Digit Identity

Digits have long been synonymized with the autopodium and the origin of limbs. However, as outlined above, the autopodial field is initiated before digit formation and does not require the presence of digits. The autopodial field seems to be a more general feature in development. Because digits always form within an autopodial field, it appears that they are a specialization of the field. The developmental mechanisms for digit development are summarized below.

Digit development normally occurs only within the autopodial field. *Hoxa-13* expression as described above is necessary for autopodial field formation and, in turn, is required for normal digit formation. *Hoxa-13* knockouts reduce digit numbers (Fromental-Ramain et al., 1996) and expand the expression domain of *Hoxa-11* into the distal limb bud (Post and Innis, 1999). *Hoxd-13* knockouts alter the growth of a normal set of digits (Dollé et al., 1993; Fromental-Ramain et al., 1996; Post and Innis, 1999). *Hoxa-13* appears to function upstream of the *AbdB*-related *HoxD* genes within the autopodial field. Although *Hoxa-13* knockouts are required for digit loss (and presumably normal autopodial field function), *AbdB*-related *HoxD* gene knockouts only result in digit malformations and polydactyly (Zákány et al., 1997). Zákány et al. (1997) suggested that *Hoxa-13* may have been incorporated earlier in evolution in the development of digits than the *AbdB*-related *HoxD* genes because of the upstream function of *Hoxa-13* and the greater severity of the effect on digit development resulting from its knockout mutant. The *AbdB*-related *HoxD* genes are hypothesized to have subsequently become utilized to limit digit numbers and assign particular digit identities. The mechanism of the *AbdB*-related *HoxD* gene signaling may be by means of subtle differences in local concentrations of the total amount of *HoxD* proteins and not the presence of any single specific *HoxD* protein (Cobb and Duboule, 2005).

The expression patterns of *HoxA* and *HoxD* genes have been well documented during limb development. The patterns of mice and chickens yield three distinct expression phases that roughly correspond to their functions during the development of the three proximodistal limb segments (Nelson et al., 1996). The third phase of expression is found only in tetrapods and is expressed as a "stepped" anteroposterior pattern of *Hoxd-13* expressed throughout the autopodium, and *Hoxd-12, -11,* and *-10* expressed postaxially. Only the first two phases are expressed in zebra fish (Sordino et al., 1995; Neumann et al., 1999), which suggests that the third phase is specific to the autopodium. One pitfall of this assumption is that the endochondral elements in the zebra fish paired appendages are not homologous to any elements in Sarcopterygii, including tetrapods (Mabee, 2000). However, the presence of a third phase of expression may still be specific to the autopodium and establishments of segments within the autopodium.

Regulation of the *AbdB*-related *HoxD* genes appears to be under a complex network that is modified temporally. A global control region (GCR) has been identified in mice, humans, and puffer fish, which is functionally responsible for the expression of *HoxD* genes (Spitz et al., 2003). When the puffer fish GCR is inserted into transgenic mice, normal expression patterns of *HoxD* genes in the genitalia and most of the central nervous system are produced. However, no expression of *HoxD* genes was found in the mouse autopodia (Spitz et al., 2003). This result suggests that the evolution of the GCR of *HoxD* genes may be one of the principal events governing the origin of the autopodium.

The regulatory control of *AbdD*-related *HoxD* genes may have at least two phases in limb development. Zákány et al. (2004) proposed a complex regulatory network that becomes modified temporally in limb development. Their model suggests that the repressive form of *Gli3*, which is present throughout the early limb bud, represses *AbdB*-related *HoxD*, *Shh,* and *dHand* expression in early developing limb buds. Soon after, *AbdB*-related *HoxD* genes are expressed, which activate *Shh* and *dHand* expression that, in turn, suppresses the repressive form of *Gli3*. Sometime later, the positive feedback loops between *AbdB*-related *HoxD, Shh,* and *dHand* maintain their expression, establishing a graded expression pattern from posterior to anterior.

The role of sonic hedgehog (*Shh*) in limb development has long been known. Summerbell et al. (1973) discovered a signaling center in the posterior margin of developing chick limb buds. The region, the zone of polarizing activity (ZPA), has been characterized by its expression and concurrent secretion of *Shh* protein (Riddle et al., 1993). Altering the position of the ZPA by either transplanting it or inducing *Shh* expression via retinoic acid leads to an increase in digit number and changes to digit identity (Summerbell, 1983). Application of *Shh* protein has yielded similar results (Lopez-Martinez et al., 1995; Yang et al., 1997). *Shh* may have two functional phases during digit development. The first phase of *Shh* expression appears to transform autopodial mesenchyme to become competent to develop digits, whereas the second phase governs digit identity (Dahn and Fallon, 2000; Drossopoulou et al., 2000). *Shh* signals via the transmembrane receptor proteins Patched and Smoothened to modify the ratio between the activator and repressor forms of *Gli3* (Litingtung et al., 2002; Welscher et al., 2002). *Gli3*, in turn, is known to form a complex with Hoxd-12 and other HoxD proteins (Chen et al., 2004). This complex converts the repressive form of *Gli3* to an activator of *Shh* target genes (Chen et al., 2004). The complex thereby regulates digit formation and digit identity.

Knockouts of *Shh* in mice lead to a complete loss of *AbdB*-related *HoxD* and *Hoxa-13* expression, ectopic expression of the repressive form of *Gli3,* and an absence of an apical ectodermal ridge (AER; Litingtung et al., 2002). The absence of *Shh* results in an abundance of the repressor form of *Gli3* that, in turn, represses the activation of *FGF* in the AER via *Gremlin* and reduces *Hox* expression (Litingtung et al., 2002). The resulting phenotype of *Shh* homozygous mutants expresses only a single digit in the hind limb and a remnant of a digit in the forelimb (Ros et al., 2003). *Shh* loss-of-function mutations do not affect *Hoxd-11* and *Hoxd-13* expression in zebra fish (Neumann et al., 1999). This suggests that the interaction between *Shh* and *HoxD* genes via *Gli3* may be specific to tetrapods, and that the origin of digits and digit identity may be tied to the two phases of *Shh* expression and its interactions with the *AbdD*-related *HoxD* genes. The mechanism of establishing digit identity by *Shh* has been hypothesized as a combination of protein concentration (digits II–III) and length of *Shh* is expressed (digits III–V; Harfe et al., 2004). Note that digit III is determined by both concentration and length of expression time, whereas digit I may be *Shh* independent. The independent nature of digit I from *Shh* is deduced from the severe reduction of the remaining digit in *Shh* knockout mice. Although the remaining digit is indeed small, I would caution against assigning an identity to a "digit" that is little more than a string of cartilaginous nodules.

Further downstream targets of *Shh* involve the expression of

Bmp4 within the autopodial field. *Bmp4* concentration within the interdigital mesenchyme is known to affect the identity of the digit adjacent to it anteriorly (Dahn and Fallon, 2000). Whether the *Bmp4* concentration gradient is localized to interdigital mesenchyme or is spread throughout the autopodial field (Drossopoulou et al., 2000) is still in question. However, digit formation does appear to be separable from digit identity (Dahn and Fallon, 2000; Drossopoulou et al., 2000). *Bmp4* may modulate digit identity via its effect on the AER (Selever et al., 2004).

In summary, the expression phases of *AbdB*-related *HoxD* and *Shh* appear to be the principal regulators of digit development and digit identity. The initiation of *AbdB*-related *HoxD* genes establishes an anteroposterior polarization involving *Shh* and *dHand*. If there are indeed two phases of *Shh* expression and function within the developing limb bud, the first makes the autopodial mesenchyme competent to form digits and may regulate digit number. The interaction of *Shh* with at least *Hoxd-11* and *Hoxd-13* via *Gli3* may be novel for tetrapods, as may be the regulation of *AbdB*-related *HoxD* genes in the autopodium by their global control region. The second phase of *Shh* expression appears to regulate digit identity through a gradient of *Gli3* activator: repressor ratios, *Bmp4* concentration, and the *AbdB*-related *HoxD* genes. Wagner and Larsson (2007) have presented the development of the autopodium as involving at least three discrete developmental steps: (1) the initiation of the autopodium, (2) digit formation (digitogenesis), and (3) digit identification.

Transformational Hypotheses of the Autopodium

The evolutionary and developmental transformations of the transition from fins to limbs can now be addressed. Previous efforts to combine the two transformations have led to at least two discrete hypotheses. The first involves paleontological and late development data, whereas the second is based primarily on molecular developmental data.

The extension of the metapterygial axis into the autopodium to form digits was first suggested by paleontologists working with sarcopterygian fish. Gegenbaur (1878), Watson (1913), Gregory et al. (1923), Steiner (1934), Gregory and Raven (1941), Westoll (1943), and Jarvik (1980) realized the importance of the metapterygial axis and hypothesized extensions of this axis into the autopodium. A schematic summary of their interpretations is presented in Figure 4.4. Gegenbaur and Gregory and Raven suggested that the metapterygial axis passed through digit V, whereas Jarvik, Steiner, and Watson interpreted the trajectory of the axis as passing through digit IV in five-fingered or -toed hands and feet. Westoll's hypothesis placed the axis toward digits I and II. Holmgren (1933) proposed an additional further interpretation in which the axis passed into the postaxial mesopodial elements (ulnare and distal carpal 4 of the forelimb), similar to those by Watson (1913), Steiner (1934), and Jarvik (1980), but the axis did not project into the digits themselves. Holmgren suggested that all remaining struc-

A

B

C

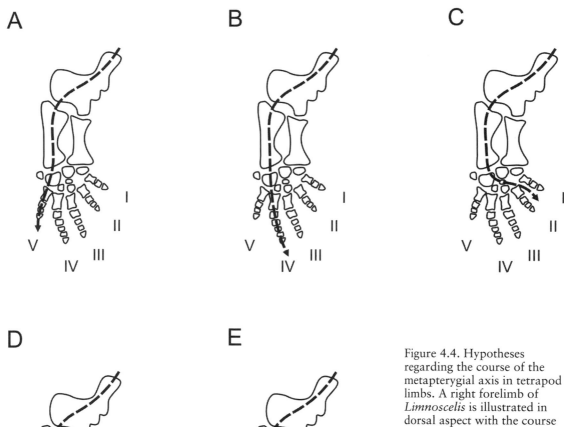

I
II
V
III
IV

I
II
V
III
IV

I
II
V
III
IV

D

E

Figure 4.4. Hypotheses regarding the course of the metapterygial axis in tetrapod limbs. A right forelimb of *Limnoscelis* is illustrated in dorsal aspect with the course of the metapterygial axis indicated with a dashed line. Hypotheses are illustrated for the following: **A**, Gegenbaur (1878), Gregory et al. (1923), and Gregory and Raven (1941); **B**, Watson (1913), Steiner (1934), and Jarvik (1980); **C**, Westoll (1943); **D**, Holmgren (1933); **E**, Burke and Alberch (1985), Shubin and Alberch (1986), and Müller and Alberch (1990).

I
II
V
III
IV

I
II
V
III
IV

tures within the autopodium were neomorphic and not related to the metapterygial axis.

The derivation of the autopodial skeleton from the metaptery-gial axis was elaborated by Burke and Alberch (1985), Müller and Alberch (1990), and Shubin and Alberch (1986). These workers hypothesized that the metapterygial axis extends into the au-topodium and bends preaxially, as previous authors had asserted, but by a novel path. After passing through the postaxial elements to the fourth digit, the axis bends preaxially and gives rise to the re-maining preaxial digits (digital arc; Fig. 4.4E). Upon bending, the axis could derive all the mesopodial and acropodial elements in a sequence that appeared to correspond with the sequence of skeletal chondrification within the autopodium. The marked third phase of *AbdB*-related *HoxD* gene expression within the autopodium was cited as supporting the bending of the metapterygial axis (Sordino et al., 1995). Evidence of some posterior limb bud growth expansion has been observed in mice and chick limb buds (Vargesson et al., 1997; Harfe et al., 2004) and is consistent with bending of the metapterygial axis.

The de novo hypothesis of the autopodium is supported by a variety of developmental and genetic data (Capdevila and Izpisúa-Belmonte, 2001). The progress zone model was proposed by Saunders (1948) to explain the establishment of each of the three developmental fields that gave rise to the stylo-, zeugo-, and autopodium. This model was further elaborated by Dudley et al. (2002) and Sun et al. (2002) to suggest that each of the three fields are specified soon after the initiation of limb bud development. The novel nonoverlapping expression domains of *Hoxa-11* and *Hoxa-13* during the early formation of the autopodial field were used by Wagner and Chiu (2001) to suggest that the autopodial field and its contents were neomorphic with respect to the rest of the limb. The lack of morphological similarities between the autopodium of tetrapods and the fins of fish-like sarcopterygians has also been used to defend a novel origin for the autopodium (Ahlberg and Milner, 1994; Coates, 1996). The latter lack any trace of the proximal nodular elements that form the mesopodium and the distal elongate elements that form the acropodium.

One difficulty in reconciling these different interpretations of the origin of the autopodium is that each hypothesis is based on different sets of data. The digital arch hypothesis is based on paleontological interpretations and chondrification patterns, whereas the de novo hypothesis is based on mechanistic interactions during development and the inability to identify homologous structures between sarcopterygian fins and tetrapod autopodia. Both hypotheses regard the origin of the autopodium and its structures as a singular or relatively limited event. With these different research routes, it seems difficult to reach consensus—or, worse, to glean novel information regarding the relationships between the evolutionary transformations and related developmental mechanisms involved during the fin-to-limb transition.

Symmetries between the evolutionary and developmental trajectories of the fin-to-limb transition may offer more useful insight into the problem. As outlined above, the development of the autopodium is initiated by the establishment of the autopodial field. This event occurs before any other developmental event relevant to the autopodium (except for the initiation of the limb bud) and has been attributed to the mutually excluding expression patterns of *Hoxa-11* and *Hoxa-13*. This event will be considered a developmental transformational unit (Fig. 4.5). Developmentally, it seems apparent that digits and digit identity are separate mechanisms.

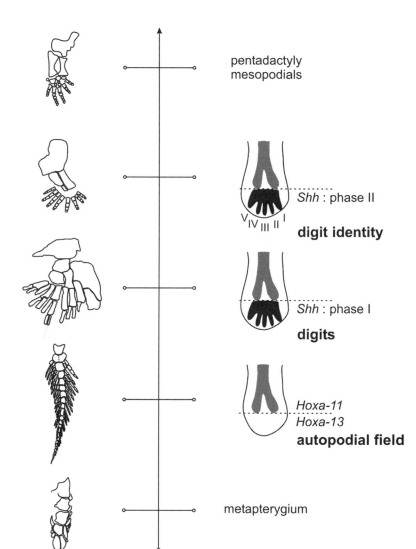

pentadactyly
mesopodials

Shh : phase II

V IV III II I

digit identity

Shh : phase I

digits

Hoxa-11
Hoxa-13

autopodial field

metapterygium

Figure 4.5. Evolutionary and developmental transformations of the fin-to-limb transition. Note the evolutionary transformations are not meant to indicate that these taxa form an ancestral-descendant sequence, but merely that their common ancestry may be best characterized by these taxa at present.

The formation of digits is also separate from the initiation of the autopodium. Thus, I will present digitogenesis and the development of digit identity as two separate developmental transformational stages (DTSs). Each of these DTSs appears to be regulated by two phases of *Shh* expression. The first initiates a cascade of events that lead to a regulation of at least *Hoxa-11* and *Hoxd-13* via modification of a *Gli3* activator-to-repressor ratio gradient. The second phase of *Shh* also acts through a *Gli3* ratio gradient to affect all *AbdB*-related *HoxD* genes that, in turn, may regulate the morphogen responsible for digit identity, *Bmp4*.

Some of the developmental data used by previous authors are based on adult skeletal morphology (topology and typology) and late embryonic development of cartilaginous and osseous elements. Although these structures do offer a wealth of morphological data, they may not provide the most useful data in this case. Gross morphological modifications may occur after the early patterning of those structures. These changes are certainly important at fine taxonomic levels of evolution, but they may not be relevant to more general modifications shared by many taxa whose ancestry is more remote, especially when the autopodium is obviously a structure that has undergone extensive repatterning.

The postaxial to preaxial chondrification of autopodial elements was used in defense of the digital arch hypothesis (Burke and Alberch, 1985; Shubin and Alberch, 1986; Müller and Alberch, 1990). Chondrification occurs after the initial patterns of the limb skeleton are established. Bone morphogenetic proteins (Bmps) are one of the primary regulators of cartilage formation (Hogan, 1996; Hall and Miyake, 2000). The high concentration of Bmps near high concentrations of *Shh* that is secreted from the postaxial ZPA may be the only explanation required for the postaxial to preaxial patterns of chondrification of the autopodial elements. Therefore, the pattern of chondrification may not represent an accurate picture of the transformation of the fin endoskeleton to the limb.

Symmetries of the Evolution and Development of the Autopodium

Possible symmetries between the sequences of the DTSs presented above with ETSs will now be explored. To begin, the initiation of the autopodial field must be identified as a discrete ETS. Molecularly, significant adaptive evolution of the sequence of *Hoxa-11* has been localized along the lineage after the common ancestry of lungfish and tetrapods but before the origin of tetrapods (G. Wagner, pers. comm.). This is interpreted as significant sequence changes to signal a genetic stabilization of the autopodial field.

Morphologically, the first indication of a paired second metapterygial mesomere is in this tetrapodomorph branch, and a distinct ulna-radius pair appears subsequently. Although this structure does not provide direct evidence for an autopodium, it does suggest that an autopodial field has been established to provide the developmental signaling to transform the ancestrally single second mesomere into a pair of bones. Distal to the paired second mesomere, the remainder of the fin exhibits a range of elaborations—colorations that were not found in fishes with a single second mesomere. Rhizodontid sarcopterygians, for instance, exhibit a wide range of distal structures that provide a radial arrangement of elongate bony elements (Jeffery, 2001). These elements have even been proposed as digits in the rhizodont *Sauripterus* (Daeschler and Shubin, 1998).

The structure of the second mesomere in extant lungfish may extend the presence of a discrete zeugopodial/autopodial boundary to include this taxon. Although fossil porolepiforms and present-day lungfish do not appear to have a paired second mesomere, the embryology of the Australian lungfish, *Neoceratodus*, offers some novel information. The adult second mesomere of *Neoceratodus* has a longitudinal groove in its midline that has been interpreted as the fusion of separate cartilaginous elements (Clack, 2002). Before ossification, the second mesomere is composed of two cartilaginous elements (Joss and Longhurst, 2001). The presence of a paired second mesomere and single distal mesomeres suggests a developmental boundary between the second and third mesomeres. Because the zeugopodium is composed of the second mesomere, it seems logical to assume that a zeugopodial field is present. Its proximal and distal boundaries appear to be defined already at the evolutionary level of *Neoceratodus*. Therefore, the beginnings of a proximal autopodial field must also be present. Thus, the presence of a boundary between zeugopodium and autopodium may be a synapomorphy of all tetrapodomorphs and may also extend to include lungfish.

The *Hd* mice mutants and *Shh* knockout mice and chicken mutants described above may provide an interesting connection here. Both maintain a stylopodium and zeugopodium but have only a single digit-like structure in the region of their presumptive autopodium. The presence of this string of elements distal to the zeugopodium is reminiscent of the pattern in lungfish, where the metapterygial axis extends past the paired second mesomere to form a string of bony elements. Although this series of bones has pre- and postaxial radials in *Neoceratodus*, the other two extant lungfish lack radials. The symmetry between the developmental and evolutionary data implies that the metapterygial axis does indeed extend into the autopodial field, but not as a digit because no digits are present yet in lungfish. This skeletal pattern suggests that the absence of digits is associated with an absence of a repressive function between *Hoxa-11* and *Hoxa-13* and a feedback interaction between *Shh* and the *AbdB*-related *HoxD* in the paired fins of lungfish.

The origin of digits must now be placed in the context of an ETS. As mentioned above, digit-like structures have been suggested for rhizodontid fish such as *Sauripterus;* and sister-taxa to Tetrapoda, the tristichopterids (such as *Eusthenopteron*), also have elongate distal structures. Although it is tempting to homologize the distal digit-like structures in rhizodontids and tristichopterids with the digits in tetrapods, the absence of endochondral elements distal to the zeugopodium in *Panderichthys* calls this homology into question. The absence of digit-like structures in *Panderichthys* may represent an independent loss within that lineage or may represent a loss of the structures in the clade *Panderichthys* + Tetrapoda, with parallel evolution of similar structures in the clade tristichopterid + Tetrapoda. The evidence is equivocal at present, although parsimony would suggest a single origin of digit-like structures. Moreover, if we assume loss of ossification or even chondrification of endochondral elements to be relatively easy compared with the origin of a novel developmental pattern, we can consider the digit-like elements of rhizodontids homologous to tetrapod digits. The origin of digits (or at least digit-like structures) can now be considered an ETS and placed with varying degrees of certainty at Tetrapodomorpha.

The origin of digits as an ETS may have symmetry in the developmental data discussed above. Digitogenesis appears to be a discrete DTS that requires the establishment of an autopodial field. I would like to propose that the origin of digit-like structures and the first phase of *Shh* expression and its subsequent regulation of *Gli3* and *AbdB*-related *HoxD* genes be related as a unit of developmental evolution (UDE).

The digit-like elements in rhizodontids and tristichopterids are clearly not entirely homologous to the digits of tetrapods. The distal radials of these fish-like sarcopterygians are nearly identical to each other, with only subtle differences in element length. These differences may have more to do with space-fitting issues than developmental control. However, even the oldest-known tetrapods, *Ichthyostega* and *Acanthostega*, have some degree of digit identity. Their digits are of unequal size and are composed of different number of phalanges. The evolutionary origin of digit identity may be associated with the developmental expression of the second phase of *Shh*, its downstream targets such as *Gli3* and *AbdB*-related *HoxD* genes, and more distally, regulation of *Bmp4* to control digit identity. This pairing will be hypothesized as another UDE.

Other UDEs surrounding the MODE of the fin-to-limb transition surely exist. These may include the reduction of the paired appendages to a metapterygial axis, stabilization at five digits, establishment of a mesopodium, and specific morphologies of each digit and associated structures (e.g., specialized musculature, epidermal structures).

New Definition of the Autopodium

The origin of the autopodium can now be examined in greater detail than before. Previously, the origin of the autopodium was identified with the origin of limbs and tetrapods. The event was claimed an evolutionary novelty because of the unique variational opportunities that the autopodium offered to tetrapods. The structure was also treated as a module because it appears to have had a great degree of internal integration over its more than 370-million-year-long history and during its development in extant tetrapods. This view of the autopodium placed its origin and maintenance in a static view where definitions of novelty and modularity struggle.

The evolution of characters, be they molecular or morphological in nature, is always dynamic. The selection for and ultimate stabilization of characters depend on a multitude of factors, such as selection gradients across adaptive landscapes, which, for the most part, cannot be gleaned from the fossil record. The maintenance of a character is under an equally varied number of factors. If one or more of these influencing factors (e.g., ecology, developmental dynamics) changes, the character itself may be modified. The evolution of characters cannot be accommodated without a dynamic definition of the character. Therefore, a dynamic sliding window is used here to assist in assessing modules of characters.

The view of the origin of the autopodium from this sliding window presents a much more complicated picture. It becomes clear that different levels within the evolution of the autopodium must be qualified. In turn, each of these levels, if they are to be justified as a module of developmental evolution, must have symmetrical states with developmental trajectories. This approach to examining the origin of the autopodium (or any other character) forces a comparison between the evolutionary patterns and developmental mechanisms. In doing so, the autopodium becomes a suite of character transformations along axes of evolution and development. The intersections produce a set of UDEs that appear to be: (1) the origin of the autopodial field, (2) the origin of digits, and (3) the origin of digit identity.

The autopodium can now be defined. From the evolutionary and developmental data, the metapterygial axis does not appear to extend into the entire autopodium. The axis instead contributes to an elongate string of elements within the autopodial field when it first evolves and is initiated developmentally. Digits later evolve/develop about this autopodial string that now becomes characterized as a digit as well. Subsequently, each digit evolves/develops its unique identity. Digits appear to be de novo structures that evolve/develop within the autopodium. They may be reiterations of the same developmental mechanisms used to create the distal radials that branch off the metapterygium of many sarcopterygians, or they may represent redeployments of the mechanisms that generate the axis itself.

Only a single digit appears to be derived from the metapterygial axis itself. This digit in pentadactyl tetrapods is probably digit IV. The transformation of the metapterygial axis into digit IV of pentadactyl autopodia fits well with the evolutionary and developmental data that suggest this digit is the most robust in terms of developmental perturbations and evolutionary reduction of digits. Interestingly, extension of the axis into digit IV but no other digit has previously been suggested by Watson (1913), Steiner (1934), and Jarvik (1980). Holmgren (1933) had a similar interpretation, but he thought that the axis went no further than distal carpal/tarsal 4 (i.e., the mesopodial base of digit IV). It seems that all these authors may have agreed, at least in part, with the hypotheses presented here. Thus, the transformational hypotheses of a digital arch or de novo autopodium may be inadequate. Instead, a hybrid of these two hypotheses may be more in line with the data at hand.

The recent discovery of the Late Devonian sarcopterygian *Tiktaalik* from Nunavut (Canada) sheds important light on the fin-to-limb transition (Daeschler et al., 2006; Shubin et al., 2006). Specimens of this taxon include nearly complete pectoral fins. The skeletal anatomy suggests that *Tiktaalik* is the closest known sister-taxon to Tetrapoda. The endoskeleton of the pectoral fins provides further evidence that the autopodium of tetrapods is derived at least in part from distal radial structures of closely related fish-like sarcopterygians. A primary axis appears to be present in *Tiktaalik* and passes from the humerus, ulna, and ulnare into three more mesomeres that each have between two and four elongate radials.

This pattern is reminiscent of the pectoral skeleton of *Sauripterus*. All articular surfaces of the pectoral endoskeleton of *Tiktaalik* appear to have had synovial joints and suggest a proximodistal series of mobile joints aligned along anteroposterior axes. These joint arrays are hypothesized to be homologous to the anteroposterior joints present in the tetrapod autopodium and are used to suggest the autopodial skeleton is derived from the primary axis of fish-like sarcopterygians (Shubin et al., 2006).

The skeletal structure of *Tiktaalik* lends further support for the contribution of the primary axis to the autopodial skeleton of tetrapods. However, each mesomere distal to the ulna exhibits multiple elongate radials and also supports the hypothesis I have presented above that suggests that the digits within the autopodial field are reiterations of proximodistal axes found more proximally and ancestrally within nontetrapod sarcopterygians. Although the well-preserved new fossils shed valuable light on the topic of the fin-to-limb transition, it remains clear that skeletal structure alone may not lead to definitive conclusions. But the combination of fossil and developmental data offers a powerful tool to address evolutionary transformations.

Conclusions

Evolutionary novelties, modules, and the evolution and development of characters are clearly complex issues. Although definitions of evolutionary novelties have become mainstays of some discussions of evolutionary biology, little attention has been paid to developmental biology. Modules are discussed in terms of phylogenetic patterns of morphological units by evolutionary biologists, whereas developmental biologists tend to refer to modules as developmental mechanisms with greater internal than external integration. I have attempted to address the intersection of evolution, development, novelties, and modules with a new set of tools. MODEs are meant to reflect a dynamic symmetry between the evolutionary and developmental transformations of a character or character set. The characters may be considered an evolutionary novelty and a module at least within the view of the sliding window proposed. The sliding window approach does not restrict us to static definitions of the beginning and end of evolutionary and developmental transformations involving the characters in question. There is clearly no discrete beginning or end for any evolutionary or developmental character.

The justification of a MODE must lie in associating symmetrical information between the evolutionary and developmental transformation series. Each UDE is composed of an ETS that has a symmetrical relationship to a DTS. Together, these tools offer a framework to examine dynamic systems. To illustrate, data are presented to outline symmetries between the evolutionary and developmental transformations of the autopodium. These UDEs help to clarify a section of the origin of the autopodium and assign discrete associations between evolutionary and developmental events. The method assists in isolating stages within each trajectory to highlight where we may have sufficient information and where we may not. Likewise, the sufficiency of information gives us some parameters to base our definition of the transformation in question. The autopodium was found to be composed of at least three UDEs that span a lengthy section of the sarcopterygian lineage and a suite of mechanisms within developing limbs. The autopodium can now be defined with respect to these UDEs. Furthermore, the symmetrical information offers novel insights into the transformation of the autopodium from ancestors without autopodia and leads to dynamic insight into the evolution and development of the autopodium.

Acknowledgments

This chapter benefited greatly from discussions with Günter Wagner. My graduate students and Robert Carroll also provided insight during various phases. An anonymous reviewer and Chi-hua Chiu-Groff provided helpful comments. I dedicate this article to Günter Wagner on the occasion of his fiftieth birthday, with gratitude for his inspiration and guidance during my initiation to developmental evolution, and to Luca Erling Larsson, whose arrival provided many sleepless nights near the completion of this chapter. This research was generously supported by NSERC, FQRNT, and the Canada Research Chairs Program.

References Cited

Ahlberg, P. E. 1991. A re-examination of sarcopterygian interrelationships, with special reference to the Porolepiformes. Zoological Journal of the Linnean Society 103:241–287.

Ahlberg, P. E., and A. R. Milner. 1994. The origin and early diversification of tetrapods. Nature 368:507–514.

Arnone, M. I., and E. H. Davidson. 1997. The hard-wiring of development: organization and function of genomic regulatory systems. Development 110:115–130.

Arthur, W. 2000. Intraspecific variation in developmental characters: the origin of evolutionary novelties. American Zoologist 40:811–818.

Atchley, W. R., and B. K. Hall. 1991. A model for development and evolution of complex morphological structures. Biological Reviews of the Cambridge Philosophical Society 66:101–157.

Baer, K. E. von. 1828. Ueber die Entwickelungsgeschichte der Thiere. Beobachtung und Reflexion. Erster Theil. Gebrüder Borntraeger, Königsberg, 271 pp.

Blanco, M. J., B. Y. Misof, and G. P. Wagner. 1998. Heterochronic differences of *Hoxa-11* expression in *Xenopus* fore- and hind limb development: evidence for a lower limb identity of the anuran ankle bones. Development Genes and Evolution 208:175–187.

Bolker, J. A. 2000. Modularity in development and why it matters to evo-devo. American Zoologist 40:770–776.

Bramble, D. M., and D. E. Lieberman. 2004. Endurance running and the evolution of *Homo*. Nature 432:345–352.

Budd, G. E. 1999. Does the evolution in body patterning genes drive morphological change—or vice versa? BioEssays 21:326–332.

Burke, A. C., and P. Alberch. 1985. The development and homology of the chelonian carpus and tarsus. Journal of Morphology 186:119–131.

Capdevila, J., and J. C. Izpisúa-Belmonte. 2001. Perspectives on the evolutionary origin of tetrapod limbs; pp. 531–558 in G. P. Wagner (ed.), The Character Concept in Evolutionary Biology. Academic Press, San Diego.

Carroll, R. L., and R. B. Holmes. 2007. Evolution of the appendicular skeleton of amphibians, pp. 185–224 in B. K. Hall (ed.), Fins into Limbs: Evolution, Development, and Transformation. University of Chicago Press, Chicago.

Chen, P.-J., Z.-M. Dong, and S.-N. Zhen. 1998. An exceptionally well-preserved theropod dinosaur from the Yixian Formation of China. Nature 391:147–152.

Chen, Y., V. Knezevic, V. Ervin, R. Hutson, Y. Ward, and S. Mackem. 2004. Direct interaction with *Hoxd* proteins reverse *Gli3*-repressor function to promote digit formation downstream of *Shh*. Development 131:2339–2347.

Clack, J. A. 2002. Gaining Ground: The Origin and Evolution of Tetrapods. Indiana University Press, Bloomington, 369 pp.

Coates, M. I. 1996. The Devonian tetrapod *Acanthostega gunnari* Jarvik: postcranial anatomy, basal tetrapod relationships and patterns of skeletal evolution. Transactions of the Royal Society of Edinburgh: Earth Sciences 87:363–427.

Coates, M. I. 2003. The evolution of paired fins. Theory in Biosciences 122:266–287.

Coates, M. I., and M. Ruta. 2007. Skeletal changes in the transition from fins to limbs; pp. 15–38 in B. K. Hall (ed.), Fins into Limbs: Evolution, Development, and Transformation. University of Chicago Press, Chicago.

Coates, M. I., J. E. Jeffery, and M. Ruta. 2002. Fins to limbs: what the fossils say. Evolution and Development 4:390–401.

Cobb, J., and D. Duboule. 2005. Comparative analysis of genes downstream of the *Hoxd* cluster in developing digits and external genitalia. Development 132:3055–3067.

Currie, P. J., and P.-J. Chen. 2001. Anatomy of *Sinosauropteryx prima* from Liaoning, northeastern China. Canadian Journal of Earth Sciences 38:1705–1727.

Daeschler, E. B., and N. H. Shubin. 1998. Fish with fingers? Nature 391:133.

Daeschler, E. B., N. H. Shubin, and F. A. Jenkins Jr. 2006. A Devonian tetrapod-like fish and the evolution of the tetrapod body plan. Nature 440:757–763.

Dahn, R. D., and J. F. Fallon. 2000. Interdigital regulation of digit identity and homeotic transformation by modulated *BMP* signaling. Science 289:438–441.

Davis, M. C., N. H. Shubin, and A. Force. 2004. Pectoral fin and girdle development in the basal actinopterygians *Polyodon spathula* and *Acipenser transmontanus*. Journal of Morphology 262:608–628.

Dollé, P., A. Dierich, M. LeMeur, T. Schimmang, B. Schubaur, P. Chambon, and D. Duboule. 1993. Disruption of the *Hoxd-13* gene induces localized heterochrony leading to mice with neotenic limbs. Cell 75:431–441.

Drossopoulou, G., K. E. Lewis, J. J. Sanz-Ezquerro, N. Nikbakht, A. P. McMahon, C. Hofmann, and C. Tickle. 2000. A model for anterioposterior patterning of the vertebrate limb based on sequential long- and short-range *Shh* signalling and *Bmp* signalling. Development 127:1337–1348.

Dudley, A. T., M. A. Ros, and C. J. Tabin. 2002. A re-examination of proximodistal patterning during vertebrate limb development. Nature 418:539–544.

Force, A., M. Lynch, F. B. Pickett, A. Amores, Y.-L. Yan, and J. Postlethwait. 1999. Preservation of duplicate genes by complementary, degenerative mutations. Genetics 151:1531–1545.

Fromental-Ramain, C., X. Warot, N. Messadecq, M. LeMeur, P. Dollé, and P. Chambon. 1996. *Hoxa-13* and *Hoxd-13* play a crucial role in the patterning of the limb autopod. Development 122:2997–3011.

Gegenbaur, C. 1878. Elements of Comparative Anatomy. Macmillan, London, 645 pp.

Gompel, N., B. Prud'homme, P. J. Wittkopp, V. A. Kassner, and S. B. Carroll. 2005. Chance caught on the wing: *cis*-regulatory evolution and the origin of pigment patterns in *Drosophila*. Nature 433:481–487.

Gould, S. J. 1977. Ontogeny and Phylogeny. Harvard University Press, Cambridge, 501 pp.

Gould, S. J., and R. Lewontin. 1978. The spandrels of San Marco and the Panglossian paradigm: a critique of the adaptationist programme. Proceedings of the Royal Society of London B 205:581–597.

Gregory, W. K., and H. C. Raven. 1941. Origin of paired fins and limbs. Annals of the New York Academy of Sciences 42:273–360.

Gregory, W. K., R. W. Miner, and G. K. Noble. 1923. The carpus of *Eryops*. Bulletin of the American Museum of Natural History 48:279–288.

Haack, H., and P. Gruss. 1993. The establishment of murine *Hox-1* expression domains during patterning of the limb. Developmental Biology 157:410–422.

Haeckel, E. 1874. Die Gastraea-Theorie, die phylogenetische Klassifikation des Tierreiches und Homologie der Keimblätter. Jenaische Zeitschrift für Naturwissenschaften 9:402–508.

Hall, B. K. 2005. Consideration of the neural crest and its skeletal derivatives in the context of novelty/innovation. Journal of Experimental Zoology, Part B: Molecular and Developmental Evolution 304:548–557.

Hall, B. K., and T. Miyake. 2000. All for one and one for all: condensations and the initiation of skeletal development. BioEssays 22:138–147.

Harfe, B. D., P. J. Scherz, S. Nissim, H. Tian, A. P. McMahon, and C. J. Tabin. 2004. Evidence for an expansion-based temporal *Shh* gradient in specifying vertebrate digit identities. Cell 118:517–528.

Harris, M. P., J. F. Fallon, and R. O. Prum. 2002. *Shh-Bmp2* signaling module and the evolutionary origin and diversification of feathers. Journal of Experimental Zoology, Part B: Molecular and Developmental Evolution 294:160–172.

Hitchcock, E. 1836. Ornithichnology: description of the foot marks of birds (Ornithichnites) on New Red Sandstone in Massachusetts. American Journal of Science and the Fine Arts 29:307–340.

Hogan, B. L. M. 1996. Bone morphogenetic proteins: multifunctinal regulators of vertebrate development. Genes and Development 10:1580–1594.

Holmgren, N. 1933. On the origin of the tetrapod limb. Acta Zoologica (Stockholm) 14:187–248.

Hopson, J. A. 1995. Patterns of evolution in the manus and pes of non-mammalian therapsids. Journal of Vertebrate Paleontology 15:615–639.

Jarvik, E. 1980. Basic Structure and Evolution of Vertebrates. Volume 1. Academic Press, London and New York, 575 pp.

Jeffery, J. E. 2001. Pectoral fins of rhizodontids and the evolution of pectoral appendages in the tetrapod stem-group. Biological Journal of the Linnean Society 74:217–236.

Ji, Q., P. J. Currie, M. A. Norell, and S.-A. Ji. 1998. Two feathered dinosaurs from northeastern China. Nature 393:753–761.

Joss, J., and T. Longhurst. 2001. Lungfish paired fins; pp. 370–376 in P. E. Ahlberg (ed.), Major Events in Early Vertebrate Evolution: Palaeontology, Phylogeny, Genetics and Development. Taylor & Francis, London and New York.

Kundrát, M. 2004. When did theropods become feathered? Evidence for pre-*Archaeopteryx* feathery appendages. Journal of Experimental Zoology, Part B: Molecular and Developmental Evolution 302:355–364.

Litingtung, Y., R. D. Dahn, Y. Li, J. F. Fallon, and C. Chiang. 2002. *Shh* and *Gli3* are dispensable for limb skeleton formation but regulate digit number and identity. Nature 418:979–983.

Long, A. D., R. F. Lyman, C. H. Langley, and T. F. C. Mackay. 1998. Two sites in the *Delta* gene region contribute to naturally occurring variation in bristle number in *Drosophila melanogaster*. Genetics 149:999–1017.

Long, M., E. Betrán, K. Thornton, and W. Wang. 2003. The origin of new genes: glimpses from the young and old. Nature Reviews Genetics 4:865–875.

Lopez-Martinez, A., D. T. Chang, C. Chiang, J. A. Porter, M. A. Ros, B. K. Simandl, P. A. Beachy, and J. F. Fallon. 1995. Limb-patterning activity and restricted posterior localization of the amino-terminal product of *Sonic hedgehog* cleavage. Current Biology 5:791–796.

Mabee, P. M. 2000. Developmental data and phylogenetic systematics: evolution of the vertebrate limb. American Zoologist 40:789–800.

Mabee, P. M., and M. Noordsy. 2004. Development of the paired fins in the paddlefish, *Polyodon spathula*. Journal of Morphology 261:334–344.

Mayr, E. 1960. The emergence of evolutionary novelties; pp. 349–380 in S. Tax (ed.), Evolution after Darwin. University of Chicago Press, Chicago.

Mayr, G., D. S. Peters, G. Plodowski, and O. Vogel. 2002. Bristle-like integumentary structures at the tail of the horned dinosaur *Psittacosaurus*. Naturwissenschaften 89:361–365.

Mercanter, N., E. Leonardo, N. Azpiazu, A. Serrano, G. Morata, C. Martinez-A, and M. Torres. 1999. Conserved regulation of proximodistal limb axis development by *Meis1/Hth*. Nature 402:425–429.

Metscher, B. D., K. Takahashi, K. Crow, C. Amemiya, D. F. Nonaka, and G. P. Wagner. 2005. Expression of *Hoxa-11* and *Hoxa-13* in the pectoral fin of a basal rayfinned fish, *Polyodon spathula*: implications for the origin of tetrapod limbs. Evolution and Development 7:186–195.

Mortlock, D. P., L. C. Post, and J. W. Innis. 1996. The molecular basis of hypodactyly (Hd): a deletion in *Hoxa 13* leads to arrest of digital arch formation. Nature Reviews Genetics 13:284–289.

Müller, G. B., and P. Alberch. 1990. Ontogeny of the limb skeleton in *Alligator mississippiensis:* developmental invariance and change in the evolution of archosaur limbs. Journal of Morphology 203:151–164.

Müller, G. B., and J. Streicher. 1989. Ontogeny of the syndesmosis tibiofibularis and the evolution of the bird hindlimb: a caenogenetic feature triggers phenotypic novelty. Anatomy and Embryology 179:327–339.

Müller, G. B., and G. P. Wagner. 1991. Novelty in evolution: restructuring the concept. Annual Review of Ecology and Systematics 22:229–256.

Nelson, C. E., B. A. Morgan, A. C. Burke, E. Laufer, E. DiMambro, L. C. Murtaugh, E. Gonzales, L. Tessarollo, L. F. Parada, and C. Tabin. 1996. Analysis of *Hox* gene expression in the chick limb bud. Development 122:1449–1466.

Neumann, C. J., H. Grandel, W. Gaffield, F. Schulte-Merker, and C. Nüsslein-Volhard. 1999. Transient establishment of anterior-posterior polarity in the zebrafish pectoral fin bud in the absence of *Sonic Hedgehog* activity. Development 126:4817–4826.

Newman, S. A., and G. B. Müller. 2000. Epigenetic mechanisms of character origination. Journal of Experimental Zoology, Part B: Molecular and Developmental Evolution 288:304–317.

Nijhout, H. F. 2001. Origin of butterfly wing patterns; pp. 511–529 in G. P. Wagner (ed.), The Character Concept in Evolutionary Biology. Academic Press, San Diego.

Ostrom, J. 1976. *Archaeopteryx* and the origin of birds. Biological Journal of the Linnean Society 8:91–182.

Post, L. C., and J. W. Innis. 1999. Altered *Hox* expression and increased cell death distinguish hypodactyly from *Hoxa13* null mice. International Journal of Developmental Biology 43:287–294.

Prum, R. O. 1999. Development and evolutionary origin of feathers. Journal of Experimental Zoology, Part B: Molecular and Developmental Evolution 285:291–306.

Prum, R. O. 2003. Dinosaurs take to the air. Nature 421:323–324.

Prum, R. O., and A. H. Brush. 2002. The evolutionary origin and diversification of feathers. Quarterly Review of Biology 77:261–295.

Prum, R. O., and S. Williamson. 2001a. Reaction-diffusion models of within-feather pigmentation patterning. Proceedings of the Royal Society of London, B 269:781–792.

Prum, R. O., and S. Williamson. 2001b. Theory of the growth and evolution of feather shape. Journal of Experimental Zoology, Part B: Molecular and Developmental Evolution 291:30–57.

Raff, R. 1996. The Shape of Life. University of Chicago Press, Chicago, 544 pp.

Richardson, M. K., and G. Keuck. 2002. Haeckel's ABC of evolution and development. Biological Reviews of the Cambridge Philosophical Society 77:495–528.

Riddle, R. D., R. L. Johnson, E. Laufer, and C. Tabin. 1993. *Sonic hedgehog* mediates the polarizing activity of the ZPA. Cell 75:1401–1416.

Rieppel, O. 1985. Ontogeny and the hierarchy of types. Cladistics 1:234–246.

Rieppel, O. 1988. Fundamentals of Comparative Biology. Birkhäuser Verlag, Basel and Boston, 202 pp.

Robertson, K. E., C. Tickle, and S. M. Darling. 1997. *Shh, Fgf4,* and *Hoxd* gene expression in the mouse limb mutant hypodactyly. International Journal of Developmental Biology 41:733–736.

Ros, M. A., R. D. Dahn, M. Fernandez-Teran, K. Rashka, N. C. Caruccio, S. M. Hasso, J. J. Bitgood, J. J. Lancman, and J. F. Fallon. 2003. The chick oligozeugodactyly (ozd) mutant lacks *Sonic hedgehog* function in the limb. Development 130:527–537.

Rosen, J. 1983. A Symmetry Primer for Scientists. John Wiley and Sons, New York, 208 pp.

Saunders, J. W. J. 1948. The proximodistal sequence of the origin of the parts of the chick wing and the role of the ectoderm. Journal of Experimental Zoology 108:363–404.

Selever, J., W. Liu, M.-F. Lu, R. R. Behringer, and J. F. Martin. 2004. *Bmp4* in limb bud mesoderm regulates digit pattern by controlling AER development. Developmental Biology 276:268–279.

Sereno, P. C. 1999a. The evolution of dinosaurs. Science 284:2137–2147.

Sereno, P. C. 1999b. A rationale for dinosaurian taxonomy. Journal of Vertebrate Paleontology 19:788–790.

Shapiro, M. D., M. E. Marks, C. L. Peichel, B. K. Blackman, K. S. Nereng, B. Jonsson, D. Schluter, and D. M. Kinglsley. 2004. Genetic and developmental basis of evolutionary pelvic reduction in threespine sticklebacks. Nature 428:717–723.

Shubin, N. H., and P. Alberch. 1986. A morphogenetic approach to the origin and basic organization of the tetrapod limb. Evolutionary Biology 20:319–387.

Shubin, N. H., E. B. Daeschler, and F. A. Jenkins, Jr. 2006. The pectoral fin of *Tiktaalik roseae* and the origin of the tetrapod limb. Nature 440:764–771.

Sordino, P., F. van der Hoeven, and D. Duboule. 1995. *Hox* gene expression in teleost fins and the origin of vertebrate digits. Nature 375:678–681.

Spitz, F., F. Gonzalez, and D. Duboule. 2003. A global control region defines a chromosomal regulatory landscape containing the *HoxD* cluster. Cell 113:405–417.

Steiner, H. 1934. Über die embryonale Hand- und Fuss-Skelett-Anlage bei den Crocodiliern, sowie über ihre Beziehungen zur Vogel-Flügelanlage und zur ursprünglichen Tetrapoden-Extremität. Revue Suisse de Zoologie 41:383–396.

Summerbell, D. 1983. The effect of local application of retinoic acid to the anterior margin of the developing chick limb. Journal of Embryology and Experimental Morphology 78:269–289.

Summerbell, D., J. H. Lewis, and L. Wolpert. 1973. Positional information in chick limb morphogenesis. Nature 244:492–496.

Sun, X., F. V. Mariani, and G. R. Martin. 2002. Functions of FGF signalling from the apical ectodermal ridge in limb development. Nature 418:501–508.

Vargesson, N., J. D. Clarke, K. Vincent, C. Coles, L. Wolpert, and C. Tickle. 1997. Cell fate in the chick limb bud and relationship to gene expression. Development 124:1909–1918.

Wagner, G. P. 1989. The origin of morphological characters and the biological basis of homology. Evolution 43:1157–1171.

Wagner, G. P., and C.-H. Chiu. 2001. The tetrapod limb: a hypothesis on its origin. Journal of Experimental Zoology, Part B: Molecular and Developmental Evolution 291:226–240.

Wagner, G. P., and H. C. E. Larsson. 2007. Fins and limbs in the study of evolutionary novelties; pp. 49–61 in B. K. Hall (ed.), Fins into Limbs: Evolution, Development, and Transformation. University of Chicago Press, Chicago.

Watson, D. M. S. 1913. On the primitive tetrapod limb. Anatomischer Anzeiger 44:24–27.

Welscher, P. T., A. Zuniga, S. Kuijper, T. Drenth, H. J. Goedemans, F. Meijling, and R. Zeller. 2002. Progression of vertebrate limb development through *SHH*-mediated counteraction of *GLI3*. Science 298:827–830.

West-Eberhard, M. J. 2003. Developmental Plasticity and Evolution. Oxford University Press, New York, 816 pp.

Westoll, T. S. 1943. The origin of the primitive tetrapod limb. Proceedings of the Royal Society of London B 131:373–393.

Xu, X., Z. Zhou, X. Wang, X. Kuang, F. Zhang, and X. Du. 2003. Four-winged dinosaurs from China. Nature 421:335–340.

Yang, Y., G. Drossopoulou, P.-T. Duprez, E. Martin, D. Brumcrot, N. Vargesson, J. Clarke, L. Niswander, A. MacMahon, and C. Tickle. 1997. Relationship between dose, distance, and time in *Sonic Hedgehog*–mediated regulation of anteroposterior polarity in the chick limb. Development 124:4393–4404.

Yokouchi, Y., S. Nakazato, M. Yamamoto, Y. Goto, T. Kameda, H. Iba, and A. Kuroiwa. 1995. Misexpression of *Hoxa-13* induces cartilage homeotic transformation and changes cell adhesiveness in chick limb buds. Genes and Development 9:2509–2522.

Zákány, J., C. Fromental-Ramain, X. Warot, and D. Duboule. 1997. Regulation of number and size of digits by posterior *Hox* genes: a dose-dependent mechanism with potential evolutionary implications. Proceedings of the National Academy of Sciences USA 94:13695–13700.

Zákány, J., M. Kmita, and D. Duboule. 2004. A dual role for *Hox* genes in limb anterior-posterior asymmetry. Science 304:1669–1672.

Zhang, J. 2003. Evolution by gene duplication: an update. Trends in Ecology and Evolution 18:292–298.

5. Incorporating Ontogeny into the Matrix: A Phylogenetic Evaluation of Developmental Evidence for the Origin of Modern Amphibians

Jason S. Anderson

ABSTRACT

The question of the evolutionary origin of modern amphibians (frogs, salamanders, and caecilians, collectively known as Lissamphibia) is among the most controversial subjects in systematic paleontology. The controversy stems from two questions: do lissamphibians share a single evolutionary (i.e., monophyletic) origin or do they descend from multiple fossil lineages, and among which of the morphologically dissimilar fossil taxa is this origin (or are these origins) found? This controversy is driving research and marshaling evidence from the cutting edge of current science, including evolutionary developmental biology (colloquially known as evo-devo). Three lines of developmental data—the pattern of vertebral development, the sequence of cranial development, and the inferred presence of a larval stage–have recently been offered as support for the independent derivation of frogs from within amphibamid temnospondyls, salamanders from within branchiosaurid temnospondyls, and caecilians from within "microsaurian" lepospondyls. To date, however, these data have not been included in a phylogenetic analysis, and thus these assertions of relationship remain untested.

This chapter evaluates these data from a phylogenetic perspective and considers their impact on the inference of the origin of lissamphibians. The described similar pattern of cranial ossification in salamanders and branchiosaurs is compared with the recently described pattern in the aistopod *Phlegethontia,* which suggests a broader distribution of these similarities than previously assumed. Finally, these data, where appropriate, are incorporated for the first time into a new matrix of morphological characters for lepospondyls and dissorophoid temnospondyls and subjected to parsimony analysis. The result of the analysis supports the polyphyly hypothesis, with frogs and salamanders (Batrachia) placed as sister-taxa to the amphibamid *Doleserpeton* (not with branchiosaurs), whereas caecilians are found to be a sister-group to brachystelechid "microsaurs," with *Rhynchonkos* placed as outgroup to this dichotomy. The enigmatic albanerpetontids are placed as outgroup to salamanders and their stem. The clade formed by ostodolepidids, pantylids, gymnarthrids, brachystelechids, *Rhynchonkos,* and *Eocaecilia* (and presumably all caecilians) is given the new name Recumbirostra. This analysis suggests that neoteny was a frequent lifestyle adaptation of dissorophoid temnospondyls, evolving in multiple, disparate instances driven by the environment, similar to the distribution of this feature in extant salamanders. It also underscores the importance of phylogenetic analysis in interpreting important features that are not known in a majority of the taxa under study.

Introduction

> Temnospondyls, nectrideans, the "reptilian microsaurs,"
> and *Tuditanus* all possess some characters in common
> with the modern amphibians, but they all have notable
> differences. Therefore we consider the problem [of
> lissamphibian origins] to be insoluble until further
> evidence on the structure of the Paleozoic amphibians, and
> especially of forms transitional to the modern amphibians,
> is found. At present only guesses are possible.
> —PARSONS AND WILLIAMS (1963:48)

Frogs (Anura), salamanders (Caudata), and caecilians (Apoda) are all that remain of the vast diversity of nonamniote tetrapods that existed since the origin of limbs with digits sometime during the Middle to Late Devonian Period, over 370 million years ago. Although all modern amphibians (lissamphibians) are small, many fossil amphibians grew to fairly large size and occupied extremely specialized ecological niches. These fossil amphibians (most of which were formerly united in the "labyrinthodonts," a group now recognized to be paraphyletic) are morphologically distinctive, and, for the most part, it is easy to distinguish between their major clades. However, none of their characteristic distinctiveness is recognizable within lissamphibians because of the latter's highly specialized anatomy and patchy fossil record. As a result, there has always been great uncertainty with respect to which fossil group, or groups, might have given rise to modern amphibians.

Throughout the literature, just about every imaginable pattern of relationship possible to propose for frogs, salamanders, and caecilians has been proposed. For the sake of brevity, this paper will focus on the hypotheses that are currently being debated (for a detailed discussion of the history of this problem, see Schoch and Milner, 2004). Three hypotheses dominate ongoing discussions: The Temnospondyl Hypothesis (TH), the Lepospondyl Hypothesis (LH), and the Polyphyly Hypothesis (PH).

History of the Lissamphibian Problem

As traditionally defined, Temnospondyli is a diverse assemblage of amphibians that lived from the Early Mississippian (350 million years ago) to the Early Cretaceous (about 100 million years ago). They are united by many features, some relevant to the origin of lissamphibians (Holmes, 2000). Dissorophoids are a subset of this larger grouping and are considered primitively terrestrially adapted on the basis of anatomical and depositional evidence (Fig. 5.1). Dissorophoidea is composed of several groups: Dissorophidae, characterized by a midline row of scutes dorsal to the vertebral neural arches (Williston, 1910; DeMar, 1966; Bolt, 1974b); Trematopidae, medium-sized, armorless amphibians with a distinctive posteriorly extended external naris (Olson, 1941; Bolt, 1974a; Berman et al., 1987; Dilkes and Reisz, 1987); Amphibamidae, small forms with a skull table of reduced length (Carroll, 1964; Bolt, 1969; Daly, 1994); and the Branchiosauridae and Micromelerpetontidae, neotenic forms of facultatively, and possibly obligatorily, aquatic amphibians (Boy and Sues, 2000).

Lepospondyls are a heterogeneous, monophyletic group of small amphibians united by many derived features (Carroll et al., 1998; Carroll, 2000a; Anderson, 2001). There are five groups of lepospondyls (only three of which are directly relevant to the question of lissamphibian origins): Nectridea, Adelospondyli, Microsauria, Aistopoda, and Lysorophia (Fig. 5.2). Nectrideans are mostly aquatic, newt-like animals with distinctive vertebrae with crenulated neural and hemal spines (Milner, 1980; Bossy and Milner, 1998). Adelospondyls are incompletely known, somewhat elongate forms known only from the Lower Carboniferous of Scotland (Andrews and Carroll, 1991; Carroll and Andrews, 1998). "Microsaurs" are small salamander-like forms, so named because they were frequently mistaken for small reptiles by early researchers (Carroll and Gaskill, 1978; Carroll, 1998a). The monophyly of this group has recently come into question (Anderson, 2001; Ruta et al., 2003). Aistopods are elongate, limbless forms, similar in vertebral regionalization with extant limbless anguid lizards (Anderson, 2002b, 2003a,b; Anderson et al., 2003). Finally, lysorophids are a group of elongate amphibians with reduced limbs and highly reduced skulls (Wellstead, 1991, 1998).

Historically, modern amphibians were not regarded to have arisen from a singular evolutionary event. Romer (1945), for instance, considered frogs descendants of such forms as the amphibamid temnospondyl, *Amphibamus*, but regarded salamanders and caecilians as lepospondyls with highly reduced (salamanders) or highly consolidated (caecilians) cranial ossification. However, not

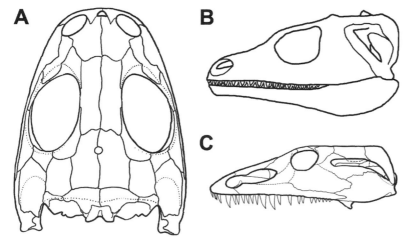

Figure 5.1. Various dissorophoid temnospondyls. **A**, *Doleserpeton*, an amphibamid (redrawn from Bolt, 1969). **B**, *Cacops*, a dissorophid (redrawn from Williston, 1910). **C**, *Acheloma*, a trematopid (redrawn from Dilkes and Reisz, 1987).

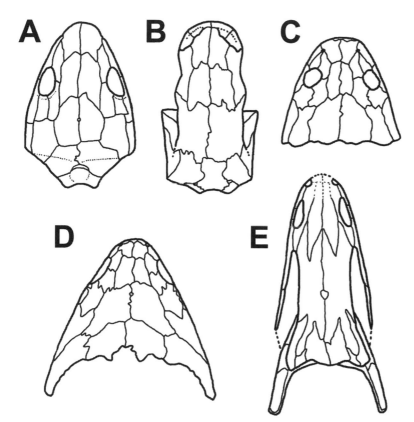

Figure 5.2. Various lepospondyls. **A**, *Cardiocephalus*, a gymnarthrid "microsaur" (redrawn from Carroll and Gaskill, 1978). **B**, *Brachydectes*, a lysorophian (redrawn from Wellstead, 1991). **C**, *Scincosaurus*, a scincosaurid nectridean (redrawn from Bossy and Milner, 1998). **D**, *Batrachiderpeton*, a diplocaulid nectridean (redrawn from Bossy and Milner, 1998). **E**, *Coloraderpeton*, an aistopod (redrawn from Anderson, 2003a).

all authors accepted this view. In 1963, Parsons and Williams published a seminal paper on lissamphibians, documenting a number of uniquely shared features that they used to argue for a single evolutionary origin (Table 5.1). Following this study, most authors accepted the monophyletic status of lissamphibians, and with the publication of the putative "protolissamphibian" *Doleserpeton* (an

TABLE 5.1

Characters cited by Parsons and Williams (1963) as evidence for the monophyletic status of Lissamphibia. Characters that are potentially observable in fossils are indicated in boldface type

---

- **Pedicellate teeth**
- **Operculum-plectrum complex**
- Papilla amphibiorum
- Green rods
- Structure of m. levator bulbi
- Fat bodies associated with gonads
- Structure of skin glands
- Cutaneous respiration
- Chromosomes and DNA content

amphibamid temnospondyl) by Bolt (1969), a consensus was seemingly reached: modern amphibians have a monophyletic origin within dissorophoid temnospondyls (Bolt, 1969; Milner, 1988; Bolt, 1991; Trueb and Cloutier, 1991; Milner, 1993; Duellman and Trueb, 1994; Gardner, 2001). This hypothesis (TH) is supported by several synapomorphies: broad openings on the palate (interpterygoid vacuities), a large embayment on the squamosal that supported the tympanic membrane ("otic notch"), and most importantly, labiolingually oriented bicuspid, pedicellate teeth (with a zone of poor ossification between tooth crown and base).

Despite the amount of evidence and seeming consensus, there are outstanding problems not explained by the TH, and it has never received universal acceptance. Of all of the characters offered as evidence of lissamphibian monophyly by Parsons and Williams (1963), few are comprised of hard tissues, which would make them observable in potential fossil sister-groups (Table 5.1). Carroll and Holmes (1980) documented differences in the arrangement of jaw adductor musculature in frogs and salamanders, which they correlated to the presence of the large otic notch in frogs. Given the differences in the dominant adductor subgroups between the two taxa, they found a temnospondyl origin for salamanders unlikely. Carroll and Holmes suggested that "microsaurs" provided a better model because they (and all lepospondyls) lack an otic notch, and thus would have the requisite arrangement of jaw adductor musculature.

Carroll and Currie (1975) made comparisons between the "microsaur" *Rhynchonkos* (then known as *Goniorhynchus*) and caecilians, demonstrating close correspondence. A "microsaurian" origin for caecilians requires that one of the uniting features of lissamphibians—a large orbitotemporal opening—is derived independently within the caecilian lineage and is not shared among all lissamphibians uniquely. This prediction was confirmed by the discovery of the Early Jurassic caecilian *Eocaecilia* (Jenkins and Walsh, 1993). It shares most of the unique caecilian anatomy, including a tentacular foramen, a long retroarticular process, and an internal process on the pseudangular (both used in the unique caecilian jaw depressor musculature). However, it retains primitive features, including a solid skull roof (thus lacking a orbitotemporal opening), limbs, and a shorter vertebral column, which demonstrate its position as an outstanding transitional form between modern caecilians and "microsaurs" (Carroll, 2000b).

A more recent fossil caecilian discovery, the Early Cretaceous *Rubricacaecilia* (Evans and Sigogneau-Russell, 2001), has reinforced the primitive status of many of these characters. Significantly for discussions of lissamphibian origins, *Eocaecilia* and *Rubricacaecilia* possess an anteriorly directed process on the atlas centrum, variously called an *odontoid process* or *tuberculum interglenoideum*. Similar processes are also present on the atlantes of both lepospondyls and salamanders (Duellman and Trueb, 1994; Anderson, 2001).

These studies represent an early formulation of the PH. More recently, a series of papers have been published drawing comparisons between patterns of development seen in modern amphibians and what can be inferred from growth series of extinct taxa (Schoch, 1992, 2002a; Carroll, 1998b, 2001, 2004; Schoch and Carroll, 2003; Carroll et al., 2004). These investigations, discussed in detail below, conclude that frogs share more developmental similarities with amphibamid temnospondyls, salamanders with branchiosaurid temnospondyls, and caecilians with "microsaurian" lepospondyls. The data from ontogeny, although exciting in their innovative approach to the question of lissamphibian origins, have not yet been incorporated into a phylogenetic analysis, and thus it cannot be said whether they represent evidence of recent shared common ancestry.

Cladistic methods of phylogenetic inference came to this subject relatively recently. Up until the mid-1980s, authors simply highlighted a few characters they felt had special importance to their ideas of relationship, resulting in the previously mentioned proliferation of phylogenetic hypotheses. By the mid-1980s, the cladistic methodology of using shared derived features as evidence of evolutionary relationship became more widely accepted, but the numbers of taxa and characters analyzed (usually by hand) were still very limited (Milner, 1988; Bolt, 1991; but see Trueb and Cloutier, 1991).

Computer-assisted phylogenetic inference allows for simultaneous consideration of large numbers of characters rather than limited numbers of key features, and it directly led to the current controversy. In 1997, Laurin and Reisz published an analysis that suggested that lissamphibians had their origins not within temnospondyls, as most researchers argued at the time, but within lepospondyls, specifically as sister-taxa to the derived lysorophians. This study was more inclusive in the number of taxa analyzed compared with earlier analyses performed by hand; however, phylogenetic analysis is highly sensitive to the number of taxa and characters included (Gauthier et al., 1988; Graybeal, 1998; Halanych, 1998; Poe, 1998; Whiting, 1998; Anderson, 2001), and Laurin and Reisz did not include a wide number of either lepospondyls or temnospondyls. This criticism was addressed to some extent by Laurin (1998), but taxonomic sampling density among temnospondyls and lepospondyls remains relatively low, even in the most recent iteration of this matrix (Vallin and Laurin, 2004). Furthermore, the evidence linking lissamphibians with lysorophians in Laurin and Reisz (1997) are absence characters rather than unique derived features, a point of some criticism (Ruta et al., 2003; Schoch and Milner, 2004).

Anderson (2001) produced what remains the most detailed analysis of lepospondyl relationships, including nearly every described lepospondyl species. He found a close relationship between *Eocaecilia* and *Rhynchonkos,* both of which were sequentially more basal to a small group of derived "microsaurs" known as Brachystelechidae. Anderson's study was not intended as a test of lissamphibian monophyly; only caecilians were included and a comparable series of temnospondyls was lacking because his primary purpose was a comprehensive examination of lepospondyl relationships. He was struck by the positive characters supporting a caecilian-"microsaur" relationship and speculated whether the PH might be true (considering the strength of the evidence supporting the TH) rather than giving full support to the LH (Anderson, 2002a). McGowan (2002) published a small analysis that found polyphyly, although in a heterodox arrangement of taxa with "microsaurs" and lissamphibians, rendering temnospondyls paraphyletic. It is the only published cladistic analysis that has supported the PH until the present study.

The most comprehensive cladistic analysis of extinct and extant amphibian-grade tetrapods to date is that of Ruta et al. (2003). Ruta et al. found an overall pattern of relationships similar to those proposed by Laurin and Reisz and by Anderson, respectively, with lepospondyls as sister-group to amniotes and a more basally placed Temnospondyli, but they recovered the traditional TH. This result has come under some criticism (Vallin and Laurin, 2004), specifically that the TH is not more statistically significant than the LH using the data of Ruta et al.; nevertheless, the 2003 paper by Ruta et al. remains the most exhaustive treatment of the subject. Despite this, a number of characters related to the origin of lissamphibians are not included in this matrix, including developmental features, so the question of lissamphibian origins is far from resolved by their study.

Molecular-based phylogenies are also uncertain. Early work that used nuclear and mitochondrial rRNA suggested that caecilians and salamanders are sister-taxa (Procera), to the exclusion of frogs (Hedges and Maxson, 1993; Hay et al., 1995; Feller and Hedges, 1998). More recent studies that used complete mitochondrial genomes and *RAG1* nuclear gene sequences have recovered the Batrachia arrangement with frogs and salamanders as sister-taxa, to the exclusion of caecilians (Zardoya and Meyer, 2001; San Mauro et al., 2004, 2005; Zhang et al., 2005). All molecular phylogenies published to date have found modern amphibians to be a monophyletic group, although this might be an artifact of a short period of divergence between modern amphibian groups and amniotes and subsequent long branches (Lee and Anderson, 2006).

Evolutionary Developmental Biology

Some of the most exciting new data to be introduced to the debate over lissamphibian origins come from the newly emerging discipline of evolutionary developmental biology (colloquially known as evo-devo). Three lines of developmental data have been offered as evidence for the shared origins of frogs, salamanders, and temnospondyls: vertebral development, cranial development, and the presence of a biphasic lifestyle.

Vertebral Development

Carroll (1998b, 2001; Carroll et al., 2004) has discussed the patterns of vertebral development in Paleozoic tetrapods and compared them with what is seen in lissamphibians. By studying growth series of branchiosaurid and amphibamid temnospondyls, Carroll documented a pattern of cranial to caudal ossification, where the neural arches would form, followed by a pause, and finally the centra would ossify. This is the same pattern Carroll documented in frogs and some basal salamanders, although the picture in salamanders turns out to be more complex (C. Boisvert, pers. comm.). On the other hand, of the lepospondyls for which growth series (or at least small individuals) exist (the "microsaurs" *Hyloplesion, Saxonerpeton,* and *Utaherpeton* and the aistopods *Oestocephalus* and *Phlegethontia*), the entire vertebra is already completely ossified at a very small size, and all elements appear simultaneously (Baird, 1965; Carroll and Gaskill, 1978; Carroll, 2001; Anderson, 2002b, 2003a). Caecilians also follow this pattern of vertebral development.

Carroll has argued that this line of evidence supports the PH; salamanders and frogs have a pattern of vertebral development identical to temnospondyls, whereas lepospondyls and caecilians are direct developers (dropping their external gills very soon after hatching, if they had them at all). However, with the description of the outgroup taxon *Eusthenopteron* as having a vertebral development pattern of arches first, then centra like that seen in frogs, basal salamanders, and temnospondyls (Cote et al., 2002), part of Carroll's interpretation has been weakened because this pattern is probably the primitive condition for all tetrapods. On the other hand, his proposed sister-group relationship between caecilians and lepospondyls is strengthened because the "all at once" pattern is a derived condition and a possible synapomorphy linking the two groups.

Until now, this character has not been included in phylogenetic analysis. The major difficulty in doing so is the limited knowledge of ontogeny in most Paleozoic tetrapods; very few of the taxa in a typical analysis will be coded for this character. However, those taxa that can be coded have been (Appendix 5.1, character 197), and as more information about ontogeny becomes available in the future for other species, the matrix will be amended.

Cranial Ossification

The pattern of cranial ossification has been offered recently as another line of evidence supporting the PH. In a series of extremely informative papers, Schoch and colleagues (Schoch, 1992, 2002a, 2004; Schoch and Carroll, 2003; Carroll, 2004; Carroll et al., 2004) have documented the pattern of ossification within a genus of branchiosaurid temnospondyl, *Apateon,* and made close comparisons with the pattern of ossification seen in modern salamanders. *Apateon* is found in very fine-grained lacustrine sediments that preserve some soft tissues. By examining a growth series of hundreds of specimens representing sizes assumed to span from "hatching" to "postmetamorphosis" (although defining metamorphosis by itself is difficult, let alone documenting it in the fossil record), Schoch was able to document the relative timing of ossification of the bones of the skull, with skull midline length used as his proxy for time. Additionally, because of the quality of the preservation, Schoch could identify bone primordia and initial centers of ossification.

When compared with the pattern of development seen in primitive salamanders (represented by *Ranodon*), close similarities were seen in the location of the initial centers of ossification (Schoch, 1992, 2002a; Schoch and Carroll, 2003). The sequence of ossification is broadly similar, with the initial appearance of the palatal and marginal tooth-bearing elements, followed by the midline bones, and ending with the circumorbital elements (Table 5.2). Initially, there is a long gap between maxilla and jaw suspension (open lower temporal bar), which is closed later in ontogeny. Although the adult *Apateon* has a fully ossified skull, like all other temnospondyls, the juvenile is similar enough to the adult salamander that it has been suggested that salamanders may have arisen through paedomorphosis, or the onset of sexual maturity and cessation of growth in a descendant lineage at a morphological stage similar to juveniles of more ancestral lineages, specifically branchiosaurid temnospondyls (Schoch, 1992, 2002a; Schoch and Carroll, 2003; Carroll, 2004; Carroll et al., 2004).

More recently, supporters of the PH have stated that "the sequence of ossification of the skull bones of conservative extant salamanders is *identical* with that of the early larval stages of the branchiosaur *Apateon*" (Carroll et al., 2004:321; emphasis mine). This statement overlooks several significant differences noted in detail in Schoch (2002a), summarized briefly here (Table 5.2). The pterygoid and prearticular bones make a later appearance in *Ranodon*. The maxilla appears late in the ontogenetic sequence of *Ran-*

TABLE 5.2

Comparison of cranial ossification sequences across several taxa

Event	*Apateon*	Salamanders	*Phlegethontia*	*Canis*
Palatal tooth-bearing elements	Early	Early	Early	NA
Marginal tooth-bearing elements	Early	Early	Early	Early
Skull roof	Early-intermediate	Early-intermediate	Early-intermediate	Early-intermediate
Circumorbitals	Late	Late	Late	Intermediate-late
Neurocranium	Late	Early	Early	Late
Gap between maxilla and suspensorium	Initial	Persistent	Initial	Initial
Maxilla	Early	Late	Early-intermediate	Early
Prearticular	Early	Late	NA	NA

odon, but early in *Apateon.* A number of early-appearing bones (e.g., the ectopterygoid and supratemporal) are absent in salamanders, as are several late-appearing bones (postfrontal, postorbital, jugal, tabular). Most importantly, the neurocranium ossifies early in salamanders, but it rarely ossifies extensively in temnospondyls.

Of even greater concern for this hypothesis is the fact that data on the sequence of ossification of cranial bones are extremely limited in fossil amphibians. Salamanders are fairly easy to study; one can raise them in a laboratory, and in very small growth increments, one can sacrifice, clear, and stain the growing larvae to document the condensation of cartilage or centers of ossification. Documenting the cranial ossification sequence in fossil taxa requires the simultaneous presence of many conditions, and this combination is rarely met in the fossil record. The organisms must have lived in an exceptional depositional environment. There must be anoxic conditions to prevent decay of soft tissues and suppress scavengers or organisms that disturb the bottom sediments (and anything contained therein). It must be an extremely low-energy depositional environment to ensure that the organisms are covered with the fine-grained sediment that is required to preserve details of soft tissue. The depositional environment must preserve a wide range of ontogenetic stages; should different-sized animals occupy differing environments, the growth series may be truncated, or it may appear to make leaps from one stage to the next. Additionally, the timing of ossification must be sufficiently slow so that discrete steps can be discerned with increased body size; this is an extremely important point and may be the limiting factor in how strongly the cranial ossification evidence described above can be taken as supporting statements of relationship.

Growth series of some other fossil amphibians have been described, mostly restricted to micromelerpetontid and amphibamid temnospondyls and seymouriamorphs (another group of Paleozoic tetrapods), but they do not inform the sequence of cranial ossification (Boy and Sues, 2000 and references therein; Witzmann and Pfretzschner, 2003). The immediate outgroup to dissorophoids, *Sclerocephalus,* is also uninformative in terms of ossification sequence (Schoch, 2003). All previously described ontogenetic series of fossil amphibians found in the literature other than the two species of *Apateon* have all elements of the skull present, even at the smallest sizes. These studies are restricted to examining changes in relative proportions of cranial regions (allometry) and ornamentation and can say nothing about ossification sequences because, presumably, the sequence of ossification takes place within too short a period for it to be documented in the fossil record.

There is, to my knowledge, only a single other described sequence of ossification in a fossil amphibian, the aistopod, *Phlegethontia* (Anderson, 2002b; Anderson et al., 2003). Despite working with a significantly smaller sample size (eight specimens rather than the hundreds used by Schoch, 1992), several details were described (Fig. 5.3). *Phlegethontia* is broadly similar to *Apateon* and salamanders in a number of respects (Table 5.2). Palatal and marginal elements ossify early in all three groups. The frontal bone is completely ossified early, and the squamosal bone begins ossifying after the frontal is complete. The circumorbital bones form late in the sequence. Finally, one of the characters that most strongly supports an origin of salamanders through an early truncation of growth, the early gap between maxilla and jaw suspension, is also seen in *Phlegethontia*. Of course, *Phlegethontia* has a highly derived cranial structure, and it, like salamanders, has lost many bones so that exact comparison with *Apateon* is impossible.

What is striking, however, is what one finds when one compares the differences in ossification sequence between *Apateon* and salamanders (outlined above) with the sequence seen in *Phlegethontia* (Table 5.2). The maxilla appears late in *Ranodon* but early in *Apateon;* it appears early in *Phlegethontia,* but is highly incomplete at that stage and only grows to its adult proportions later in ontogeny. In this feature, *Phlegethontia* appears to be intermediate between the two conditions. *Phlegethontia* shares the absence of many bones with *Ranodon,* but it is difficult to establish the homology of these loss characters. On the positive side, *Ranodon* shares with *Phlegethontia* the early ossification of the neurocranium. In many respects, salamanders appear to share more of the cranial development sequence with aistopods than with *Apateon*.

This comparison underscores the importance of phylogenetic constraints on evolutionary developmental scenarios. The fact that such phylogenetically distant species as aistopods and branchiosaurs (Laurin and Reisz, 1997; Anderson, 2001; Ruta et al., 2003) share so many similarities suggests that the sequence of cranial ossification is stereotyped within tetrapods—if not vertebrates—as a whole. This observation is supported when one considers a broader sample of vertebrates.

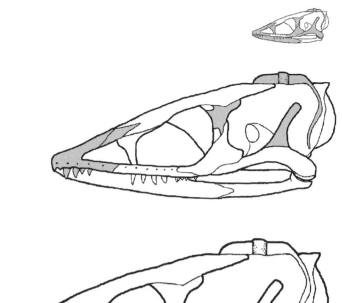

Figure 5.3. Sequence of cranial ossification in the aistopod *Phlegethontia,* drawn to scale. Modified from Anderson et al. (2003).

Frogs have a pattern of cranial development that is strongly influenced by larval feeding adaptations and is not comparable (Duellman and Trueb, 1994; Carroll, 2004). Little is known about the pattern of development of the direct developing caecilian or salamander skull (Rose and Reiss, 1993). The dog, however, is a well-studied vertebrate because of its economic and emotional importance to people, and, unlike for many other vertebrates, there are ample developmental data available. Many studies of ossification have been conducted, summarized by Evans (1993). When the pattern of ossification is examined, one sees many of the same patterns observed in modern and fossil amphibians (Table 5.2), although the high degree of bone fusion or loss of individual elements in the mammalian skull obscures details. The first bones to ossify are the tooth-bearing elements (maxilla, dentary) and the frontal. These are followed by the parietal, palatine, jugal (zygomatic), premaxilla (incisive), and nasal. Next to ossify are the temporal, pterygoid, and shortly thereafter the lacrimal (representing, in part, the circumorbitals). In the dog, as in the temnospondyl, *Apateon,* the neurocranium is a late-forming structure. Is this the primitive or derived condition? It is impossible to say from the developmental sequence data alone, but the preponderance of the sequence is shared and thus is presumably primitive. Given the vast phylogenetic separation between salamanders and dogs, it is reasonable to conclude that much of the sequence of cranial ossification is highly conserved in tetrapods, if not vertebrates.

From this, it seems the majority of the shared sequence of ossification between salamanders, branchiosaurs, and aistopods may not be shared derived features—evidence of recent common ancestry—but primitive features that are evident only because of extraordinary depositional conditions and a shared slowed rate of cranial development. That is not to say there is no potential phylogenetic signal within these developmental sequence data; there are, after all, numerous differences between all three that support various groupings of the taxa. However, it is impossible to say which are derived and which are primitive without a robust phylogeny on which to map these characters, and without more complete sampling of sequences within other extinct and extant species.

That being said, incorporating developmental sequence data, such as the sequence of cranial ossification, into phylogenetic analysis is seriously problematic. Unlike larval characters, which can be added to a matrix with adult characters by using a total-evidence approach (Steyer, 2000), sequence data are confounded by other considerations. One cannot compare features between species at the same chronological age or size, as both correlate poorly with developmental maturity (Bininda-Edmonds et al., 2002; Jeffery et al., 2002). For this reason, developmental biologists have relied on ontogenetic stages, which are established by the occurrence of a key developmental event (or events). However, it is extremely difficult to compare two species at the same developmental stage because shifts in the timing of developmental events (*sequence heterochrony;* Smith, 2001) used to delineate developmental stages may transpose events from species to species or cause them to occur simultaneously (Richardson et al., 2001). Stages are also problematic because although the two landmarks establishing the beginning and end of a stage may be present, the intervening steps from one to another may be completely different, which also calls into question whether the end state is truly homologous (Bininda-Edmonds et al., 2002). Ultimately, the decision to use a particular developmental event (or events) to define a stage is arbitrary, with little biological meaning. The selection of defining stages is complicated further by intraspecific variation (Bininda-Edmonds et al., 2002). Finally, during development, some characters undergo transformation and may completely disappear (Richardson et al., 2001).

Despite these difficulties, a few methods have been proposed to analyze developmental sequence data in a phylogenetic context (reviewed in Smith, 2001; Bininda-Edmonds et al., 2002). The most common current approach is *event-pairing* (Fig. 5.4a), which breaks down a developmental sequence into paired events (say, event 1, then event 2), and the order of their occurrence in different species can then be compared when mapped on a phylogenetic tree. Should event 2 precede event 1 in taxa within a subclade on the tree

A

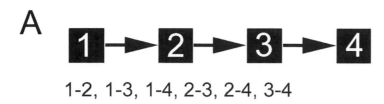

1-2, 1-3, 1-4, 2-3, 2-4, 3-4

B

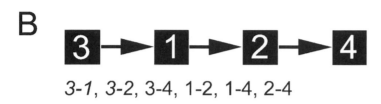

3-1, *3-2*, *3-4*, 1-2, 1-4, 2-4

C

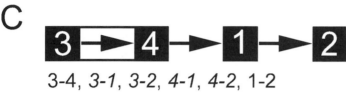

3-4, *3-1*, *3-2*, *4-1*, *4-2*, 1-2

Figure 5.4. Sequence heterochrony. **A**, generalized developmental sequence, with each event indicated by a number. The event pairs are given below. **B**, a sequence heterochrony where a single event (3) is shifted to earlier in the sequence. Below in italics are the alternate event pairs. A single-sequence heterochrony produces two event pair changes in this example. **C**, a sequence heterochrony where a single event (3) is shifted earlier, causing a shift in another event (4) related to the former by induction or pleiotropy, with the change in event pairs in italics below. A single-sequence heterochrony in this example produces four changes in event pairs. The more intricate the interactions in a developmental sequence, the more likely a single-sequence heterochrony will cause a large number of nonindependent changes in event pairs, which will bias phylogenetic analysis.

in an instance of sequence heterochrony, it may be considered a shared derived developmental change. A modified version of this method, *event-pair cracking,* has recently been used in finding the phylogenetic tree (Jeffery et al., 2002), rather than being restricted to a posteriori character analysis using a previously generated tree.

However, there are concerns with these methods—concerns primarily centering on the lack of character independence (Schulmcister and Wheeler, 2004). Shifting the timing of an event to an earlier position in the sequence will make it demonstrate a shift with respect to many of the events with which it pairs, when only a single sequence heterochrony actually occurred (Fig. 5.4b). Additionally, some developmental events depend on other events occurring earlier for them to be expressed; thus, a single shift may cause multiple features related to it through pleiotropy or induction also to shift (Fig. 5.4c). This would be reflected in changes within numerous event pairs and would bias a phylogenetic analysis.

A new method has recently been proposed that avoids these issues by considering the entire developmental sequence as a single character. This single character can have changes occur after an edit cost function, where the event pair changes are weighted a priori by a procedure similar to a step matrix (Schulmeister and Wheeler, 2004). This new method shows promise. However, given the highly restricted distribution of taxa for which information about cranial ossification is available and the intensive programming and computation required by this method, I have decided to forego including these data in my phylogenetic analysis.

Metamorphosis

The third new piece of developmental evidence is the distribution of metamorphosis (Carroll, 2001, 2004; Schoch and Carroll, 2003). The inference of the presence of a larval stage in fossil tetrapods comes from the slow sequence of vertebral ossification and "clear differences in the degree of ossification between young and adults" (Carroll, 2004:159) and the presence of external gills. Because no lepospondyls are known to have fossilized external gills despite coming from deposits that preserve them, it is argued that they are direct developers and can have nothing to do with the ancestry of frogs and salamanders. Caecilians do possess external gills, but they rapidly drop off after hatching, so this is not seen to be an impediment to the PH (Carroll, 2004).

But what exactly is metamorphosis? There are many definitions for metamorphosis, from a discretely different mechanism of development at a particular stage, to a phase of development where there is a shift in ecological habit and feeding (Rose and Reiss, 1993). Rose and Reiss observe that although most workers are explicit about restricting metamorphosis to a change from larval to juvenile stages of life history, other periods of life history can also show dramatic change. They conclude that metamorphosis is a metaphorical construct used to describe a period with a concentration of developmental events.

Schoch (2002b) explicitly restricts his discussion of the evolution of metamorphosis in temnospondyls to the ecological definition; Carroll (2004) uses this same definition implicitly. Recognizing that metamorphosis cannot be directly observed in the fossil record, Schoch, building on the work of previous researchers, offered a list of seven morphological correlates to metamorphosis that could be observed: (1) terrestrial adaptations seen in the postcranial skeleton, (2) disappearance of lateral line sulci, (3) loss of branchial denticles in larger specimens, (4) changes in the morphology of the choana, (5) disappearance of ceratohyals, (6) changes in the structure and size of the septomaxilla, and (7) closure of the nasolacrimal canal. Many of these characters are included in the following phylogenetic analysis (presence of branchial denticles and further elaboration of these into gill rakers, presence of lateral line canal grooves during ontogeny, ossified hyoids; Table 5.3) but without ontogenetic discrimination; if the state occurs at any stage of development, it is scored present. This is in part due to the very limited number of taxa with described ontogeny; without knowledge of the polarity of the ontogenetic change, it is impossible to say that they are shared derived features. On the other hand, the complete absence of these features in any known stage is derived and can support the clustering of taxa on the basis of shared recent common ancestry. From the distribution of characters and the tree

TABLE 5.3
Characters used in phylogenetic analysis that have developmental component

Character	Description
14	Quadratojugal-jugal
15	Quadratojugal-maxilla
47	Dermal sculpturing
60	Teeth
63	Labyrinthine infolding
73	Palatal teeth
80	Palatal teeth
111	Ossified hyoids
159	Scapulocoracoid ossification
162	Deltopectoral crest
163	Supinator process
166	Olecranon process
167	Podial ossification
173	Femoral shaft
174	Femur
196	Gill osteoderms
197	Vertebral development

they produce, the interested investigator may infer the evolutionary process explaining the distribution.

I believe that simply documenting derived changes in morphology is the most conservative approach to best optimize ontogenetic information and phylogenetic signal. This is in part due to the distinction drawn between pattern and process in modern phylogenetics; metamorphosis is recognized by possession of correlates to metamorphosis and is inferred rather than observed in the fossil record. More importantly, the correlates themselves can evolve, and heterochrony can further obscure interpretations of metamorphosis by eliminating part, or all, of the metamorphic event (Rose and Reiss, 1993). This further justifies my decision to deal with Schoch's anatomical correlates of metamorphosis as individual characters; the shared loss of one correlate among a group of taxa is a potential synapomorphy. The evolutionary scenario should be read from the phylogenetic distribution of character transformations, and not vice versa, because the individual characters which might be used to define metamorphosis can change independently of the others. As Rose and Reiss (1993:306) put it:

> This diversity of ontogenetic patterns within metamorphic groups well illustrates the evolutionary lability of metamorphosis, as well as the fallacy of defining metamorpho-

sis on the basis of character-specific criteria alone. All patterns of cranial remodelling are relative in their scope and intensity, and how one defines the set of events constituting a "complete" or "true" metamorphosis is essentially a taxonomic or cladistic issue.

This approach also allows for differing interpretations of anatomical evidence. The "microsaur" *Microbrachis,* for instance, was previously described as having branchial denticles associated with external gills (Carroll and Gaskill, 1978), despite the lack of direct evidence of gills being preserved in any of the known specimens. However, this interpretation of the evidence seems to have been abandoned (Carroll, 2004), although it is still plausible. Including the presence of small ossicles in a position likely to be related with gills does not render a judgment that *Microbrachis* might have been a metamorphic or neotenic lepospondyl; that is achieved by the presence of several correlates of metamorphosis (i.e., remnants of the lateral line canal system also present) within the context of a phylogeny. I have coded branchial denticles as present in *Microbrachis* pending detailed analysis of this feature.

Carroll (2004) also distinguished between the branchial denticles of branchiosaurids and all other temnospondyls. Branchiosaurs, he argued, have narrowly triangular osteoderms that interdigitate, forming a seal necessary for gape-and-suck feeding, whereas other temnospondyls have oval disks with multiple denticles "that could not have served to seal the gill slits" (Carroll, 2004:169). Other authors have called all branchial denticles *gill rakers,* and it appears that at least one saurerpetontid temnospondyl from Mazon Creek might also have narrowly triangular branchial osteoderms (Milner, 1982). I have maintained Carroll's distinction in my matrix, which biases this analysis toward finding a sister-group relationship between salamanders and branchiosaurs.

New Character Evidence

Pedicellate teeth with two labiolingually oriented cusps at some life stage are only known in lissamphibians, branchiosaurs, and some amphibamids, making this feature the strongest evidence for the TH. However, recently described lepospondyls are starting to challenge this view by demonstrating the presence of bicuspid teeth with labiolingually oriented cuspules in taxa closely associated with the origin of some lissamphibians.

Brachystelechid "microsaurs" are united with caecilians in Anderson's (2001) matrix by a similar degree of cranial consolidation and by having multicuspid teeth, which was thought to be a unique feature among lepospondyls until more recent fossil discoveries. Unlike the multicuspid teeth of lissamphibians, where the cuspules are oriented labiolingually, the cusps of brachystelechid "microsaurs" are oriented anteroposteriorly, which suggests that multicuspid teeth are independently derived in brachystelechids and lissamphibians.

Recently, however, the new gymnarthrid *Bolterpeton* was described by Anderson and Reisz (2003) as having weakly bicuspid teeth (sensu Bolt, 1977) with labiolingually oriented cusps. Reinvestigation of a lower jaw of the gymnarthrid *Cardiocephalus peabodyi* originally described by Gregory et al. (1956: fig. 8c,f,g) demonstrated that it also had cusps with identical structure to *Bolterpeton* on its anteriormost teeth (Anderson and Reisz, 2003). Given the tree shape established by Anderson (2001), it can be predicted that *Rhynchonkos* also had weakly bicuspid teeth, but this has not yet been confirmed.

The newly observed, weakly bicuspid tooth structure of *Cardiocephalus* has been incorporated into the following phylogenetic analysis. The orientation of the cusps is included as a separate character to increase the rigor of the test of lissamphibian origins. Additionally, multicuspidality and pedicely are separate characters in this matrix because in modern amphibians, multicuspid teeth can occur without pedicely, and pedicely can be present in monocuspid teeth (Tesche and Greven, 1989).

Lepospondyl and dissorophid temnospondyl skulls are very small and are frequently three-dimensionally preserved, making them ideal candidates for detailed investigation of in situ, delicate features of anatomy. One of the characters of Parsons and Williams (1963) that is based on hard tissue and thus potentially observable in fossil taxa is the presence of an operculum, a small bone that is articulated with, or in some extant species fused to, the footplate of the stapes (Duellman and Trueb, 1994). This bone is connected with the pectoral girdle by a muscle and is thought to serve to transmit low-frequency sound waves to the inner ear. It has been suggested that the aistopod *Phlegethontia* had a similar system, although the operculum, if present at all, is indistinguishably fused to the footplate of the stapes (Anderson, 2002b), as is seen in some salamanders (Duellman and Trueb, 1994). The only fossil amphibians previously described as having accessory ossifications related to the stapes are the "microsaurs" (Carroll and Gaskill, 1978), but these accessory ossicles have never been described in detail because of their small size and inaccessibility. These tantalizing elements represent an exciting line of new research but are not yet appropriate for inclusion in a phylogenetic matrix as homologous to the operculum.

Phylogenetic Analysis

As the previous section emphasized, potentially new lines of evidence are extremely difficult to evaluate independent of a well-substantiated phylogeny. However, even the most comprehensive analysis of amphibian relationships (Ruta et al., 2003) suffers from the lack of revision of the species relevant to the various hypotheses of lissamphibian origins. The most densely sampled single taxonomic grouping of the various competing large-scale analyses (Anderson, 2001; Ruta et al., 2003; Vallin and Laurin, 2004) is Anderson's Lepospondyli; thus it is the focus of expansion over the next several years. Previous coding errors have been corrected (e.g., the number of coronoids in *Limnoscelis,* now revealed after new preparation to be two [Reisz, this volume]; number of sacral vertebrae in *Utaherpeton* changed to one). Vallin and Laurin (2004) have discussed these, but many of their additional charges of error are based on either the literature rather than specimens or on differing interpretations of specimen drawings. Temnospondyl taxa have been added on the basis of direct observation where possible, but this is a time-consuming process that is still ongoing (and will continue until they are as densely sampled as the lepospondyls). Thus, this analysis should be considered preliminary. Despite this, the present analysis is among the more thorough tests of lissamphibian origins and allows for the possibility of frogs and salamanders to be placed independently as sister-taxa to amphibamids and branchiosaurs, as suggested by some to be the case (Carroll, 2004; Carroll et al., 2004). Albanerpetontids have also been added to Anderson's matrix to test whether they are more closely related to lissamphibians than to any other fossil amphibian group, as has been long assumed (Gardner, 2001; McGowan, 2002).

Developmental Characters

To summarize, the three lines of developmental evidence cited in the debate over lissamphibian origins are patterns of vertebral ossification, patterns of cranial ossification, and the presence of metamorphosis in larvae. Vertebral ossification has been included in this analysis (character 197), with the arches first, then centra pattern coded primitive (Cote et al., 2002). The persistent presence of a gap between the maxilla and suspensorium is reflected in characters 14 and 15. Various dental characters, like the later-occurring bicuspid condition (Bolt, 1977; Duellman and Trueb, 1994), are also included (Table 5.3).

As discussed at length above, the pattern of cranial ossification is highly conserved within tetrapods, and too little is known to determine whether variation held in common between the few taxa for which we have data is primitive or derived. Therefore, it is inappropriate to include these data in a phylogenetic analysis at present, methodological difficulties aside.

The presence of metamorphosis in fossils is an inference, not a character, and so it is not included in this matrix. This is similar to the inference of terrestriality in Paleozoic tetrapods, which I discussed elsewhere (Laurin and Anderson, 2004). However, many of the characters mentioned previously from which the presence of metamorphosis is inferred are included in some form or another. Terrestrial adaptations in the postcranial skeleton in this matrix include many characters related to large muscle insertions or degrees of ossification (e.g., characters 159, 162, 163, 166, 167, 173, and 174). Character 47, dermal ornamentation, is an ontogenetic character at least in part, where the primitive state, high ridges and pits, is the last feature to develop (Boy and Sues, 2000). Paedomorphosis will lead to the derived adult retention of a lesser degree of ornamentation, so it is phylogenetically informative. However, I believe this character deserves detailed study in the future.

The disappearance of lateral line sulci with growth is not accommodated in this matrix, except for their complete loss in non-aquatic lineages because I coded the presence of lateral line canals at any point in ontogeny as primitive. This is justified for the present by noting that very few Paleozoic tetrapods have known growth series from which the late loss of lateral line canals can be documented; most fossil species lack associated juveniles and adults. The same can be said for the characters "changes in the morphology of the choana" and "closure of the nasolacrimal canal." It is to be hoped that future discoveries will permit eventual inclusion of these characters.

The remainder of the characters can be accommodated to a limited degree. The correlate "disappearance of ceratohyals" has been included in this matrix as "ossified hyoids present" (character 111), where hyoids are considered present if they are known from any stage in growth. The identity of a particular hyoid element, other than the basihyoid, is difficult to establish with a high degree of confidence without the presence of all elements. Again, considering the paucity of growth series with respect to the fossils included in this matrix, greater specification is not presently phylogenetically informative. Changes in the structure and size of the septomaxilla are partially accounted for by character 20. Absence of branchial denticles in larger specimens is not as phylogenetically informative as their presence (which is derived) and morphology, so character 196 was included, which considers both the derived presence of branchial denticles and the uniquely derived raker morphology of branchiosaurs and salamanders.

Not all of these decisions are ideal, but they are pragmatic. Most of these detailed correlates of metamorphosis can only be documented in salamanders, branchiosaurids, and micromelerpetontids. This distribution leaves us with a classic three-taxon statement and no ability to polarize the character states in a phylogenetically meaningful way. In discussions of similarities between branchiosaurs and salamanders, when attempting to infer a metamorphosis in the former, this is acceptable, but it is not appropriate for phylogenetic analysis. It is hoped that as the study of development of more fossil taxa progresses, these characters can be included, because presumably they will permit fine degrees of taxonomic discrimination. All developmental characters are listed in Table 5.3 and are indicated in italics in Appendix 5.1.

Results

Results of this analysis of 197 characters and 62 taxa are presented in Figures 5.5 to 5.7. Figure 5.5 is a strict consensus of 56 most parsimonious trees (each with a length of 1,113 steps, Consis-

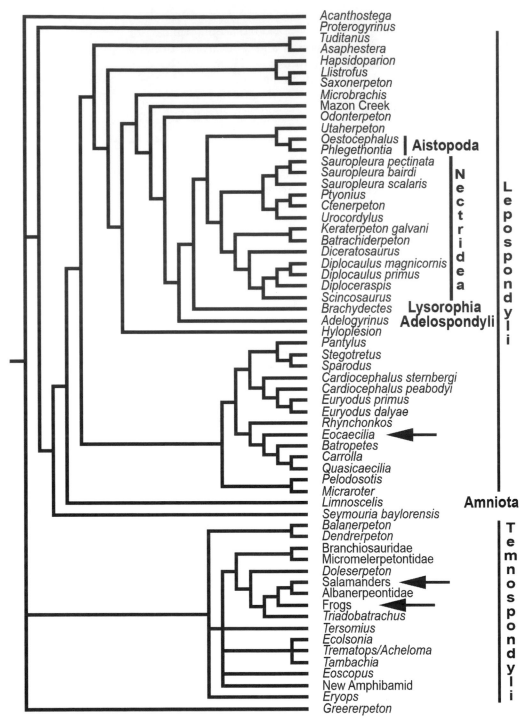

Figure 5.5. Strict consensus tree of 56 most parsimonious trees found in the phylogenetic analysis. Trees differ in the placement of various amphibamids. See text for further discussion.

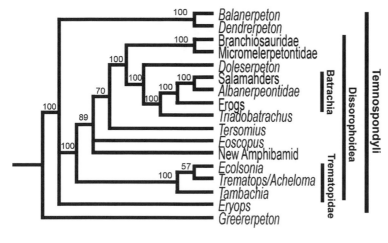

Figure 5.6. Placement of frogs and salamanders among dissorophoid temnospondyls. This is from the majority-rule consensus of 56 trees; frequency of occurrence among the most parsimonious trees is given at each node. Inclusion of dissorophoids may stabilize the position of *Ecolsonia*, which is controversially considered either a trematopid or dissorophid. Additional instability caused by *Tersomius* may be due to its being a composite of multiple taxa.

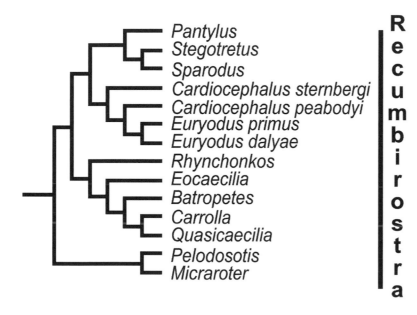

Figure 5.7. Placement of caecilians among recumbirostrine lepospondyls from the strict consensus tree.

tency Index = 0.23, Retention Index = 0.59) found after a heuristic search (1,000 random replicates). It is the most comprehensive analysis to support the PH. *Eocaecilia* maintains its relationship with brachystelechid "microsaurs" and *Rhynchonkos*, and this group in turn is the sister-taxon to the gymnarthrids (Fig. 5.7), as found by Anderson (2001). Frogs and salamanders (Batrachia; Milner, 1988), however, are placed within temnospondyls, with albanerpetontids as the sister-group to salamanders. *Triadobatrachus*, the putative stem anuran (Rage and Rocek, 1989), is the sister-taxon to Batrachia, which itself is the sister-group to the amphibamid *Doleserpeton*, as Bolt (1969) suggested for the entire Lissamphibia (Fig. 5.6). Micromelerpetontidae and Branchiosauridae form a monophyletic Phyllospondyli, which is the sister-group to the clade (*Doleserpeton*

(*Triadobatrachus* (Batrachia))). The remaining amphibamids are poorly resolved; in the strict consensus tree, *Tersomius* and a new form from the Lower Permian of Germany currently under study (Anderson et al., 2004) fall to a polytomy of "dissorophoids" including a fully resolved Trematopidae and *Eoscopus*. In the majority-rule consensus tree (Fig. 5.6), Trematopidae is removed to the next node below this polytomy, thus establishing the monophyly of Amphibamidae (including Batrachia).

Discussion

This is the first large-scale analysis to support the PH. McGowan's (2002) analysis was limited in the number of taxa it considered, which might explain the heterodox arrangement of taxa in his tree. It bears emphasizing again that this present study is a work in progress; as other taxa become better characterized, they will be added to the matrix. Work on amphibamid, trematopid, and dissorophid temnospondyls is currently under way, and some preliminary data have been incorporated in this study (Anderson et al., 2004). Additionally, many more basal temnospondyls remain to be considered. Because the goal of this study is to test lissamphibian monophyly, however, amphibamid and branchiosaurid temnospondyls are most critical, and they are represented well enough in the present study to refute the LH. It remains quite possible that when temnospondyls are as well represented as lepospondyls, the caecilians will be pulled away from the "microsaurs."

With this caveat, the results of this analysis are intriguing with respect to my inference of development in dissorophoid temnospondyls. The discovery that frogs and salamanders are more closely related to *Doleserpeton* than to branchiosaurs (Fig. 5.6), despite including the developmental evidence supporting a branchiosaur-salamander relationship (indeed, intentionally biasing this study toward finding this pattern), suggests that patterns of development are highly plastic in dissorophoids. True gill rakers (as defined by Carroll, 2004) are found in some salamanders and branchiosaurs, but not in other intervening taxa (micromelerpetontids, *Doleserpeton*, albanerpetontids, and frogs). This supports the idea that certain groups of amphibians repeatedly exploit the same vacant niche through a fairly simple alteration of their development. Neoteny may be an example of this because it is presumed to be present in some branchiosaurs sensu lato (Boy and Sues, 2000), and it is repeatedly acquired throughout salamanders (Duellman and Trueb, 1994). New research shows that neoteny tends to cluster taxa on similar morphology, and a total evidence analysis of morphological and molecular data supports an even patchier distribution of neoteny among salamanders than previously thought (Wiens et al., 2005). In this light, it is quite possible biologically that neoteny is independently derived in branchiosaurs and salamanders, and ontogenetic plasticity is characteristic of dissorophoids in general, if not temnospondyls as a whole. A possible explanation for this patchy distribution is that it is a means for exploiting ephemeral bodies of water (Boy and Sues, 2000). Frogs as a clade have taken niche exploitation by larvae to an extreme (Duellman and Trueb, 1994).

This broad distribution of repeated acquisition of a similar larval feeding strategy in dissorophoids suggests that other temnospondyls may also share the gill raker structure. Milner (1982) described a saurerpetontid temnospondyl from the nodules of Mazon Creek as possessing gill rakers. The quality of preservation of the specimen makes it difficult to describe the fine details of the pattern that Carroll (2004) used to distinguish true rakers—triangular ossicles that intermesh across the pharyngeal slit—from that present in other temnospondyls with a disk with multiple projections. However, the structure in the Mazon Creek specimen appears to conform more closely with true rakers in shape and arrangement (Milner, 1982: fig. 1). If this is the case, true rakers are widely distributed across temnospondyls, present in branchiosaurs, salamanders, and saurerpetontids (basal member of Dvinosauria). On the other hand, the possible presence of ossicles associated with the ceratohyals in *Microbrachis* does not appear to affect the placement of modern amphibians. It does emphasize how frequently neoteny appears to have taken place among nonamniote tetrapods, and it strengthens the inference of multiple acquisitions of this ecological strategy within the more closely related temnospondyls.

This analysis finds Amphibamidae, a family established by Daly (1994) for *Doleserpeton, Amphibamus, Tersomius,* and *Eoscopus,* paraphyletic (Fig. 5.6). *Doleserpeton* is the sister-taxon to *Triadobatrachus* + Batrachia, whereas *Eoscopus, Tersomius,* and a new form from the Lower Permian of Germany currently under study (Anderson et al., 2004) are placed in a polytomy with trematopids at the basal node of Dissorophoidea. This lack of resolution may be the result of a lack of recent revision and bears closer scrutiny; such work is currently underway. The polyphyly of amphibamids was previously suggested by Milner (1993).

The enigmatic Albanerpetontidae are found to be the sister-group to salamanders. Because my coding for salamanders comprises a composite of basal crown taxa, it cannot be determined at present whether albanerpetontids are the sister-group either to stem salamanders (Caudata) or salamanders (Urodela). Milner (1988) demonstrated they could be either, with only a few conflicting characters supporting either hypothesis. Gardner (2001) and McGowan (2002) placed albanerpetontids slightly lower, as the sister-group to Batrachia, whereas Ruta et al. (2003) had difficulty resolving their position, variously placing them as outgroup to Batrachia and crown Lissamphibia.

Caecilians remain closely associated with brachystelechid "microsaurs" and *Rhynchonkos* (Fig. 5.7), as suggested by Carroll (2000b) and previously found by Anderson (2001). They are part of a previously recognized but unnamed clade of "microsaurs" with an overturned snout and medially inclined dental laminae that is here named **Recumbirostra** (defined as a node-based taxon: the clade descended from the most recent common ancestor of *Panty-lus, Cardiocephalus sternbergi, Rhynchonkos,* and *Micraroter,* but not including *Tuditanus* or *Microbrachis*). Unfortunately, some of the characters shared between caecilians and brachystelechids are loss features (shared fusion of bones). One exception is the presence of multicuspid teeth. The issue of multicuspid teeth, and tooth pedicely, would bear closer investigation in the future. Recognition of tooth pedicely is very difficult in fossil forms and is usually hypothesized when all tooth cusps are broken off at the same level—something that is very likely in small animals with delicate teeth. It is also complicated by the fact that it is ontogenetically variable in its presence in modern amphibians, as is, in fact, multicuspidality (Tesche and Greven, 1989; Duellman and Trueb, 1994).

However, a bladed tooth crown appears to be present in adults of frogs and salamanders irrespective of the number of cusps, and bladed teeth are present in the primary teeth of anurans (Tesche and Greven, 1989). Bladed teeth are slightly compressed labiolingually, with a cutting edge extending along the medial edge, a feature also present in *Bolterpeton* and *Cardiocephalus* (Anderson and Reisz, 2003), which suggests that the "weakly bicuspid" condition described by Bolt (1980) is another term for a homologous structure.

One of the outstanding criticisms of the PH is the fact that no molecular phylogeny has placed frogs, salamanders, or caecilians closer to amniotes than the rest of the extant forms (Laurin, 2002). It is widely recognized that phylogenetic analysis, even analysis based on molecular data, is highly sensitive to taxonomic sampling (e.g., Gauthier et al., 1988; Donoghue et al., 1989; Novacek, 1992; Graybeal, 1998; Poe, 1998; Anderson, 2001), so this is not considered a strong criticism. Additionally, should the divergence between caecilians, batrachians, and amniotes have been sufficiently rapid, it may not have been captured by slowly evolving nuclear DNA, whereas mitochondrial DNA evolved so rapidly that the signal may have been subsequently saturated (Lee and Anderson, 2006). Future studies should add molecular data to a total evidence matrix to explore this question further.

Conclusions

Developmental data provide interesting new lines of investigation into evolutionary problems as long as they are placed within the proper phylogenetic context. When possible, the presence of derived developmental states should be coded and included in phylogenetic analyses. Developmental sequence data, like the sequence of ossification of the vertebrate skull, are best analyzed by character mapping, although even this is plagued by problems of nonindependence.

Several lines of developmental evidence have recently been applied to the question of the origins of modern amphibians, which were considered in turn and included for the first time in a phylogenetic analysis when appropriate. The pattern of vertebral development was coded and included in the analysis while the pattern of cranial ossification was discussed but not included for several reasons. Metamorphosis was incorporated as a series of discrete characters because it is through the cumulative presence of these correlates that metamorphosis is inferred to be present in fossil taxa.

The results from the analysis support an origin for caecilians within lepospondyls; specifically, lepospondyls were found to be the sister-group to brachystelechid "microsaurs" and *Rhynchonkos*. Frogs and salamanders, on the other hand, form a monophyletic grouping, with albanerpetontids as the sister-group to salamanders at a currently unknown level. These taxa were found to be the sister-group to *Doleserpeton*, and not separately with amphibamids and branchiosaurids, as has been suggested. This topology, if correct, demonstrates that even among the Paleozoic precursors to frogs and salamanders, changes in development were important to the exploitation of empty niches.

Acknowledgments

I thank Bob Carroll for providing an environment of excitement and discovery while I was a student at McGill University, and for many stimulating conversations that continue to this day. Some of the data presented in this paper come from ongoing work, so I would like to thank my collaborators David Berman, John Bolt, Amy Henrici, and Stuart Sumida. I also thank Andrew Milner and Jenny Clack for many helpful comments that improved this paper, and my coeditor Hans-Dieter Sues for exclusively handling the editorial duties of this chapter.

References Cited

Anderson, J. S. 2001. The phylogenetic trunk: maximal inclusion of taxa with missing data in an analysis of the Lepospondyli. Systematic Biology 50:170–193.

Anderson, J. S. 2002a. Use of well-known names in phylogenetic nomenclature: a reply to Laurin. Systematic Biology 51:822–827.

Anderson, J. S. 2002b. Revision of the aïstopod genus *Phlegethontia* (Tetrapoda: Lepospondyli). Journal of Paleontology 76:1029–1046.

Anderson, J. S. 2003a. Cranial anatomy of *Coloraderpeton brilli*, postcranial anatomy of *Oestocephalus amphiuminus*, and reconsideration of Ophiderpetontidae (Tetrapoda: Lepospondyli: Aistopoda). Journal of Vertebrate Paleontology 23:532–543.

Anderson, J. S. 2003b. A new aistopod (Tetrapoda: Lepospondyli) from Mazon Creek, Illinois. Journal of Vertebrate Paleontology 23:79–88.

Anderson, J. S., and R. R. Reisz. 2003. A new microsaur (Tetrapoda: Lepospondyli) from the Lower Permian of Richards Spur (Fort Sill) Oklahoma. Canadian Journal of Earth Sciences 40:499–505.

Anderson, J. S., R. L. Carroll, and T. B. Rowe. 2003. New information on *Lethiscus stocki* (Tetrapoda: Lepospondyli: Aistopoda) from high-resolution computed tomography and a phylogenetic analysis of Aistopoda. Canadian Journal of Earth Sciences 40:1071–1083.

Anderson, J. S., S. S. Sumida, D. S Berman, A. C. Henrici, and T. Martens. 2004. The dissorophoid temnospondyls from the Early Permian of Bromacker, Germany. Journal of Vertebrate Paleontology 24 (Suppl. to 3):34A.

Andrews, S. M., and R. L. Carroll. 1991. The order Adelospondyli: Carboniferous lepospondyl amphibians. Transactions of the Royal Society of Edinburgh: Earth Sciences 82:239–275.

Baird, D. 1965. Paleozoic lepospondyl amphibians. American Zoologist 5:287–294.

Berman, D. S, R. R. Reisz, and D. A. Eberth. 1987. A new genus and species of trematopid amphibian from the Late Pennsylvanian of North-Central New Mexico. Journal of Vertebrate Paleontology 7:252–269.

Bininda-Edmonds, O. R. P., J. E. Jeffery, M. I. Coates, and M. K. Richardson. 2002. From Haeckel to event-pairing: the evolution of developmental sequences. Theory in Biosciences 121:297–320.

Bolt, J. R. 1969. Lissamphibian origins: possible protolissamphibian from the Lower Permian of Oklahoma. Science 166:888–891.

Bolt, J. R. 1974a. Osteology, function, and evolution of the trematopsid (Amphibia: Labyrinthodontia) nasal region. Fieldiana 33:11–30.

Bolt, J. R. 1974b. Armor of dissorophids (Amphibia: Labyrinthodontia): an examination of its taxonomic use and report of a new occurrence. Journal of Paleontology 48:135–142.

Bolt, J. R. 1977. Dissorophoid relationships and ontogeny, and the origin of the Lissamphibia. Journal of Paleontology 51:235–249.

Bolt, J. R. 1980. New tetrapods with bicuspid teeth from the Fort Sill locality (Lower Permian, Oklahoma). Neues Jahrbuch für Geologie und Paläontologie, Monatshefte 1980:449–459.

Bolt, J. R. 1991. Lissamphibian origins; pp. 194–222 in H.-P. Schultze and L. Trueb (eds.), Origins of the Higher Groups of Tetrapods: Controversy and Consensus. Comstock Publishing Associates, Ithaca and London.

Bossy, K. A., and A. C. Milner. 1998. Order Nectridea; pp. 73–132 in R. L. Carroll (ed.), Lepospondyli. Handbuch der Paläoherpetologie. Part 1. Verlag Dr. Friedrich Pfeil, Munich.

Boy, J. A., and H.-D. Sues. 2000. Branchiosaurs: larvae, metamorphosis and heterochrony in temnospondyls and seymouriamorphs; pp. 1150–1197 in H. Heatwole and R. L. Carroll (eds.), Amphibian Biology. Volume 4, Palaeontology. Surrey Beatty & Sons, Chipping Norton.

Carroll, R. L. 1964. Early evolution of the dissorophid amphibians. Bulletin of the Museum of Comparative Zoology, Harvard University 131:161–250.

Carroll, R. L. 1998a. Order Microsauria; pp. 1–71 in R. L. Carroll (ed.), Lepospondyli. Handbuch der Paläoherpetologie. Part 1. Verlag Dr. Friedrich Pfeil, Munich.

Carroll, R. L. 1998b. Vertebral development and amphibian relationships. Journal of Vertebrate Paleontology 18(Suppl. to 3):31A.

Carroll, R. L. 2000a. Lepospondyls; pp. 1198–1269 in H. Heatwole and R. L. Carroll (eds.), Amphibian Biology. Volume 4, Palaeontology. Surrey Beatty & Sons, Chipping Norton.

Carroll, R. L. 2000b. *Eocaecilia* and the origin of caecilians; pp. 1402–1411 in H. Heatwole and R. L. Carroll (eds.), Amphibian Biology. Volume 4, Palaeontology. Surrey Beatty & Sons, Chipping Norton.

Carroll, R. L. 2001. The origin and early radiation of terrestrial vertebrates. Journal of Paleontology 75:1202–1213.

Carroll, R. L. 2004. The importance of branchiosaurs in determining the ancestry of the modern amphibian orders. Neues Jahrbuch für Geologie und Paläontologie, Abhandlungen 232:157–180.

Carroll, R. L. (ed.). 1998. Lepospondyli. Handbuch der Paläoherpetologie. Part 1. Verlag Dr. Friedrich Pfeil, Munich, 216 pp.

Carroll, R. L., and S. M. Andrews. 1998. Order Adelospondyli; pp. 149–162 in R. L. Carroll (ed.), Lepospondyli. Handbuch der Paläoherpetologie. Part 1. Verlag Dr. Friedrich Pfeil, Munich.

Carroll, R. L., and P. J. Currie. 1975. Microsaurs as possible apodan ancestors. Zoological Journal of the Linnean Society 57:229–247.

Carroll, R. L., and P. Gaskill. 1978. The Order Microsauria. Memoirs of the American Philosophical Society 126:1–211.

Carroll, R. L., and R. Holmes. 1980. The skull and jaw musculature as guides to the ancestry of salamanders. Zoological Journal of the Linnean Society 68:1–40.

Carroll, R. L., C. Boisvert, J. R. Bolt, D. M. Green, N. Philip, C. Rolian, R. R. Schoch, and A. Tarenko. 2004. Changing patterns of ontogeny from osteolepiform fish through Permian tetrapods as a guide to the early evolution of land vertebrates; pp. 321–343 in G. Arratia, M. V. H. Wilson, and R. Cloutier (eds.), Recent Advances in the Origin and Early Radiation of Vertebrates. Verlag Dr. Friedrich Pfeil, Munich.

Cote, S., R. L. Carroll, R. Cloutier, and L. Bar-Sagi. 2002. Vertebral development in the Devonian sarcopterygian fish *Eusthenopteron foordi* and the polarity of vertebral evolution in non-amniote tetrapods. Journal of Vertebrate Paleontology 22:487–502.

Daly, E. 1994. The Amphibamidae (Amphibia: Temnospondyli),with a description of a new genus from the Upper Pennsylvanian of Kansas. University of Kansas Museum of Natural History, Miscellaneous Publications 85:1–59.

DeMar, R. E. 1966. The phylogenetic and functional implications of the armor of the Dissorophidae. Fieldiana: Geology 16:55–88.

Dilkes, D. W., and R. R. Reisz. 1987. *Trematops milleri* Williston, 1909 identified as a junior synonym of *Acheloma cumminsi* Cope, 1882, with a revision of the genus. American Museum Novitates 2902:1–12.

Donoghue, M. J., J. A. Doyle, J. Gauthier, A. G. Kluge, and T. Rowe. 1989. The importance of fossils in phylogeny reconstruction. Annual Review of Ecology and Systematics 20:431–460.

Duellman, W. E., and L. Trueb. 1994. Biology of Amphibians. Johns Hopkins University Press, Baltimore, 670 pp.

Evans, H. E. 1993. Miller's Anatomy of the Dog. Third Edition. W. B. Saunders, Philadelphia, 1113 pp.

Evans, S. E., and D. Sigogneau-Russell. 2001. A stem-group caecilian (Lissamphibia: Gymnophiona) from the Lower Cretaceous of North Africa. Palaeontology 44:259–273.

Feller, A. E., and S. B. Hedges. 1998. Molecular evidence for the early history of living amphibians. Molecular Phylogenetics and Evolution 9:509–516.

Gardner, J. D. 2001. Monophyly and affinities of albanerpetontid amphibians (Temnospondyli; Lissamphibia). Zoological Journal of the Linnean Society 131:309–352.

Gauthier, J., A. G. Kluge, and T. Rowe. 1988. Amniote phylogeny and the importance of fossils. Cladistics 4:105–209.

Graybeal, A. 1998. Is it better to add taxa or characters to a difficult phylogenetic problem? Systematic Biology 47:9–17.

Gregory, J. T., F. E. Peabody, and L. I. Price. 1956. Revision of the Gymnarthridae: American Permian microsaurs. Bulletin of the Peabody Museum of Natural History, Yale University 10:1–77.

Halanych, K. M. 1998. Lagomorphs misplaced by more characters and fewer taxa. Systematic Biology 47:138–146.

Hay, J. M., I. Ruvinsky, S. B. Hedges, and L. R. Maxson. 1995. Phylogenetic relationships of amphibian families inferred from DNA sequences of mitochondrial 12S and 16S ribosomal RNA genes. Molecular Biology and Evolution 12:928–937.

Hedges, S. B., and L. R. Maxson. 1993. A molecular perspective on lissamphibian phylogeny. Herpetological Monographs 7:27–42.

Holmes, R. 2000. Palaeozoic temnospondyls; pp. 1081–1120 in H. Heatwole and R. L. Carroll (eds.), Amphibian Biology. Volume 4, Palaeontology. Surrey Beatty & Sons, Chipping Norton.

Jeffery, J. E., M. K. Richardson, M. I. Coates, and O. R. P. Bininda-Emonds. 2002. Analyzing developmental sequences within a phylogenetic framework. Systematic Biology 51:478–491.

Jenkins, F. A., Jr., and D. M. Walsh. 1993. An early Jurassic caecilian with limbs. Nature 365:246–249.

Laurin, M. 1998. The importance of global parsimony and historical bias in understanding tetrapod evolution. Part I. Systematics, middle ear evolution and jaw suspension. Annales des Sciences Naturelles 1:1–42.

Laurin, M. 2002. Tetrapod phylogeny, amphibian origins, and the definition of the name Tetrapoda. Systematic Biology 51:364–369.

Laurin, M., and J. S. Anderson. 2004. Meaning of the name Tetrapoda in the scientific literature: an exchange. Systematic Biology 53:68–80.

Laurin, M., and R. R. Reisz. 1997. A new perspective on tetrapod phylogeny; pp. 9–59 in S. S. Sumida and K. L. M. Martin (eds.), Amniote Origins. Academic Press, San Diego.

Lee, M. S. Y., and J. S. Anderson. 2006. Molecular clocks and the origin of modern amphibians. Molecular Phylogenetics and Evolution 40:635–639.

McGowan, G. 2002. Albanerpetontid amphibians from the Lower Cretaceous of Spain and Italy: a description and reconsideration of their systematics. Zoological Journal of the Linnean Society 135:1–32.

Milner, A. C. 1980. A review of the Nectridea; pp. 377–405 in A. L. Panchen (ed.), The Terrestrial Environment and the Origin of Land Vertebrates. Academic Press, London and New York.

Milner, A. R. 1982. Small temnospondyl amphibians from the Middle Pennsylvanian of Illinois. Palaeontology 25:635–664.

Milner, A. R. 1988. The relationships and origin of living amphibians; pp. 59–102 in M. J. Benton (ed.), The Phylogeny and Classification of the Tetrapods. Volume 1, Amphibians, Reptiles, Birds. Systematics Association Special Volume No. 35A. Clarendon Press, Oxford.

Milner, A. R. 1993. The Paleozoic relatives of lissamphibians. Herpetological Monographs 7:8–27.

Novacek, M. J. 1992. Fossils as critical data for phylogeny; pp. 46–88 in M. J. Novacek and Q. D. Wheeler (eds.), Extinction and Phylogeny. Columbia University Press, New York.

Olson, E. C. 1941. The Family Trematopsidae. Journal of Geology 49:149–176.

Parsons, T. S., and E. E. Williams. 1963. The relationship of modern Amphibia: a re-examination. Quarterly Review of Biology 38:26–53.

Poe, S. 1998. Sensitivity of phylogeny estimation to taxonomic sampling. Systematic Biology 47:18–31.

Rage, J.-C., and Z. Rocek. 1989. Redescription of *Triadobatrachus massinoti* (Piveteau, 1936), an anuran amphibian from the Early Triassic. Palaeontographica A 206:1–16.

Richardson, M. K., J. E. Jeffery, M. I. Coates, and O. R. P. Bininda-Emonds. 2001. Comparative methods in developmental biology. Zoology 104:278–283.

Romer, A. S. 1945. Vertebrate Paleontology. Second Edition. University of Chicago Press, Chicago, 687 pp.

Rose, C. S., and J. O. Reiss. 1993. Metamorphosis and the vertebrate skull: ontogenetic patterns and developmental mechanisms; pp. 289–346 in B. K. Hall (ed.), The Skull. Volume 1, Development. University of Chicago Press, Chicago.

Ruta, M., M. I. Coates, and D. L. Quicke. 2003. Early tetrapod relationships revisited. Biological Reviews of the Cambridge Philosophical Society 78:251–345.

San Mauro, D., D. J. Gower, O. V. Oommen, M. Wilkinson, and R. Zardoya. 2004. Phylogeny of caecilian amphibians (Gymnophiona) based on complete mitochondrial genomes and nuclear RAG1. Molecular Phylogenetics and Evolution 33:413–427.

San Mauro, D., M. Vences, M. Alcobendas, R. Zardoya, and A. Meyer. 2005. Initial diversification of living amphibians predated the breakup of Pangaea. American Naturalist 165:590–599.

Schoch, R. R. 1992. Comparative ontogeny of Early Permian branchiosaurid amphibians. Developmental stages. Palaeontographica A 222:43–83.

Schoch, R. R. 2002a. The early formation of the skull in extant and Paleozoic amphibians. Paleobiology 28:278–296.

Schoch, R. R. 2002b. The evolution of metamorphosis in temnospondyls. Lethaia 35:309–327.

Schoch, R. R. 2003. Early larval ontogeny of the Permo-Carboniferous temnospondyl *Sclerocephalus*. Palaeontology 46:1055–1072.

Schoch, R. R. 2004. Skeleton formation in the Branchiosauridae: a case study in comparing ontogenetic trajectories. Journal of Vertebrate Paleontology 24:309–319.

Schoch, R. R., and R. L. Carroll. 2003. Ontogenetic evidence for the Paleozoic ancestry of salamanders. Evolution and Development 5:314–324.

Schoch, R. R., and A. R. Milner. 2004. Structure and implications of theories on the origin of lissamphibians; pp. 345–377 in G. Arratia, M. V. H. Wilson, and R. Cloutier (eds.), Recent Advances in the Origin and Early Radiation of Vertebrates. Verlag Dr. Friedrich Pfeil, Munich.

Schulmeister, S., and W. C. Wheeler. 2004. Comparative and phylogenetic analysis of developmental sequences. Evolution and Development 6:50–57.

Smith, K. K. 2001. Heterochrony revisited: the evolution of developmental sqeuences. Biological Journal of the Linnean Society 73:169–186.

Steyer, J. S. 2000. Ontogeny and phylogeny in temnospondyls: a new method of analysis. Zoological Journal of the Linnean Society 130:449–467.

Tesche, M., and H. Greven. 1989. Primary teeth in Anura are nonpedicellate and bladed. Zeitschrift für zoologische Systematik und Evolutionsforschung 1989:326–329.

Trueb, L., and R. Cloutier. 1991. A phylogenetic investigation of the inter- and intrarelationships of the Lissamphibia (Amphibia: Temnospondyli);

pp. 223–313 in H.-P. Schultze and L. Trueb (eds.), Origins of the Higher Groups of Tetrapods: Controversy and Consensus. Comstock Publishing Associates, Ithaca and London.

Vallin, G., and M. Laurin. 2004. Cranial morphology and affinities of *Microbrachis,* and a reappraisal of the phylogeny and lifestyle of the first amphibians. Journal of Vertebrate Paleontology 24:56–72.

Wellstead, C. F. 1991. Taxonomic revision of the Lysorophia, Permo-Carboniferous lepospondyl amphibians. Bulletin of the American Museum of Natural History 209:1–90.

Wellstead, C. F. 1998. Order Lysorophia; pp. 133–148 in R. L. Carroll (ed.), Lepospondyli. Handbuch der Paläoherpetologie. Part 1. Verlag Dr. Friedrich Pfeil, Munich.

Whiting, M. F. 1998. Long-branch distraction and the Strepsiptera. Systematic Biology 47:134–138.

Wiens, J. J., R. M. Bonett, and P. T. Chippindale. 2005. Ontogeny discombobulates phylogeny: paedomorphosis and higher-level salamander relationships. Systematic Biology 54:91–110.

Williston, S. W. 1910. *Cacops, Desmospondylus;* new genera of Permian vertebrates. Bulletin of the Geological Society of America 21:249–284.

Witzmann, F., and H.-U. Pfretzschner. 2003. Larval ontogeny of *Micromelerpeton credneri* (Temnospondyli, Dissorophoidea). Journal of Vertebrate Paleontology 23:750–768.

Zardoya, R., and A. Meyer. 2001. On the origin of and phylogenetic relationships among living amphibians. Proceedings of the National Academy of Sciences USA 98:7380–7383.

Zhang, P., H. Zhou, Y.-Q. Chen, Y.-F. Liu, and L.-H. Qu. 2005. Mitogenomic perspectives on the origin and phylogeny of living amphibians. Systematic Biology 54:391–400.

PLATES

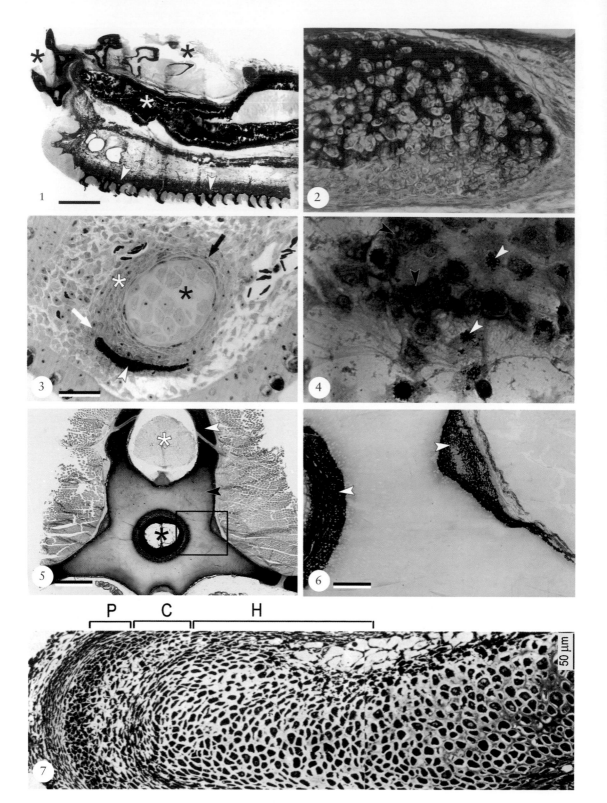

P   C   H

PLATES 1–7

(*Opposite page*)

PLATE 1. Sagittal section through the lower jaw of a juvenile spiny dogfish, *Squalus acanthias*, showing dermal skeletal elements consisting of teeth in different developmental stages (black asterisks) and placoid scales (white arrowheads). All elements are made from enamel (enameloid) and dentine and are attached to acellular bone of attachment. The lower jaw (white asterisk) consists of calcified cartilage. Kossa–Van Gieson, scale bar = 120 μm.

PLATE 2. This very cellular bone arose when maxillary preosteogenic mesenchyme from a 7-day-old chick embryo was combined with mandibular arch epithelium and maintained in organ culture for 7 days. HBQ.

PLATE 3. Meckel's cartilage (black asterisk) and early bone of the dentary (white arrowhead) in the lower jaw of Atlantic salmon are surrounded by condensed cells that constitute condensations (white asterisk) that give rise to chondroblasts (black arrow) and osteoblasts (white arrow). Despite the fact that the bone of Atlantic salmon later contains osteocytes (and so is cellular bone), bone initially is acellular (white arrowhead). Toluidine blue, scale bar = 25 μm.

PLATE 4. Continuous labeling with ³H-thymidine of prechondrogenic cells in vitro shows that chondroblasts, which are surrounded by extracellular matrix (black arrowheads) and prechondroblasts, continue to synthesize DNA, shown as black granules over the nuclei (white arrowheads). Autoradiography, counterstained with Alcian blue.

PLATE 5. In the spiny dogfish, *Squalus acanthias*, the vertebral body (black arrowhead) surrounding the notochord (black asterisk) and the neural arches (white arrowhead) surrounding the spinal cord (white asterisk) are entirely made of cartilage. Some areas mineralize (the area within the inset is enlarged in Plate 6). Toluidine blue, scale bar = 200 μm.

PLATE 6. A higher magnification of the area within the inset in Plate 5. Kossa stain displays mineralizing cartilage matrix (white arrowheads) but no bone formation. Kossa–Van Gieson; scale bar = 40 μm.

PLATE 7. Condylar cartilage from the mandible of 20-day-old mouse embryo to show organization into a proliferative zone (**P**) closest to the joint surface (left), a chondrogenic zone (**C**) in which prechondroblasts become chondroblasts, and a zone of chondrocyte hypertrophy (**H**) in which chondroblasts/chondrocytes become hypertrophic as they undergo terminal differentiation. Modified from Silbermann et al. (1983). Hematoxylin and eosin, scale bar = 50 μm.

(*Opposite page*)

PLATE 8. The secondary cartilage at the posterior hook of the quadratojugal (a membrane bone) from a 16-day-old chick embryo is being transformed into bone. Masson trichrome.

PLATE 9. Histogenesis at the fracture site 19 days after the quadratojugal of a newly hatched chick was fractured to show the formation of cartilage (white asterisk) and the transition between cartilage and bone (black asterisks). Cartilage matrix stains blue due its high glycosaminoglycan content. Hall-Brunt quadruple.

PLATE 10. A nodule of chondrocytes (white asterisk) formed after the quadratojugal of a newly hatched chick was fractured. Note the integration of chondrogenic and osseous matrices (black asterisk). Hall-Brunt quadruple.

PLATE 11. Growth of the dentary in a juvenile Atlantic salmon, *Salmo salar,* involves chondroid bone (black asterisk) and the transformation of chondroid bone into "regular" bone (white asterisk). The cartilaginous component of the chondroid bone derives from the periosteum (white arrowheads) and is unrelated to Meckel's cartilage (black arrowhead). Masson trichrome; scale bar = 150 μm.

PLATE 12. The transitional zone between chondroid bone and compact bone at the tip of the dentary of an adult male Atlantic salmon, *Salmo salar,* displays several tissues in close approximation: bone (white asterisk), cartilage (black asterisk), chondroid bone (black arrowheads), and tissue types intermediate between all three categories. Masson trichrome, scale bar = 40 μm.

PLATE 13. Ectopic ossicles composed of bone and chondroid bone form in yellow perch, *Perca flavescens,* that have been infected with metacercaria of the trematode parasite, *Apophallus brevis.* **A,** cross-section of a maturing ossicle to show the central chamber (asterisk) that houses the metacercaria and the ports (arrowheads) that allow the metacercaria to move in and out of the ossicle. **B,** cross-section of an ossicle from a yellow perch given three intraperitoneal injections of tetracycline. Note the lamellar bone deposited at the periphery (white arrowhead). **C,** scanning electron micrograph of an EDTA-etched ossicle to show the inner cellular zone of chondroid bone (white asterisk) and the outer zone of lamellar bone (white arrowhead). Modified from Taylor et al. (1994). **A,** scale bar = 250 μm, **B** and **C,** scale bar = 150 μm.

PLATE 14. The acellular bone of the dentary in the Nile tilapia, *Oreochromis niloticu*s, is devoid of bone cells and bone cell processes (white asterisks). Typical for bony fish, the bone marrow contains fat tissue (black asterisks) but no hematopoietic tissue. Hematoxylin and eosin, scale bar = 150 μm.

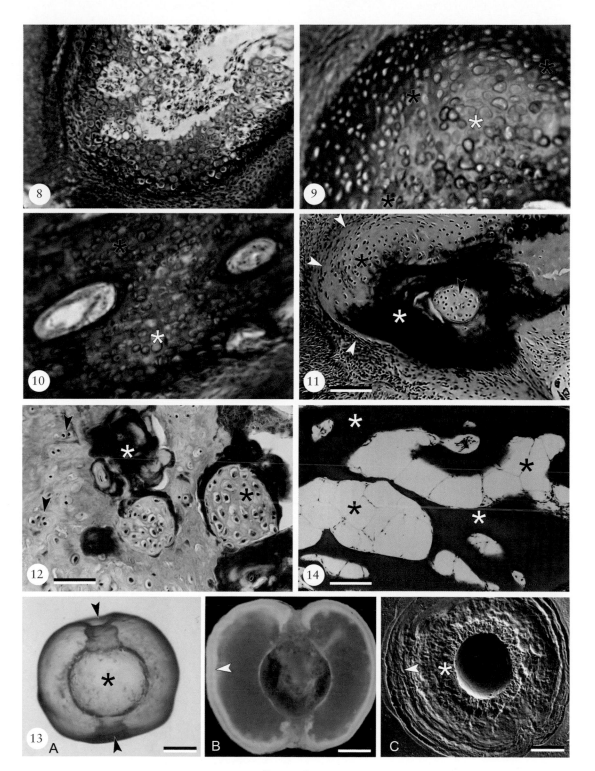

PLATES 8–14

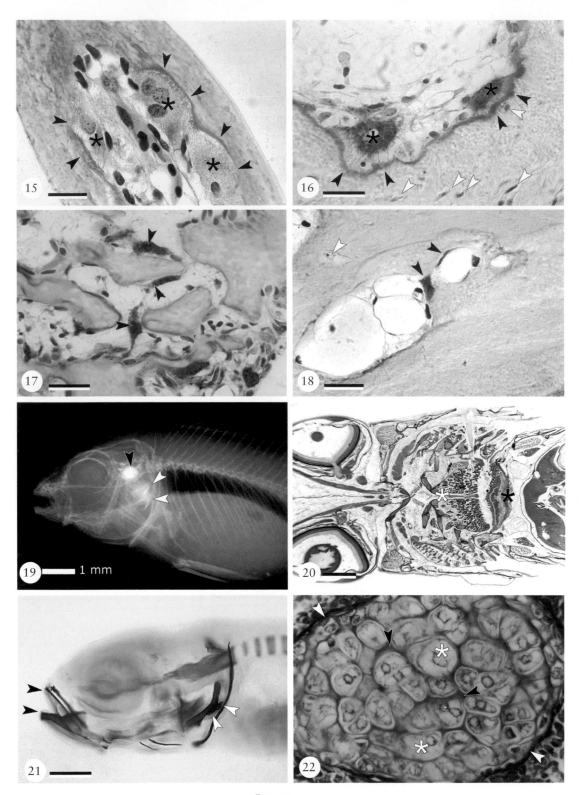

PLATES 15–22

(*Opposite page*)

PLATE 15. Large multinucleated odontoclasts (osteoclasts; black asterisks) inside the pulp cavity resorb a tooth in the gray bichir (*Polypterus senegalus*). The cells create pronounced resorption lacunae (black arrowheads) and segregate tartrate-resistant acid phosphatase (TRAP) into the subcellular space. Visualization of TRAP counterstained with hematoxylin, scale bar = 20 μm.

PLATE 16. Large multinucleated osteoclasts (black asterisks) resorb the bone of the dentary in a juvenile carp, *Cyprinus carpio*. The osteoclasts create pronounced resorption lacunae (black arrowheads) and secrete tartrate-resistant acid phosphatase (TRAP) into the subcellular space. The white arrowheads point to the abundant osteocytes. Modified from Witten et al. (2000). Visualization of TRAP counterstained with hematoxylin, scale bar = 20 μm.

PLATE 17. In contrast to the carp (Plate 16), bone resorption in the Nile tilapia, *Oreochromis niloticus*, is accomplished by abundant mononucleated cells (black arrowheads). Similar to multinucleated osteoclasts, the cells contain tartrate-resistant acid phosphatase (TRAP) and secrete TRAP onto the bone surface. Notice the complete lack of osteocytes (see also Plate 14). Visualization of TRAP counterstained with hematoxylin, scale bar = 20 μm.

PLATE 18. Small mononucleated bone-resorbing cells (black arrowheads) on the vertebral column of the gray bichir, *Polypterus senegalus*, are identified by demonstration of the osteoclast marker enzyme tartrate-resistant acid phosphatase (TRAP). The white arrowhead points to an osteocyte. Visualization of TRAP counterstained with hematoxylin, scale bar = 40 μm.

PLATE 19. A lateral radiograph of a juvenile cichlid, *Oreochromis niloticus*, showing the location of the radiodense upper and lower pharyngeal jaws (white arrowheads). The second radiodense structure is the otoliths (black arrowhead). Scale bar = 1 mm.

PLATE 20. A horizontal section through the pharyngeal region of the cichlid, *Oreochromis niloticus*, showing the well-developed lower (white asterisk) and upper (black asterisk) pharyngeal jaws, used for processing food. Azan, scale bar = 500 μm.

PLATE 21. A juvenile carp, *Cyprinus carpio*, to show the presence of teeth on the lower pharyngeal jaw (white arrowheads). The mandibular jaws are toothless (black arrowheads). A similar situation is found in the related cyprinid species, *Danio rerio*. Alizarin red, scale bar = 500 μm.

PLATE 22. Typical chondroid, consisting of large chondrocytes (white asterisks), sparse extracellular matrix (black arrowheads), and a minimal perichondrium (white arrowheads). Mallory trichrome.

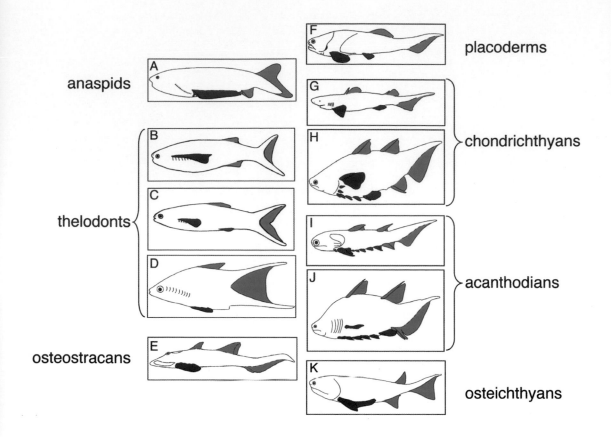

PLATE 23. Suggested homologies for the paired fins of "agnathans" and gnathostomes. Green, median fins; red, pectoral fins, prepectoral fin spines, and suprabranchial paired fins; blue, pelvic fins, prepelvic fin spines and finlets, and ventrolateral paired fins.

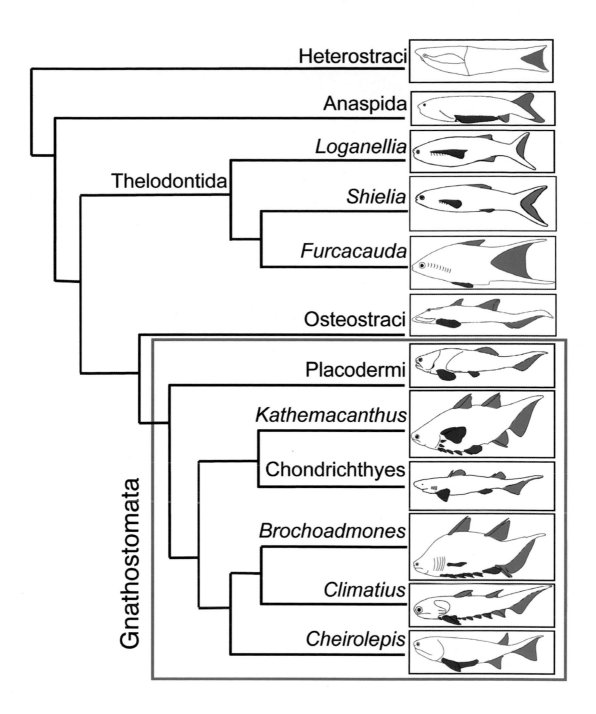

PLATE 24. Simplified phylogenetic distribution of paired fins in fossil and extant vertebrates across the "agnathan"-gnathostome transition. Of the taxa shown, only Chondrichthyes and Osteichthyes (*Cheirolepis* illustrated here) have extant representatives. Paired fins with possible homologies to both pectoral and pelvic fins evolved before the origin of jaws; loss of paired fins was common among fossil "agnathans."

5 cm

PLATE 25. Skull of *Limnoscelis paludis* (YMP 811) in *(from top to bottom)* lateral, dorsal, and ventral views. Courtesy of Dr. David S Berman. Scale = 5 cm.

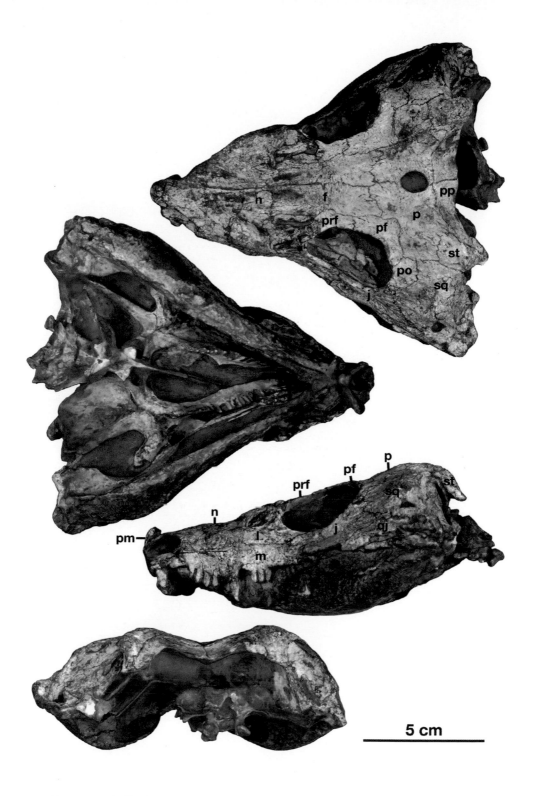

5 cm

PLATE 26. Skull of *Tseajaia campi* (UCMP 59012) in *(from top to bottom)* dorsal, ventral, lateral, and occipital views. Scale = 5 cm.

PLATE 27. Anterior part of skeleton of the diadectid *Orobates pabsti*
(Museum der Natur Gotha, MNG 10181) in *(top)* ventral and *(bottom)* dorsal views.
Courtesy of Dr. David S Berman. Scale = 5 cm.

Characters and states used in the phylogenetic analysis. Italic text indicates developmental characters; boldface text, new characters added to the matrix of Anderson (2001)

Character	State 0	State 1	State 2	State 3
1 Intertemporal	Present	Absent		
2 Supratemporal	Present	Absent		
3 Supratemporal exposure on occiput	Absent	Present		
4 Tabular-postorbital	Absent	Present		
5 Tabular-postfrontal	Absent	Present		
6 Squamosal-tabular	Absent	Present	Fused	
7 Squamosal-temporal articulation	Weakly overlapping	Sutural		
8 Lacrimal-prefrontal articulation	Mobile	Sutural		
9 Lacrimal	Present	Absent		
10 Lacrimal forms narial margin	Present	Absent		
11 Lacrimal forms orbital margin	Absent	Present		
12 Lacrimal orbital processes	Only ventral	Dorsal and ventral	Neither present	
13 Quadratojugal	Present	Absent		
14 *Quadratojugal-jugal*	*Articulate*	*No contact*		
15 *Quadratojugal-maxilla*	*Articulate*	*No contact*		
16 Frontals	Paired	Fused		
17 Frontal in orbital margin	Present	Absent		
18 Nasals	Present	Absent		
19 Alary process of premaxilla	Absent	Present		
20 Septomaxilla	Ossified	Unossified		
21 Prefrontal to external naris	Distant from	Near	Contacts	
22 External naris in dorsal view	Visible	Not exposed		
23 Dorsal exposure of premaxilla	Broad	Narrow	None	
24 Dorsal shape of skull	Triangular	Diamond		
25 Posterior skull margin	Concave	Straight	Convex	Undulating

(continued)

Character	State 0	State 1	State 2	State 3
26 Snout	Blunt	Pointed		
27 Snout	Short	Long		
28 Quadrate-internal flange of squamosal	Absent	Present		
29 Otic notch	Present	Absent		
30 Raised orbital rim	Absent	Present		
31 Postorbital	Present	Absent		
32 Jugal-postorbital interfingered processes	Absent	Present		
33 Postorbital in orbital margin	Present	Absent		
34 Parietal-postorbital contact	Absent	Present		
35 Parietal-squamosal contact	Absent	Present		
36 Parietal-tabular contact	Absent	Present		
37 Postparietals	Paired	Fused	Absent	
38 Parietal foramen	Present	Absent		
39 Postparietals	Moderate	Large		
40 Postparietal-squamosal contact	Absent	Present		
41 Squamosal-jugal contact	Present	Absent		
42 Tabular	Present	Absent		
43 Tabular horns	Absent	Present		
44 Tabular horns	Parallel	Widely divergent		
45 Squamosal forms base of tabular horn	Absent	Present		
46 Lateral line canal grooves	Present	Absent		
47 *Dermal sculpturing*	*Circular pits*	*Shallow ridges*	*Little to none*	
48 Premaxilla anterior margin	Vertical	Overturned		
49 Maxilla into orbital margin	No	Yes		
50 Maxilla into external naris	Present	Absent		
51 Maxilla entire ventral narial margin	Absent	Present		
52 Maxilla	Longer than palatine	Shorter than palatine		

	State 0	State 1	State 2	State 3
53 Marginal teeth	Vertical	Turned medially		
54 Marginal teeth largest anterior	Absent	Present		
55 Marginal teeth	Pointed pegs	Blunt pegs	Large cones	
56 Number of premaxillary teeth	≥10	5–9	<5	
57 Number of maxillary teeth	≥30	20–29	<20	
58 Teeth laterally compressed	No	Yes		
59 Enlarged teeth mid–tooth row	Absent	Present		
60 *Teeth*	*Simple points*	*Multiple cusps*		
61 Multiple cusp orientation	Labiolingual	Anteroposterior		
62 Enamel fluting	Absent	Present		
63 *Labyrinthine infolding*	*Present*	*Absent*		
64 Occipital profile	Low and wide	High and wide	High and narrow	
65 Supraoccipital	Absent	Present		
66 Occipital condyle	Concave	Convex		
67 Occipital condyle	Single	Double		
68 Jugular foramen	Between opisthotic and exoccipital	Through exoccipital		
69 Opisthotic connection to skull roof	Strongly integrated	Loosely articulated		
70 Jaw articulation	Posterior to occiput	Even with occiput	Anterior to occiput	Far anterior (>20% basal skull length)
71 Internal nares	Widely separated	Narrowly separated		
72 Palatal teeth	Present	Absent		
73 *Palatal teeth*	*Single pit-pairs*	*Multiple in rows*	*Multiple random*	
74 Vomerine teeth	Present	Absent		
75 Denticles on pterygoid	Present	Absent		
76 Teeth on pterygoid	Absent	Present		
77 Denticles on vomers	Present	Absent		
78 Denticles on palatines	Present	Absent		
79 Denticles on parasphenoid	Present	Absent		

(continued)

APPENDIX 5.1
Continued

Character	State 0	State 1	State 2	State 3
80 *Palatal teeth*	*Larger than marginals*	*Equal to marginals*	*Smaller than marginals*	
81 Parasphenoid	Medial of stapes	Under footplate of stapes		
82 Basicranial articulation	Loose	Sutured or fused		
83 Stapes	Perforated stem	Imperforate stem	No stem	
84 Footplate of stapes	Oval	Round	Palmate	
85 Dorsal process of stapes	Absent	Present		
86 Accessory ossicle	Absent	Present		
87 Pleurosphenoid	Unossified	Ossified		
88 Sphenethmoid	Ossified	Unossified		
89 Interpterygoid vacuities	Narrow ("closed")	Wide	Fused at midline	
90 Pterygoids contact anteriorly	Present	Absent		
91 Pterygoid-exoccipital contact	Absent	Present		
92 Ectopterygoid	Present	Absent		
93 "Pharyngeobranchial pouches"	Absent	Present		
94 Dentary	Long	Short		
95 Dentary forms coronoid process	Absent	Present		
96 Surangular	Normal	Reduced	Absent	
97 Angular	Narrow	Deep		
98 Number of splenials	2	1	0	
99 Splenial exposed laterally	Present	Absent		
100 Meckelian fossae	2 or more	1	0	
101 Ventral border of Meckel's fossa	Splenial	Angular		
102 Retroarticular process	Absent	Present		
103 Retroarticular process	Straight	Hooked		
104 Articulation to tooth row	Above	Equal	Below	

No.	Character	Posterior tooth row	Middle of tooth row	
105	Angular anterior extent			0
106	Number of coronoids	3	2	1
107	Coronoid teeth	Present	Absent	
108	Coronoid teeth	Larger than marginals	Equal to marginals	Smaller than marginals
109	Symphysis	Dentary and splenal	Dentary alone	
110	Jaw sculpture	Present	Absent	
111	*Ossified byoids*	*Present*	*Absent*	
112	Number of accessory vertebral articulations	0	1	2 or more
113	Strong lateral ridge between zygapophyses	Absent	Present	
114	Caudal processes between depression	Absent	Present	
115	Trunk intercentra	Present	Absent	
116	Atlas-axis intercentra	Present	Absent	
117	Trunk neural arch to centrum	Loosely articulated	Sutured	Fused
118	Base of neural to hemal spine	Equal to	Smaller than	
119	Height of neural spines	Even	Alternating	
120	Dorsal neural spine	Narrow and smooth	Laterally broad and sculpted	
121	Neural spine shape lateral	Rectangular	Triangular	
122	Neural spine lateral surface	Smooth	Crenulated	
123	Hemal arches	Present	Absent	
124	Hemal arches	Intercentral	Fused to midlength of centrum	
125	Hemal arches to neural spine length	Longer	Shorter	
126	Hemal accessory articulations		One	Two
127	Hemal arch shape	Triangular	Rectangular	
128	Tail	Tapers	Deep with sudden end	
129	Trunk arches	Paired	Fused	
130	Spinal nerve foramina	Absent	Present	
131	Extended transverse processes	Absent	Present	

(continued)

APPENDIX 5.1
Continued

Character	State 0	State 1	State 2	State 3
132 Transverse process	On arch pedicle	On centrum		
133 Atlas anterior centrum	Same size as posterior	Laterally expanded		
134 Atlas centrum	Multipartite	Single notochordal	Single odontoid	
135 Atlas neural arch	Loosely articulated	Sutured to centrum	Fused to centrum	
136 Atlas parapophyses	On centrum	On transverse process	Absent	
137 Atlas neural arch	Paired	Sutured at midline	Fused at midline	
138 Atlas accessory articulation	Absent	Zygosphene	Zygantra	
139 Proatlantes	Present	Absent		
140 Second cervical arch to more posterior	Expanded	Equal	Shorter	
141 Atlas ribs	One pair	Two pairs	Absent	
142 Cervical rib distal shape	Spatulate	Pointed		
143 Ribs anterior to sacrum	Short	Long		
144 Costal process at rib head	Absent	Present		
145 Number of sacrals	1	2	3	
146 Sacral parapophysis	On centrum	On transverse process		
147 Number pairs of caudal ribs	5 or more	4	3	2 or fewer
148 Interclavicle posterior stem	No or short	Long		
149 Interclavicle posterior stem	Wide	Narrow		
150 Interclavicle	Diamond-shaped	T-shaped		
151 Interclavicle anterior plate	Broad	Narrow		
152 Interclavicle shape–diamond	Broad	Narrow		
153 Interclavicle anterior fimbriation	Present	Absent		
154 Interclavicle sculpture	Present	Absent		
155 Cleithrum dorsal shape	Rounded or pointed	T- or Y-shaped		
156 Proximal clavicle blades	Widely separate	Articulate medially	Interdigitate	
157 Supraglenoid foramen	Present	Absent		
158 Number coracoid foramina	0	1	2	

	State 1	State 2	State 3	State 4
159 *Scapulocoracoid ossification*	*Both*	*Scapula only*	*Absent*	
160 *Entepicondylar foramen*	*Present*	*Absent*		
161 *Torsion in humerus*	*Absent*	*Less than 80 degrees*	*More than 80 degrees*	
162 *Deltopectoral crest*	*Weak*	*Intermediate*	*Prominent*	
163 *Supinator process*	*Absent*	*Present*		
164 *Humerus length*	*Long (>4 trunk centra)*	*Short*		
165 *Humerus/radius ratio*	*≥0.7*	*0.5–0.7*	*<0.5*	
166 *Olecranon process*	*Unossified*	*Ossified*		
167 *Podial ossification*	*Fully or partially*	*Unossified*		
168 *Number of digits, manus*	*5 or more*	*4*	*3*	
169 *Pelvis*	*Fused*	*Sutured*	*Poorly ossified*	
170 *Iliac blade*	*2 dorsal processes*	*Narrowly bifurcate*	*Single blade*	
171 *Puboischiadic plate spans*	*6 vertebrae*	*5 vertebrae*	*4 vertebrae*	*3 vertebrae*
172 *Intertrochanteric fossa*	*Distinct*	*Absent*		
173 *Femoral shaft*	*Robust*	*Slender*		
174 *Femur*	*Long*	*Short*		
175 *Distal tibia expansion*	*Present*	*Absent*		
176 *Tarsals*	*Ossified*	*Unossified*		
177 *Number of distal tarsals*	*6*	*5 or fewer*		
178 *Astragalus*	*Absent*	*Present*		
179 *Number of digits, pes*	*5 or more*	*4 or fewer*		
180 *Number of presacrals*	*25–35*	*<25*	*>35*	
181 *Basal skull length*	*>70 mm*	*50–70 mm*	*30–50 mm*	*<30 mm*
182 *Skull/trunk ratio*	*≥0.45*	*0.30–0.45*	*0.20–0.29*	*<0.20*
183 **Tooth pedicely**	**Absent**	**Present**		
184 **Lateral process of palatal pterygoid**	**Absent**	**Present**		
185 **LEP**	**Absent**	**Present**		
186 **Semilunar flange of supratemporal**	**Absent**	**Present**		
187 **Supratympanic flange**	Absent	Present "trematopid"	Present "dissorophid"	

(continued)

APPENDIX 5.1
Continued

Character	State 0	State 1	State 2	State 3
188 Supratympanic shelf	Absent	Present		
189 Parasphenoid-vomer articulation	Narrow	Broad		
190 Otic notch posterior margin	Open	Closed		
191 Lacrimal-jugal contact	Present	Absent		
192 Shape of postorbital	Irregular trapezoid	Triangular, apex caudal		
193 Internarial vacuity	Absent	Present		
194 External naris shape	Circular	Extended along lateral-prefrontal suture	Excavation of lateral only	
195 Pterygoid-vomer contact	Present	Absent		
196 *Gill osteoderms*	*Absent*	*Present, noninterdigitating*	*Interdigitating rakers*	
197 *Vertebral development*	*Arches, then centra*	*Centra and arches*		

Taxon	1	2	3	4	5	6	7	8	9	10	11	12	13	14	15	16	17	18	19	20	21	22	23	24	25	26	27
Acanthostega	1	0	0	0	0	1	0	0	0	0	0	0	?	0	0	0	0	0	0	0	0	0	0	0	0	0	0
Proterogyrinus	0	0	0	0	0	0	0	0/1	0	0	0	1	?	0	0	1	0	0	0	?	0	0	0	0	0	0	0
Balanerpeton	0	0	0	0	0	0	1	1	0	0	1	0	0	0	0	0	0	0	1	1	0	0	0	0	0	0	0
Dendrerpeton	0	0	0	0	0	0	1	1	0	0	1	0	0	0	0	0	0	0	1	0	0	0	0	0	0	0	0
Tuditanus	1	1	-	1	1	1	1	1	0	0	1	1	0	0	1	0	0	0	0	0	0	0	0	0	0	0	0
Asaphestera	1	1	-	1	1	1	1	1	0	0	1	0	0	0	1	0	0	0	0	1	0	0	0	0	3	0	0
Hapsidopareion	1	1	-	0	1	1	0	1	0	0	1	0	1	-	-	0	0	0	0	0	1	0	0	0	1	0	0
Llistrofus	1	1	-	1	1	1	0	1	0	0	1	0	0	1	1	0	1	0	0	0	1	0	0	0	1	0	0
Saxonerpeton	1	1	-	1	1	1	0	1	0	0	1	0	0	0	1	0	1	0	0	1	1	0	0	0	1	0	0
Pantylus	1	1	-	1	1	1	1	1	0	0	1	0	0	0	1	0	0	0	0	0	0	1	1	0	1	1	0
Cardiocephalus sternbergi	1	1	-	1	1	1	1	1	0	0	1	1	0	0	1	0	0	0	0	0	1	1	0	0	2	1	0
Cardiocephalus peabodyi	1	1	-	1	1	1	1	1	0	0	1	1	0	0	1	0	0	0	0	0	1	1	1	1	2	1	0
Euryodus primus	1	1	-	1	1	1	1	1	0	0	1	0	0	0	0	0	0	0	0	0	1	1	2	1	1	1	0
Euryodus dalyae	1	1	-	1	1	1	1	1	0	0	1	1	0	0	1	0	0	0	0	0	1	1	2	0	1	1	0
Pelodosotis	1	1	-	1	1	1	1	1	0	0	1	1	0	1	1	0	0	0	0	0	1	0	0	0	3	1	0
Micraroter	1	1	-	1	1	1	1	1	0	0	1	0	0	1	1	0	0	0	0	0	0	0	0	0	3	0	0
Rhynchonkos	1	1	-	1	1	1	1	1	0	0	1	0	0	0	1	0	0	0	0	0	1	2	1	2	1	0	0
Eocaecilia	1	1	-	-	0	1	1	-	1	0	0	-	-	0	0	0	0	0	0	1	2	1	1	1	2	1	0
Microbrachis	1	1	-	1	0	1	1	1	0	1	1	0	0	0	1	0	0	0	0	1	0	0	0	0	1	0	0
Hyloplesion	1	1	-	0	0	1	1	1	0	0	1	0	0	0	1	0	0	0	0	0	0	0	0	0	1	0	0
Mazon Creek	?	?	?	?	?	?	?	?	?	?	?	?	?	?	?	?	?	?	?	?	?	?	?	0	?	0	0
Batropetes	1	1	-	1	0	1	1	1	0	0	1	0	?	?	?	0	1	0	0	1	2	1	2	1	3	0	0
Carrolla	1	1	-	?	0	?	?	1	0	0	1	1	1	-	-	0	1	0	0	0	2	0	0	1	3	0	0
Quasicaecilia	1	1	-	-	-	-	?	1	0	0	1	0	1	-	-	0	1	0	?	1	1	1	2	1	3	0	0
Odonterpeton	1	1	-	-	-	-	1	1	0	0	1	0	0	0	1	0	0&1	0	0	1	1	0	0	0/1	2	0	0
Utaherpeton	?	?	?	?	0	?	?	1	0	0	1	1	?	?	?	0	0	0	0	?	2	0	0	0	?	0	0
Stegotretus	1	1	-	1	1	1	1	1	0	0	1	1	0	0	0/1	0	0	0	0	0	1	1	1	0	1	0	0
Sparodus	1	1	-	1	1	1	1	?	0	?	1	1	0	0	1	?	0	?	?	?	?	1	?	1	2	?	0
Sauropleura pectinata	1	0	0	0	0	1	0	0	0	0	1	1	0	0	0	1	0	0	0	1	2	0	0	0	0	1	1
Sauropleura scalaris	1	0	0	0	0	1	0	0	0	0	1	1	0	0	0	1	0	0	0	1	2	0	0	0	0	1	1
Sauropleura bairdi	?	?	?	?	?	?	?	0	0	0	1	?	?	?	?	1	?	0	0	1	2	0	0	0	?	1	1
Ptyonius	1	0	1	0	0	1	0	1	0	0	1	1	0	0	1	0	0	0	0	1	2	0	0	0	1	1	0
Ctenerpeton	1	?	?	?	?	?	?	?	?	?	?	?	?	?	?	?	?	?	?	?	?	?	?	?	?	?	?
Urocordylus	1	0	?	?	0	?	0	0	0	0	1	1	0	0	?	?	?	?	?	?	?	0	0	0	?	1	0
Keraterpeton galvani	1	1	-	0	0	1	1	1	0	0	1	0	0	0	0	1	0	0	0	1	2	0	0	0	0	0	0
Batrachiderpeton	1	1	-	0	0	1	1	1	0	0	1	0	0	0	1	0	0	0	0	1	1	0	0	0	0	0	0
Diceratosaurus	1	1	-	0	0	1	1	1	0	0	1	0	0	0	1	0	0	0	0	1	2	0	0	0	0	0	0
Diplocaulus magnicornis	1	1	-	0	0	1	1	1	0	0	0	-	0	0	1	1	1	1	0	1	1	1	0	0	0	0	0
Diplocaulus primus	1	1	-	0	0	1	1	1	0	0	0	-	0	0	1	1	1	1	0	1	1	0	0	0	0	0	0
Diploceraspis	1	1	-	0	0	1	1	-	1	-	-	-	0	0	1	1	1	1	0	1	2	0	0	0	0	0	0
Scincosaurus	1	1	-	1	1	1	1	1	0	0	1	0	0	0	1	0	0	0	0	1	2	0	0	0	1/3	0	0
Brachydectes	1	1	-	0	0	1	1	1	0	0	1	1	1	-	-	0	0	0	0	1	2	0	0	-	2	0	0
Adelogyrinus	1	1	-	1	0	2	1	-	1	-	-	-	0	0	0	0	0	0	0	1	1	0	0	0	3	0	0
Oestocephalus	1	0	0	-	0	1	0	1	0	0	1	0	0	1	1	0	0	0	0	1	2	1	1	0	2	0	0
Phlegethontia	1	1	-	-	-	-	1	-	1	-	-	-	-	-	1	1	0	0	1	2	0	0	0	2	1	0	0
Greererpeton	1	0	0	0	0	1	0	0	0	1	1	0	0	0	0	0	0	0	1	2	0&1	0	0	0	3	0	0
Seymouria baylorensis	0	0	0	0	0	0	1	0	0	1	0	0	0	0	0	0	0	0	0	0	0	0	0	0	1	0	0
Limnoscelis	1	0	1	0	0	1	0	1	0	0	1	0	0	0	0	0	0	0	0	0	0	0	0	0	0	1	0
Branchiosauridae	1	0	0	0	0	0	1	0	0	0	1	2	0	0	0	0	0	0	1	0	1	0	0	0	1	0	0
Micromelerpetontidae	1	0	0	0	0	0	1	0	0	0	1	0	0	0	0	0	1	0	1	0	0	0	0	0	0&1	0	0
Tersomius	1	0	0	0	0	0	1	1	0	0	1	0	0	0	0	0	1	0	1	0	0	0	0	0	1	0	0
Ecolsonia	1	0	0	0	0	1	1	1	0	0	1	1	0	0	0	0	1	0	1	0	2	0	0	0	0	0	0
Trematops/Acheloma	1	0	0	0	0	1	1	1	0	0	?	?	0	0	0	0	1	0	?	0	2	0	0	0	0	0	0
Eryops	1	0	0	0	0	1	?	1	0	0	0	-	0	0	0	0	1	0	?	0	0	0	0	0	0	0	1
Doleserpeton	1	0	0	0	0	1	1	1	0	0	1	0	0	0	0	0	1	0	1	0	0	0	0	0	1	0	0
Salamanders	1	1	-	-	-	-	0	0	0	0	1	2	1	-	-	1	0	1	0	0	0	0	1	1/2	0	0	0
Frogs	1	1	-	-	-	-	0	-	1	-	-	-	1	-	-	0/1	1	0	1	0	-	0	0	0	1	0	0
Albanerpetontidae	1	1	-	-	-	-	0	1	0	0	1	2	1	-	-	1	0	0	?	2	0	0	0	1	1	0	0
Eoscopus	1	0	0	0	0	0/1	1	1	0	0	1	0	0	0	0	0	1	0	1	?	0	0	0	0	0	0	0
Tambachia	1	0	0	0	0	1	1	1	0	0	1	0	0	0	0	0	1	0	1	0	2	0	0	0	0	0	0
new amphibamid	0	0	0	0	?	1	1	1	0	0	1	0	0	0	0	0	1	0	1	0	0	0	0	0	0	0	0
Triadobatrachus	1	1	-	-	-	-	-	?	?	?	?	?	1	-	-	1	1	0	?	?	?	?	?	0	1	?	?

Taxon	28	29	30	31	32	33	34	35	36	37	38	39	40	41	42	43	44	45	46	47	48	49	50	51	52	53	54
Acanthostega	0	0	0	0	0	0	1	0	0	0	0	0	0	0	0	0	-	-	0	0	0	0	0	0	0	0	0
Proterogyrinus	0	0	0	0	0	0	0	0	1	0	0	0	0	0	0	0	-	-	0	0	0	0	0	0	0	0	0
Balanerpeton	0	0	0	0	0	0	0	0	0	0	0	0	0	0	0	0	-	-	1	0	0	0	0	0	0	0	0
Dendrerpeton	0	0	0	0	0	0	0	0	0	0	0	0	0	0	0	0	-	-	1	0	0	0	0	0	0	0	0
Tuditanus	0	1	1	0	0	0	0	0	1	?	0	0	0	0	0	0	-	-	1	1	0	0	0	0	?	0	0
Asaphestera	0	1	1	0	0	0	0	0	1	0	0	0	0	0	0	0	-	-	0/1	1	0	0	0	0	0	0	1
Hapsidopareion	0	1	0	0	0	0	0	0	1	0	0	0	0	0	1	0	-	-	1	2	0	0	0	1	0	0	0
Llistrofus	0	1	0	0	0	0	0	0	1	0	0	1	1	1	0	0	-	-	1	1	?	0	0	1	?	0	0
Saxonerpeton	0	1	1	0	0	0	0	0	1	0	0	0	0	0	0	0	-	-	0	2	0	0	0	1	0	0	0
Pantylus	0	1	0	0	0	0	0	0	1	0	1	1	0	0	0	0	-	-	1	0	1	0	0	0	0	0	0
Cardiocephalus sternbergi	0	1	1	0	0	0	0	0	1	0	0&1	1	0	0	0	0	-	-	1	2	1	1	0	0	0	1	0
Cardiocephalus peabodyi	0	1	1	0	0	0	0	0	1	0	0&1	1	0	0	0	0	-	-	1	2	1	0	0	0	?	1	0
Euryodus primus	0	1	1	0	0	0	0	0	1	0	1	1	0	0	0	0	-	-	?	?	1	0	0	0	0	1	0
Euryodus dalyae	0	1	1	0	0	0	0	0	1	0	0	0	1	0	0	0	-	-	1	2	1	0	0	0	0	1	0
Pelodosotis	0	1	0	0	0	0	0	0	1	0	1	1	0	1	0	0	-	-	1	2	1	0	1	-	0	1	0
Micraroter	0	1	1	0	0	0	0	0	1	0	1	0&1	0	1	0	0	-	-	1	2	1	1	0	0	0	1	0
Rhynchonkos	0	1	1	0	0	0	0	0	1	0	0	1	0	0	0	0	-	-	1	2	1	0	0	0	0	1	0
Eocaecilia	0	1	1	1	1	-	-	-	1	1	0	1	1	1	0	0	-	-	1	2	0	1	0	0	0	0	0
Microbrachis	0	1	0	0	0	0	1	1	1	0	0	0	0	0	0	0	-	-	0	1	0	0	0	0	0	0	0
Hyloplesion	0	1	0	0	0	0	1	1	1	0	0	0	1	0	0	0	-	-	1	2	0	0	0	0	1	0	0
Mazon Creek	?	?	?	?	?	?	?	?	?	?	?	?	?	?	?	?	?	?	?	?	?	?	?	?	?	?	0
Batropetes	0	1	0	0	0	0	1	1	1	2	0	-	-	0	0	0	-	-	1	2	1	0	1	-	?	1	0
Carrolla	0	1	0	0	0	0	1	1	?	2	0	-	-	?	?	0	-	-	1	2	1	0	0	?	0	1	0
Quasicaecilia	0	1	1	0	0	0	1	?	-	2	0	-	-	?	1	0	-	-	1	2	1	0	?	?	?	?	?
Odonterpeton	0	1	1	0	0	0	1	1	-	1	0	0	0	0	1	-	-	-	1	2	0	0	0	0	0	0	0
Utaherpeton	0	1	0	0	0	0	1	?	?	?	0	?	?	0	?	0	-	-	1	2	0	1	0	?	?	0	0
Stegotretus	0	1	1	0	0	0	0	0	1	0	0	1	0	0	0	0	-	-	1	1	1	0	0	1	0	1	0
Sparodus	0	1	1	0	0	0	0	0	1	0	0	1	0	0	0	0	-	-	1	1	?	0	?	?	?	?	0
Sauropleura pectinata	0	1	0	0	1	0	1	0	1	0	0	0	0	0	0	0	-	-	1	1	0	1	1	-	0	0	1
Sauropleura scalaris	0	1	0	0	1	0	1	0	1	0	0	0	0	0	0	0	-	-	1	1	0	1	0&1	0	0	0	1
Sauropleura bairdi	?	?	0	?	?	?	?	?	?	?	?	?	?	?	?	?	?	?	1	1	0	1	1	-	?	0	1
Ptyonius	0	1	0	0	1	0	1	0	1	0	0	0	0	0	0	0	-	-	1	1	0	0	0	0	0	0	0
Ctenerpeton	?	?	?	?	?	?	?	?	?	?	?	?	?	?	?	?	?	?	?	?	?	?	?	?	?	?	?
Urocordylus	0	1	0	0	1	0	1	?	?	?	?	?	?	0	?	0	-	-	1	1	0	0	?	?	?	0	1
Keraterpeton galvani	?	1	0	0	0	0	1	1	1	0	0	1	0	0	0	1	0	1	0	1	0	0	0	0	?	0	0
Batrachiderpeton	1	1	0	0	0	0	1	1	1	0	0	1	0	0	0	1	1	0	0	1	0	1	0	0	1	0	0
Diceratosaurus	1	1	0	0	0	0	1	1	1	0	0	1	0	0	0	1	0	0	0	1	0	1	0	0	1	0	0
Diplocaulus magnicornis	1	1	0	0	0	1	1	1	1	0	0	1	0	0	0	1	1	1	0	1	0	0	0	0	1	0	0
Diplocaulus primus	1	1	0	0	0	1	1	1	1	0	0	1	0	0	0	1	1	1	0	1	0	0	0	0	?	0	0
Diploceraspis	0	1	0	0	0	1	1	1	1	0	0	1	0	0	0	1	1	1	0	1	0	0	0	0	1	0	0
Scincosaurus	1	1	0	0	0	0	0	0	1	2	0	-	-	0	0	0	-	-	0	1	0	1	0	0	0	0	0
Brachydectes	0	1	0	1	-	-	-	1	1	0	1	1	1	-	0	0	-	-	1	2	0	1	0	0	0	0	1
Adelogyrinus	0	0	0	0	0	1	1	1	1	0	0	0	0	0	0/1	0	-	-	0	1	0	1	?	?	?	0	0
Oestocephalus	0	1	0	1	-	-	-	0	0	0	0	1	0	1	0	0	-	-	1	2	0	1	0	0	1	0	0
Phlegethontia	0	1	0	1	-	-	-	-	-	2	1	-	-	0	1	0	-	-	1	2	0	1	0	1	-	0	0
Greererpeton	0	1	0	0	0	0	1	0	0	0	0	0&1	0	0	0	0	-	-	0	0	0	0	0	0	0	0	0
Seymouria baylorensis	0	0	0	0	0	0	0	1	0	0	0	0	0	0	0	0	-	-	0	0	0	0	0	0	0	0	0
Limnoscelis	0	1	0	0	0	0	1	0	1	1	0	0	0	0	0	0	-	-	1	2	1	0	0	0	0	0	1
Branchiosauridae	0	0	0	0	0	0	0	0	0	0	0	0	0	0	0	0	-	-	1	1	0	1	0	0	0	0	0
Micromelerpetontidae	0	0	0	0	0	0	0	0	1	0	0	0	0	0	0	0	-	-	0	1	0	0	0	0	0	0	0
Tersomius	0	0	0	0	0	0	0	0	0	0	1	0	0	0	0	0	-	-	1	1	0	0/1	0	0	0	0	0
Ecolsonia	0	0	0	0	0	0	0	0	0	0	1	0	0	0	0	0	-	-	1	0	0	0	0	0	0	0	0
Trematops/Acheloma	0	0	0	0	0	?	0	0	0	0	1	0	0	0	0	0	-	-	1	0	0	0	0	0	0	?	0
Eryops	0	0	0	0	0	0	0	0	0	0	0	0	0	0	0	0	-	-	0	0	0	0	0	0	0	0	0
Doleserpeton	0	0	0	0	0	0	0	0	0	0	0	0	0	0	0	0	-	-	1	1	0	0	0	0	0	0	0
Salamanders	0	1	0	1	-	-	-	1	-	2	1	-	-	-	1	0	-	-	1	2	0	1	0	0	-	0	0
Frogs	0	0	0	1	-	-	-	0	-	2	0	-	-	-	1	0	-	-	1	2	0	1	0	0	-	0	0
Albanerpetontidae	0	1	0	1	-	-	-	0	-	2	1	-	-	0	1	0	-	-	1	1	0	1	0	0	-	0	0
Eoscopus	0	0	0	0	0	0	0	0	0	0	0	0	0	0	0	0	-	-	1	1	0	1	0	0	0	0	0
Tambachia	0	0	0	0	0	0	0	0	0	0	0	0	0	0	0	0	-	-	1	1	0	0	0	0	0	0	0
new amphibamid	0	0	0	0	0	0	0	0/1	?	0	?	0	0	0	0	0	-	-	1	2	0	1	1	0	0	0	0
Triadobatrachus	?	0	0	1	-	-	-	1	-	2	0	-	-	-	1	0	-	-	1	2	?	1	?	?	?	?	?

Taxon	55	56	57	58	59	60	61	62	63	64	65	66	67	68	69	70	71	72	73	74	75	76	77	78	79	80
Acanthostega	0	0	0	0	0	0	-	0	0	0	0	-	-	0	0	0	0	0	0	0	0	0	0	0	0	0
Proterogyrinus	0	1	0	0	0	0	-	0	0	2	0	0	0	0	0	0	0	0	0	?	0	0	?	1	0	0
Balanerpeton	0	0	0	0	0	0	-	0	0	0	0	0	?	0	0	0	0	0	0	0	0	0	0	0	0	0
Dendrerpeton	0	0	0	0	0	0	-	0	0	0	?	?	0	?	?	0	0	0	0	0	0	0	0	0	0	0
Tuditanus	0	1	1	0	0	0	-	0	1	1	?	?	0	?	?	1	0	?	?	?	0	0	0	0	0	?
Asaphestera	0	1	0	0	0	0	-	0	1	2	1	0	0	0	1	2	0	0	1	0	0	1	0	0	?	0
Hapsidopareion	0	1	1	0	0	0	-	0	1	0	1	1	0	0	1	2	0	0	1	1	0	0	1	1	0	1
Llistrofus	0	?	1	0	0	0	-	0	1	1	?	?	?	?	?	2	?	0	1	0	?	?	?	?	?	1
Saxonerpeton	0	1	1	0	0	0	-	0	1	1	0	?	0	?	1	2	0	1	-	1	0	0	0	0	0	-
Pantylus	2	2	2	0	1	0	-	0	1	0	1	0	0	0	0	1	1	0	2	0	0	0	0	0	1	1
Cardiocephalus sternbergi	2	2	2	1	1	?	?	1	1	0	0	0	0	0	0	3	0	0	1	0	1	1	1	1	1	2
Cardiocephalus peabodyi	2	2	2	1	1	1	0	1	1	0	0	0	0	0	0	3	?	?	?	?	?	?	?	?	?	?
Euryodus primus	2	2	2	0	1	0	-	1	1	0	0	0	0	1	1	2	0	0	1	0	0	0	0	0	1	2
Euryodus dalyae	2	2	2	0	1	0	-	0	1	0	0	0	0	0	1	2	0	0	1	0	0	0	0	0	0	2
Pelodosotis	0	1	2	0	1	0	-	0	1	2	1	0	0	0	1	2	?	0	1	?	0	0	?	?	0	2
Micraroter	1	1	2	0	1	0	-	0	1	0	1	0	0	1	1	2	0	0	1	0	0	0	0	0	0	2
Rhynchonkos	0	1	2	0	0	0	-	0	1	0	1	1	0	1	0	3	1	0	1	0	1	0	1	1	0	1
Eocaecilia	0	0	0	0	0	1	0	0	1	0	0	1	0	-	0	3	0	0	1	0	0	0	1	1	0	1
Microbrachis	0&1	1	1	0	0	0	-	0	1	0	?	0	0	?	0	2	0	1	-	1	0	0	0	0	0	-
Hyloplesion	0	1	1	0	1	0	-	0	1	0	1	0	0	?	?	1&2	0	1	-	1	0	0	0	0	0	-
Mazon Creek	?	?	?	0	0	0	-	0	1	?	?	?	?	?	?	?	?	?	?	?	?	0	?	0	0	?
Batropetes	0	1	2	0	0	1	1	0	1	1	1	1	0	-	0	3	?	?	?	?	1	0	1	1	0	-
Carrolla	0	2	2	1	0	1	1	0	1	1	0	1	0	1	0	3	0	1	-	1	1	0	1	1	1	-
Quasicaecilia	?	?	?	?	?	?	?	?	?	2	0	?	0	-	0	3	?	?	?	?	1	0	?	?	1	?
Odonterpeton	0	1	2	0	0/1	0	-	0	1	1	?	?	?	0	?	2	1	0	1	1	0	0	0	0	0	1/2
Utaherpeton	1	?	1	0	0	0	-	0	1	?	?	?	?	?	?	1/2	0	?	?	1	0	0	?	0	?	?
Stegotretus	2	2	2	0	1	0	-	0	1	0	1	0	0	0	0	2	1	0	2	0	0	1	0	0	0	1
Sparodus	2	2	2	0	1	0	-	1	1	?	?	?	?	?	?	3	?	0	1	?	0	1	?	0	?	1
Sauropleura pectinata	0	0	1	0	0	0	-	0	0&1	2	?	?	0	?	?	0	0	0	1	0	0	1	0	1	1	2
Sauropleura scalaris	0	1	2	0	0	0	-	0	1	1	?	?	0	?	0	0	0	0	1	1	0	1	1	1	1	2
Sauropleura bairdi	0	1	?	0	?	0	-	0	1	1	?	?	?	?	?	0	?	?	?	?	1	?	?	?	?	?
Ptyonius	0	1	2	0	0	0	-	0	1	1	?	?	0	?	0	0	0	0	1	0	1	0	0	1	1	2
Ctenerpeton	?	?	?	?	?	0	?	?	?	?	?	?	?	?	?	?	?	?	?	?	?	?	?	?	?	?
Urocordylus	0	?	?	0	0	0	-	0	1	?	?	?	0	?	0	0	0	?	1	?	0	0	?	1	?	2
Keraterpeton galvani	0	2	2	0	0	0	-	0	1	0	?	1	1	?	?	2	0	0	1	1	0	0	0	1	1	1
Batrachiderpeton	0	1	2	0	0	0	-	0	1	1	?	1	1	?	0	2	0	0	1	0	0	0	0	1	1	1
Diceratosaurus	?	?	?	?	0	0	-	0	1	0	?	1	1	?	?	2	?	?	?	?	?	?	?	?	?	?
Diplocaulus magnicornis	0	1	2	0	0	0	-	0	1	0	0	1	1	0/1	0	3	0	0	1	0	1	0	1	1	1	1
Diplocaulus primus	0	1	2	0	0	0	-	0	1	0	?	1	1	?	0	3	0	0	1	0	1	0	1	1	1	1
Diploceraspis	0	1	2	0	0	0	-	0	1	0	0	1	1	0	0	3	0	0	1	1	1	0	1	1	1	1
Scincosaurus	0	1	2	0	0	0	-	0	1	?	?	1	1	?	0	3	0	1	-	1	1	0	1	0	1	-
Brachydectes	0	1	2	0	0	0	-	0	1	2	1	0	0	0	0	3	1	1	1	0	1	0	1	1	1	1
Adelogyrinus	0	?	0	0	0	0	-	0	1	0	0	0	0	1	?	2	?	1	-	1	0	?	0	0	1	-
Oestocephalus	0	1	0	0	0	0	-	0	1	2	0	0	0	0	0	0	0	0	1	?	1	0	?	1	0	1
Phlegethontia	0	2	2	0	0	0	-	0	1	2	0	0	0	-	0	2	?	?	?	?	1	0	?	?	1	?
Greererpeton	0	0	0	0	0	0	-	0	0	0	0	0	0	?	1	0	0	0	0&1	0	0	0	1	1	0	0&1
Seymouria baylorensis	0	1	2	0	1	0	-	0	0	0	0	0	0	0	0	0	0	0	0	0	0	0	0	0	0	1
Limnoscelis	0	2	2	0	1	0	-	0	0	0	1	0	0	0	0	1	1	1	-	1	0	1	0	0	0	1
Branchiosauridae	0	1	0	0	0	0	-	0	0	0	0	?	?	?	?	0&1	0	0	0&1	0	0	0	0	0	0	1
Micromelerpetontidae	0	0	0	0	0	0	-	0	0	0	0	?	?	?	?	?	0	0	0&1	0	0	0	0	0	0	0
Tersomius	0	0	0	0	0	0&1	0	0	0&1	0	0	0	1	0	1	0	0	0	0	0	0	0	0	0	0	0
Ecolsonia	0	1	0/1	0	1	0	-	0	0	0	0	0	0	?	?	0	0	0	0	0	0	0	0	0	0	0
Trematops/Acheloma	0	1	1	0	1	0	-	0	0	0	0	?	?	?	?	0	0	0	0	0	?	0	?	?	?	0
Eryops	0	0	0	0	1	0	-	0	0	0	0	0	0	1	?	0	0	0	0	0	?	0	?	?	?	0
Doleserpeton	0	0	0	0	0	1	0	0	0	1	0	0	0	1	1	1	0	0	0	1	0	0	0	0	0	0&1
Salamanders	0	0&1	0	0	0	1	0	0	1	0	0	1	1	?	0	2&3	0	-	1&2	0	1	0	0	-	0	1
Frogs	0	0&1	0	0	0	1	0	0	1	0	0	1	1	1	0	0	0	-	1&2	0	1	0	0	-	0	1
Albanerpetontidae	0	?	?	1	0	1	1	0	1	1	0	1	1	-	0	2	?	?	?	?	?	?	?	?	?	?
Eoscopus	0	0	0	0	0	0	-	0	0	0	0	?	?	?	?	0	0	0	0	0	0	0	0	0	0	0
Tambachia	0	1	1	0	1	0	-	0	0	0	0	?	?	?	?	0	0	0	0	0	0	0	0	1	1	0
new amphibamid	0	1	0	0	0	0	-	0	?	0	0	1	1	?	1	0	0	0	0	0	?	0	?	?	?	0
Triadobatrachus	?	?	?	?	?	?	?	?	?	0	0	?	1	?	1	0	?	?	?	?	?	?	?	?	?	?

APPENDIX 5.2
Continued

Taxon	81	82	83	84	85	86	87	88	89	90	91	92	93	94	95	96	97	98	99	100	101	102	103	104	105	106	107	108	109
Acanthostega	0	0	0	0	0	0	0	0	0	0	0	0	0	0	0	0	0	0	0	0	0	0	-	0	1	0	0	0	0
Proterogyrinus	0	0	?	?	?	0	0	0	0	0	0	0	0	0	0	0	0	0	0	?	0	0	-	0	0	0/1/2	0	0	1
Balanerpeton	0	0	0	0	0	0	0	1	1	0	0	0	0	0	0	0	1	0	0	0	0	0	-	0	0	0	1	-	1
Dendrerpeton	0	0	?	?	?	?	0	?	1	1	0	0	0	0	0	0	0	0	1	1	1	0	-	0	0	?	?	?	1
Tuditanus	0	0	2	0	0	0	0	?	0	0	0	?	0	1	0	0	1	0	0	2	-	0	-	2	1	?	?	?	0
Asaphestera	?	0	?	?	?	0	0	1	0	1	0	?	0	0	?	0	1	0	0	?	?	?	?	2	0	?	?	?	?
Hapsidopareion	0	0	1	1	0	1	0	?	0	1	0	0	0	0	1	1	0	1	1	2	-	0	-	2	0	?	?	?	1
Llistrofus	?	?	?	?	?	?	?	?	?	?	?	?	?	0	1	1	0	?	1	?	?	1	0	1	0	?	?	?	?
Saxonerpeton	?	0	?	?	?	?	1	0	1	0	0	0	0	0	1	1	0	0/1	1	?	?	1	0	2	0	?	?	?	?
Pantylus	1	0	1	0	1	1	?	0	0	0	0	1	0	0	1	0	0	0	1	1	0&1	1	0	2	0	2	0	1	0
Cardiocephalus sternbergi	1	0	1	0	0	1	0	0	0	0	0	0	0	0	1	?	0	0	1	2	-	0	-	2	0	2	0	2	1
Cardiocephalus peabodyi	?	?	?	?	?	?	?	?	?	?	0	?	?	0	1	1	1	?	1	?	?	0	-	2	0	?	?	?	?
Euryodus primus	1	0	?	?	?	0	?	0	0	0	0	0	0	0	1	1	1	1	1	2	-	1	1	2	0	3	-	-	1
Euryodus dalyae	1	0	0	0	0	1	0	0	0	0	0	0	0	0	1	1	1	1	1	2	-	0	-	2	0	3	-	-	1
Pelodosotis	1	0	1	2	1	1	0	0	0	1	0	?	0	0	0/1	0	1	1	1	1	1	1	0	2	0	?	0	2	1
Micraroter	0	0	0	1	0	1	0	?	0	0	0	0	0	0	1	0	1	0	1	?	?	1	0	2	0	?	?	?	1
Rhynchonkos	1	0	0	0	0	1	1	0	0	1	0	0	0	0	1	0	0	0	0	1	1	1	0	1	0	1	0	1	1
Eocaecilia	1	0	0	0	0	?	?	0	0	1	0	1	0	0	-	-	-	2	-	?	?	1	0	1	0	3	-	-	1
Microbrachis	1	0	0	0	0	0	0	1	0	0	0	0	0	1	0	1	0	0	1	1	1	0	0	0	0	0	0	2	1
Hyloplesion	0	0	1	1	0	0	?	?	0	1	0	0	0	0	1	?	0	0	0	?	?	0	-	2	0	?	?	?	0
Mazon Creek	?	0	?	?	?	?	?	?	?	0	?	?	?	?	?	?	?	?	?	?	?	?	?	?	?	?	?	?	?
Batropetes	1	0	1	0	0	0	0	0	?	0	?	0	0	1	?	?	1	1	?	?	?	1	0	1	?	?	?	?	1
Carrolla	0	0/1	0	0	0	0	1	0	0	1	0	0	0	?	?	-	?	2	?	?	?	0	?	2	0	3	-	-	1
Quasicaecilia	0	0	1	0	1	?	1	0	?	1	0	?	?	?	?	?	?	?	?	?	?	?	?	?	?	?	?	?	?
Odonterpeton	1	0	2	0	0	1	0	1	0	0	0	0	0	0	1	?	?	0	1	?	?	0	-	2	1	1	1	-	1
Utaherpeton	?	?	?	?	?	?	?	?	?	?	?	?	?	0	0	?	1	0	?	?	?	0	-	2	1	?	?	?	?
Stegotretus	1	0	?	?	?	0	0	0	0	0	1	0	0	1	?	0	0	1	1	1	1	0	-	0	0	?	0	0&1	0
Sparodus	?	?	?	?	?	?	?	?	?	?	0	?	0	?	0	?	?	?	0	2	-	1	0	2	0	?	0	0	?
Sauropleura pectinata	0	0	?	?	?	?	0	0	0	?	0	1	0	0	0	0	0	1	0	0	0	1	0	0	0	3	-	-	1
Sauropleura scalaris	0	0	?	?	?	?	0	0	0	?	0	1	0	0	0	0	0	1	0	0	0	1	0	0	0	3	-	-	1
Sauropleura bairdi	?	?	?	?	?	?	?	?	?	?	?	?	?	?	?	?	?	?	?	?	?	?	?	?	?	?	?	?	1
Ptyonius	0	0	?	?	?	?	0	1	1	0	0	0	0	0	0	1	?	1	?	?	1	1	0	1	?	?	?	?	1
Ctenerpeton	?	?	?	?	?	?	?	?	?	?	?	?	?	?	?	?	?	?	?	?	?	?	?	?	?	?	?	?	?
Urocordylus	?	0	?	?	?	?	0	0	0	0	?	?	0	0	0	?	0	?	0	?	?	1	1	0	0	?	?	?	1
Keraterpeton galvani	?	?	?	?	?	?	0	0	2	0	?	?	0	1	0	0	0	1	0	1	1	1	1	1	1	?	?	?	0
Batrachiderpeton	?	1	?	?	?	?	0	0	2	0	0	1	0/1	1	0	0	0	1	0	1	1	1	1	1	1	2	0	1&2	0
Diceratosaurus	?	1	?	?	?	?	0	0	0	0	1	?	0	0	0	?	1	1	0	1	1	1	0	2	?	2	0	1	1
Diplocaulus magnicornis	0	1	?	?	?	?	0	0	1	1	1	1	1	0	0	0	2	0	1	0	0&1	0	-	2	0	2	0	1	0
Diplocaulus primus	0	1	?	?	?	?	0	0	1	1	1	?	1	0	0	0	?	1	?	1	?	?	?	2	?	?	?	?	?
Diploceraspis	0	1	?	?	?	?	0	0	1	1	1	0	1	0	1	0	1	1	0	1	1	1	0	2	0	2	0	1	0
Scincosaurus	0	1	?	?	?	?	0	0	0	1	0	1	0	?	0	?	?	?	?	?	?	?	?	?	?	?	0	1	?
Brachydectes	1	?	1	1	0	0	1	0	0	1	0	1	0	1	1	0	0	1	1	1	1	1	0	2	0	3	-	-	1
Adelogyrinus	0	0	?	?	?	0	0	?	?	0	0	0	0	0	0	?	0	0	?	1	0	0	0	0&1	?	?	?	?	1
Oestocephalus	?	0	?	?	?	0	0	0	0	0	0	1	0	0	0	0	0	0	1	1	1	0	-	0	0	3	-	-	1
Phlegethontia	0	0	2	0	0	0	0	0	?	0	1	0	0	0	?	?	2	-	2	-	0	-	0	-	?	3	-	-	1
Greererpeton	0	0	0	0	0	0	0	0	0	0	0	0	0	0	0	0	0	0	0	1	0	0	-	0	0	1	0	1	0
Seymouria baylorensis	0	0	1	?	0	1	0	2	0	0	0	0	0	0	0	0	0	0	0	0	0&1	0/1	-	2	0	0	0	2	0
Limnoscelis	0	0	?	?	0	0	0	0	1	0	0	0	0	0	0	0	0	0	1	0	0&1	0	-	0	0	1	0	2	0
Branchiosauridae	?	0	?	?	?	?	?	1	1	1	0	0	0	?	?	?	?	?	?	?	?	0	-	?	?	?	?	?	?
Micromelerpetontidae	?	0	?	?	?	?	?	1	1	1	0	0	0	?	?	?	?	?	?	?	?	?	?	?	?	?	?	?	?
Tersomius	?	0	1	0	0	?	0	0	1	1	0	0	0	0	0	0	0	0	0	0	0&1	0	-	0	1	0	0	2	1
Ecolsonia	?	0	?	?	0	0	?	1	1	0	0	0	0	0	0	0	0	0	?	0	0&1	0	-	0	?	0	0	2	0
Trematops/Acheloma	?	0	?	?	?	?	?	1	1	?	0	0	0	0	?	0	1	0	0	?	?	0	-	0	1	?	?	?	0
Eryops	?	1	?	?	?	?	0	1	1	?	0	0	0	0	?	0	0	?	?	?	0	?	?	?	?	?	?	?	?
Doleserpeton	?	0	?	?	?	?	0	0	1	1	0	1	0	0	0	0	0	0	0	0	0	0	-	0	1	0	0	2	?
Salamanders	0	0	0&1	0/1	0	1	?	1	1	1	0	1	0	0	0	2	-	2	-	2	-	0	-	1	0	3-Feb	0/1	1	0
Frogs	0	0	1	0/1	0	1	?	0	1	1	0	1	0	0	0	2	-	2-Jan	1	2	-	0	-	1	1	3	-	-	0
Albanerpetontidae	?	?	?	?	?	?	?	?	?	?	?	?	?	0	0	2	-	2	-	1	1	1	0	2	1	3	-	-	0
Eoscopus	?	0	0	0	0	?	?	1	1	1	0	0	0	0	0	0	0	0	?	0/1	?	0	-	?	0	?	0	2	0
Tambachia	?	0	0	0	0	?	?	?	1	1	0	0	0	0	?	?	1	0/1	1	?	?	0	-	0	1	?	?	?	0
new amphibamid	?	0	?	?	?	?	?	1	1	0	?	0	0	?	?	?	0	?	?	0	-	?	?	?	0	?	?	?	0
Triadobatrachus	0	0	1	0	0	0	0	0	1	1	-	1	0	?	?	?	?	?	?	1	2	?	?	?	?	?	?	?	?

Taxon	110	111	112	113	114	115	116	117	118	119	120	121	122	123	124	125	126	127	128	129	130	131	132	133	134	135	136	137	138	139	140	141	142	143	144
Acanthostega	0	0	0	0	0	0	0	0	0	0	0	0	0	0	0	0	0	0	0	0&1	0	0	0	0	0	0	0	0	0	0	0	0	0	0	0
Proterogyrinus	0	0	0	0	0	0	0	0	0	0	0	0	0	0	0	0	0	1	?	1	0	0	0	0	0	0	0	0	0	0	0	0	0	?	0
Balanerpeton	0	1	0	0	0	0	0	0	0	0	0	0	0	0	0	0	0	0	0	0	0	0	0	0	0	0	?	0	0	0	0	2	0	0	0
Dendrerpeton	0	1	0	0	0	0	0	0	0	0	0	0	0	0	0	0	0	1	?	1	0	0	0	0	0	0	?	0	0	?	0	?	0	0	0
Tuditanus	0	1	0	0	0	1	1	2	-	0	0	1	0	1	-	-	-	-	?	1	0	0	0	1	2	2	0	2	0	1	1	0	1	0	0
Asaphestera	1	1	0	0	?	1	1	2	?	0	0	1	0	?	?	?	?	?	?	1	0	0	0	1	2	2	?	?	0	1	1	?	?	0	0
Hapsidopareion	1	0	0	0	?	?	1	0	?	?	0	?	0	?	?	?	?	?	?	?	0	0	0	1	2	1	2	0	0	1	?	0	1	?	0
Llistrofus	1	1	0	0	?	1	?	1	?	0	0	1	0	?	?	?	?	?	?	1	0	0	0	?	?	?	?	?	?	?	?	?	?	?	0
Saxonerpeton	1	1	0	0	0	1	1	1	-	0	0	1	0	1	-	-	-	-	?	1	0	0	0	1	2	1	?	2	0	1	1	0	0	1	0
Pantylus	0	0	0	0	0	1	0	1	0	1	0	1	0	0	0	0	0	0	0	1	0	0	0	1	2	2	0	2	0	1	1	1	0	1	0
Cardiocephalus sternbergi	1	1	0	?	?	?	?	?	?	?	?	?	0	?	?	?	?	?	?	?	?	?	?	?	?	?	?	?	?	?	?	?	?	?	?
Cardiocephalus peabodyi	1	1	0	0	?	0	0	1	?	0	0	1	0	?	?	?	?	?	?	1	0	0	0	1	2	2	0	?	0	1	1	0	0	0	0
Euryodus primus	?	1	0	0	?	0	?	2	?	0	0	1	0	?	?	?	?	?	?	1	0	0	0	?	?	?	?	?	?	?	?	?	?	?	?
Euryodus dalyae	1	1	0	0	?	0	0	2	?	0	0	1	0	?	?	?	?	?	?	1	0	0	0	1	2	2	0	0	0	1	0	0	?	?	0
Pelodosotis	1	0	0	0	?	0	1	1	?	1	0	1	0	?	?	?	?	?	?	1	0	0	0	1	2	1	?	1	0	1	1	0	0	1	0
Micraroter	1	0	0	0	0	0	0	1	0	1	0	1	0	0	0	0	0	1	0	1	0	0	0	1	2	2	0	0	0	1	1	0	0	0	0
Rhynchonkos	1	1	0	0	?	0	0	2	?	0	0	1	0	?	?	?	?	?	?	1	0	0	0	1	2	2	-	2	0	1	1	2	1	0	0
Eocaecilia	1	1	0	0	0	0	0	2	?	0	0	-	0	?	?	?	?	?	0	1	1	0	1	1	2	2	-	2	0	1	1	2	1	0	0
Microbrachis	0	0	0	0	0	1	1	1	0	0	0	1	0	0	0	0	0	0	0	1	0	1	0	1	2	1	2	1	0	1	1	0	0	0	0
Hyloplesion	1	1	0	0	0	1	1	1	?	0	0	1	0	?	?	?	?	?	0	0&1	0	1	0	1	2	1	2	2	0	1	1	0	0	0	0
Mazon Creek	?	?	0	0	0	1	1	?	?	?	0	?	0	?	?	?	?	?	0	1	0	1	0	1	2	?	?	0	0	1	1	0	0	0	0
Batropetes	1	1	0	0	0	1	1	2	0	0	0	1	0	0	0	1	0	0	0	1	0	0	0	1	2	2	0	?	0	1	1	0	0	0	0
Carrolla	1	?	?	?	?	?	?	?	?	?	?	?	?	?	?	?	?	?	?	?	?	?	?	1	2	?	?	?	?	?	?	?	?	?	?
Quasicaecilia	?	?	?	?	?	?	?	?	?	?	?	?	?	?	?	?	?	?	?	?	?	?	?	?	?	?	?	?	?	?	?	?	?	?	?
Odonterpeton	1	1	0	0	?	1	1	1	?	0	0	1	0	?	?	?	?	?	?	0	0	0	0	0	1	1	?	1	0	1	1	1	0	?	0
Utaherpeton	1	1	0	0	0	1	1	2	0	0	0	0	0	0	0	0	0	0	?	0&1	0	?	0	?	?	?	?	?	?	?	?	?	0	0	0
Stegotretus	?	1	0	0	?	1	1	1	?	0	0	1	0	?	?	?	?	?	?	1	0	0	0	1	2	1	?	?	?	?	?	?	?	?	0
Sparodus	1	1	0	0	?	1	1	1	?	0	0	1	0	?	?	?	?	?	?	1	0	0	0	?	?	?	?	?	0	1	1	?	0	0	0
Sauropleura pectinata	0	0	1	1	0	1	1	2	0	0	1	1	1	0	1	0	0	0	0	1	0	0	0	0	1	2	2	2	2	1	1	2	1	0	0
Sauropleura scalaris	0	0	1	0	0	1	1	2	0	0	0	1	1	0	1	0	1	0	?	1	0	1	0	0	1	2	2	2	1	1	1	2	1	0	0
Sauropleura bairdi	0	?	1	1	?	?	?	?	?	?	1	?	1	?	?	?	?	?	?	?	1	?	0	?	?	?	?	?	?	?	?	?	?	?	?
Ptyonius	0	0	2	0	1	1	1	2	1	0	0	0&1	1	0	1	0	2	1	0	1	0	1	0	0	2	2	2	2	2	1	1	2	1	1	0
Ctenerpeton	?	?	2	0	1	1	1	2	1	0	0	1	1	0	1	0	2	1	1	1	0	1	0	?	?	?	?	?	?	?	?	?	?	1	0
Urocordylus	0	0	2	0	1	1	1	2	1	0	0	1	1	0	1	1	1	0&1	1	1	0	1	0	0	1	2	2	2	2	1	?	2	1	0	0
Keraterpeton galvani	0	0	1	0	0	1	1	2	0	0	0	1	1	0	1	1	0	0	0	1	0	1	0	0	1	2	2	2	1	1	1	2	1	0	0
Batrachiderpeton	0	0	1	0	0	1	1	2	0	0	0	0&1	1	0	1	1	0	0	0	1	0	1	0	1	1	2	2	2	1	1	2	2	?	0	0
Diceratosaurus	0	0	1	0	0	1	1	2	0	0	1	0	1	0	1	0	0	0	0	1	0	1	0	0	1	2	2	2	1	1	1	2	1	0	0
Diplocaulus magnicornis	0	0	1	0	0	1	1	2	0	0	1	0	1	0	1	0	0	0	0	1	0	1	1	1	1	2	2	2	1	1	2	2	0	0	0
Diplocaulus primus	0	0	1	0	?	1	1	2	?	0	1	0	1	0	1	?	0	?	?	1	0	1	1	1	1	2	2	2	?	?	2	2	0	?	0
Diploceraspis	0	0	2	0	0	1	1	2	0	0	1	0	1	0	1	0	1	0	0	1	0	1	1	1	2	2	2	2	1	1	2	2	?	?	0
Scincosaurus	0	0	1	0	0	1	1	2	1	0	0	0	1	0	1	1	1	1	0	1	0	0	0	0	1	2	2	2	?	1	1	2	1	0	0
Brachydectes	1	0	0	0	0	1	1	1	0	0	0	1	0	0	0	0	0	1	0	0	0	1	0	0	2	1	2	0	0	1	1	2	1	0	0
Adelogyrinus	0	0	0	0	0	1	1	0	?	0	0	1	0	?	?	?	?	?	?	0	0	0	0	0	2	0	2	2	0	1	1	2	?	?	0
Oestocephalus	1	1	1	0	0	1	1	2	-	0	0	0	0	1	-	-	-	-	0	1	1	1	0	0	1	2	2	2	0	0	1	2	1	-	1
Phlegethontia	1	1	1	0	0	1	1	2	-	0	0	0	0	1	-	-	-	-	0	1	1	1	0	0	1/2	2	2	2	0	0	1	2	1	-	1
Greererpeton	0	1	0	0	0	0	0	0	0	0	0	0	0	0	0	0	0	0	?	1	0	1	0	0	0	0	?	0	0	0	0	2	1	1	0
Seymouria baylorensis	0	0	0	0	0	0	0	0	0	0	0	0&1	0	0	0	0	0	1	0	1	0	0	0	0	0	0	0	0	0	1	1	0	0	0	0
Limnoscelis	1	1	0	0	0	0	0	0	0	0	1	0	0	0	0	0	0	0	0	1	0	0	0	0	0	0	0	0	0	1	0	0	0	0	0
Branchiosauridae	?	0	0	0	0	0	0	0	?	0	0	?	0	0	0	?	0	?	0	0	0	0	0	?	?	?	?	?	0	?	1	2	0	0	0
Micromelerpetontidae	?	0	0	0	0	0	0	0	?	0	0	?	0	0	0	?	0	?	0	0	0	0	0	?	?	?	?	?	0	?	1	2	1	0	0
Tersomius	0	1	0	?	?	?	?	?	?	?	?	?	?	?	?	?	?	?	?	?	?	?	?	?	?	?	?	?	?	?	?	?	?	?	?
Ecolsonia	0	?	0	0	0	0	?	0	1	0	0	0	0	0	0	0	0	1	0	1	0	0	0	?	?	?	?	?	?	?	?	?	?	?	0
Trematops/Acheloma	0	?	0	0	0	0	?	0	0	0	0	0	0	0	0	0	0	0	?	1	0	0	0	?	?	?	?	?	?	?	?	?	0	0	?
Eryops	0	?	0	0	0	0	?	0	0	0	0	0	0	0	0	0	0	0	?	1	0	0	0	?	?	?	?	?	?	?	?	?	0	0	?
Doleserpeton	0	1	0	0	0	0	0	1&2	?	0	0	0	0	0	0	?	0	?	?	1	0	0	0	?	1	2	2	2	0	?	?	2	?	?	0
Salamanders	1	0	0	0	0	1	1	2	0	0	0	0	0	0	1	0	0	1	0	1	1	1	0	1	2	2	2	2	0	1	1	2	0	0	0
Frogs	1	0	0	0	0	1	1	2	-	0	0	0	0	1	-	-	-	-	1	1	0	1	0	1	1	2	2	2	0	1	1	2	0	0	0
Albanerpetontidae	1	?	0	0	0	1	1	2	?	0	0	0	0	?	?	?	?	?	?	1	1	0	0	1	2	2	2	2	0	1	2	2	0	0	0
Eoscopus	0	?	0	0	0	0	?	0	0	0	0	0	0	0	0	0	0	1	1	1	0	0	0	?	?	0	2	0	0	?	0&1	2	0	0	0
Tambachia	0	1	0	0	0	?	?	0	?	0	0	0	0	?	?	?	?	?	?	1	?	0	?	?	?	?	?	?	?	?	?	?	0	0	0
new amphibamid	1	0	?	?	?	?	?	?	?	?	?	?	?	?	?	?	?	?	?	?	?	?	?	?	?	?	?	?	?	?	?	?	?	?	?
Triadobatrachus	?	0	0	0	0	1	0	2	-	0	0	?	0	1	-	-	-	-	0	1	0	1	0	1	0	?	1	2	0	0/1	0	0	0	0	0

Taxon	145	146	147	148	149	150	151	152	153	154	155	156	157	158	159	160	161	162	163	164	165	166	167	168	169	170	171	172	173	174	175	176	177	178	179
Acanthostega	0	0	0	0	0	0	0	0	0	0	0	0	0	0	0	0	0	0	0	0	0	0	0	0	0	0	0	0	1	0	0	0	0	0	0
Proterogyrinus	0	0	0	0	0	0	0	0	1	0	0	0	0	2	0	0	1	0	0	?	1	0	1	0	1	1	0	1	1	0	0	0	1	0	0
Balanerpeton	0	0	0	0	0	0	0	0	0	0	0	0	?	?	0	0	2	2	0	0	1	1	0	1	?	2	0	1	1	0	0	0	1	0	0
Dendrerpeton	0	0	0	0	0	0	0	0	0	0	0	0	0	1	0	0	2	2	0	0	1	1	0	1	0	2	1	?	0	0	?	?	?	?	0
Tuditanus	1	0	2	1	1	1	1	-	?	1	0	1	-	-	2	0	2	2	0	0	1	0	0	1	1	0	0	0	1	0	0	0	0	1	0
Asaphestera	?	?	?	1	1	1	0	-	?	0	?	1	0	2	0	0	2	1	0	0	?	1	?	?	2	0	?	?	1	0	?	?	?	?	?
Hapsidopareion	?	?	?	?	?	?	?	?	?	?	?	?	?	?	?	?	?	?	?	?	?	0	0	?	2	2	?	?	?	?	?	?	?	?	?
Llistrofus	?	?	?	?	?	?	?	?	?	?	?	?	?	?	?	?	?	?	0	?	?	0	0	?	?	?	?	?	?	?	?	?	?	?	?
Saxonerpeton	0	?	2	1	1	1	1	-	1	1	0	1	?	?	0	0	2	0	0	0	2	1	0	1	1	2	1	0	1	0	0	0	1	0	0
Pantylus	0	?	0	1	1	1	1	-	1	0	0	2	0	1	0	0	2	2	0	0	2	1	0	1	1	2	0	1	1	0	0	0	1	1	0
Cardiocephalus sternbergi	?	?	?	?	?	?	?	?	?	?	?	?	?	?	?	?	?	?	?	?	?	?	?	?	?	?	?	?	?	?	?	?	?	?	?
Cardiocephalus peabodyi	1	0	?	?	?	?	?	?	?	?	0	?	0	-	1	1	2	1	0	1	2	?	0	?	2	2	1	0	1	0	0	0	?	?	?
Euryodus primus	?	?	?	1	1	1	1	-	1	?	?	1	?	?	0	0&1	2	1	0	1	?	1	0	?	1	2	?	0	1	0	1	?	?	?	?
Euryodus dalyae	?	0	?	?	?	?	?	?	?	?	?	?	?	?	?	?	?	?	?	?	?	1	?	?	?	?	?	?	?	?	?	?	?	?	?
Pelodosotis	1	0	?	0	1	1	1	-	1	1	0	1	1	-	1	0	2	2	0	0	2	1	?	?	0	2	?	0	1	0	0	0	?	?	?
Micraroter	2	?	?	?	?	?	?	?	?	?	0	?	0	-	1	?	?	?	0	?	1	2	0	1	1	0	?	?	?	?	?	?	?	?	?
Rhynchonkos	1	?	3	?	?	?	?	?	?	?	0	?	?	?	?	1	2	2	0	1	2	1	0	1	1	2	2	0	1	0	1	0	1	0	?
Eocaecilia	0	1	3	?	?	?	?	?	?	?	?	?	?	?	0	1	2	2	0	1	2	1	0	2	?	?	?	0	1	1	?	0	1	?	1
Microbrachis	0	0	2	0	1	1	0	-	0	0	0	0	0	-	1	0	2	0	0	1	1	0	0	2	1	1	3	1	1	1	1	0	?	0	0
Hyloplesion	0	0	2	0	?	?	0	?	0	?	0	0	?	?	0	1	2	0	0	1	2	1	0	2	1	2	2	1	1	0	0	0	1	0	0
Mazon Creek	0	?	3	?	?	?	?	?	?	?	0	?	?	?	?	0	2	0	0	1	1	0	0	?	2	2	3	1	1	1	0	1	-	-	0
Batropetes	0	0	0	1	1	1	0	-	1	1	0	1	0	2	0	0	2	2	0	1	2	1	0	1	0	2	3	0	1	1	1	0	1	1	0
Carrolla	?	?	?	?	?	?	?	?	?	?	?	?	?	?	?	?	?	?	?	?	?	?	?	?	?	?	?	?	?	?	?	?	?	?	?
Quasicaecilia	?	?	?	?	?	?	?	?	?	?	?	?	?	?	?	?	?	?	?	?	?	?	?	?	?	?	?	?	?	?	?	?	?	?	?
Odonterpeton	?	?	?	?	?	?	?	?	?	?	0	?	?	?	2	1	2	0	0	1	1	0	0	2	?	?	?	?	?	?	?	?	?	?	?
Utaherpeton	1	?	?	0	0	-	0	0	0&1	?	?	?	?	-	1	0	2	2	0	1	0	1	1	1	1	1	1	?	1	0	0	1	-	-	0
Stegotretus	?	?	?	1	1	1	1	-	?	?	?	0	1	1	0	1	2	2	0	0	2	1	0	?	2	0	0	0	1	0	0	0	?	?	?
Sparodus	?	0	?	1	1	1	0	-	0	?	0	1	?	?	0	1	?	2	0	0	1	1	0	?	1	0	1	?	?	0	0	?	?	?	?
Sauropleura pectinata	0	0	1	0	-	0	0	1	0	0	0	0	?	?	?	1	0	0	0	1	?	?	1	1	2	2	?	1	1	1	0	1	-	-	?
Sauropleura scalaris	0	0	1	0	-	0	0	1	0	0	0	0	?	?	0	1	0	0	1	1	0	0	0	1	2	2	3	1	1	1	0	0	?	1	0
Sauropleura bairdi	?	?	?	?	?	?	?	?	?	?	?	?	?	?	?	?	?	?	?	?	?	?	?	?	?	?	?	?	1	?	?	?	?	?	?
Ptyonius	0	0	2	0	-	0	0	0	0	0	0	0	?	?	?	1	0	0	0	1	?	0	1	1	2	2	3	1	1	1	0	1	-	-	1
Ctenerpeton	0	0	2	0	-	0	0	?	?	0	?	?	?	?	?	?	?	?	?	?	?	?	1	1	0	2	3	1	1	1	0	1	-	?	1
Urocordylus	0	0	3	0	-	0	0	0	?	0	0	0	?	?	0	1	0	2	0	1	1	1	1	?	1	2	3	1	0	1	0	1	-	-	0
Keraterpeton galvani	0	0	3	0	-	0	0	0	?	0	1	2	?	?	?	1	0	0	0	1	0	1	1	1	2	2	?	1	1	1	0	1	-	-	0
Batrachiderpeton	0	?	1	0	-	0	0	0	?	0	1	2	?	?	?	?	?	?	?	?	?	?	?	?	2	?	?	?	?	?	?	?	?	?	?
Diceratosaurus	0	0	3	0	-	0	0	0	?	0	1	2	-	-	2	1	0	0	0	1	0	1	1	1	2	2	?	1	1	1	0	1	-	-	0
Diplocaulus magnicornis	0	1	3	0	0	0	0	0	1	0	1	2	?	?	2	0	0	0	0	1	1	0	1	1	2	?	?	?	1	1	0	1	-	-	0
Diplocaulus primus	?	?	?	?	0	0	0	0	1	0	1	?	?	?	?	?	?	?	?	1	?	?	1	?	?	?	?	?	?	?	?	?	?	?	?
Diploceraspis	?	1	?	0	-	0	0	0	1	0	1	2	?	?	?	?	?	?	?	?	?	?	?	?	2	?	?	?	?	?	?	?	?	?	?
Scincosaurus	0	?	2	0	-	0	0	0	?	0	0	2	?	?	0	0	2	2	0	0	1	0	0	0	1	2	2	0	1	0	0	0	?	1	0
Brachydectes	0	0	3	1	1	0	0	0	1	0	0	0	1	0	0	1	1	0	0	1	0&1	1	1	1	2	2	3	1	1	1	0	0	?	0	0
Adelogyrinus	?	?	?	0	-	0	0	0	0	0	0	?	?	?	2	?	?	?	?	?	?	?	?	?	2	?	?	?	?	?	?	?	?	?	?
Oestocephalus	-	-	0	-	-	-	-	-	-	-	0	-	-	-	-	-	-	-	-	-	-	-	-	-	-	-	-	-	-	-	-	-	-	-	-
Phlegethontia	-	-	0	-	-	-	-	-	-	-	0	-	-	-	-	-	-	-	-	-	-	-	-	-	-	-	-	-	-	-	-	-	-	-	-
Greererpeton	0	0	0	0	-	0	1	0	0	0	0	1	1	1	0	0	1	1	0	1	1	1	1	1	0	2	2	0	1	0	0	1	-	-	0
Seymouria baylorensis	1	0	0	1	1	0	0	0	1	0	0	1	0	2	0	0	2	2	1	0	0	0	0	0	0	1	0	1	1	1	0	0	?	0	0
Limnoscelis	1	0	0	1	1	0	0	0	?	?	0	?	0	?	0	0	2	2	1	0	1	0	1	0	1	1	0	1	1	0	0	0	1	1	0
Branchiosauridae	0	?	0	0	-	0	0	0	?	?	?	?	?	?	2	1	0	0	0	0	1	0	1	1	2	2	-	1	1	0	1	1	-	-	0
Micromelerpetontidae	0	?	0	0	-	0	0	0	?	?	?	?	?	?	2	1	0	0	0	1	1	0	1	1	2	2	-	1	1	0	0	1	-	-	0
Tersomius	?	?	?	?	?	?	?	?	?	?	?	?	?	?	?	?	?	?	?	?	?	?	?	?	?	?	?	?	?	?	?	?	?	?	?
Ecolsonia	0	?	?	?	?	?	?	?	?	?	0	?	0	1	0	1	?	1	1	?	?	0	?	?	1	0	0	1	1	0	?	?	?	?	?
Trematops/Acheloma	?	?	?	?	?	?	?	?	?	?	?	?	?	?	0	1	?	?	?	?	?	?	?	?	?	?	?	?	?	?	?	?	?	?	?
Eryops	?	?	0	0	-	0	0	0	?	0	0	0	?	?	0	?	?	?	?	?	?	2	0	1	0	1	0	?	?	1	0	0	0	0	0
Doleserpeton	0	?	?	?	?	?	?	?	?	?	?	?	?	?	?	?	?	?	?	?	?	?	?	?	?	?	?	?	?	?	?	?	?	?	?
Salamanders	0	1	3	-	-	-	-	-	-	-	-	-	1	1	1	1	1	2	0	0	1	0	0	1	2	2	?	1	1	0	1	0	0/1	0	0
Frogs	0	1	3	-	-	-	-	-	-	-	0	1	1	0	0	1	1	1	0	0	1	0	0	1	0	2	-	1	1	0	1	0	1	0	0
Albanerpetontidae	0	1	?	?	?	?	?	?	?	?	?	?	?	?	0	1	?	?	?	0	1	?	?	?	?	?	?	?	1	0	?	?	?	0	?
Eoscopus	0	1	0	0	-	0	0	0	0	0	?	?	?	?	0	1	1	1	0	?	2	1	0	1	0&1	1	?	0	1	0	0	0	1	0	0
Tambachia	?	?	?	?	?	?	?	?	?	?	?	?	?	?	0	1	1	0	0	1	?	?	?	?	?	?	?	?	?	?	?	?	?	?	?
new amphibamid	?	?	?	?	?	?	?	?	?	?	?	?	?	?	?	?	?	?	?	?	?	?	?	?	?	?	?	?	?	?	?	?	?	?	?
Triadobatrachus	0	1	1/2	-	-	-	-	-	-	-	0	?	?	?	0	1	?	0	0	0	1	1	0	?	1&2	2	-	-	1	0	0	0	?	0	?

Taxon	180	181	182	183	184	185	186	187	188	189	190	191	192	193	194	195	196	197
Acanthostega	0	0	0	0	0	0	0	0	0	0	0	0	0	0	0	0	0	?
Proterogyrinus	0	0	1	0	0	0	0	0	0	0	0	0	0	0	0	0	0	?
Balanerpeton	1	0	0	0	1	0	0	0	0	0	0	0	1	0	0	0	0	0
Dendrerpeton	1	0	0	0	0	0	0	0	0	0	0	0	1	1	0	0	0	?
Tuditanus	0	3	1	0	0	0	0	0	0	0	-	0	0	0	0	0	0	?
Asaphestera	?	2	2	0	0	0	0	0	0	0	-	0	0	0	0	0	0	?
Hapsidopareion	?	3	?	0	0	0	0	0	0	0	-	0	0	0	0	0	0	?
Llistrofus	?	3	?	0	?	0	0	0	0	0	-	0	0	0	0	0	0	?
Saxonerpeton	0	3	1	0	0	0	0	0	0	0	-	0	0	0	0	0	0	1
Pantylus	1	1	0	0	0	0	0	0	0	0	-	0	0	0	0	0	0	?
Cardiocephalus sternbergi	?	3	?	0	0	0	0	0	0	0	-	1	0	0	0	0	0	?
Cardiocephalus peabodyi	2	3	2	0	?	0	0	0	0	0	-	0	0	0	0	?	0	?
Euryodus primus	?	2	?	0	0	0	0	0	0	0	-	0	0	0	0	0	0	?
Euryodus dalyae	?	2	?	0	0	0	0	0	0	0	-	0	0	0	0	0	0	?
Pelodosotis	2	1	3	0	0	0	0	0	0	0	-	0	0	0	0	0	0	?
Micraroter	0	1	3	0	0	0	0	0	0	0	-	1	0	0	0	0	0	?
Rhynchonkos	2	3	3	0	0/1	0	0	0	0	0	-	0	0	0	0	0	0	?
Eocaecilia	2	3	3	1	0/1	0	0	0	0	0	-	-	-	0	0	1	0	?
Microbrachis	2	3	2	0	0	0	0	0	0	0	-	0	0	0	0	0	1	1
Hyloplesion	0	3	2	0	0	0	0	0	0	0	-	0	0	0	0	0	0	1
Mazon Creek	0	3	1	0	?	?	?	?	?	?	-	?	?	?	?	?	0	?
Batropetes	1	3	2	0	0	0	0	0	0	0	-	0	0	0	0	?	0	?
Carrolla	?	3	?	0	0	0	0	0	0	0	-	0	0	?	0	0	0	?
Quasicaecilia	?	3	?	?	?	0	0	0	0	0	-	0	1	?	0	?	0	?
Odonterpeton	?	3	?	0	0	0	0	0	0	0	-	0	0	0	0	0	0	?
Utaherpeton	1	3	1	0	?	0	0	0	0	0	-	1	?	0	0	?	0	?
Stegotretus	?	3	?	0	0	0	0	0	0	0	-	0	0	0	0	0	0	?
Sparodus	1	2	1	0	?	0	?	?	?	?	-	?	?	?	?	?	0	?
Sauropleura pectinata	0	2	1	0	0	0	0	0	0	0	-	1	1	0	0	0	0	?
Sauropleura scalaris	0	2	1	0	0	0	0	0	0	0	-	1	1	0	0	0	0	?
Sauropleura bairdi	?	?	2	0	?	?	?	?	?	?	1	2	0	0	?	?	?	?
Ptyonius	1	1	1	0	0	0	0	0	0	0	-	0	1	0	0	0	0	?
Ctenerpeton	?	?	?	?	?	?	?	?	?	?	?	?	?	?	?	?	?	?
Urocordylus	1	2	2	0	?	0	?	?	?	?	-	0	0	?	?	?	?	?
Keraterpeton galvani	1	3	2	0	?	0	0	0	0	0		0	0	0	0	?	0	?
Batrachiderpeton	1	?	?	0	0	0	0	0	0	0	-	0	0	0	0	0	0	?
Diceratosaurus	1	?	1	0	?	0	0	0	0	0	-	1	0	0	0	?	0	?
Diplocaulus magnicornis	1	0	1	0	0	0	0	0	0	0	-	1	0	0	0	1	0	?
Diplocaulus primus	?	?	?	?	?	?	?	?	?	?	?	?	?	?	?	?	?	?
Diploceraspis	?	0	?	0	0	0	0	0	0	0	-	0	0	0	0	1	0	?
Scincosaurus	1	3	3	0	0	0	0	0	0	0	-	0	0	0	0	0	0	?
Brachydectes	2	3	3	0	0	0	0	0	0	0	-	-	-	0	0	0	0	?
Adelogyrinus	?	2	?	0	0	0	0	0	0	0	-	0	0	0	0	?	0	?
Oestocephalus	2	2	3	0	0	-	0	0	0	-	-	1	-	0	0	0	0	?
Phlegethontia	2	2	3	0	0	-	-	0	0	-	-	-	-	0	0	?	0	1
Greererpeton	2	0	2	0	0	0	0	0	0	0	0	0	0	0	0	0	0	?
Seymouria baylorensis	1	0	1	0	0	0	0	0	0	0	0	0	0	0	0	0	1	?
Limnoscelis	0	0	1	0	0	0	0	0	0	0	0	0	0	0	0	0	0	?
Branchiosauridae	1	3	1	0/1	0	0	?	?	?	0	0	1	0&1	1	0	1	2	0
Micromelerpetontidae	0	2&3	1	0	1	1	?	?	?	0&1	0	1	1	1	0	1	1	0
Tersomius	?	2	?	0	1	1	1	1	1	0	0	0/1	1	1	0	1	0	?
Ecolsonia	?	0	?	0	1	1	1	2	1	0	1	1	0	0	1	0	0	?
Trematops/Acheloma	?	0	?	0	?	1	1	1	1	?	1	0	0	1	1	0	0	?
Eryops	1	0	1	0	1	0	0	0	0	0	0	0	0	0	0	0	0	?
Doleserpeton	?	3	?	1	1	1	1	0	0	0	0	0	1	1	0	1	0	?
Salamanders	0&1	?	2	1	0	0	0	0	0	1	-	-	-	1	0	1	1&2	0&1
Frogs	1	?	1	1	0	0	0	0	1	0&1	0	-	-	1	0	1	0	0
Albanerpetontidae	0&1	?	?	0	?	0	0	0	0	?	-	1	-	0	0	?	0	?
Eoscopus	1	2	?	0	1	0	1	2	1	0	0	1	1	1	0	1	0	0
Tambachia	?	2	?	0	1	1	1	1	1	0	0	0	0	0	1	0	0	?
new amphibamid	?	2	?	0	1	0/1	?	?	?	0	0	1	1	1	2	1	0	?
Triadobatrachus	1	?	1	?	0	?	-	?	1	?	0	?	-	?	?	?	0	?

6. The Cranial Anatomy of Basal Diadectomorphs and the Origin of Amniotes

Robert R. Reisz

ABSTRACT

The cranial anatomy of the diadectomorphs *Limnoscelis* and *Tseajaia* is reevaluated in the context of cotylosaur history and relationships. Previous phylogenetic analyses of tetrapods confirm the monophyly of Cotylosauria, with Diadectomorpha as the sister-group of Amniota, but the pattern of relationships within diadectomorphs has not been examined in detail. A recent phylogenetic analysis supports the hypothesis that *Limnoscelis, Tseajaia,* and Diadectidae form a monophyletic group, with diadectids and *Tseajaia* sharing a more recent common ancestor than either does with *Limnoscelis*. The pattern of diadectomorph relationships and the sister-group relationship between Diadectomorpha and Amniota result in long ghost lineages between the minimum divergence time of these taxa and the earliest appearance of diadectomorphs in the latest Pennsylvanian (Late Carboniferous). Reevaluation of the fossil record implicated in the "anamniote"-amniote transition, the known Permo-Carboniferous diadectomorphs and the Carboniferous amniotes, indicates that the origins of amniotes remains unresolved, and that many aspects of the above transition merit more detailed investigation.

Introduction

Diadectomorph tetrapods have attracted a lot of attention since the nineteenth century, and the composition of the clade Diadectomorpha, with the exception of the type genus *Diadectes*, has varied extensively. Case (1911) was the first paleontologist to evaluate the composition and relationships of what he called the Diadectosauria. However, he included in the larger Cotylosauria (Cope, 1880) not only the diadectids, but also *Bolosaurus,* pareiasaurs, captorhinids, procolophonids, *Seymouria,* and even the microsaur *Pantylus.* At that time, *Seymouria* was thought to be the most primitive amniote, and *Pantylus* was believed to be an early reptile. Watson (1917) proposed Diadectomorpha for the reception of diadectids and attempted to provide a definition. However, he included pareiasaurs and procolophonids in the group. Although *Pantylus* was later recognized to be a microsaurian lepospondyl, all other taxa were variably included in Diadectomorpha. Moss (1972) described *Tseajaia* in some detail and argued that it was closely related to diadectids, but included the latter in Seymouriamorpha together with *Seymouria.* Williston (1911) first described *Limnoscelis* on the basis of an essentially complete, articulated skeleton. Romer (1946) redescribed the skull and interpreted *Limnoscelis* as a reptile, close to the ancestry of all later groups. This relationship was reiterated by him (Romer, 1956) and by Carroll (1969), who allied Limnoscelidae with the reptilian group Captorhinomorpha as early amniotes.

Heaton (1980) was the first to provide a phylogeny of the clade; he regarded Diadectomorpha as a clade of cotylosaurs that included Limnoscelidae, Tseajaiidae, and Diadectidae. This pattern has come to be generally accepted (e.g., Berman et al., 1992) after many decades of debate.

Our understanding of Diadectomorpha has increased greatly in the last two decades (Berman and Sumida, 1990; Berman et al., 2004) and the composition of the clade has been clarified (Laurin and Reisz, 1995), but there is still some uncertainty about diadectomorph phylogeny and its relationship to amniotes. Phylogenetic analyses (Gauthier et al., 1988; Laurin and Reisz, 1995, 1997; Lee and Spencer, 1997) have shown that Diadectomorpha is the sister-taxon of Amniota, with Diadectomorpha + Amniota constituting Cotylosauria, but none of these analyses included *Tseajaia* or considered in detail the interrelationships of diadectomorphs. The only study that undertook such an analysis (Berman et al., 1992) used a data matrix of seven taxa and only nine characters, and interestingly indicated on that restricted data set that diadectomorphs formed a clade with synapsids within Amniota. The most recent diadectomorph phylogeny (Kissel and Reisz, 2004) confirms its status as a monophyletic group, with *Limnoscelis* as the sister-taxon of *Tseajaia* + Diadectidae.

I present here a reconsideration of the cranial anatomy of *Limnoscelis* and *Tseajaia* and review the interrelationships of diadectomorphs, with particular emphasis on the available evidence for diadectomorph monophyly and the relationship of diadectomorphs to amniotes within the framework of the "anamniote"-amniote transition.

Cranial Anatomy of *Limnoscelis*

The Limnoscelidae included the type genus *Limnoscelis* and a number of incompletely known forms (Langston, 1966; Carroll, 1969; Berman and Sumida, 1990), but only two taxa, *L. paludis* and *L. dynatis,* have remained as valid after the most recent taxonomic review (Wideman, 2002).

The material pertaining to *Limnoscelis paludis,* the type species of the genus, was collected by David Baldwin from El Cobre Canyon, New Mexico, in 1879. According to Fracasso (1980), the specimen now designated the holotype of *Limnoscelis paludis* was first mentioned in a letter written by Baldwin to O. C. Marsh (Yale University) and dated September 19, 1879. However, as is all too common in early vertebrate paleontology, this significant find (Yale University, Peabody Museum of Natural History, YMP 811) was overlooked for more than three decades. Since its original description by S. W. Williston in a series of publications (Williston, 1911, 1912), *Limnoscelis* has been considered one of the most important Paleozoic tetrapods, closely associated with the origin of amniotes. Subsequently several paleontologists have restudied its cranial anatomy. Notable among these are A. S. Romer (1946), who reinterpreted the structure of the skull on the basis of the growing knowledge of early reptiles, and M. A. Fracasso (1983, 1987), who restudied the skull in great detail, but only published a redescription of the braincase (Fracasso, 1983). Despite its long history of study, extensive handling, and repeated attempts at preparation and description, the skull of *Limnoscelis paludis* (Fig. 6.1, Plate 25) remains inadequately known. The following description emphasizes those aspects of the cranial structure of this genus that distinguish it from other diadectomorphs.

General Features

The skull of *Limnoscelis* is massively built but low in profile, with a large snout that appears slender because of the lateral expansion of the antorbital region. The massiveness of the snout region is exemplified by the size of the premaxilla, but the skull is robust, as indicated by the transverse thickness of the cranial elements, the thickness of the circumorbital bones, and the width of the ventral edges of the skull. The supraorbital region and the skull table are unusually broad, and the temporal part of the skull roof is low, only slightly taller than the snout. The external narial opening and the orbit face laterally and are only slightly visible in dorsal view. The dorsal surface of the skull roof is relatively flat throughout the length of the skull, extending from the level of the external naris to the posterior edge of the skull table. This dorsal surface is gently domed transversely throughout the snout region, but becomes nearly flat in the orbital and postorbital portions of the skull roof, bending down gently only in the postorbital region. The pineal foramen is oval in outline, slightly wider transversely than anteroposteriorly. The occipital area of the skull is

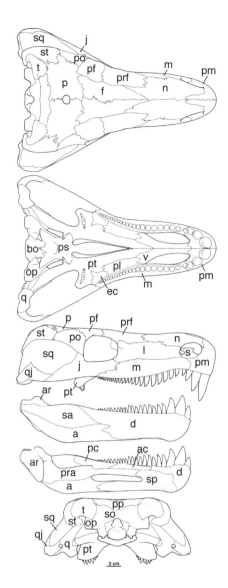

Figure 6.1. Reconstruction of skull and mandible of *Limnoscelis paludis*. Dorsal, ventral, lateral views of skull; lateral and medial views of mandibular; occipital view of skull. **a**, angular; **ac**, anterior coronoid; **ar**, articular; **bo**, basioccipital; **ec**, ectopterygoid; **d**, dentary; **ec**, ectopterygoid; **f**, frontal; **j**, jugal; **l**, lacrimal; **m**, maxilla; **n**, nasal; **op**, opisthotic; **p**, parietal; **pc**, posterior coronoid; **pf**, postfrontal; **pl**, palatine; **pm**, premaxilla; **po**, postorbital; **pp**, postparietal; **pra**, prearticular; **prf**, prefrontal; **ps**, parasphenoid; **pt**, pterygoid; **q**, quadrate; **qj**, quadratojugal; **s**, septomaxilla; **sa**, surangular; **sp**, splenial; **sq**, squamosal; **st**, supratemporal; **t**, tabular; **v**, vomer.

nearly vertical, but is sufficiently tilted anteriorly to make the occipital plate and foramen magnum visible in dorsal view.

In palatal view, the anteroposterior length of the paired internal nares, which occupy nearly 30% of the skull length, reflects the elongation of the snout. The well-developed suborbital foramen is located in an anteriorly oriented, elongate groove that straddles the suture between the pterygoid and ectopterygoid. The interpterygoid vacuity is quite elongate and occupies the central region of the palatal midline. The occipital region of the skull is broad and low, with a large occipital plate that attaches to skull roof elements sloping onto the occiput.

The mandible is also massively built despite being relatively low. It is markedly shorter than the skull, indicating that at full closure the large anterior mandibular teeth fitted into the deep anterior pits of the internal nares. The symphyseal region is large.

Dental Arcade and Circumorbital Bones

The massiveness of the snout is probably related to the large size of the premaxillary teeth, and their deep implantation into the body of this element. The first tooth is the largest of the series, and the largest of the entire marginal dentition. The other premaxillary teeth gradually decrease in size posteriorly, and the size difference between the first and last premaxillary teeth is dramatic, greater than in all other Permo-Carboniferous cotylosaurs, with the possible exception of such derived forms as *Dimetrodon*. Of the three processes that are typically present in all cotylosaurs (diadectomorphs and amniotes), the maxillary process is the most distinct in *Limnoscelis*. It has a long lateral exposure that borders most of the ventral margin of the external naris, then continues posteriorly for a considerable distance as a tall blade along the medial surface of the maxilla and forms the sockets for the premaxillary teeth. This process is taller than in other cotylosaurs, exceeding the height of the maxilla along its length, and not only forms the deep anterolateral border of the internal naris, but also provides supports to the base of the septomaxilla.

The dorsal process of the premaxilla is sutured along the midline to its fellow. It projects slightly anteriorly as it extends anterodorsally, and then curves strongly posterodorsally, forming the tip of the snout. The suture between the premaxillae is strongly interdigitated along the anterior surface of the bone, and remains gently interdigitated on the posterior, narrow part of the dorsal process.

The maxilla is the longest element of the skull roof, forming a thick alveolar shelf and little or no vertical lamina. Its ventral edge is slightly convex in lateral view along most of the length of the bone. Anteriorly, however, the maxilla curves slightly ventrally just beneath the external naris, where it overlaps the premaxilla. The maxillary teeth are set in very deep sockets, but tooth implantation cannot be considered thecodont because the lateral edge of the maxilla extends ventrally well below the level of the alveolar shelf, creating a pleurodont-like topology. In addition, there is clear evidence along the maxillary tooth row (tooth 6) that tooth resorption commences medially. The tooth row extends along most of the ventral edge of the maxilla, and there is space for a total of 23 teeth along the length of the bone. The previous attempts at preparation have resulted in loss of fine detail. However, the fourth maxillary tooth on the left side was protected between the mandibular rami, allowing careful preparation in medial view. Two surprising features were discovered: the marginal teeth are gently recurved, and there are posterior and anterior cutting edges on the distal third of the tooth crown. One of the maxillary teeth was cut near its base, slightly below the edge of the maxillary bone, revealing the presence of a labyrinthodont-like infolding of the tooth. However, the infolding involves mainly the dentine, and there is no enamel layer at this level.

The lacrimal is an elongate element, extending as a sheet of bone from the posterior border of the external naris to the orbit. However, most of its external surface is not flat, but rather slightly convex in transverse section, so that it is partly visible in dorsal view between the essentially vertical maxilla and the ventrally curving nasals. The overall proportions of the lacrimal of *Limnoscelis* are different from those in other diadectomorphs, including *Tseajaia* and diadectids, the bone being relatively more slender in lateral view and only slightly taller than the maxilla throughout its length. In strong contrast, the lacrimal in other diadectomorphs is tall, occupying much of the side of the snout.

Skull Roof and Occiput

The nasals are large elements that form the dorsal surface of the snout. In contrast to other diadectomorphs, they are the longest paired midline elements of the skull roof, in accordance with the obvious elongation of the snout in *Limnoscelis paludis*.

As in all diadectomorphs, the frontal is a relatively narrow, flat element that is excluded from the orbital margin by broad, plate-like postfrontals and prefrontals. In contrast to the condition in other diadectomorphs, where the frontals are approximately rectangular in outline, the frontals of *Limnoscelis* are narrow anteriorly where they are attached to the nasals and wide posteriorly where they contact the parietals, resulting in an approximate triangular outline on either side of the midline. The prefrontal is longer in *Limnoscelis* than in other diadectomorphs; this is possibly a reflection of the snout elongation that characterizes the former.

In contrast to condition seen in other diadectomorphs, the posterior border of the bone is in line with the suture between the frontal and parietal bones, rather than extending deeply into the parietal. The sutural contact between the postfrontal and parietal is therefore significantly shorter than in other diadectomorphs. The jugal is relatively long in *Limnoscelis*, extending far posteriorly from the level of the anterior orbital margin. Its free ventral edge is much longer than in other diadectomorphs, partly because of the large size of the posterior process and partly because of the sutural pattern between the jugal and the quadratojugal. Thus, the jugal not only contributes to the ventral edge of the skull in the orbital region, but also to more than half of the edge in the temporal region.

As in other cotylosaurs, the quadratojugal has a dorsal process that extends upward under the squamosal at the posterolateral corner of the dermal skull roof. This process contacts the lateral edge of the tall dorsal process of the quadrate along approximately half its height.

The squamosal forms most of the temporal region, contacting the elements of the skull table above (supratemporal and postorbital), and the two other elements in this region, the jugal and the quadratojugal below. Over the years, the squamosal of *Limnoscelis* has attracted a lot of attention because of its sutural relationship to the postorbital and the supratemporal. This special relationship has been interpreted as a so-called cardinal line, or line of weakness between the skull roof and the temporal region (Kemp, 1980). This is because both sides of the skull show that the dorsal part of the squamosal is not sculptured near the postorbital and the supratemporal, nor does it attach to the lateral edges of these bones, but at least 3–5 mm medial to them. There is thus no possible line of weakness in this area of the skull because this is a complex, massive suture, and in addition, the anterior junction between the postorbital, jugal, and squamosal is a strongly overlapping sutural area. Posteriorly, the squamosal has a large medial flange that extends onto the occipital surface of the skull, forms a strong overlapping suture with the dorsal process of the quadrate, and extends far ventrally, to the level of the quadrate foramen. This flange therefore extensively overhangs the occipital surface of the quadratojugal and is separated from the temporal region by a clearly defined ridge.

The skull table is composed mainly by the large parietals, which have the typical diadectomorph configuration, contacting the frontal and the postfrontal anteriorly, the postorbital and the supratemporal laterally, and the tabular and the postparietal posteriorly. The median parietal suture is interrupted halfway by a relatively small pineal foramen, relatively smaller than in diadectids or *Tseajaia*. As in other diadectomorphs, the single median postparietal straddles the dorsal and occipital surfaces of the skull, rather than being located dorsally as in noncotylosaurian tetrapods, or occipitally as in amniotes (Berman, 2000).

The large supratemporal forms the posterolateral corner of the skull table. In addition to contributing to the skull table, this bone curves strongly ventrolaterally and posteroventrally to overhang the squamosal, both in the temporal and occipital regions of the skull. The supratemporal is strongly attached to the posterolateral flange of the parietal, has a short anterior contact with the postorbital, overlies the squamosal, contacts the tabular posteromedially, and clearly underlies it on the occiput. As in other diadectomorphs, this bone has a complex relationship to the other elements of the skull, one that appears to be quite distinctive in a number of features. Its contribution to the posterolateral skull table has dermal sculpturing and forms an important structural component of this region of the skull, rather than being the thin, superficial bone more commonly found in most basal amniotes. The supratemporal is strongly curved ventrolaterally in *Limnoscelis,* overhanging the squamosal, and has a free ventral surface that lacks sculpturing. In addition, a large, sheet-like process of the supratemporal extends ventrally beyond the tabular and curves medially to form the lateral and part of the ventral border of a large occipital notch. This notch has been interpreted as the posttemporal fenestra (Berman, 2000). Careful study of this area has revealed that, contrary to Berman's preliminary interpretation of this area, the supratemporal does not contact the paroccipital process of the occiput. Instead of having a posttemporal fenestra completely surrounded by bone, *Limnoscelis* probably had a fenestra that was open ventrally, as in *Varanosaurus* and *Ophiacodon* (Berman et al., 1993), and was probably completed by a cartilaginous extension of the paroccipital process of the opisthotic.

Perhaps the most striking feature of the supratemporal is its relationship to the occipital flange of the squamosal. The well-developed posteroventral flange of the supratemporal does not appear to contact the underlying squamosal. Instead, as indicated on both sides of the skull, the occipital flange of the supratemporal lies approximately 2 mm above the surface of the squamosal, creating a slight gap between this wing-like posteroventral extension of the skull table and the occipital surface. Interestingly, the supratemporal also overhangs the occiput in diadectids and in *Tseajaia* (Moss, 1972), and a similar overhang is present in caseasaurian synapsids (Reisz, 1986). However, the separation between the posteroventral process of the supratemporal and the occiput is less pronounced in *Limnoscelis* and the aforementioned cotylosaurs.

Like the postparietal medial to it, the paired tabulars straddle the posterior edge of the skull table and the occiput and attach anteriorly to the parietal. In addition, however, the tabular has a large contribution to the occipital plate. Medially, the tabular contacts not only the unpaired median postparietal as a narrow, thin sheet of bone, but also forms a massive abutting contact with the lateral process of the supraoccipital. The supraoccipital has been variously interpreted in *Limnoscelis,* and Berman and Sumida (1990) have argued that this bone was paired and arranged on either side of the midline in *L. dynatis.* There is no evidence for such an arrangement in *Limnoscelis paludis.*

Palate

In most Paleozoic cotylosaurs, the vomers and the anterior region of the palate are difficult to study in detail because the lower jaws are frequently preserved in place. Thus, we know relatively little about the structure of the vomer in other diadectomorphs. This paired median bone is long in *Limnoscelis,* extending for almost 30% of the total skull length. Anteriorly, it covers most of the palatal process of the premaxilla, making it appear as if *Limnoscelis* lacks such a process. Posteriorly, the vomer becomes more plate-like rather than rounded in transverse section, even though it continues to form the medial border of the internal naris. This plate-like sheet of bone is tilted posterodorsally and extends posterolaterally on either side of the wedge-shaped anterior process of the pterygoid. This laterally expanded portion of the vomer contacts the palatine along a bifurcating suture and slightly overlaps the pterygoid.

The palatine is a surprisingly robust element that braces the rest of the palate against the cheek. It therefore contacts the other three palatal elements in a series of massive butt joints: the vomer anteriorly, the pterygoid medially, and the ectopterygoid and pterygoid posteriorly. In addition, however, the palatine also has a dorsal sheet of bone that overlies the pterygoid and extends slightly over the vomer and ectopterygoids. Laterally, the palatine forms a massive, tall, and long contact with the cheek. Because it has a dorsoventrally tall sutural contact with the alveolar shelf of the maxilla, it extends far ventrally from its suture with the posterolateral process of the vomer, creating a deep posterior border of the internal naris. On the ventral surface of the palatine, only a relatively small posteromedial area is covered by fine denticles, and these are confluent with a similarly denticulated area of the pterygoid. In this particular regard, the palatine is different in *Limnoscelis* from those of other diadectomorphs in retaining teeth, the primitive condition for cotylosaurs.

Although much smaller than the palatine, the ectopterygoid is a massive element that also braces the skull roof in the orbital region to the palate. It is edentulous and contacts the palatine anteriorly, the pterygoid medially and posteriorly, and the maxilla laterally. Like the palatine, its lateral sutural surface is so deep dorsoventrally that the ectopterygoid probably not only contacted the medial surface of the alveolar shelf of the maxilla but also the jugal.

As in amniotes, the pterygoid has three major structural components: the anterior palatal process, the transverse flange, and the posterior quadrate process. In most respects, the structure of this bone is closest to that in early amniotes, rather than other diadectomorphs, suggesting that this represents the primitive condition for cotylosaurs. The long anterior process of the pterygoid is wedged between the vomers and forms a long, strongly overlapping suture with the palatine laterally. Much of the ventral surface of palatal process is covered by fine denticles. The denticular field terminates posteriorly near the base of the transverse flange and is separated by a gap from the dentition on the flange. In contrast to the condition found in the other diadectomorphs, where a single row of large teeth extends along the medial edge of the entire palatal process, the pterygoid of *Limnoscelis* is edentulous along this edge. Instead, the medial edge of the palatal process of the pterygoid curves dorsally and forms a smooth edge of the elongate interpterygoid vacuity. The transverse flange of the pterygoid is massively constructed in *Limnoscelis* and forms the most robust part of the palate. This large structure bears a row of large teeth along its posteroventral edge, and it also has a narrow field of small denticles medial, lateral, and anterior to the large teeth.

The teeth vary in size, as indicated by the transverse dimensions of their bases, with the largest teeth being those in the middle of the row. Between the ventrolaterally projecting transverse flange and the posterolaterally projecting quadrate process, the pterygoid forms a waist-like constriction. This area is quite robust, although relatively more slender than in diadectids, with a well-developed ridge that supports the transverse flange laterally, and a dorsally curved medial edge that extends the margin of the interpterygoid vacuity posterior to the level of the transverse flange. At the base of the quadrate ramus, the pterygoid expands medially to form the posterolateral edge of the interpterygoid vacuity and forms a fork-like recess that encloses the basicranial articulation. Contrary to previous interpretations (Fracasso, 1987), the pterygoid does not actually form the articulating surface with the braincase. Although the paired fork-like recesses of the pterygoid wrap around the basal tubera, the actual articular surface is formed by the epipterygoid. The same condition is seen in *Diadectes* and *Orobates* as well as in basal amniotes.

The structure of the quadrate probably represents the primitive condition for diadectomorphs, and possibly all cotylosaurs, in having a large ventral condyle for articulation with the mandible, and a dorsal process that forms the posterior wall of the adductor chamber. In *Limnoscelis,* the dorsal process of the quadrate is not emarginated as it is in the other diadectomorphs. Instead, it follows the slightly curved posterior edge of the skull roof and is partly covered posteriorly by the supratemporal, the occipital flange of the squamosal, and the quadratojugal.

Braincase

The braincase and the occiput have been described in detail in previous papers (Fracasso, 1987; Berman et al., 1992) and need not be described here. One issue of contention is the anatomy of the supraoccipital, which has been described as a double or paired ossification (Berman, 2000), and with an anterior ossification (Fracasso, 1987). Although these are potentially significant interpretations of the supraoccipital because the presence of a well-ossified supraoccipital is a synapomorphy of diadectomorphs and amniotes, neither interpretation appears to be well supported by the evidence provided by the skull of *Limnoscelis paludis*. At this time, I prefer to take the conservative approach and regard the controversial anatomical features as artifacts of preservation. There is no evidence of a midline division of the supraoccipital in YMP 811, as has been suggested for *Limnoscelis dynatis* (CM 47653; Berman and Sumida, 1990). Thus, the supraoccipital is interpreted here as a flat, single, plate-like ossification. Another controversial feature of the occipital anatomy is related to the medial extent of the ventral process of the supratemporal on the posterior surface of the skull. This process curves medially underneath the tabular and contributes to the border of the posttemporal fenestra. However, contrary to a previous interpretation of this area (Berman, 2000), there is no direct evidence that it contacts the paroccipital process of the opisthotic.

Mandible

The massiveness of the mandible is reflected by the shape and the size of the dentary. The dentary carries the marginal tooth row and occupies a large proportion of the lateral surface of the mandibular ramus, covering more than two-thirds of its length. Anteriorly, the dentary forms most of the massive symphysis. It is deep dorsoventrally, probably to accommodate the roots of the large anterior teeth, but is also wide mediolaterally. Posterior to the symphyseal area, the dentary is noticeably constricted along its ventral edge and its lateral surface, creating a slight waist in this anterior part of the mandible. Throughout the anterior part of the jaw, the dentary is strongly rounded in transverse section. The dentary increases in height only slightly as it extends posteriorly. As in all other cotylosaurs, the dentary narrows posteriorly toward the posterior end of the ramus and extends dorsally as the angular and surangular border it posteroventrally. The posterior tip of the dentary lies at the dorsal margin of the jaw, just anterior to the peak of the coronoid eminence. There is space for 23 teeth on the dentary, the largest being the second and third teeth.

The splenial is large in *Limnoscelis*. It forms the ventral part of the mandibular symphysis, contributes to the medial surface of the mandibular ramus, and forms most of the ventral surface of the lower jaw. In contrast to the condition seen in most amniotes, where the splenial tends to be a slender, sheet-like element, this bone matches in transverse section the thickness of the dentary along most of its length. Two posterior processes of the splenial are unequal in size, the ventral process being much longer than the dorsal one. The short posterodorsal process of the splenial overlies the prearticular at the anterodorsal corner of the fenestra and contacts the anterior coronoid, whereas the more massive, long posteroventral process overlies the angular and forms more that half the ventral border of the infra-Meckelian fenestra.

Two coronoids are present in *Limnoscelis:* a slender, elongate anterior element, and a slightly wider, deeper posterior bone. Both coronoids have small denticles on their medially facing surfaces, medial to the marginal tooth row. The posterior coronoid bridges the medial and lateral walls of the mandible, and forms the anterior edge of the adductor fossa. However, the coronoid does not extend far dorsally, and the mandibular ramus does not have a well-developed coronoid eminence, in contrast to the condition seen in diadectids.

The prearticular is a long, slender element that extends along the medial surface of the mandibular ramus. Posteriorly, the prearticular contributes significantly to the medial (pterygoideus) process of the jaw. Surprisingly, the angular is the longest element of the lower jaw. This does not appear to be the case when the complete mandibular ramus is considered, but transverse sections (not illustrated) along the length of the jaw demonstrate that the angular not only forms the massive posteroventral region of each ramus, but also extends anteriorly as a long, slender ventromedial process that forms much of the floor of the Meckelian canal. The exposed part of the angular contributes to the lateral, medial, and ventral surfaces of the posterior half of the mandibular ramus.

Although not completely exposed because of the close apposition of the mandible to the skull, the available evidence shows that the surangular is a relatively massive bone that forms the lateral wall of the adductor fossa and most of the dorsolateral edge of the jaw posterior to the dentary. Anteriorly, the surangular is attached to the dentary and the posterior coronoid, ventrally to the angular, and posteriorly, it braces the articular above the angular. Although most of the surangular is a sheet-like, flat bone, it is curved slightly medially as it attaches to the articular posteriorly, and its dorsal edge is somewhat thickened in transverse section. This free dorsal edge of the surangular is only gently domed, so that the coronoid eminence is represented by slight convexity. This is in contrast to the condition in diadectids where the coronoid eminence is pronounced. Consequently, the medial and lateral edges of the adductor fossa in *Limnoscelis* are not as widely separated transversely or dorsoventrally, as in other diadectomorphs.

The articular bone is only partly exposed on both rami because it has retained its contact with the quadrate bone. This massive, relatively compact element is braced by the surangular medially, the angular ventrally, and by the surangular laterally. Thus, the articular forms the posterior corner of the lower jaw as well as the posterior wall of the adductor fossa. It has three processes: a large medial process, a short posterior knob that can be interpreted as a retroarticular process similar to that in some amniotes, and a slender anterior sheet-like process. Transverse sections of the mandibular ramus indicate that this anterior process did not extend far anteriorly as in diadectids but probably terminated somewhere along the medial wall of the adductor fossa, attached to the medial surface of the prearticular.

Cranial anatomy of *Tseajaia*

Tseajaia campi was originally described by Vaughn (1964) on the basis of an articulated skeleton (University of California at Berkeley, Museum of Paleontology, UCMP 59012) from the Lower Permian of San Juan County, Utah. He characterized it as a cotylosaur with diadectid-like otic notches. It was subsequently studied and described in detail by Moss (1972), who compared it to various taxa and concluded that it was closely related to *Diadectes* (and hence diadectids), but who also argued that *Tseajaia* should be included in the Seymouriamorpha.

A second, nearly complete skeleton of *Tseajaia campi* (Carnegie Museum of Natural History, CM 38033) was recovered in Arroyo de Agua, New Mexico, in 1979. The skull in this specimen was strongly fractured but has added important new information on the dentition and the posterior end of the skull roof.

The skull of *Tseajaia* (Plate 26) was relatively well described by Moss (1972), but his emphasis of certain features was misplaced. In the context of diadectomorph morphology, the following anatomical features are relevant.

Dental Arcade and Circumorbital Bones

The marginal dentition is generally similar to that of *Limnoscelis,* with the largest upper teeth located anteriorly on the premaxilla. A large tooth is located near the midpoint of the maxillary tooth row. It can be designated as a caniniform. The teeth are relatively simple and conical, and because of damage, it is not possible to determine whether there were any cutting edges on the crowns. However, broken surfaces on the teeth of CM 38033 reveal the presence of the typical diadectomorph infolding of the dentine near the base of the crown.

The lacrimal is different from that in *Limnoscelis,* and similar to that in diadectids in being a tall bone that occupies much of the cheek in lateral view. This is an obvious synapomorphy of diadectids and *Tseajaia,* the primitive condition being present in *Limnoscelis* and amniotes.

The suborbital ramus of the jugal is deep in *Tseajaia,* as in all diadectomorphs, and widely separates the maxilla and quadratojugal. As in diadectids, the free ventral edge of the jugal is shorter than in *Limnoscelis,* and this is because the quadratojugal is relatively short in the latter genus. In contrast to the condition in *Limnoscelis,* the quadratojugal is a long bone in *Tseajaia* and diadectids, extending slightly beyond the posterior edge of the orbit.

The postorbital is unusually short and not in contact with the supratemporal, and there is an extensive contact between the parietal and squamosal bones. In contrast, *Limnoscelis* has a long posterior process that contacts the supratemporal, widely separating the parietal and squamosal. In diadectids, the postorbital contacts the supratemporal, as in *Limnoscelis.*

Skull Roof and Occiput

On the parietal, the pineal foramen is relatively larger than in *Limnoscelis,* and similar to that in diadectids. However, in contrast to the condition in all other diadectomorphs, the opening is located near the posterior edge of the parietal in *Tseajaia,* which is an autapomorphy of this genus.

An otic notch or emargination is clearly present in *Tseajaia,* and in all respects is similar to the emargination seen in all diadectids.

As in all diadectomorphs, the supratemporal extends far ventrally onto the occiput. In *Tseajaia,* this posteroventral process extends over the otic emargination, similar to the condition in diadectids, and is widely separated from the posterior edge of the skull.

Palate

On the palate, the pterygoid bears a row of large teeth that extends anteriorly along the medial edge of the bone, as in diadectids. In contrast, the dentition on the anterior portion of the bone in *Limnoscelis* comprises a broad field of denticles, the presumably primitive condition for cotylosaurs. Thus, the presence of the single row of large teeth on the pterygoid is a synapomorphy of *Tseajaia* and diadectids. The well-developed, anteroposteriorly broad transverse flange of the pterygoid appears to bear numerous small teeth in *Tseajaia*. As in diadectids, and in contrast to the condition seen in *Limnoscelis* and amniotes, there are no teeth on the palatine of *Tseajaia*, which constitutes another synapomorphy for the latter genus and diadectids.

The internal naris is very long in *Tseajaia*, extending posteriorly to the level of the anterior orbital region of the skull. The internal naris is also elongated in *Limnoscelis*, but it does not extend as far posteriorly as in *Tseajaia*. This difference is related to the reduced size of the palatine bone in the latter genus, another clear autapomorphy.

The overall structure of the lower jaw in *Tseajaia* is similar to that in diadectids, with a long, large medial mandibular fenestra and a deep mandibular ramus.

Cranial Anatomy of Diadectidae

Numerous taxa of diadectids were erected in the last 125 years. However, after significant taxonomic revisions (Case, 1911; Olson, 1947), only seven diadectid genera are currently recognized: *Phanerosaurus* Meyer, 1860; *Diadectes* Cope, 1878; *Stephanospondylus* Geinitz and Deichmüller, 1882; *Desmatodon* Case, 1908; *Diasparactus* Case, 1910; *Orobates* Berman et al., 2004; and *Ambedus* Kissel and Reisz, 2004. The cranial anatomy is reasonably well known in only two taxa, *Diadectes absitus* (Berman et. al, 1998) and *Orobates pabsti* (Berman et al., 2004; Plate 27). A revision of this clade and detailed redescription of all taxa is in progress and will not be addressed here. However, some anatomical features of diadectids are worth mentioning within the context of diadectomorph relationships.

In all known diadectids, the dentition is heterodont, with procumbent anterior premaxillary and dentary teeth, and transversely expanded, molariform cheek teeth. These cheek teeth have a central cusp flanked by labial and lingual shoulders and wear facets developing on the lingual and labial shoulders of the maxillary and dentary cheek teeth, respectively. This pattern is seen in all diadectids. As in all diadectomorphs, the teeth are deeply implanted, with delicate fluting near the base. This fluting represents infolding of the dentine and extends onto the root of the tooth (Reisz and Sutherland, 2001). The roots are significantly longer than in other diadectomorphs. On the palate, the transverse flange of the pterygoid is reduced in size. In addition, in all diadectids where this region of the skull is preserved, there is row of large palatal teeth that extends anteriorly on the pterygoid on either side of the midline. Unfortunately, five of the seven diadectid genera are poorly known, making general statements about diadectid cranial anatomy difficult. It is nevertheless reasonable to argue that all diadectids possessed an otic emargination, even though this area of the skull is unknown in some members of the clade. Less certain is the presence of a tuberculate sculpturing pattern that is seen in several diadectid taxa, but it may be absent in the basal form *Ambedus*. Other morphological features will be mentioned in the next section.

Phylogeny of Diadectomorpha

Heaton (1980) proposed that Diadectomorpha represents a monophyletic grouping (Limnoscelidae (Tseajaiidae + Diadectidae)). Subsequent studies (e.g., Gauthier et al., 1988; Laurin and Reisz, 1995, 1997, 1999; Lee and Spencer, 1997) indicated that Diadectomorpha is the sister-taxon of Amniota, with Diadectomorpha + Amniota constituting Cotylosauria. This has now become the generally accepted pattern of relationships, with diadectomorphs as the closest known relatives of the crown group Amniota. However, *Tseajaia* was not incorporated in these analyses, leaving Diadectomorpha as a monophyletic group consisting of *Limnoscelis* and diadectids. The only study since that of Heaton (1980) to consider the interrelationships of diadectomorphs is that of Berman et al. (1992). On the basis of a data matrix of seven taxa and only nine characters of the temporal and occipital region, the analysis of Berman et al. (1992) supported the conclusions of Heaton (1980), with Diadectomorpha consisting of (*Limnoscelis* (*Tseajaia* + *Diadectes*)), but argued for inclusion of this clade within Amniota as the sister-taxon of Synapsida. Similarly, Sumida et al. (1992) also argued for the monophyly of Diadectomorpha on the basis of the atlas-axis complex.

In order to evaluate the phylogenetic position of the new diadectid *Ambedus pusillus*, a new phylogenetic analysis of Diadectomorpha has recently been undertaken (Kissel and Reisz, 2004; Fig. 6.2). It comprised nine taxa, including two outgroups, and 37 cranial, dental, and postcranial characters. The analysis incorporated a number of characters that were derived from the studies by Gauthier et al. (1988), Berman et al. (1992), Laurin and Reisz (1995, 1997), Lee and Spencer (1997), Berman et al. (1998), and Berman (2000).

Amniota, the sister-group of Diadectomorpha, and Lepospondyli were selected as outgroup taxa for the phylogenetic analysis. Because previous studies have strongly supported the monophyly of Cotylosauria, it was unnecessary to test the sister-group relationship between Amniota and Diadectomorpha. Although *Solenodonsaurus* has been hypothesized as the sister-taxon of Cotylosauria (Gauthier et al., 1988; Laurin and Reisz, 1999), the specimens referred to that genus are fragmentary and lack much anatomical information; thus, Lepospondyli, the sister-group of *Solenodonsaurus* + Cotylosauria, was selected as the outgroup in that study. Within Diadectomorpha, *Limnoscelis, Tseajaia, Ambedus, Orobates, Desmatodon, Diasparactus,* and *Diadectes* were analyzed. Because of their incomplete nature and uncertain affinities, *Phanerosaurus, Stephanospondylus,* and the Richards Spur diadectid were not included in the analysis. Because *Tseajaia* remains the only genus assigned to Tseajaiidae, and all valid limnoscelid taxa have been referred to the genus *Limnoscelis* (Wideman, 2002), both Tseajaiidae and Limnoscelidae represent

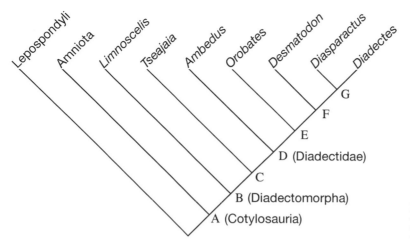

Figure 6.2. Diadectomorph phylogeny. Modified from Kissel and Reisz (2004).

monogeneric taxa, and these family-level names were formally abandoned in that study.

The resulting tree (based on a branch-and-bound search; Swofford, 2002) supported the previous hypothesis that *Limnoscelis,* *Tseajaia,* and diadectids form a monophyletic group, with diadectids and *Tseajaia* sharing a more recent common ancestor than either does with *Limnoscelis.*

The following osteological characters were found to diagnose Cotylosauria (Kissel and Reisz, 2004; numbers correspond to characters and character states used by those authors): postparietal and tabular restricted to the occiput (5/1, 6/2); neural arches of dorsal vertebrae swollen (33/1); sacrum composed of at least two vertebrae (35/1). Additional potential diagnostic characters not part of this list because of uncertain evolutionary history include the following: presence of large, ossified supraoccipital; palatal dentition divided into fields, with an anterior field extending along the medial edge of the pterygoid, and another field extending laterally on the transverse flange. This is because these characters could not be coded for *Solenodonsaurus,* the sister-taxon of Cotylosauria (Laurin and Reisz, 1999).

The monophyly of Diadectomorpha is well supported by the following set of characters that diagnose this clade (Kissel and Reisz, 2004): lateral parietal lappet present (2/1); single median postparietal (4/1); otic trough in ventral flange of opisthotic (18/1); infolding of dentine (25/1); deep marginal tooth roots with root length less than crown height (26/1); anterior process of axial intercentrum-atlantal pleurocentrum complex (32/1); lateral shelf on iliac blade (36/1); humerus short and robust, without a distinct shaft (37/1). Additional diadectomorph characters that were not included in previous analyses include the presence of an elongate medial mandibular fenestra and the presence of large unossified areas in the tarsus and carpus that were probably occupied by cartilage in life (Berman et al., 2004).

The clade of *Tseajaia* and Diadectidae is diagnosed by the following derived characters (Kissel and Reisz 2004): parietal foramen diameter 33% or greater than the anteroposterior length of the parietal midline suture (3/1); vertical, shallow temporal notch present (8/1). Additional characters that diagnose node C (Fig. 6.2) include tall lacrimal, occupying most of the lateral surface of the antorbital region; absence of denticles on palatine bone; and presence of a single row of large teeth on the anterior process of the pterygoid, nearly as large as the marginal dentition.

The monophyly of Diadectidae was also well supported, with diadectids consisting of a series of nested clades that terminates with *Diasparactus* + *Diadectes* (Kissel and Reisz, 2004). *Ambedus* was found to be the sister-taxon to all other members of Diadectidae. Because currently known remains of this genus are restricted to maxilla and dentary, only the following characters were found to be synapomorphies of *Ambedus* and other diadectids (Kissel and Reisz, 2004): node D (Diadectidae)—heterodont dentition, characterized by the presence of transversely expanded cheek teeth (27/1); presence of molarization of largest preserved, midseries dentary cheek teeth (30/1); labial and lingual cusps of cheek teeth represented by shoulders (31/1).

Although the known remains of *Ambedus* lack many of the structures considered in that analysis, the absence of a labial parapet of the dentary, deep tooth implantation, deep lower jaw, and well-developed molariform teeth with lateral and lingual cusps supports the position of *Ambedus* as the sister-group of all other diadectids.

All other diadectids are characterized by presence of a labial parapet of the dentary and the presence of marginal teeth with roots longer than the height of the crown. The presence of a secondary palatal shelf, a surface that may have provided an occlusal surface for the dentary cheek teeth (Olson, 1947; Berman et al., 1998), and the presence of a deep lower jaw diagnoses node F; the presence of a jaw articulation located below the occlusal plane, the presence of a high degree of molarization of the cheek teeth, and the presence of well-developed labial and lingual cusps of the cheek teeth diagnoses node G (Fig. 6.2).

As indicated above, the analyses of Gauthier et al. (1988), Laurin and Reisz (1995, 1997, 1999), and Lee and Spencer (1997) established that Amniota and Diadectomorpha are sister-taxa. Berman et al. (1992) and Berman (2000) argued for a different pattern of relationships, suggesting that Synapsida is the sister-taxon of Diadectomorpha, thus making the latter a member of the crown group Amniota. Berman et al. (1992) united synapsids and diadectomorphs on the basis of the presence of three synapomorphies: (1) the posterolateral corner of the skull table formed entirely or almost entirely by the supratemporal; (2) long posterior expansion of postorbital contacting supratemporal to exclude the parietal lappet from contacting the squamosal; and (3) presence of an otic trough. These three characters are problematic. Although the first character state is present in basal synapsids, recently described remains of *Diadectes* indicate that the posterolateral corner of the skull table of *Diadectes* is formed subequally by the supratemporal and tabular (Berman et al., 1998), a condition shared with all other diadectids in which this region of the skull is known. In *Limnoscelis paludis*, the posterolateral corner of the skull table is also characterized by subequal contributions of the supratemporal and tabular. Only in *Tseajaia* does the tabular only contribute slightly to the posterolateral corner of the skull table, with the supratemporal being the dominant element of this region (Moss, 1972). However, the size of the tabular is uncertain in *Tseajaia*. Thus, the presence of a skull table on which the posterolateral corner is formed entirely or nearly entirely by the supratemporal is not shared by synapsids. As indicated by Laurin and Reisz (1995), the second character of Berman et al. (1992) is present in diadectomorphs and all Amniota, whereas the third probably evolved convergently in both diadectomorphs and Synapsida because several basal synapsids (such as *Eothyris*, *Cotylorhynchus*, *Varanops*, and *Aerosaurus*) have no otic trough.

A more recent analysis that used eight characters of the occipital region (Berman, 2000) is equally problematic. With most of the characters used by Berman et al. (1992) and Berman (2000) to link diadectomorphs and synapsids either refuted or called into question (Kissel and Reisz, 2004), the hypothesis of diadectomorphs and synapsids as sister-taxa is not supported. Instead, in the more inclusive analyses of Gauthier et al. (1988), Laurin and Reisz (1995, 1997, 1999), and Lee and Spencer (1997), which used large data matrices, Diadectomorpha is placed as the sister-taxon of Amniota rather than among Amniota.

Diadectomorphs and the Origin of Amniota

The sister-taxon relationship between Diadectomorpha and Amniota emphasizes the importance of the former to our understanding the origin of the latter. The composition and definition of Amniota have not varied significantly since this taxon was named by Haeckel (1866). Its currently accepted definition (the most recent common ancestor of synapsids, turtles, and diapsids, and all of its descendants) is a crown-group definition (Laurin and Reisz, 1995). The osteologically based phylogeny clearly places diadectomorphs outside the crown group Amniota. However, the available evidence does not address the very important question of whether diadectomorphs laid amniote eggs.

Lee and Spencer (1997) suggested that diadectomorphs laid amniotic eggs, basing this tentative conclusion on two lines of evidence. First, they argued that Cotylosauria was diagnosed by a much larger number of synapomorphies than Amniota, indicating that perhaps some kind of ecological shift had occurred at the base of Cotylosauria. Second, they suggested that diadectomorphs reproduced on land. More recent work (Laurin and Reisz, 1999; Kissel and Reisz, 2004) has shown that most of Lee and Spencer's list of putative synapomorphies for amniotes and diadectomorphs have a wider distribution than they suggested, and hence their conclusion about the ecological shift at the base of Cotylosauria appears to be invalid. Lee and Spencer (1997) also presented some ecological interpretations of the diadectomorph morphology in order to argue that they may have had obligate terrestrial habits and laid amniotic eggs. Although they recognized the dubious nature of this evidence, they suggested that the relatively long ribs of diadectomorphs indicate that they could probably have had coastal ventilation of the lungs, and that their relatively large size precluded the use of cutaneous respiration as a major means of gas exchange. Although this line of reasoning may be valid, their conclusion that diadectomorphs had obligate terrestrial habits and terrestrial reproduction is questionable. Many extant amphibians are terrestrial as adults, but lay their anamniotic eggs in water.

A possible counterargument to their ecological interpretation is based on the structure of the shoulder girdle of diadectomorphs. Although the diadectomorph shoulder girdle resembles that of amniotes in some regards, it retains a large cleithrum and is dramatically different from those of amniotes in being tilted in such a manner that the clavicles and interclavicle extended far anteriorly and were close to the skull. This condition is also seen in many Paleozoic anamniotes like *Seymouria*, embolomeres, and temnospondyl amphibians. Consequently, the gap between the head and the shoulder girdle was significantly narrower in diadectomorphs than in amniotes, where the cleithrum is very slender or completely absent, and the shoulder girdle is widely separated from the head and essentially perpendicular to the long axis of the skeleton. It is therefore reasonable to conclude that diadectomorphs, like their anamniote relatives, were probably less efficient than early synapsids and reptiles during terrestrial locomotion. These differences between the shoulder girdles of diadectomorphs, other groups of Paleozoic anamniotes, and early amniotes have not been explored from a functional perspective, and future work may contribute to our understanding of the biology of these early tetrapods.

Other lines of investigation may provide clues that would allow us to envision diadectomorphs as archaic amniotes. For example, detailed studies of growth and development in diadectomorphs may allow us to determine whether they lacked larval stages or whether they went through metamorphosis. Unfortunately, both *Limnoscelis* and *Tseajaia* are represented by a few adult specimens. However, the fossil record of diadectids is extensive, and growth series may be available for study. Another possible line of investigation involving diadectids would examine the acquisition of high-fiber herbivory. Diadectids like *Diadectes, Desmatodon,* and *Orobates* have dental and skeletal features that indicate they fed on a diet of high-fiber vegetation (Sues and Reisz, 1998; Reisz and Sues, 2000). Modesto (1992) argued that diadectomorphs may have been amniotes because the endosymbionts required for high-fiber herbivory could only be acquired by nest-building amniote tetrapods. This argument is based on evidence provided by modern analogues, like extant iguanid lizards. In these amniotes, the juveniles of each generation acquire the requisite microbes for endosymbiotic fermentation by consuming the droppings of adult conspecifics (Troyer, 1982). Although this hypothesis is attractive because it draws on the most obvious distinction between amniotes and anamniotes (Lee and Spencer, 1997), it is not compelling because amphibian hatchlings and juveniles could just as easily have picked up the endosymbiotic microbes in a terrestrial environment by feeding at or near sites where adults defecated. The apparent inability of present-day amphibians to acquire, or at least maintain, cellulytic endosymbionts is of considerable interest and merits detailed investigation. However, the ability of diadectids to process high-fiber vegetation does not necessarily indicate that they laid amniotic eggs. As the closest relative of amniotes, diadectids may have shared with the latter certain anatomical and physiological attributes that allowed them to maintain cellulytic endosymbionts and feed on plants without being amniotes in reproductive terms.

A novel approach to the study of the "amphibian-reptilian transition" (Carroll, 1970) involved considerations of reproductive patterns of extant lizards and plethodontid salamanders. Carroll suggested that the "anamniote"-amniote transition involved intermediate stages in which anamniotic eggs were laid on land. According to his interpretation, these intermediate forms required a small body size (less than 100 mm snout-vent length). This proposed bottleneck of small size may have been recognizable in the skeletal anatomy of the descendant forms because it would have required reorganization in the otic region to accommodate the relatively larger semicircular canals in miniaturized early amniotes. In this way, Carroll argued that forms like *Seymouria* and *Gephyrostegus* retain the primitive condition of the braincase with a dorsally oriented opisthotic that could not accommodate for the great lateral expansion of the otic capsule. However, on the basis of information on the cranial anatomy of *Limnoscelis* available at that time, he also argued that these diadectomorphs had also not achieved amniote status. More recent studies indicate that the otic region of *Limnoscelis* (Fig. 6.1; Berman, 2000) is intermediate between those of well-known "anamniotes" and amniotes, but closer to the condition seen in early synapsids and reptiles than those in the anamniotic lepospondyls, anthracosaurs, and temnospondyls. Most significantly, the semicircular canals of *Limnoscelis* are enclosed in the laterally projecting paroccipital process of the opisthotic, closely resembling the condition seen in basal amniotes. Thus, the line of reasoning used by Carroll would suggest that diadectomorphs, like crown-group amniotes, went through a bottleneck of small size and laid amniotic eggs.

Carroll's hypothesis of transformation includes a number of evolutionary modifications of the reproductive process, like internal fertilization, abbreviation and elimination of larval stages, laying of eggs in terrestrial locations, and development of extraembryonic membranes and shell. Although this hypothesis is attractive, it is greatly weakened by the assumptions associated with it. For example, Carroll (1970) suggested that all the above changes are interdependent and must have occurred concurrently. In addition, he assumed that the evolution of the amniote reproductive pattern proceeded by an intermediate stage in which anamniotic eggs were laid in damp places on land. Alternative interpretations of the sequence and timing of various parts of the transformation are also possible. There is also no reason to assume that evolution of the amniotic egg could not have occurred in an aquatic environment. It is hoped that future experimental work that uses extant amphibians may be able to test this interesting and potentially important hypothesis, but for the time being, the reproductive strategy of diadectomorphs cannot be established by using the available fossil evidence.

Conclusions

Redescription of the skulls of *Limnoscelis* and *Tseajaia* adds valuable new data on the cranial anatomy of diadectomorphs, the closest known relatives of amniotes. Phylogenetic analysis supports previous hypotheses that *Limnoscelis, Tseajaia,* and Diadectidae form a monophyletic group Diadectomorpha, with diadectids and *Tseajaia* sharing a more recent common ancestor than either does with *Limnoscelis.* With the recognition of Limnoscelidae as a monogeneric taxon (Wideman, 2002), the Permo-Carboniferous clade Diadectomorpha consists of (*Limnoscelis* (*Tseajaia* + Diadectidae)). Only two species of *Limnoscelis, L. paludis* Williston, 1911, and *L. dynatis* Berman and Sumida, 1990, are currently recognized, and *Tseajaia* is a monotypic genus, with *T. campi* Vaughn, 1964 as its type and only known species.

Both *Tseajaia* and *Limnoscelis* appear to be restricted to the Permo-Carboniferous strata (Wolfcampian [Lower Permian] and upper Virgilian [Upper Pennsylvanian], respectively) of the southwestern United States. This restricted geographic and temporal distribution is in strong contrast to that of diadectids, which have an extensive fossil record and a wide geographic distribution. *Desmatodon* is a Late Pennsylvanian taxon known primarily from craniodental remains from the Missourian Red Knob Formation of Pennsylvania (Case, 1908; Berman and Sumida, 1995), the Missourian Sangre de Cristo Formation of Colorado (Vaughn, 1969; Berman and Sumida, 1995), and Virgilian Cutler Formation of New Mexico (Berman, 1993). *Diasparactus* remains have been recovered from the Upper Pennsylvanian Virgilian Cutler and Ada formations of New Mexico and Oklahoma, respectively (Berman, 1993; Kissel and Lehman, 2002). The best-known diadectid, *Diadectes,* is known from the Lower Permian of Colorado, New Mexico, Ohio, Oklahoma, Texas, Utah, West Virginia, Prince Edward Island, and Germany (Olson, 1947, 1967; Langston, 1966; Lewis and Vaughn, 1965; Berman, 1971; Berman et al., 1998). The youngest known diadectid remains have been recovered from the Upper Leonardian (Lower Permian) sediments in the Dolese Quarry near Richards Spur (Reisz and Sutherland, 2001). Thus, if the number of species and the geographic distributions associated with the sister-taxa Diadectidae and *Tseajaia* are considered, it is clear that members of Diadectidae constitute the majority of species within that clade and the much wider pattern of dispersal. This asymmetry in species richness and geographic distribution suggests that Diadectidae represents an evolutionary radiation within Diadectomorpha, possibly associated with the acquisition of high-fiber herbivory (Sues and Reisz, 1998).

The oldest-known diadectomorphs are the diadectids *Desmatodon* and *Diasparactus,* both of which are significantly older than *Tseajaia* and *Limnoscelis.* Diadectomorph phylogeny requires that ghost lineages for the latter two taxa extend well into the Carboniferous (Fig. 6.3). In addition, as the sister-group of amniotes, the early history of diadectomorphs appears to be rather poorly represented in the fossil record. The oldest-known amniotes have been recovered from Middle Pennsylvanian sediments of Joggins, Nova Scotia, where both the eureptile *Hylonomus* and the probable synapsid *Protoclepsydrops* have been recovered. Thus, cotylosaur phylogeny with amniotes and diadectomorphs as sister-taxa and of equal age necessitates inference of long ghost lineages between the minimum divergence time of these two taxa (Fig. 6.3) and the earliest appearance of the diadectomorphs in the Late Pennsylvanian (Late Carboniferous). This indicates that we still know relatively little about the early evolutionary history of diadectomorphs. It is to be hoped that future discoveries of early members of this clade will clarify the early history of this important group of amniote relatives.

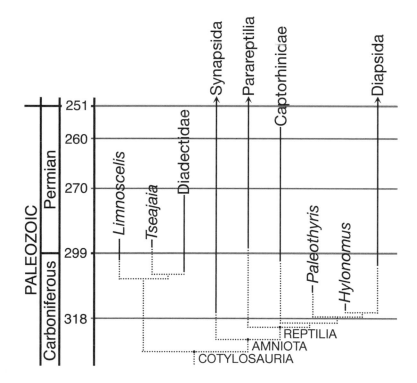

Figure 6.3. Cotylosaur phylogeny within a temporal framework. Modified from Reisz (1997) and Reisz and Sues (2000).

Acknowledgments

I thank Dr. David Berman for the photographs of Limnoscelis *and* Orobates. *I also wish to thank him, Richard Kissel, and Diane Scott for many informative discussions on diadectid and diadectomorph anatomy and relationships. I am grateful to Diane Scott for photographs of* Tseajaia *and assistance with the figures. This research was supported by a Discovery grant from the Natural Sciences and Engineering Research Council of Canada.*

References Cited

Berman, D. S. 1971. A small skull of the Lower Permian reptile *Diadectes* from the Washington Formation, Dunkard Group, West Virginia. Annals of Carnegie Museum 43:33–46.

Berman, D. S. 1993. Lower Permian vertebrate localities of New Mexico and their assemblage; pp. 11–21 in S. G. Lucas and J. Zidek (eds.), Vertebrate Paleontology in New Mexico. New Mexico Museum of Natural History and Science Bulletin No. 2.

Berman, D. S. 2000. Origin and early evolution of the amniote occiput. Journal of Paleontology 74:938–956.

Berman, D. S, and S. S. Sumida. 1990. A new species of *Limnoscelis* (Amphibia, Diadectomorpha) from the Late Pennsylvanian Sangre de Cristo Formation of central Colorado. Annals of Carnegie Museum 59:303–341.

Berman, D. S, and S. S. Sumida. 1995. New cranial material of the rare diadectid *Desmatodon hesperis* (Diadectomorpha) from the Late Pennsylvanian of central Colorado. Annals of Carnegie Museum 64:315–336.

Berman, D. S, S. S. Sumida, and R. E. Lombard. 1992. Reinterpretation of the temporal and occipital regions in *Diadectes* and the relationships of diadectomorphs. Journal of Paleontology 66:481–499.

Berman, D. S, S. S. Sumida, and T. Martens. 1998. *Diadectes* (Diadectomorpha: Diadectidae) from the Early Permian of central Germany, with description of a new species. Annals of Carnegie Museum 67:53–93.

Berman, D. S, A. C. Henrici, R. A. Kissel, S. S. Sumida, and T. Martens. 2004. A new diadectid (Diadectomorpha), *Orobates pabsti*, from the Early Permian of central Germany. Bulletin of Carnegie Museum of Natural History 35:1–36.

Carroll, R. L. 1969. A middle Pennsylvanian captorhinomorph and the interrelationships of primitive reptiles. Journal of Paleontology 43:151–170.

Carroll, R. L. 1970. Quantitative aspects of the amphibian-reptilian transition. Forma et Functio 3:165–178.

Case, E. C. 1908. Description of vertebrate fossils from the vicinity of Pittsburgh, Pennsylvania. Annals of Carnegie Museum 4:234–241.

Case, E. C. 1910. New or little known reptiles and amphibians form the Permian of New Mexico. Bulletin of the America Museum of Natural History 38:163–196.

Case, E. C. 1911. A revision of the Cotylosauria of North America. Carnegie Institution of Washington Publication 145:1–121.

Cope, E. D. 1878. Descriptions of extinct Batrachia and Reptilia from the Permian Formation of Texas. Proceedings of the American Philosophical Society 17:505–530.

Cope, E. D. 1880. Second contribution to the history of the Vertebrata of the Permian formation of Texas. Proceedings of the American Philosophical Society 19:38–58.

Fracasso, M. 1980. Age of the Permo-Carboniferous Cutler Formation vertebrate fauna from El Cobre Canyon, New Mexico. Journal of Paleontology 41:1256–1261.

Fracasso, M. 1983. Cranial osteology, functional morphology, systematics and paleoenvironment of *Limnoscelis paludis* Williston. Ph.D. dissertation, Yale University, New Haven, 624 pp.

Fracasso, M. 1987. Braincase of *Limnoscelis paludis* Williston. Postilla 201:1–22.

Gauthier, J. A., A. G. Kluge, and T. Rowe. 1988. The early evolution of Amniota; pp. 105–155 in M. J. Benton (ed.), The Phylogeny and Classification of Tetrapods. Volume 1, Amphibians, Reptiles, Birds. Systematics Association Special Volume, 35A. Clarendon Press, Oxford.

Geinitz, H. B., and J. Deichmüller. 1882. Die Saurier der unteren Dyas von Sachsen. Palaeontographica 9:1–46.

Haeckel, E. 1866. Generelle Morphologie der Organismen. Allgemeine Grundzüge der organischen Formen-Wissenschaft, mechanisch begründet durch die von Charles Darwin reformirte Descendenz-Theorie. Volumes 1 and 2. Georg Reimer, Berlin.

Heaton, M. J. 1980. The Cotylosauria: a reconsideration of a group of archaic tetrapods; pp. 497–551 in A. L. Panchen (ed.), The Terrestrial Environment and the Origin of Land Vertebrates. Systematics Association Special Volume 15. Academic Press, London.

Kemp, T. S. 1980. Origin of mammal-like reptiles. Nature 283:378–380.

Kissel, R. A., and T. M. Lehman. 2002. Upper Pennsylvanian tetrapods from the Ada Formation of Seminole County, Oklahoma. Journal of Paleontology 76:529–545.

Kissel, R. A., and R. R. Reisz. 2004. *Ambedus pusillus,* new genus, new species, a small diadectid (Tetrapoda: Diadectomorpha) from the Lower Permian of Ohio, with a consideration of diadectomorph phylogeny. Annals of Carnegie Museum 73:197–212.

Langston, W., Jr. 1963. Fossil vertebrates and the late Palaeozoic red beds of Prince Edward Island. Bulletin of the National Museum of Canada 187:1–36.

Langston, W., Jr. 1966. *Limnosceloides brachycoles* (Reptilia: Captorhinomorpha), a new species from the Lower Permian of New Mexico. Journal of Paleontology 40:690–695.

Laurin, M., and R. R. Reisz. 1995. A reevaluation of early amniote phylogeny. Zoological Journal of the Linnean Society 113:165–223.

Laurin, M., and R. R. Reisz. 1997. A new perspective on tetrapod phylogeny; pp. 9–59 in S. S. Sumida and K. L. M. Martin (eds.), Amniote Origins. Academic Press, San Diego.

Laurin, M., and R. R. Reisz. 1999. A new study of *Solenodonsaurus janenschi*, and a reconsideration of amniote origins and stegocephalian evolution. Canadian Journal of Earth Sciences 36:1239–1255.

Lee, M. S. Y., and P. S. Spencer. 1997. Crown-clades, key characters and taxonomic stability: when is an amniote not an amniote; pp. 61–84 in S. S. Sumida and K. L. M. Martin (eds.), Amniote Origins. Academic Press, San Diego.

Lewis, G. E., and P. P. Vaughn. 1965. Early Permian vertebrates from the Cutler Formation of the Placerville area of Colorado. U.S. Geological Survey Professional Paper 503-C:1–50.

Meyer, H. von. 1860. *Phanerosaurus naumanni* aus dem Sandstein des Rothliegenden in Deutschland. Palaeontographica 7:248–252.

Modesto, S. 1992. Did herbivory foster early amniote diversification? Journal of Vertebrate Paleontology 12(Suppl. to 3):44A.

Moss, J. L. 1972. The morphology and phylogenetic relationships of the Lower Permian tetrapod *Tseajaia campi* Vaughn (Amphibia: Seymouriamorpha). University of California Publications in Geological Sciences 98:1–72.

Olson, E. C. 1947. The family Diadectidae and its bearing on the classification of reptiles. Fieldiana: Geology 11:1–53.

Olson, E. C. 1967. Early Permian vertebrates. Oklahoma Geological Survey Circular 74:1–111.

Reisz, R. R. 1986. Pelycosauria. Handbuch der Paläoherpetologie, Teil 17A. Gustav Fischer Verlag, Stuttgart, 102 pp.

Reisz, R. R. 1997. The origin and early evolutionary history of amniotes. Trends in Ecology and Evolution 12:218–222.

Reisz, R. R., and H.-D. Sues. 2000. Herbivory in late Paleozoic and Triassic terrestrial vertebrates; pp. 9–41 in H.-D. Sues (ed.), Evolution of Herbivory in Terrestrial Vertebrates. Cambridge University Press, Cambridge.

Reisz, R. R., and T. E. Sutherland. 2001. A diadectid (Tetrapoda: Diadectomorpha) from the Lower Permian fissure fills of the Dolese Quarry, near Richards Spur, Oklahoma. Annals of Carnegie Museum 70:133–142.

Romer, A. S. 1946. The primitive reptile *Limnoscelis* restudied. American Journal of Science 244:149–188.

Romer, A. S. 1956. The Osteology of Reptiles. University of Chicago Press, Chicago, 772 pp.

Sues, H.-D., and R. R. Reisz. 1998. Origins and early evolution of herbivory in tetrapods. Trends in Ecology and Evolution 13:141–145.

Sumida, S. S., R. Lombard, and D. S Berman. 1992. Morphology of the atlas-axis complex of the late Paleozoic tetrapod suborders Diadectomorpha and Seymouriamorpha. Philosophical Transactions of the Royal Society of London B 336:259–273.

Swofford, D. L. 2002. PAUP: phylogenetic analysis using parsimony, Version 4.0b10. Sinauer Associates, Inc., Sunderland, Mass.

Troyer, K. 1982. Transfer of fermentative microbes between generations in a herbivorous lizard. Science 216:540–542.

Vaughn, P. P. 1964. Vertebrates from the Organ Rock Shale of the Cutler Group, Permian of Monument Valley and vicinity, Utah and Arizona. Journal of Paleontology 40:603–612.

Vaughn, P. P. 1969. Upper Pennsylvanian vertebrates from the Sangre de Cristo Formation of central Colorado. Los Angeles County Museum of Natural History Contributions in Science 164:1–28.

Watson, D. M. S. 1917. A sketch classification of the pre-Jurassic tetrapod vertebrates. Proceedings of the Zoological Society of London 1917:167–186.

Wideman, N. K. 2002. The postcranial anatomy of the late Paleozoic Family Limnoscelidae and its significance for diadectomorph taxonomy. Journal of Paleontology 22(Suppl. to 3):119A.

Williston, S. W. 1911. A new family of reptiles from the Permian of New Mexico. American Journal of Science (4)31:378–398.

Williston, S. W. 1912. Restoration of *Limnoscelis,* a cotylosaur reptile from New Mexico. American Journal of Science (4)34:457–468.

7. Snake Phylogeny, Origins, and Evolution: The Role, Impact, and Importance of Fossils (1869–2006)

Michael W. Caldwell

ABSTRACT

The role of the fossil record in shaping hypotheses on the relationships, evolution, and origin of snakes is reviewed. The earliest hypothesis concerning the relationships of snakes within Squamata was given by Cope (1869); he concluded that snakes and mosasaurs, among all squamates, were most closely related to each other. The ensuing debate on snake-lizard relationships and origins, where fossil lizards were at the center of debate, continued to Bolkay (1925) with the description of one of only two known species of Cretaceous marine snake. From 1925 onward, the evidence used to infer snake phylogeny, and secondarily origins, moved away from the fossil record of snakes and lizards and focused on the evidence obtained from the study of extant snakes and their soft tissues. The description of *Pachyrhachis problematicus* Haas, 1979, the first reported snake with hind limbs, reignited a fossil-based focus on the collection and interpretation of evidence bearing on snake phylogeny and origins. The end result of the last decade of fossil-based studies has led to the description of several new species of legged snakes, and the collection and characterization of numerous specimens of previously described taxa. The history of ideas from Cope (1869) to the present is briefly reviewed, followed by an overview of the fossil record of *Pachyrhachis, Eupodophis, Haasiophis, Pachyophis, Mesophis, Najash, Dinilysia, Wonambi,* and *Yurlungurr.* The character delimitation arguments for the jugal, postorbital, postfrontal and ectopterygoid are reviewed, followed by an overview of phylogenetic arguments presented since Caldwell and Lee (1997).

Introduction

Since Cope (1869) presented the first hypothesis of snake relationships, the phylogeny and origins of these reptiles have been a vexing and passionately debated problem in the study of vertebrate evolution. The difficulty in satisfactorily addressing these questions arises from the highly modified anatomy of snakes (e.g., Estes et al., 1988) and the resulting difficulty of delimiting characters (and thus primary homologies; de Pinna, 1991) for phylogenetic inference. For example, all extant snakes share the absence of limbs (although some retain vestiges of a pelvic girdle and perhaps a femur); absence of eyelids; a unique structure of the eye; absence of external ear openings; a reduced osteological complexity of the middle ear (while at the same time evolving a unique inner ear re-entrant fluid circuit supported by a bony crest [crista circumfenestralis]); a highly kinetic, or uniquely akinetic, skull; in many taxa, specialized teeth to deliver venoms; and only a single lung (and, in many cases, an additional, tracheal "lung"). These characteristic features of snakes have made easy answers to questions concerning phylogeny and origins elusive, but they have also been the inspiration for a long history of scientific research on the topic.

The debate on snake origins and relationships dates back to Cope (1869), making it the second-oldest debate of its kind in the history of the study of vertebrate evolution; the debate concerning bird origins, initiated by Huxley (1868, 1870), is merely a single year older. Since Cope (1869) first proposed a close relationship between present-day snakes and extinct mosasaurs, many fossil lizards and snakes from all over the world have been added to the 2,900 species of extant snakes and 3,700 species of extant lizards known to date. As powerful a source of evidence as these data might be, it also means that there is inadequate space in this review to exhaustively examine every detail, even though such a treatment is long overdue.

Because the fossil record of snakes is consistently underappreciated in the literature (for instance, Greene [1997] and Pianka and Vitt [2003] devote a mere 9 pages out a total exceeding 700 to a brief review of fossil snakes and lizards), I focus here on the role that fossils have played in our understanding of their evolution, phylogeny, and origins. Space constraints make it impossible to consider the role of extinct lizards.

I begin with a brief overview of the fossil record of snakes in the order of the age of the deposits and follow this with a chronological review of the data and hypotheses that fueled the first part of the fossil-driven debates concerning snake origins, commencing with Cope (1869) and ending with Bolkay (1925). I then review the record of the most recently described fossil snakes in order to identify characters and structures that have placed them at the center of the modern phase of the debate. I will also explore in detail one set of characters (jugal, postorbital, postfrontal, and ectopterygoid) that has been pivotal in the recent debate; it is also an excellent example of the problem of character delimitation. Finally, I review the most recent phase of fossil-based phylogenetic hypotheses by comparing and contrasting ideas and conclusions proposed from Caldwell and Lee (1997) to Apesteguía and Zaher (2006).

Institutional Abbreviations

HUJ-PAL, Hebrew University of Jerusalem, Palaeontology Collections; MACN, Museo Argentino de Ciencias Naturales "Bernardino Rivadavia," Buenos Aires, Argentina; MLP, Museo de La Plata, La Plata, Argentina; MPCA-PV, Museo de la Ciudad de Cipoletti, Cipoletti, Río Negro, Argentina; MSNM V, Museo di Storia Naturale di Milano, Milan, Italia; MUCP, Museo de Geología y Paleontología, Universidad Nacional del Comahue, Neuquén, Argentina; NHM-BiH, Zemaljski Musej, Sarajevo, Bosnia-Herzegovina; NMW, Naturhistorisches Museum, Vienna, Austria; Rh-E.f., Museum of Gannat, France; QMF, Queensland Museum, Brisbane, Australia; SAM, South Australian Museum, Adelaide, Australia.

A Brief History of the Fossil Record of Snakes

The geologically oldest snakes known to date are represented only by a few isolated fossil vertebrae from North Africa and North America. Cuny et al. (1990) assigned the African material (late Albian) to the family Lapparentophiidae (see the description of *Lapparentophis defrennei* by Hoffstetter, 1959) and to a snake taxon *incertae sedis*. The slightly younger North American material (late Albian–early Cenomanian), which is equal in age to *Lapparentophis defrennei* Hoffstetter, 1959, was assigned to *Coniophis* by Gardner and Cifelli (1999), a genus first described and uncertainly attributed to the modern Aniliidae by Marsh (1892). An older snake dating from the Barremian of Spain described by Rage and Richter (1994) was subsequently reidentified by Rage and Escuillié (2003) as a lizard. Unfortunately, *Lapparentophis* and *Coniophis* are known exclusively from vertebrae and present little useful information on snake origins other than providing a date of origin that predates the late Albian.

It is the early to middle Cenomanian record of snakes (*Pachyrhachis problematicus* Haas, 1980, *Haasiophis terrasanctus* Tchernov, Rieppel, Zaher, Polcyn and Jacobs, 2000, *Eupodophis descouensi* Rage and Escuillié, 2000), found in marine carbonates, that is truly informative on the anatomy of early snakes. Among many fascinating anatomical features, both plesiomorphic and apomorphic, these snakes retain small but well-developed hind limbs, show aquatic adaptations, and possess skulls that are macrostomate in structure (Caldwell and Lee, 1997; Lee and Caldwell, 1998; Rage and Escuillié, 2000; Tchernov et al., 2000; Rieppel and Head, 2004).

The most recent addition to the fossil record of snakes is a terrestrial snake with hind limbs, *Najash rionegrina* (Apesteguía and Zaher, 2006), found in Turonian clastic strata in the Neuquén Basin of Argentina (Apesteguía et al., 2001; Leanza et al., 2004). This taxon, together with the somewhat younger (Santonian) snake *Dinilysia patagonica* Woodward, 1901 (Estes et al., 1970; Caldwell and Albino, 2002), represents our entire record of well-preserved Late Cretaceous terrestrial snakes to date.

The fossil record of snakes at the end of the Cretaceous and into the early Cenozoic is again largely restricted to isolated vertebrae (e.g., *Coniophis* ranges throughout the Cretaceous and into the Palaeogene [Rage, 1984], and two vertebrae of another enigmatic snake, *Pouitella pervetus* [Rage, 1988]). What is clear is that the family-level groups of snakes, such as the Aniliidae, Madtsoiidae, and Boidae, are present in many faunas around the world (Rage, 1984). However, because of the nature of their preservation, this fossil record remains largely phylogenetically uninformative and provides little more than insights on the timing of snake family-level radiations. It is not until the Eocene that well-preserved snake fossils, specifically booids, are again found in North America (Marsh, 1871; Rage, 1984; Holman, 2000) and western Europe (Schaal and Ziegler, 1992; Baszio, 2004; Schaal and Baszio, 2004). The Eocene record of snakes also includes the earliest scolecophidians and a number of relatively obscure taxa, represented by numerous vertebrae of giant aquatic snakes, assigned to the family Palaeophiidae (for reviews see Rage, 1984, and Holman, 2000).

By the end of the Oligocene, the modern snake fauna is represented in the fossil record by fragmentary specimens assignable to scolecophidians and most of the present-day groups of alethinophidians (Holman, 2000). At the present time, there are 2,978 named species of extant snakes (Uetz, 2004), of which 342 are scolecophidians; the bulk of alethinophidian diversity is within the Caenophidia (Greene, 1997). The overall data set on snake anatomy, temporal distributions, and paleobiogeography is surprisingly good and detailed, although at important points in its temporal coverage (specifically the Cretaceous Period), the available fossil record is scant and difficult to interpret. What data are informative on snake origins and sister-group relationships are extremely good because they are represented by the even more diverse "lizards," which have been grouped with snakes in Squamata since Oppel (1811).

Paleontological Roots of a Debate

Cope (1869)

The debate concerning snake origins was inadvertently initiated by Cope (1869) when he proposed to classify the then known families of mosasaurs, the Clidastidae and Mosasauridae, into the Order Pythonomorpha, which was elevated to equivalency with the Order Ophidia and Order Lacertilia. Cope (1869:254–255) listed 20 characters of mosasaurs, 10 of which he considered similar to or shared with snakes, 5 shared with lacertilians (lizards), and 5 shared with sauropterygians. His creation of the Order Pythonomorpha, and its inherent nomenclatural reference to snakes, was intended to reflect what he concluded was the presence of a significant number of snake characters in mosasaurs. Cope (1869:258) defended his choice of nomenclature in terms of shared characters leading to a hypothesis of phylogeny, saying, "On account of the ophidian part of their affinities, I have called this order the Pythonomorpha" (Cope, 1869:258).

It is clear that Cope (1869) was not attempting to address the question of snake origins, but rather was trying to illuminate the relationships of mosasaurs. Interpreting his writing indicates he was supporting a squamate origin for mosasaurs and was also supporting a closer affinity of mosasaurs to snakes than to all other lizards. To this end, he created three orders of equal rank: Ophidia (snakes), Pythonomorpha (mosasaurs), and Lacertilia (lizards). As was noted by Rieppel (1988), Cope's approach was to consider relationships in terms of pattern rather than process, thus arguing for phylogenetic pattern based on character support as opposed to ancestor-descendant relationships that were based on the process of transformation. More importantly, however, Cope's real interest in proposing the Pythonomorpha was to understand more about mosasaurian relationships. His use of the name Pythonomorpha, reiterated in a later publication (Cope, 1875), initiated a lively nineteenth-century debate focused on the squamate relationships of mosasaurs. Almost without exception, Cope's critics assumed he was arguing that snakes evolved from mosasaurs, when in fact he was arguing for a closer relationship between snakes and mosasaurs within Squamata (Cope, 1895a,b), and not an ancestor-descendant relationship where snakes evolved from mosasaurs (Fig. 7.1).

Owen (1877)

The first respondent to Cope's (1869, 1875) concept of Pythonomorpha, and the first to misinterpret Cope's major innovation of searching for characters to support a relationship between fossil and extant forms, was Owen (1877), who followed Cuvier (1824) in classifying mosasaurs within the Order Lacertilia. Owen reviewed squamate morphology in great detail and concluded that most, if not all, of the features cited by Cope (1869) as shared be-

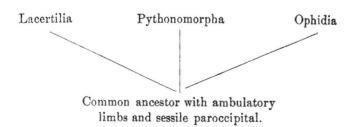

Figure 7.1. Phylogeny of Lacertilia, Ophidia, and Pythonomorpha from Cope (1895:858). Note that Cope indicated at the base of his phylogeny that all three share a common ancestor, but that one is not derived from the other, only more closely related.

tween Pythonomorpha and Ophidia were incorrectly identified. Owen's (1877:713) contribution to the issue of the origin of snakes was as follows: "If I were to hazard a guess as to any antecedent form leading toward the earliest certainly known *Ophidian,* viz. the pythonic and perhaps marine forms of the Eocene period, it would be the *Dolichosaurus* of the white Chalk of Kent that would suggest itself." In relationship to the modern debate on squamate sister-group relationships, Owen's statement is interesting as regards dolichosaurs but still remains problematic concerning his search for antecedents and his criticisms of Cope.

Baur versus Cope

Baur (1892) reviewed the work of Cope (1869, 1875), Owen (1877), Marsh (1880), and others, and concluded that Cope was in error in raising mosasaurs to an ordinal rank equivalent to Lacertilia and Ophidia. He noted that mosasaurs were most clearly derived from the lacertilian group Varanidae and showed no meaningful similarities to the Ophidia, which he ranked as a suborder (p. 18). He concluded by saying (p. 21), "To express this affinity, I placed the Varanidae and Mosasauridae in a superfamily, the Varanoidea. By this I wanted to say that the Mosasauridae cannot be separated from the true Lacertilia, to which the Varanoidea belong; in other words, that they cannot be placed as a sub-order of the Squamata, but have to placed among the sub-order Lacertilia. In this opinion I have nothing to change."

In a series of rejoinders, Cope (1895a,b, 1896a) and Baur (1895, 1896), and additionally Cope (1896b) and Boulenger (1893), exchanged opinions and criticisms of each other's assessments of squamate cranial structure and the resulting definitions as to whether or not mosasaur skulls were more snake-like or more lizard-like. Again, however, this debate was focused on the squamate relations of mosasaurs rather than snakes.

Williston (1904)

Williston (1904:43) dealt with mosasaur relationships in his paper entitled "The Relationships and Habits of the Mosasaurs," where he noted that neither snakes nor lizards should be treated as orders within a Superorder Squamata sensu Osborn (1899). He agreed with the rationale and arguments of Baur (1892) concerning the varanoid affinities of mosasaurs, but stated concerning snake origins that "snakes originated from the lizards much earlier than we have definite knowledge would seem very probable" (p. 47), and "it is not at all unreasonable to suppose that the snakes had branched off as early as, or earlier than, the beginning of the Lower Cretaceous" (p. 48). Thus, although Williston was convinced that snakes are squamates with their origin somewhere among the lizard families, he gave no clue as to which lizard group is "ancestral" to the snakes, although he did suggest a deeper geological time of origin of snakes than that implied by either Cope (1869) or Owen (1877).

Nopcsa (1903)

In four notable papers whose publication dates span three decades, Nopcsa (1903, 1908, 1923, 1925) examined and expanded on Cope's ideas concerning the "Pythonomorpha." In the first of these papers, Nopcsa (1903) reviewed all of the "*Varanus*-like" lizards from Istria (eastern Italy, Slovenia, and Croatia) and the English Chalk that had previously been described by Kornhuber (1873, 1893, 1901), Kramberger (1892), Owen (1850), and Seeley (1881). Nopcsa assigned these various aquatic lizards to two major groups, dolichosaur-like and aigialosaur-like forms, and concluded that the latter were much more like varanids then the former. He then addressed Cope's (1869) Order Pythonomorpha in relation to the characters of dolichosaurs and aigialosaurs and noted a large number of characters shared by aigialosaurs and pythonomorphs, but few between pythonomorphs (at this time, Pythonomorpha was synonymous with the modern definition of Mosasauridae) and dolichosaurs. Nopcsa also reviewed the work of Williston (1898) and Osborn (1899), who both noted, using anatomical features as evidence, that there were similarities between mosasaurs and varanids, but not remarkable ones—in other words, they were insufficient to merit consideration of the placement of varanids within Pythonomorpha. From his assessment of the studies by Baur (1892), Boulenger (1893), Williston (1898), and Osborn (1899), Nopcsa (1903:41) concluded, "The only characters which can be offered as differentiating the aigialosaurids from the varanids are those points in which the Aigialosauridae approach the pythonomorphs, while on the other hand, it is typical varanid characters which separate the aigialosaurids from the pythonomorphs."

Janensch (1906)

Janensch (1906) redescribed the unusual Eocene aquatic snake *Archaeophis proavus* Massalongo, 1859, providing an excellent and detailed anatomical description and a detailed scenario for the terrestrial origin of snakes. He presented anatomical counterarguments as to why there was no certain phylogenetic relationship between snakes and the dolichosaurs, aigialosaurs, and mosasaurs. Unfortunately, rather than positing a close relative among squamates, he simply finished by stating that "they probably developed from an unknown terrestrial lizard, not adapted to aquatic life."

Although Janensch (1906:27) presented anatomical counterarguments for Kornhuber's (1901) and Nopcsa's (1903) phylogenetic hypotheses, he presented only logical counterarguments for his terrestrial origins and relations hypotheses: "If we thus see that the living snake-like lizards are exclusively land animals, we must assume that their predecessors gained their body form on the land, but not in water." Janensch (1906:27–28) continued with a number of interesting and topical points:

Of the forerunners of snake, it may hence with great probability be assumed that the acquisition of the snake type is to be attributed to adaptation to residence on ground covered with thick vegetation, perhaps also a burrowing way of life, or both conditions at once. However, we certainly do not have to think of such marked fossorial forms as today's typhlopids and glauconids, which are quite extremely differentiated in this direction. . . . The possibility may also not be ignored that those recent fossorial groups could have developed secondarily from snakes living above ground.

Nopcsa (1908, 1923, 1925)

Nopcsa (1908) presented the first set of explicit statements addressing the problem of the aquatic origin of snakes as inferred from his phylogenetic hypotheses. Near the end of his 1908 paper, having examined new specimens of aquatic fossil lizards made available after his 1903 study, he addressed the issue of the long neck of dolichosaurs, and more specifically, of *Dolichosaurus longicollis* Owen, 1850, as it related to the "neck" of snakes. Counting the number of ophidian precloacal vertebrae that bear hypapophyses, and taking the position of the heart and lungs as an index of the end of the neck (Nopcsa, 1908:28–29), he noted, "[S]o we see ourselves compelled to count a quite considerable number of vertebrae in the snakes as cervical vertebrae, which decidedly suggests a dolichosaur-like elongation of the neck achieved through increase in the number of cervical vertebrae."

Thus Nopcsa (1908) concluded that snakes and dolichosaurs both have long necks and on this basis hypothesized a close phylogenetic relationship between them as opposed to either being closer to varanids, aigialosaurs, and mosasaurs (pythonomorphs) (Fig. 7.2). However, he went further in discussing his ideas on snake origins by indicating that the specialized snake skull did not compare well with that of *Adriosaurus suessi* Seeley, 1881, nor with any other dolichosaur, other than by size (Lee and Caldwell, 2000). Nopcsa further stated that he felt that the question of the origin of snakes could only be solved definitively with new information on the cranial structure of mid-Mesozoic snakes.

Following his 1908 treatise, Nopcsa (1923) described two new aquatic squamates from Istria: *Eidolosaurus trauthi* (a lizard) and *Pachyophis woodwardi* (which he considered a snake-like reptile). In assessing the relationships of *Eidolosaurus*, he revised his view that aigialosaurs and dolichosaurs were distantly related and instead concluded that Dollo's (1903) Dolichosauridae (Dolichosaurinae, Aigialosaurinae, and Mesoleptinae) was a classification reflecting real relationships. Additionally, Nopcsa (1923) also reaffirmed, with inclusion of the ideas of Fejérváry (1918), that mosasaurs were "descended" from the Aigialosaurinae (contra Dollo, 1903).

The second description by Nopcsa (1923) included the first detailed examination of the systematics of *Pachyophis woodwardi* and the origins and relationships of snakes. Here he discussed and

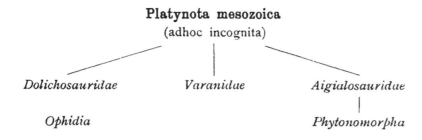

Platynota mesozoica

(adhoc incognita)

Dolichosauridae   Varanidae   Aigialosauridae

Ophidia   Phytonomorpha

Figure 7.2. Phylogeny of Mesozoic Platynota from Nopcsa (1908).

dismissed Janensch's (1906) arguments for a terrestrial origin. Nopcsa's (1923) paper is significant because this was the first time in the scientific literature that consideration was given to the origin and relationships of both extant and extinct snakes. Most importantly, Nopcsa compared snakes to a group of fossil squamates, the dolichosaurs, which he considered closely related to aigialosaurs, mesoleptines, and thus Cope's (1869) mosasaurian Pythonomorpha.

Nopcsa (1923) listed 10 anatomical characters that he found to be variously shared between *Pachyophis*, "primitive snakes," aigialosaurs, and dolichosaurs; in particular, he noted that the number of cervical vertebrae shared by snakes and dolichosaurs remained a compelling line of evidence. Interestingly, he concluded that the oarlike tail of dolichosaurs, which he considered absent in aigialosaurs, indicated that *Pachyophis* was descended from aigialosaur-like forms. Nopcsa (1923:139) stated that the teeth and ribs suggest that *Pachyophis* was a "true snake," and in the end, he assigned the taxon to a new family of primitive snakes, Pachyophiidae. Nopcsa concluded that *Pachyophis* helped answer questions on the origin of snakes but also clearly indicated that snakes were not most closely related to mosasaurs within Squamata: "That the snakes have no close relationship with the mosasaurs seems finally proven [cf. the works of Janensch and most recently of J. G. von Fejérváry]" (Nopcsa, 1923:140n). His revised phylogeny (Fig. 7.3) built on but heavily modified his previous hypothesis (Fig. 7.2), with the central point being that although mosasaurs are still closely related to snakes, there are other groups of lizards that are even more closely related to snakes.

Nopcsa (1923) continued the expansion of his phylogenetic hypothesis by defining the Proplatynota, which he placed at the base of his phylogeny and from which he derived the Varanidae and Aigialosaurinae. The Aigialosaurinae gave rise to three groups, the pelagic Mosasauridae, the littoral and benthic Mesoleptinae (ancestors to the Dolichosaurinae), and the marine and benthic Cholophidia (inclusive of the Pachyophiidae) from which the modern lineages of snakes are derived, the Angiostomata (=Scolecophidia) and Alethinophidia. The topology of Nopcsa's (1923) phylogeny is rather similar to recent phylogenies to be discussed below, but what is more important is the taxic composition of his Cholophidia, Angiostomata, and Alethinophidia.

The primitive (basal) cholophidians included the Palaeophiidae,

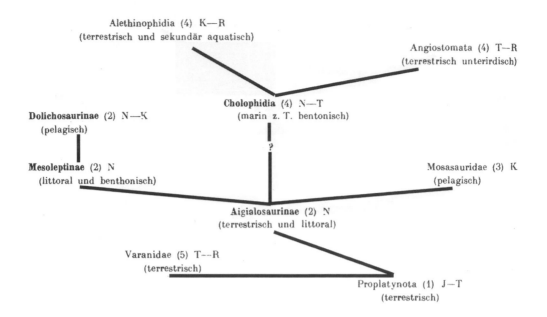

Alethinophidia (4) K—R
(terrestrisch und sekundär aquatisch)

Angiostomata (4) T—R
(terrestrisch unterirdisch)

Cholophidia (4) N—T
(marin z. T. bentonisch)

Dolichosaurinae (2) N—K
(pelagisch)

Mesoleptinae (2) N
(littoral und benthonisch)

Mosasauridae (3) K
(pelagisch)

?

Aigialosaurinae (2) N
(terrestrisch und littoral)

Varanidae (5) T—R
(terrestrisch)

Proplatynota (1) J—T
(terrestrisch)

Figure 7.3. Phylogeny by Nopcsa (1923).

Pachyophiidae, *Simoliophis* (Nopcsa, 1925), and Archaeophiidae. Nopcsa stated that the Angiostomata (Typhlopidae + Glauconiidae [=Leptotyphlopidae]) are derived from within the Cholophidia and possess a number of specialized snake characteristics as well as a number of primitive features. Nopcsa's (1923) Alethinophidia, a name still in use, comprised at least two groups: the more primitive Xenopeltidae, Ilysiidae, Uropeltidae, and Boidae; and the more specialized Colubridae, Viperidae, and Amblycephalidae (slug-eating snakes). Nopcsa (1923:143) stated, "The Angiostomata and Alethinophidia must both in any case be regarded as direct descendants of the cholophidians, and indeed it seems that the split leading to these two groups already takes place within the cholophidians. It seems to me that *Archaeophis*, on account of its skull structure, is decidedly closer to the alethinophidians, but *Simoliophis* and *Palaeophis* are closer to the angiostomatans."

Nopcsa's (1923) scenario for the aquatic origin of snakes was as detailed and comprehensive as his phylogeny and classification (Fig. 7.3). He argued for a scenario deriving aquatic cholophidian snakes from an aquatic dolichosaurid, specifically an aigialosaurine, and outlined an ecological scenario for the ancestral cholophidian whereby the neck is elongated and limbs are reduced in the preophidian stage. The early cholophidian snake was reconstructed as a benthic aquatic burrower that later moved back onto land, from whence its descendants radiated and evolved into the Angio-stomata and Alethinophidia.

Nopcsa's (1923) scenario of snake origins was the first thorough argument to account for all snake characteristics (those that seemed derived from an aquatic ancestry and those that seemed derived from a burrowing ancestry) in the literature. He incorporated the earlier considerations of Cope (1869), Owen (1850, 1877), Kornhuber (1873, 1893, 1901), Kramberger (1892), Baur (1892, 1895, 1896), Boulenger (1893), Osborn (1899), Dollo (1903), Janensch (1906), and Fejérváry (1918) concerning mosasaurs, varanids, basal platynotans, dolichosaurs, aigialosaurs, and other lizards, but most importantly, he contributed new data and new ideas on snakes.

Bolkay (1925)

Shortly after Nopcsa's (1923) innovative synthesis, and with a definite knowledge of Janensch's (1906) broadly innovative and thoughtful work, Bolkay (1925) described a second taxon of Istrian snake-like squamate as *Mesophis nopcsai* and assigned it to the Cholophidia Nopcsa, 1923, family Pachyophiidae Nopcsa, 1923. Bolkay's contribution to the debate on snake origins was to contradict Nopcsa's (1923) assertion of a benthic ecology for the primitive Cholophidia (which for Nopcsa was also the source of the observed pachyostosis in these animals), but he still retained the framework of the aquatic origin of snakes. Bolkay argued that, contra Nopcsa (1923), snake-like reptiles such as *Mesophis* and *Pachyophis* had evolved in similar ways to those postulated by Nopcsa (1923) for dolichosaurs and nothosaurs. He interpreted early snakes as living in shallow water, feeding both at the surface and at moderate depth (roughly 3–4 meters). Bolkay further expanded his scenario from the ecological to the phylogenetic, stating that all extant snakes could be derived from booids, and that the latter are closest to the original four-legged ancestor of snakes.

Pachyrhachis: Hind Limbs in Snakes

During the decades between Bolkay's (1925) description of *Mesophis* and the descriptions by Haas (1979, 1980a,b) of *Pachyrhachis problematicus* and *Ophiomorphus colberti* (=*Estesius colberti* [Wallach, 1984]), the only significant work on a Mesozoic snake was the redescription of *Dinilysia patagonica* Woodward, 1901 by Estes et al. (1970). With this exception, the snake fossils described during this time remain as rather uninformative form taxa based mostly on isolated vertebrae (Rage, 1984; Holman, 2000) compared with the information obtained from the study of extant snakes (e.g., Underwood, 1967). This is an unfortunate reality reflecting the stochastic nature of fossil discoveries; it is made doubly unfortunate because this intervening period, from 1925 to 1979, witnessed a transformative phase in the study of the evolution, ecology, and anatomy of extant lizards and snakes (for overviews, see Rieppel, 1988; Greene, 1997; and Pianka and Vitt, 2003). The infusion of new fossil-based data into the substantial neontological evidence sets from this phase would have proven very useful in shaping hypotheses generated during the mid-twentieth century (Walls, 1940, 1942; Bellairs and Underwood, 1951; Bellairs, 1972). For example, McDowell and Bogert's (1954) examination of the affinities of *Lanthanotus* would have been greatly assisted by new material of aigialosaurs; those authors had to rely on the specimen descriptions by Kornhuber (1901) and Kramberger (1892) as well as the review by Nopcsa (1908).

Even the descriptions by Haas (1979, 1980a,b) of *Pachyrhachis problematicus* and *Ophiomorphus colberti* (now synonymized with *P. problematicus*; Caldwell and Lee, 1997) were slow to enter the debate because of Haas's identification of these forms as dolichosaurs or aquatic varanoids (Haas, 1980a:100): "The fact that also other reptiles of at least ophidian relationship, like *Pachyophis* and *Simoliophis*, are marine or estuarine, recalls Nopcsa's old ideas about marine snake origins (in which I do not believe, in view of many characters of the most primitive recent snakes)." However, Haas's views are extremely interesting and important in the history of this debate because he explored various lines of platynotan relationship for both dolichosaurs and snakes, which tended in parallel ways toward an "ophidian end-stage." On the basis of his earlier, detailed work on scolecophidians (e.g. Haas, 1968) as well as that of others (e.g., List, 1966), it is clear that Haas viewed this group of animals to be the most primitive or basal snakes and that the ancestor of all snakes had evolved under the adaptive constraints of a burrowing lifestyle. The anatomy of *Pachyrhachis* (sensu Haas, 1979) was not scolecophidian and thus not commensurate with Haas's primitive snake type or ancestral snake.

It is curious in this regard that Zaher and Rieppel (2000:73) attempted to revise Haas's (1979, 1980a,b) perspectives when they wrote, "In his initial description of *Pachyrhachis*, Haas (1979) placed *Pachyrhachis* in the Varanoidea but outside Serpentes, because he realized that those characters which this fossil taxon does share with snakes are macrostomatan characters, i.e., features characteristic of relatively advanced snakes (Rieppel, 1988; 1994). It therefore seemed prudent to interpret *Pachyrhachis* as a varanoid, convergent upon macrostomatans in some cranial characters (Haas, 1979)." The Zaher and Rieppel (2000) version of Haas's view is further contradicted by an earlier statement from Rieppel (1988:98), where he stated, "This vague information in Prof. G. Haas' last publication reflects the dilemma created for him by the discovery of these snakelike marine fossils, which seemed to contradict his earlier work favoring a burrowing ancestry of snakes." If Haas (1979, 1980a,b) had regarded *Pachyrhachis* as sharing many features with macrostomatan snakes, to the exclusion of the dozens of characters viewed by him as merely snake-like, his assignment of *Pachyrhachis* to the Varanoidea would have been even less plausible. Notwithstanding the interpretation of Haas (1979) by Zaher and Rieppel (2000), the convergence discussed by Haas was convergence with snakes in general, but certainly not with macrostomatan snakes in particular.

Credit for the first certain consideration of the macrostomatan characteristics of *Pachyrhachis* must be given to McDowell (1987), followed by Carroll (1988), who referenced McDowell. Cuny et al. (1990) provided the next suggestion that *Pachyrhachis* might be a primitive snake belonging to a divergent lineage of uncertain affinity. Finally, Rieppel's (1994) brief discussion of *Pachyrhachis* noted that the supratemporals (the squamosals of Haas, 1979) are excluded from the braincase as in advanced (macrostomatan) snakes. Rieppel (1994:31) concluded in reference to *Pachyrhachis,*

> Again, however, major uncertainties prevail in the interpretation of Mesozoic ophidians. *Pachyrhachis* is a snakelike reptile from the lower Cenomanian of Israel [Haas, 1979; 1980b] that shows a curious mixture of varanoid and ophidian characteristics. . . . The fossil was used by Haas [1979] in his argument that simoliopheids are, in fact, varanoid lizards. Rage [1984] did not follow Haas [1979, 1980b] in the inclusion of *Pachyrhachis* in the Simoliopheidae.

From my perspective, Haas's interpretations highlight the critical issues at the very center of the debate: the determination of ancestral conditions, the determination of polarities of character transformation, and the determination of character and character state delimitations (Rieppel and Kearney, 2001, 2002; Kluge, 2003). His perspectives were clear and keen regarding the important and informative data set represented by scolecophidian snakes and remain perfectly in line with the conclusions of many other students of snake evolution (Walls, 1940; Underwood, 1967, 1970; Rieppel, 1988; Apesteguía and Zaher, 2006)—that snakes had a burrowing or terrestrial ancestry and origin.

New and Redescribed Snake Fossils—1997 to the Present

Since 1997, the year of publication for Caldwell and Lee's redescription of the early Cenomanian snake *Pachyrhachis problematicus,* the debate on snake phylogeny, evolution, and origins has been further invigorated by the descriptions of several new taxa of fossil snakes with legs (e.g., Rage and Escuillié, 2000, 2001; Tchernov et al., 2000; Rieppel and Head, 2004; Apesteguía and Zaher, 2006); snakes without hind limbs, including *Pachyophis* (Lee et al., 1999); new specimens of *Dinilysia* (Caldwell and Albino, 2002; Budney et al., 2006); and the madtsoiids *Wonambi* (Scanlon and Lee, 2000; Scanlon, 2005) and *Yurlunggur* (Scanlon, 2006) from the Neogene of Australia. There are now so many new and pivotal fossil taxa, both aquatic and nonaquatic, that it is important to give a brief overview of the record of these animals.

Pachyrhachis problematicus Haas, 1979 (Fig. 7.4), is a longbodied snake with a small and narrow head, distinct cervical region, small pelvis, and small hind limbs. The frontal is very long and narrow. The quadrate is sheet-like, being greatly expanded anteroposteriorly. The coronoid forms a tall rectangular process with a long anterior flange. The splenial-angular joint is located anteriorly. The posterior trunk vertebrae and ribs are pachyostotic (Lee and Caldwell, 1998). The specimen was collected from limestone-dolomite quarries at Ein Jabrud, near the West Bank town of Ramallah, from units of rock dated as early Cenomanian (Late Cretaceous).

Caldwell and Lee (1997) and Lee and Caldwell (1998) synonymized *Ophiomorphus colberti* Haas, 1980b (HUJ-PAL 3775; Figs. 7.4C,D) with *Pachyrhachis problematicus* Haas, 1979 (HUJ-PAL 3659; Fig. 7.4A,B), creating a single taxon represented by two complete and articulated specimens.

Collected at nearly the same time and from the same quarries as the specimens now assigned to *Pachyrhachis,* the more recently described *Haasiophis terrasanctus* Tchernov, Rieppel, Zaher, Polcyn, and Jacobs, 2000 (HUJ-PAL 695; Fig. 7.5), is a somewhat smaller and anatomically distinct marine snake with hind limbs. As described, *Haasiophis* possesses a small, narrow, edentulous premaxilla, 24 tooth positions on the maxilla, 8 on the palatine, 15 to 17 on the pterygoid, and 26 on the dentary. The enamel surface of the teeth is distinctly striated. There is a mandibular nerve foramen underlapped by a distinct prootic flange. The quadrate is slender and oriented vertically. The coronoid process on the mandible is small and formed by only the coronoid bone. There are 154 precloacal vertebrae, five distally expanded and bifurcated lymphapophyses in the cloacal region, and distally expanded hemapophyses present on caudal vertebrae (Rieppel et al., 2003).

A number of new fossils of snakes with legs from late Cenomanian rocks in Lebanon (type locality is near Al Nammoura, Valley of Nahr Ibrahim, northern Lebanon; Rage and Escuillié, 2000; Rieppel and Head, 2004), representing a third species from the Middle East, have recently been described and assigned to the taxon *Eupodophis descouensi* (Rage and Escuillié, 2000, 2001) (Fig. 7.6). The holotype (Rh-E.f. 9001, 9002, 9003) and referred specimens (MSNM V 3660, 3661, 4014) characterize yet another small-bodied snake with short but well-developed hind limbs. The skull has been characterized as macrostomatan. As in *Pachyrhachis,* the coronoid process is very tall, the body is laterally compressed, and the tail is short and paddle-like. Most of the vertebrae and ribs are pachyostotic, and the chevrons articulate with the caudal vertebrae.

The most recently described snake with hind limbs is *Najash rionegrina* Apesteguía and Zaher, 2006, from the late Turonian. The holotype (MPCA 390–398) is a short but articulated section of the axial skeleton that includes the sacral region, and a loosely articulated and associated series of limb and girdle elements. The referred material (MPCA 380–383, 385) includes a second and much longer string of articulated vertebrae with an associated dentary (with at least 11 large tooth sockets), and an isolated and fragmentary skull with two associated presacral vertebrae; the skull compares well with that of *Dinilysia* (cf. Caldwell and Albino, 2002).

In an interesting departure from what has become the "typical" Cretaceous marine snake, i.e., a snake with hind limbs, there

Figure 7.4 *(opposite page). Pachyrhachis problematicus.* **A,** dorsal view of holotype (HUJ-PAL 3659), as photographed during the late 1970s (from file photograph courtesy of Eitan Tchernov). **B,** ventral view of the skull of holotype (HUJ-PAL 3659) (photograph by author). **C,** dorsal view of referred specimen (HUJ-PAL 3775) (photograph by author). **D,** pelvis and hind limb of referred specimen (HUJ-PAL 3775). Photograph by author.

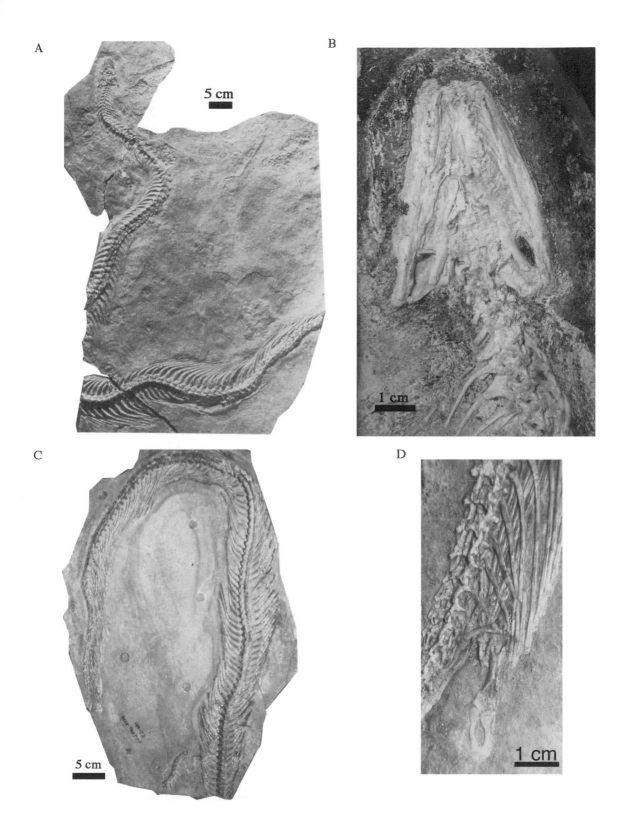

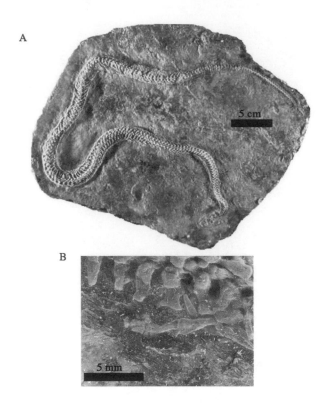

A

5 cm

B

5 mm

Figure 7.5. *Haasiophis terrasanctus.* **A,** dorsal view of holotype (HUJ-PAL 659). **B,** close-up of limb and girdle (HUJ-PAL 659). Photographs courtesy of Michael Polcyn.

are two limbless taxa known from limestone quarries located near Selista (Selisca, Selisce), an eastern suburb of Bilek (Bileca) in eastern Bosnia-Herzegovina about 40 km northeast of Dubrovnik: *Pachyophis woodwardi* Nopcsa, 1923 (Fig. 7.7A), and *Mesophis nopcsai* Bolkay, 1925 (Fig. 7.7B). The age of the rocks at Bilek has been debated for some time, ranging from the "Neocomian" (Lower Cretaceous) to Cenomanian-Turonian (Langer, 1961), but it now seems to be established as late Cenomanian (Sliskovic, 1970). The holotype of *Pachyophis woodwardi* (NMW A3919; Fig. 7.7A) was recently recharacterized by Lee et al. (1999b) as an elongated marine squamate with a small skull and slender neck. It is smaller than *Pachyrhachis problematicus,* and it shows more exaggerated pachyostosis in the mid-dorsal ribs and vertebrae. In this region, the proximal and middle portions of the ribs are very thick, so that the intercostal space is almost obliterated. The distal ends of the ribs are round rather than flat in cross section, and the body appears to have been laterally compressed. The neural arches and bases of the neural spines are very swollen. There are also differences in the dentition: although smaller, *Pachyophis* has many more teeth on the dentary (approximately 23) than *Pachyrhachis,* which has 12.

In contrast, the holotype and only known specimen of *Mesophis nopcsai* (NHM-BiH 2039) (Fig. 7.7B), which until recently was thought to be lost but has been rediscovered in the col-

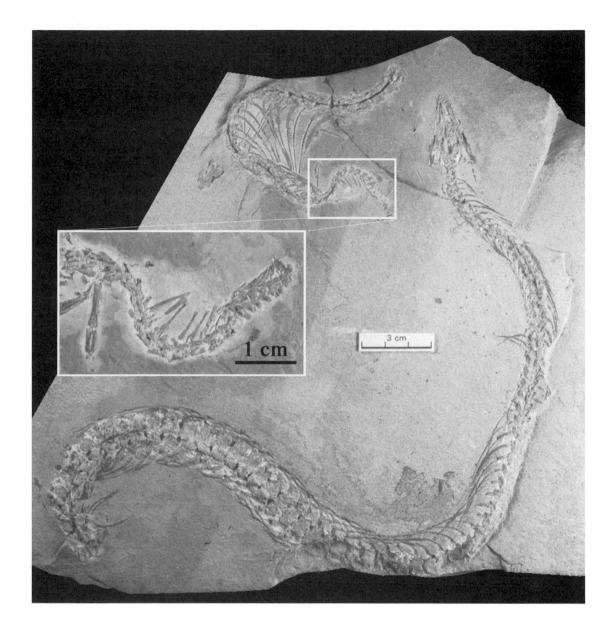

Figure 7.6. *Eupodophis descouensi.* Dorsal view of holotype (Rh-E.f. 9001, 9002, 9003) with inset box magnifying limb and girdle from counterpart specimen. Photographs courtesy of Jean-Claude Rage.

lections of the Natural History Museum of Bosnia-Herzegovina, Sarajevo, is much less pachyostotic than *Pachyophis*. A description of this snake fossil is in progress.

To date, there is only a single well-known nonmarine and probably nonlegged Cretaceous snake: *Dinilysia patagonica* Woodward, 1901 (Fig. 7.8). The holotype (MLP 26-410) and all new specimens of *Dinilysia* were found at three localities in or near Neuquén, Neuquén Province, Argentina. The fossil-producing units are red- to white-weathering, coarse-grained sandstones of the Bajo de la Carpa Member of the Rio Colorado Formation, recently assigned to the Santonian (Upper Cretaceous) by Leanza et al. (2004) and cur-

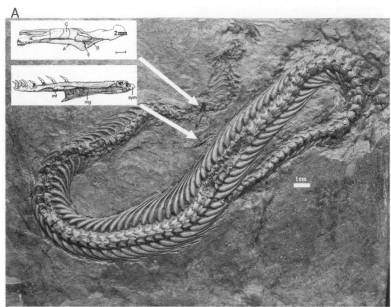

Figure 7.7. **A**, photograph of holotype of *Pachyophis woodwardi* (NMW A3919) (from Lee et al., 1999b). **Inset boxes,** line drawings of left dentary as preserved in medial view, and jaw fragment tentatively identified as part of the right mandibular ramus, preserved in lateral view. **B**, photograph of holotype (counterpart) of *Mesophis nopcsai* (NHM-BiH 2039). **mg**, Meckelian groove; **mf**, medial flange of dentary; **sym**, symphysis; **ij**, intramandibular joint; **A**, putative right angular; **B**, putative right splenial; **C**, putative right compound bone.

rently interpreted as deposited in aeolian environments (Caldwell and Albino, 2001, 2002). The holotype (Woodward, 1901; Estes et al., 1970) and two fragmentary skulls were found at Boca del Sapo, north of Neuquén. Another fragmentary skull was collected in the 1980s from outcrops just north of the Universidad Nacional del Comahue in Neuquén. The most recent collection of skulls and articulated postcranial skeletons is from outcrops of the Rio Colorado Formation exposed at the Tripailao Farm Locality (Caldwell and Albino, 2002).

Dinilysia is a medium-sized snake that possesses a complex, interdigitating frontal-parietal suture. The maxilla has a deep, anterolaterally directed trough on the suborbital surface, and the pre-

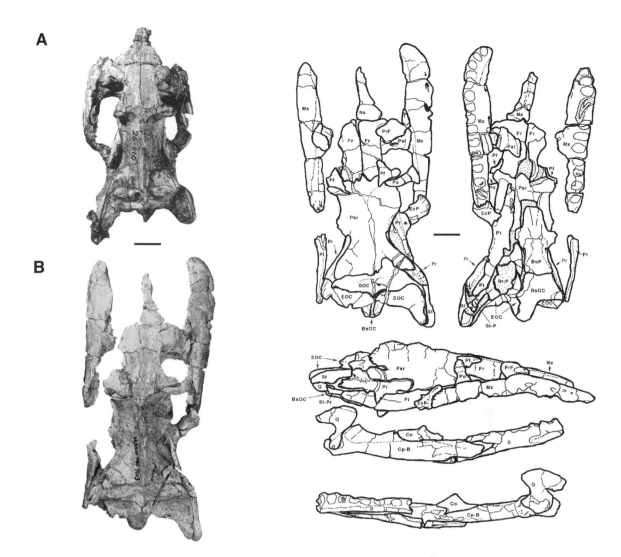

Figure 7.8. Skulls of *Dinilysia patagonica*. **A**, dorsal view of skull of holotype, MLP 26-410. Scale = 1 cm. **B**, dorsal view of skull of MACN-RN-1013; **inset box**, line drawings of MACN-RN-1013 in dorsal, ventral, and right lateral views, and lateral and medial views of right mandibular ramus (from Caldwell and Albino, 2002: fig. 1). **BsOC**, basioccipital; **cen**, centrum; **ns**, neural spine; **Co**, coronoid; **CpB**, compound bone; **D**, dentary; **EcP**, ectopterygoid; **ex-col**, extracolumella; **Fr**, frontal; **fld**, foramen for the lacrimal duct; **fPal-Mx**, palatine-maxillary foramen; **fRO**, fenestra rotunda; **Mx**, maxilla; **Na**, nasal; **Pal**, palatine; **Par**, parietal; **PrF**, prefrontal; **Po**, postorbital; **Pf**, postfrontal; **Pr**, prootic; **Pt**, pterygoid; **Q**, quadrate; **SOC**, supraoccipital; **BsP**, basisphenoid; **St**, supratemporal; **St-F**, stapedial footplate; **St-P**, stapedial shaft/process; **VI**, N. abducens (VI); **X**, N. vagus (X).

frontal forms a broad ventral facet for articulation with the maxilla and palatine and also forms the choanal groove. There is a ventro-medial process of coronoid that contacts the angular. Dorsally, the postfrontal has a triradiate process that overlaps the frontal-parietal suture. The postorbital (jugal of Estes et al., 1970) closes the posterior orbital margin with a distinct "foot" that articulates with the maxillary trough. The quadrate is a large "?"-shaped element, and the stapes/columella is robust and bears a large and expanded stapedial footplate. There is an extracolumella-intercalary element that contacts the suprastapedial process of the quadrate. The anterior precloacal (cervical?) vertebrae are characterized by the presence of unfused intercentra on the hypapophyses of third and fourth precloacals. The next six precloacals possess fused hypapophyses/intercentra; the anterior third of the known precloacals bears prominent ventral hypapophyseal keels (Caldwell and Albino, 2002).

Relatively recent additions to the debate are numerous new specimens of Gondwanan relic taxa from the Australian Neogene, *Wonambi naracoortensis* Smith, 1976, and *Yurlungurr camfieldensis* Scanlon, 1992 (Scanlon, 2005, 2006). *Yurlungurr* is the older of the two taxa (articulated braincase QMF 45111; Scanlon, 1992, 2006), ranging from the late Oligocene to the Pleistocene of northeastern and central Australia; *Wonambi* (holotype SAM P16168) also lived during the Late Pleistocene and is considered to have overlapped with the human occupation of Australia (Scanlon and Lee, 2000). Both *Wonambi* and *Yurlungurr* were large to gigantic snakes showing madtsoiid characteristics (Hoffstetter, 1961) in the axial skeleton (Scanlon and Lee, 2000: fig. 1; Scanlon, 2005, 2006: fig. 1). The Madtsoiidae show an important paleogeographic (Europe, Africa, Madagascar, South America, and Australia) and temporal distribution (mid-Cretaceous to Pleistocene) that links them to the South American Gondwanan snakes *Dinilysia* and *Najash*. Not too surprisingly, *Wonambi* and *Yurlungurr* show a number of cranial features in common with *Dinilysia* (Scanlon and Lee, 2000; Caldwell and Albino, 2002; Scanlon, 2005, 2006), although regarding *Wonambi,* these similarities are disputed by Rieppel et al. (2003). It is also important to note that the diagnosis for *Najash rionegrina* by Apesteguía and Zaher (2006) is virtually identical to that of the Madtsoiidae as revised by Scanlon and Lee (2000).

Matters of Character: Jugals, Ectopterygoids, Postorbitals, or Unidentified Elements?

I want to review the character delimitation debate arising from Caldwell and Lee's (1997) identification of a jugal in *Pachyrhachis,* but I also want to point out that many other anatomical features of lizards and snakes have been reevaluated through the course of this debate. Important features not examined in this review, but worthy of consideration nonetheless, include the following:

1. The anatomy, implantation, attachment, and histology of the teeth in snakes and lizards (Lee, 1997a; Lee and Scanlon, 2001; Scanlon and Lee, 2002; Caldwell et al., 2003; Budney et al., 2006; versus Zaher and Rieppel, 1999a; Rieppel and Kearney, 2001).
2. The structure of the intramandibular joint and the evolution of the macrostomatan feeding system (Lee and Caldwell, 1998; Lee et al., 1999a; Caldwell, 1999a; Lee and Scanlon, 2002; versus Rieppel and Zaher, 2000a, 2001a; Rieppel and Kearney, 2001).
3. The braincase of squamates, in particular the otic region and the crista circumfenestralis of snakes (Caldwell and Albino, 2002; Lee and Scanlon, 2002; Scanlon, 2005; versus Rieppel and Zaher, 2000b; 2001b).
4. Inferences concerning the paleoecology of early snakes and pythonomorph sister-groups (Caldwell, 1999a,b; Scanlon et al., 1999).
5. The orientation of the quadrate (Caldwell and Lee, 1997; Caldwell, 2000a; versus Zaher and Rieppel, 1999b; Polcyn et al., 2005; Apesteguía and Zaher, 2006).
6. The limb and its pelvic or sacral/lymphapophyseal features (Caldwell and Lee, 1997; Lee and Caldwell, 1998; Caldwell, 2003; versus Zaher and Rieppel, 1999b, 2002; Tchernov et al., 2000; Rieppel et al., 2003).
7. The neck and the identity of cervical vertebrae in snakes (Lee and Caldwell, 1998; Caldwell, 1999a, 2000b, 2003; Caldwell and Albino, 2002; Scanlon, 2004).

Extant Snakes and Lizards

In all extant snakes, the jugal is considered absent, as is the postfrontal; the only bone in the postcircumorbital series that is identified in some, but not all, snakes is the postorbital. In this latter case, the only portion of the postorbital that is present is the descending process that would have articulated with the jugal; the posteriorly directed squamosal ramus of the postorbital is absent, as is the squamosal (Fig. 7.9A–C,E,F). In nearly all present-day snakes, the ectopterygoid articulates, usually in an overlapping kinetic joint, with the posterior ramus of the maxilla where the maxilla frames the ventral and posterior margin of the orbit. It is at this position that the descending process of the postorbital contacts or overhangs the maxilla-ectopterygoid contact. Thus the identification of a postorbital in all extant snakes is problematic: (1) the element does not contact the postfrontal, which is missing; (2) instead of the frontoparietal suture, it contacts only the frontal or just the parietal; (3) there is no posterior ramus; and (4) ventrally, it makes contact with the maxilla.

In nonophidian squamates (Fig. 7.9G,H), the postorbital contacts the postfrontal (and is sometimes fused with it), has a posterior ramus contacting the squamosal, and contacts the jugal where that element rises to meet the posterior margin of the postfrontal.

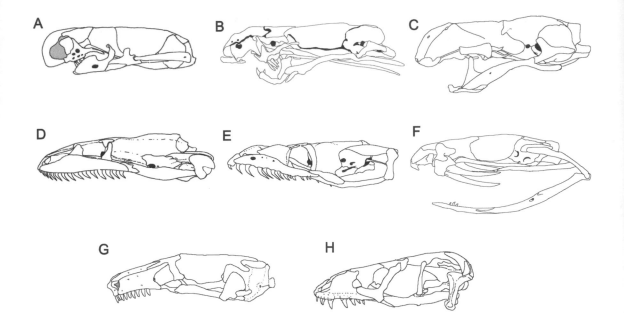

A   B   C

D   E   F

G   H

Figure 7.9. Comparison of osteological features of scolecophidian skulls with those of extinct and extant snakes and lizards. **A,** skull of *Leptotypholops;* **B,** skull of *Anomalepsis* (redrawn from Haas, 1968); **C,** skull of *Typhlops* (redrawn from Mattison, 1995); **D,** reconstructed skull of *Dinilysia;* **E,** skull of *Python* (redrawn from Rage, 1984); **F,** skull of *Atractaspis* (redrawn from Mattison, 1995); **G,** skull of *Dibamus* (redrawn from Rieppel, 1984); **H,** skull of *Heloderma* (redrawn from Rieppel, 1980).

In contrast, the jugal variably articulates with the postfrontal or postorbital (or the jugal process of the fused postorbitofrontal), articulates with the maxilla behind and beneath the orbit, and articulates with the lacrimal if present. The jugal of the varanoid lizard *Lanthanotus* (Rieppel, 1980; Estes et al., 1988; Conrad, 2004) articulates with the postfrontal, maxilla-ectopterygoid, and lacrimal (the postorbital being absent in *Lanthanotus*). The remaining articulation, i.e., with the ectopterygoid, is variable among lizards and snakes (e.g., the postorbital contacts the maxilla above the ectopterygoid-maxilla articulation in *Python*).

Fossil Snakes: Juggling the Jugal

I will begin with the description by Caldwell and Albino (2002) of the cranial structure of new specimens of *Dinilysia* (Fig. 7.10A–C) and proceed from there to *Eupodophis* (Fig. 7.11A–E), *Haasiophis* (Fig. 7.12A–B), and finally to *Pachyrhachis* (Figs. 7.13A–D, 7.14A–H).

Caldwell and Albino (2002) disagree with the identification by Estes et al. (1970) of a postfrontal, postorbital, and a tentative jugal in the holotype of *Dinilysia patagonica* (Fig. 7.8B). Caldwell and Albino (2002:864) identified both a postfrontal and postorbital, but not a jugal, and did so with a caveat concerning the postorbital: "Our identification of a postorbital in *Dinilysia* follows the convention for snakes of identifying the element that descends ventrally from the area of the frontal-parietal suture, and framing the posterior margin of the orbit, as the postorbital. Though we follow this convention for *Dinilysia,* we do find this identification problematic due to the presence of the postfrontal and the jugal-like articulations

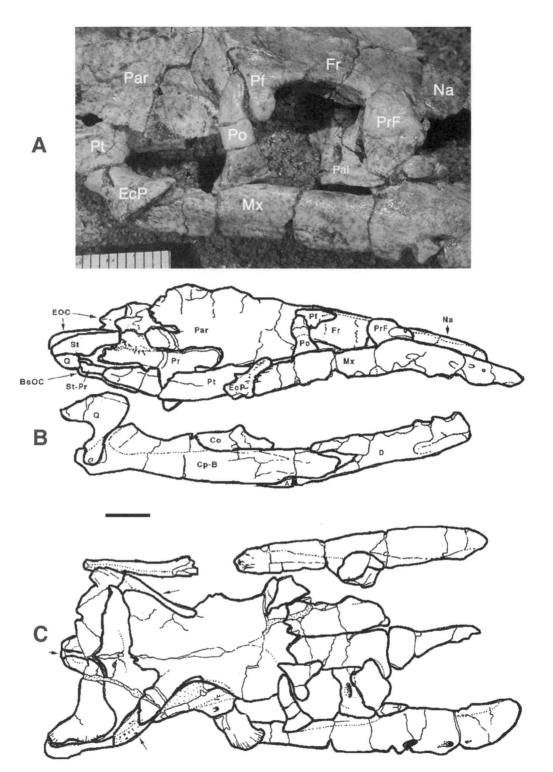

Figure 7.10. *Dinilysia patagonica*, MACN-RN-1013. **A**, left lateral view of skull, with close-up of orbital region. Specimen drawings of **B**, lateral, and **C**, dorsal views of skull of MACN-RN-1013. Scale bar = 1 cm. Abbreviations as in Figure 7.8.

of the postorbital of *Dinilysia.*" Estes et al. (1970) had identified the distal portion of the postorbital identified by Caldwell and Albino (2002) as a possible jugal, and although Caldwell and Albino (2002) contradicted Estes et al. (1970), they did so noting the problematic nature of this element in *Dinilysia* specifically and snakes generally. The identification of a postfrontal for *Dinilysia* by Estes et al. (1970) (despite the element being present in these topological relations only in scleroglossan lizards) was also confirmed by the new specimens described by Caldwell and Albino (2002:864), who noted that it "passes the positional and relational criteria of the test of similarity (i.e., element clasping the fronto-parietal suture), and the test of composition (i.e., dermatocranial bone)."

What can be distilled from the accounts by Estes et al. (1970) and Caldwell and Albino (2002) is that at least one fossil snake taxon, *Dinilysia patagonica,* retains the scleroglossan lizard character of a V-shaped postfrontal clasping both the parietal and frontal across the suture between those elements; in some anguimorph scleroglossans, such as *Heloderma* or *Lanthanotus* (Conrad, 2004), both of which have lost the postorbital, the element is closely similar to *Dinilysia.* Scanlon (2006) goes even further regarding *Yurlungurr* and identifies an element with topological relations similar to those of *Dinilysia* as the postorbitofrontal. The question regarding the presence or absence of the jugal in *Dinilysia* (and, by extension, in all snakes) is simple but important: If a postfrontal (or postorbitofrontal) is identified, then where would one expect to find either the jugal or postorbital? This is particularly true of the conservative identification by Caldwell and Albino (2002) of a postorbital in *Dinilysia:* an element lacking all topological relations of a typical squamate postorbital (e.g., no squamosal ramus) but possessing the topological connections of a lizard jugal. Scanlon (2006) gave this same rationale in his identification of a jugal and postorbitofrontal in *Yurlungurr.*

It is clear from the examination of any lizard taxon (e.g., *Heloderma;* Fig. 7.9H) that the jugal contacts the lateral and ventral (jugal) process of the V-shaped postfrontal, toward the posterior half of the postfrontal. The jugal also frames the posterior margin of the orbit, descending to contact the maxilla above the ectopterygoid; at this contact, the jugal possesses both an anterior and posterior process. The only topological and connective difference between the jugal of *Heloderma* (Fig. 9H) and the postorbital of *Dinilysia* (Figs. 7.8C,D, 7.9A–C) is that in the former, the anterior process forms the lateroventral margin of the orbit and contacts the lacrimal. Thus it is possible that the postorbital (jugal) of *Dinilysia,* and thus perhaps the postorbitals of extant snakes that still retain this bone, might be better identified as the primary homologues of the lizard jugal.

Eupodophis descouensi *(Rage and Escuillié, 2000)*

When Rage and Escuillié (2000, 2001) described this snake with legs, they only incorporated the details of the part (Fig. 7.11D) but not the counterpart (Fig. 7.11E) of the holotype; unfortunately, even if they had illustrated both halves of the specimen, the details of elements such as the postorbital and ectopterygoid, and possible elements such as the jugal and postfrontal, would not have been easily extracted. A recent description by Rieppel and Head (2004) of three new specimens from Lebanon, two of which have moderately well-preserved skulls (Fig. 7.11A,B), includes these authors' identifications of right and left ectopterygoids and right and left postorbitals (Fig. 7.11A) and thus illuminates aspects of the holotype. Rieppel and Head (2004:12) stated:

> The ectopterygoid of specimen MSNM V 3661 closely resembles that of *Pachyrhachis* (Rieppel & Zaher, 2000, fig. 1B). It is a horizontally positioned lamina of bone that slightly expands toward its anterior end. . . . The anterior end is not exposed as it underlaps (morphologically: overlaps) the posterior end of the maxilla. The somewhat narrower posterior end has a complete, rounded margin, and underlies the ventral tip of the postorbital. The complete preservation of the ectopterygoid in specimen MSNM V 3661 refutes the claim made by Rieppel & Zaher (2000) that the ectopterygoid is broken across the posterior end of the maxilla in *Pachyrhachis*. This claim was made to account for the shortness of the ectopterygoid, and for its position almost entirely in front of the ventral tip of the postorbital. These are characteristics which specimen MSNM V 3661 shares with *Pachyrhachis*, and which led to the identification of the ectopterygoid as a jugal in *Pachyrhachis* (Lee & Caldwell, 1998; Lee & Scanlon, 2002). However, as will be discussed in more detail below, the position of the ectopterygoid almost entirely in front of the postorbital results from the unusually elongate frontals that carry the postorbital backwards.

This description for the ectopterygoids and postorbitals of *Eupodophis* is remarkably different from the claims and arguments presented from Zaher and Rieppel (1999b) to just before Rieppel and Head (2004). In my opinion, it also falls short of being an accurate delimitation. Rieppel and Head (2004) argued forcefully that their ectopterygoid in *Eupodophis*, and the jugal of *Pachyrhachis* as interpreted by Caldwell and Lee (1997), are the same element: a thin lamina of bone, positioned along the maxillary groove, immediately below the postorbital. I agree with this assessment. I also agree that the earlier argument (Zaher and Rieppel, 1999b) about symmetrical right and left breakage of the ectopterygoid as a means of accounting for the required length of an ectopterygoid (which must span the distance from the lateral margin of the pterygoid to posterior ramus of the maxilla) was problematic. However, the length and structure of what Rieppel and Head (2004) identified as ectopterygoids for *Eupodophis* (Fig. 7.11A) and interpretation of Zaher and Rieppel (1999b) of *Pachyrhachis* (Fig. 7.14C; compare

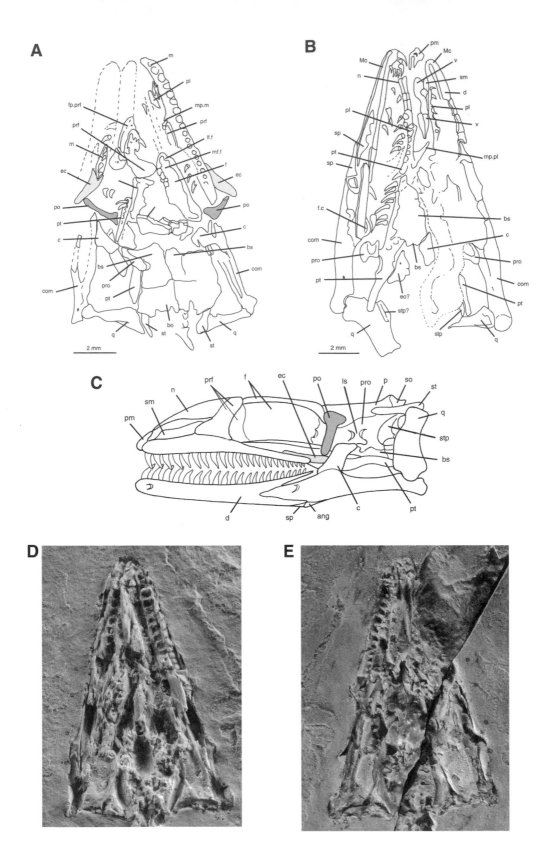

both with Rieppel and Zaher, 2000a: fig. 15B) remain a critical issue. The identification as an ectopterygoid, and the explanation of how Lee and Caldwell (1998) and Lee and Scanlon (2002) were misled by unique structures masquerading as plesiomorphic ones do not address the question of just how this putative ectopterygoid actually made contact with the ectopterygoid process of the pterygoid. The element is too short. In their reconstruction of *Eupodophis* (Fig. 7.11C), the bone identified as the ectopterygoid by Rieppel and Head (2004) is 1.4 times longer than their available fossil evidence indicates (cf. Fig. 7.11A).

Although Rieppel and Head (2004) refuted the identifications of Zaher and Rieppel (1999b) and Rieppel and Zaher (2000a,b), I dispute their identification of the right and left laminac of bone appressed to the maxillae in MSNM V 3661 as ectopterygoids. I also question their mechanistic argument of modified topological relations due to the posteriorly translocated postorbitals and elongated frontals. In every known lizard and snake, extinct or extant, the posterior margin of the orbit is aligned with one of, or a combination of, (1) the mesokinetic joint formed by the frontal-parietal suture dorsally and its articular contacts laterally along the margins of the decensus frontalis and parietalis, (2) in the middle of which is located the foramen for passage of cranial nerve II (N. opticus), and the position of the (3) postfrontal, (4) postorbital, (5) postorbitofrontal, and (6) jugal. As Rieppel and Head (2004) have reconstructed *Eupodophis*, this series of topological relations exists (Fig. 7.11C), but it is not the one they claim represents a posterior translocation of the postorbitals. Not a single positional feature of their reconstructed postorbital appears to bc modified. Instead, the skull of *Eupodophis* appears to be anteriorly enlarged so that the orbit is very elongate, as is the snout, while the braincase is proportionally shortened similar to the reconstruction (Fig. 7.13C,D) provided by Caldwell and Lee (1997) for *Pachyrhachis*. They argued that the postorbital is posteriorly translocated so as to accommodate what is seen as the artificial length of their reconstructed ectopterygoid.

Haasiophis terrasanctus *Tchernov et al., 2000*
(Figs. 7.5A,B and 7.12A,B)

In their original description of *Haasiophis*, the authors stated only that the specimen possessed a nearly complete postorbital arch (which is puzzling because there is only a single, slender, and boomerang-shaped element on either side of the frontal-parietal suture). The putative postorbital of *Haasiophis* appears to be slender dorsally and then expanded ventrally into an anteriorly elongate process. This element does not appear to possess the structure that would ally it with the putative postorbitals of both *Dinilysia* (Figs. 7.8D, 7.9D, 7.10A–C) and *Eupodophis* (Fig. 7.11A), or the postorbitofrontal of *Pachyrhachis* as identified by Caldwell and Lee (1997) (see also Lee and Scanlon, 2002).

The ectopterygoids were identified in Tchernov et al. (2000:

Figure 7.11. *(opposite page)* Skulls of *Eupodophis descouensi*. **A**, skull of specimen MSNM V 3661 as preserved in ventral view (from Rieppel and Head, 2004: figs. 6 and 10, respectively). **B**, skull of specimen MSNM V 4014 as preserved in ventral view (from Rieppel and Head, 2004). **C**, reconstruction (from Rieppel and Head, 2004: fig. 13). **D**, photograph of holotype, part (Rh-E.f. 9001, 9002, 9003) (photograph by author). **E**, photograph of holotype, counterpart (Rh-E.f. 9001, 9002, 9003) (photograph by author). **ang**, angular; **bo**, basioccipital; **bs**, basisphenoid; **c**, coronoid; **com**, compound bone; **d**, dentary; **ec**, ectopterygoid; **eo**, exoccipital; **f**, frontal; **f.c**, facet on compound bone for coronoid; **fp,prf**, foot process of prefrontal; **lf.f**, lateral frontal flange; **ls**, laterosphenoid; **m**, maxilla; **Mc**, Meckel's canal; **mf.f**, medial frontal flange; **mp.m**, medial (palatine) process of maxilla; **mp.pl**, medial (choanal) process of palatine; **n**, nasal; **p**, parietal; **pl**, palatine; **po**, postorbital; **pm**, premaxilla; **prf**, prefrontal; **pro**, prootic; **pt**, pterygoid; **q**, quadrate; **r.prl**, recess for paralymphatic system; **sm**, septomaxilla; **so**, supraoccipital; **sp**, splenial; **stp**, stapes; **st**, supratemporal; **v**, vomer.

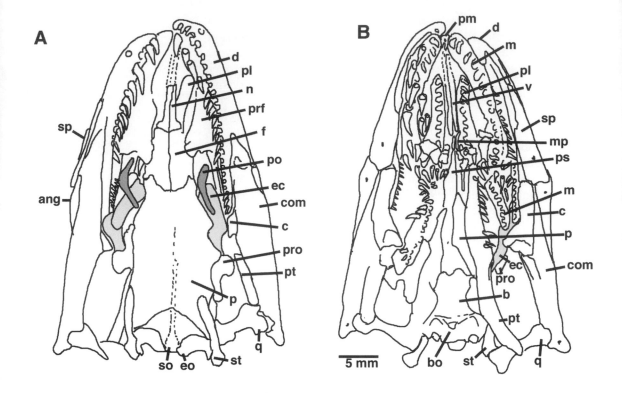

A

- d
- pl
- n
- prf
- f
- po
- ec
- com
- c
- pro
- pt
- p

sp

ang

so eo — st
q

B

- pm
- d
- m
- pl
- v
- sp
- mp
- ps
- m
- c
- p
- com
- ec
- pro
- b
- pt

5 mm   bo   st   q

Figure 7.12. Drawing of the skull of *Haasiophis terrasanctus* (HUJ-PAL 659; from Tchernov et al., 2000: fig. 1) in A, dorsal, and B, ventral views. **ang,** angular; **bo,** basioccipital; **bs,** basisphenoid (parabasisphenoid); **c,** coronoid; **com,** compound bone; **d,** dentary; **ec,** ectopterygoid; **eo,** exoccipital; **f,** frontal; **m,** maxilla; **mp,** medial process of palatine; **n,** nasal; **p,** parietal; **pl,** palatine; **pm,** premaxilla; **po,** postorbital; **prf,** prefrontal; **pro,** prootic; **ps,** parasphenoid rostrum; **pt,** pterygoid; **q,** quadrate; **so,** supraoccipital; **sp,** splenial; **st,** supratemporal; **v,** vomer. Modified from Rieppel et al., 2003.

fig. 1) in dorsal view but showed right and left asymmetries, unusual forms and articulations, and, in the illustration on the left side of the skull underneath the unidentified left ectopterygoid, the presence of an unidentified lamina of bone (Fig. 7.12A). The information for assessing the illustrations of the ectopterygoids by Tchernov et al. (2000) was provided by Rieppel et al. (2003: figs. 2, 4, 5). In that latter work, they presented photos, line drawings, and radiographs of a bone, which appear to show an anteriorly flattened, spatulate element with a distinct medial bend where it twists to articulate with a bone that they identify as the pterygoid. If Rieppel et al. (2003) were correct regarding their similarity criteria for the ectopterygoid of *Haasiophis,* following Tchernov et al. (2000), the ectopterygoid is approximately 1.2 cm in length with the anterior process significantly overlapping the maxilla (at least eight or nine tooth positions) and is a large L- or J-shaped bone.

Pachyrhachis problematicus *Haas, 1979*

Caldwell and Lee (1997) and Lee and Caldwell (1998) identified right and left jugals (Fig. 7.14A), right and left postorbitofrontals (both clasp the frontal-parietal suture and possess ventral processes that completely frame the posterior margin of the orbit), and large, elongate, spatulate ectopterygoids that were best observed as complete elements in the referred specimen (HUJ-PAL 3775; Fig. 7.13B). Similar ideas were also discussed by Scanlon (1996), although without the benefit of firsthand observation of *Pachyrhachis.*

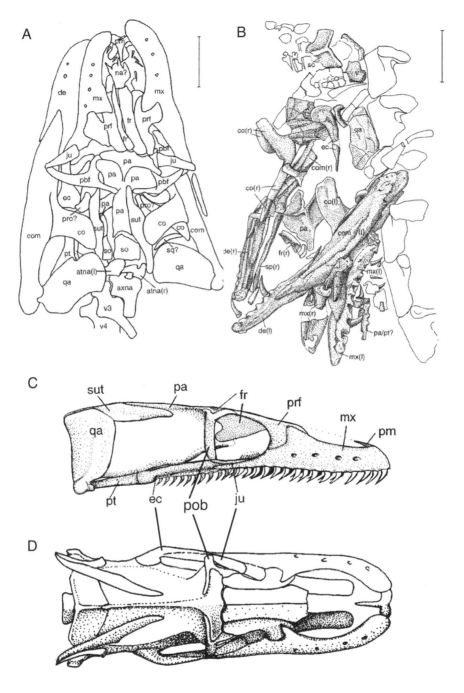

Figure 7.13. *Pachyrhachis problematicus.* **A**, drawing of the dorsal surface of the skull of holotype (HUJ-PAL 3659; from Lee and Caldwell, 1998: fig. 3). **B**, drawing of ventral surface of the region of referred specimen HUJ-PAL 3775 containing skull fragments (from Lee and Caldwell, 1998: fig. 6). Reconstruction of the skull of *Pachyrhachis* in **C**, lateral and **D**, dorsal views (from Lee and Caldwell, 1998: fig. 7). **an**, angular; **atna**, atlas neural arch; **ax**, axis neural arch; **bo**, basioccipital; **bs**, basisphenoid; **co**, coronoid; **com**, compound bone; **de**, dentary; **ec**, ectopterygoid; **ex**, exoccipital; **fr**, frontal; **ju**, jugal; **mx**, maxilla; **na**, nasal; **pa**, parietal; **pl**, palatine; **pbf**, postorbitofrontal; **pm**, premaxilla; **prf**, prefrontal; **pro**, prootic; **pt**, pterygoid; **qa**, quadrate; **sm**, septomaxilla; **so**, supraoccipital; **sp**, splenial; **sq**, squamosal; **sta**, stapes; **sut**, supratemporal; **v**, vertebra; **vo**, vomer (as listed by Lee and Caldwell, 1998; the small "r" or "l" indicates right or left element).

Zaher and Rieppel (1999b) challenged the identification of the jugal and argued against Caldwell and Lee's (1997) identification of the anterior and posterior rami as well as the postorbitofrontal bone (compare Fig. 7.14B and 7.14C). They further argued that Caldwell and Lee (1997) had mistakenly modified the margins of the parietal, frontals, and the sizes of the reidentified jugals (now ectopterygoids). However, Zaher and Rieppel (1999b) ignored the element that Caldwell and Lee (1997) did identify as an ectopterygoid (Fig. 7.14C, left side, just anterior to the left coronoid bone). In ventral view (Fig. 7.14E), the holotype skull revealed the presence and position of the paired pterygoids. The actual position of the ectopterygoid process of the pterygoid—in other words, the distance separating the ectopterygoid identified by Zaher and Rieppel (1999b) from its articulation with the pterygoid—is nearly 2 cm. These problems are now highlighted by the refutation by Rieppel and Head (2004) of arguments by Zaher and Rieppel (1999b) and Rieppel and Zaher (2000a) that the jugal identified by Caldwell and Lee (1997) was a misinterpretation of broken elements. Furthermore, the ectopterygoids of *Haasiophis* (Rieppel et al., 2003; Fig. 7.12A,B) clearly indicate that similar-sized and -shaped ectopterygoids in *Pachyrhachis* were accurately identified by Caldwell and Lee (1997) and are not the result of some unusually symmetrical postmortem fracturing.

Figure 7.14. *(opposite page) Pachyrhachis problematicus* Haas, 1979. Comparison of anatomical identifications of the dorsal view, and one ventral of the skull of the holotype (HUJ-PAL 3659), and computer reconstructions derived from computed tomographic data. **A**, from Haas (1979: fig. 3). **B**, from Caldwell and Lee (1997: fig. 1). **C**, from Zaher and Rieppel (1999: fig. 1). **D–H**, from Polcyn et al. (2005: figs. 5, 10). **D**, dorsal view of compressed skull of holotype. **E**, ventral view of compressed skull of holotype. **F**, lateral view of computer-generated skull reconstruction. **G**, dorsal view of computer-generated skull reconstruction. **H**, oblique anterior view of computer-generated skull reconstruction. **A**, Haas (1979): **ang**, angular; **atna**, atlas neural arch; **axna**, axis neural arch; **com**, compound bone; **co**, coronoid; **d**, dentary; **ect**, ectopterygoid; **f**, frontal; **mx**, maxilla; **p**, parietal; **pmx**, premaxilla; **porb**, postorbital; **prf**, prefrontal; **ptf**, postfrontal; **pro**, prootic; **squ**, squamosal; **qu**, quadrate; **socc**, supraoccipital; **spt**, septomaxilla; **porb**, postorbital; **squ**, squamosal; **st**, stapes. **B**, see abbreviations for Figure 7.12. **C**, Rieppel and Zaher (2000a): **at**, atlas neural arch; **axna**, axis neural arch; **com**, compound bone; **co**, coronoid; **d**, dentary; **ect**, ectopterygoid; **f**, frontal; **mx**, maxilla; **n**, nasal; **p**, parietal; **pm**, premaxilla; **po**, postorbital; **prf**, prefrontal; **q**, quadrate; **spt**, septomaxilla; **porb**, postorbital; **squ**, squamosal; **st**, supratemporal; **stp**, stapes. **D–E**, from Polcyn et al. (2005): **a**, angular; **atn**, atlas neural arch; **ax**, axis vertebra; **bs**, basisphenoid; **bo**, basioccipital; **c**, compound bone; **cor**, coronoid; **d**, dentary; **ect**, ectopterygoid; **ex**, exoccipital; **f**, frontal; **m**, maxilla; **p**, parietal; **pm**, premaxilla; **pob**, postorbital; **pf**, prefrontal; **pl**, palatine; **pt**, pterygoid; **q**, quadrate; **s**, stapes; **so**, supraoccipital; **sp**, splenial; **st**, supratemporal, **ue1–5**; unidentified elements; **v3–v4**, cervical vertebra 3 and 4. Scale bar = 1 cm.

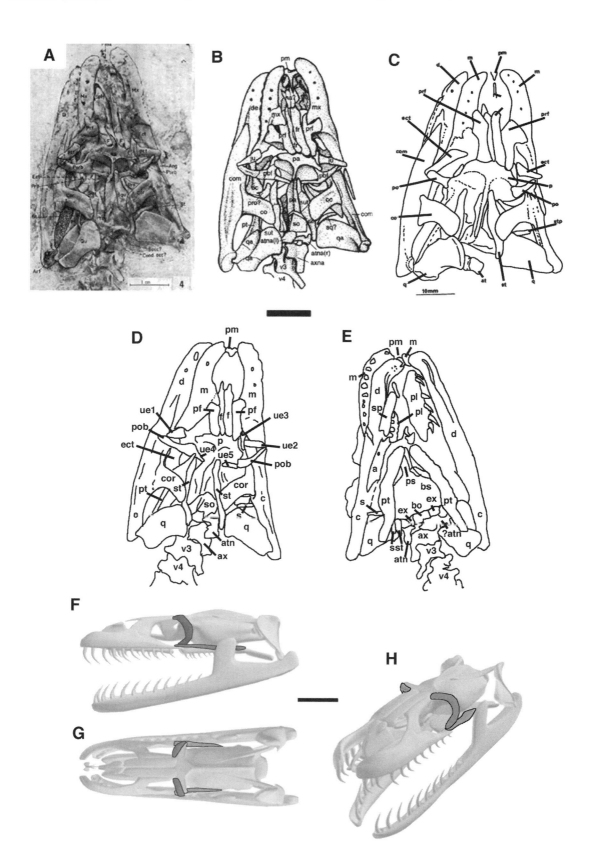

It is also interesting that Polcyn et al. (2005) recently published their analysis of computed tomographic (CT) data, using surface images, internal images, and computer reconstructions for the skull of *Pachyrhachis*. Although they were far from full agreement on all interpretations, Polcyn et al. (2005) concurred with Caldwell and Lee (1997) regarding the identity, size, and shape of the ectopterygoid, thus refuting the arguments for this bone as presented by Zaher and Rieppel (1999b), Rieppel and Zaher (2000a), Tchernov et al. (2000), Rieppel et al. (2003), and Rieppel and Head (2004).

Polcyn et al. (2005) also identified a postorbital that appears to be long enough to completely bridge the posterior margin of the orbit; in an earlier work, Tchernov et al. (2000) attempted to reconstruct a postorbital that was a very large element, interpreted here as a composite of the postorbitofrontal and jugal of Caldwell and Lee (1997). What is problematic about this aspect of the study by Polcyn et al. (2005) is that they illustrate five unidentified elements (UEs) (Fig. 7.14D–E): UE 1 (left jugal of Caldwell and Lee, 1997); UE 2 (right jugal of Caldwell and Lee, 1997); UE 3 (anteriormost process of postorbitofrontal of Caldwell and Lee, 1997); and UE 4 and UE 5 (left and right posterior rami of the postorbitofrontal of Caldwell and Lee, 1997). As such, Polcyn et al. (2005) neither disagreed with Caldwell and Lee (1997) nor agreed with Zaher and Rieppel (1999b), Tchernov et al. (2000), and Rieppel et al. (2003). The striking feature of illustrations of *Pachyrhachis* (Fig. 14D,E) provided by Polcyn et al. (2005) is that they are virtually identical to the illustration by Caldwell and Lee (1997) (Fig. 7.14B) of the orbital bones and ectopterygoid of the holotype. The very different structure illustrated and argued for by Zaher and Rieppel (1999b) and Rieppel and Zaher (2000a) was not confirmed by CT scan data in the study by Polcyn et al. (2005).

Although the use of UEs by Polcyn et al. (2005) demonstrates an otherwise conservative stance on the identification of primary homologies, they stated in their abstract, "There is no jugal." This firm statement concerning primary homology is puzzling when contrasted with their hesitation to identify problematic elements that Caldwell and Lee (1997) identified as the jugals.

Summary

The nonidentification of five bones or bone fragments by Polcyn et al. (2005), labeled as distinct UEs, does not imply that these elements are not jugals or fragments of the postorbitofrontal processes. It only means that Polcyn et al. (2005) chose not to provide an identification for these bones while at the same time ascertaining the absence of the jugal. In addition, Polcyn et al. (2005) also identified a large ectopterygoid on the left side of the skull of the holotype, as did Caldwell and Lee (1997). Thus all of the elements identified as separate bones by Caldwell and Lee (1997) and refuted as absent by Zaher and Rieppel (1999b) and Rieppel and Zaher (2000a) are verified as being present by Polcyn et al. (2005), regardless of their identity.

The structure of the ectopterygoid of *Haasiophis* is identical to that illustrated by Lee and Caldwell (1998) in the referred specimen of *Pachyrhachis* (Fig. 7.13B) and as reconstructed (Fig. 7.13C–D); the same elongate ectopterygoid is reconstructed by Polcyn et al. (2005) for *Pachyrhachis* (Fig. 7.14F–H). The position taken by Rieppel and Head (2004) on these elements in *Eupodophis* further supports the growing body of evidence indicating that these elements are not broken ectopterygoids; however, there is also no evidence to suggest these elements are extremely short ectopterygoids (the position they take now). Illustrations by Polcyn et al. (2005) agree with those of Caldwell and Lee (1997) and Lee and Caldwell (1998) that the ectopterygoid is present in the holotype; they also concur that the size and shape of the ectopterygoid as reconstructed by Lee and Caldwell (1998), and presented twice in the skull of the referred specimen, is accurate. Therefore, both Lee and Caldwell (1998) and Polcyn et al. (2005) refuted the "short ectopterygoid" of Rieppel and Head (2004) for *Pachyrhachis,* and, by extension, *Eupodophis.*

From *Pachyrhachis* to *Dinilysia,* and actually in all snakes, dermatocranial evolution has resulted in considerable modifications of the circumorbital bones, perhaps related to the reduction and complete loss of the squamosal and postorbital. What also seems useful, despite the rhetoric, is the debate surrounding the delimitation of characters and character states. Such argumentation is essential to the testability of synapomorphies.

Although phylogenies arise from parsimony analysis of overall character distributions in cladograms, the testable element of the hypothesis remains the individual character, its state, and its resultant distribution. Debate on the nature, identity, and delimitation of a character, as pedantic as that debate might become, is the only mechanism of falsification available for testing cladistic statements. Phylogenies are impossible to falsify, but the same is not true of cladograms because their character distributions are very much dependent on the limits of the character concept. If that concept is erroneous in its delimitation (identity or taxic assignment), it can be falsified.

The debate on the jugal, ectopterygoid, and postorbital thus represents an essential debate concerning character delimitation on the sister-group relationships of snakes. It is not over yet because agreement has not been reached on the identity of the element, even in *Pachyrhachis.* However, the field of possibilities is being narrowed with each new specimen and each new study.

Squamate Sister-Groups and Snake Ingroup Relations

The post-1997 debate on snake sister-group relationships and ingroup phylogeny has raised yet again the question of primitive versus derived characters in snakes, the role and value of fossil taxa in the phylogenetic analysis of squamates, which extant snake taxon (Alethinophidia or Scolecophidia) best represents the ances-

Figure 7.15. *(opposite page)*
Phylogenetic hypotheses of
snake sister-group and ingroup
phylogeny. **A**, phylogeny of Lee
and Caldwell (1998),
concerning the relationships of
Pachyrhachis with
Scolecophidia and
Alethinophidia, and between
snakes and other nonsnake
squamates. Gray box indicates
the major result of this
phylogenetic hypothesis:
Scolecophidia are not placed as
basal snakes. **B**, Phylogenetic
topology (single-lined portion)
as illustrated by Rieppel and
Zaher (2000); double-lined
portion represents not
illustrated portion as given in
the text: (Anguidae
(Xenosauridae ((*Heloderma*
(Mosasauroidea (*Lanthanotus,
Varanus*)) ((*Sineoamphisbaenia*,
Dibamidae) (Scolecophidia
(*Dinilysia* (Anilioidea
(*Pachyrhachis*,
Macrostomata)))))))))). **C**, Snake
ingroup phylogeny published by
Apesteguía and Zaher (2006);
note large-mouthed, large-
bodied snake as representing
ancestral snake morphology,
consistent with Lee et al.
(1999b), Caldwell and Albino
(2002), and Scanlon (2006).

tral snake condition, the contrasting of a nonmacrostomate skull as
primitive (allied with the burrowing origin scenario) versus a
macrostomate skull as primitive (allied with the aquatic origin scenario), the identity of the proximate sister-group of snakes among
squamates, and a lively debate concerning method and philosophy
in the test of similarity (Zaher and Rieppel, 2000; Rieppel and
Kearney, 2001, 2002; versus Kluge, 2003).

Attempts at collating the enormous data set delimited by fossil
and extant squamates began with Cope's (1869) phylogeny (Fig.
7.1) and continues to the most recent molecular phylogenies of
Vidal and Hedges (2004) and Townsend et al. (2004), and the total
evidence analysis by Lee (2005). All of these authors have struggled
to identify the closest squamate relative of snakes, i.e., their sister-
group, and to work out relationships among snakes.

For example, Camp's (1923) phylogenetic hypothesis of all
squamates underwent its first rigorous test (rigor identified in this
case as a discrete listing of characters and alternative states of each
character) when Estes et al. (1988) conducted their cladistic analysis of squamate phylogeny. Of importance to note for both of these
hypotheses is that neither of them included fossil taxa in their data
sets, and further, that neither author considered he had resolved the
relationships of snakes, amphisbaenians, or dibamids (although
both found more support for varanoid relationships of snakes than
any other alternative).

The first cladistic reevaluation of Estes et al. (1988) was that by
Caldwell and Lee (1997), who used an edited and abridged version of
matrix compiled by Estes et al. that included a smaller and revised
version of their characters (95 as opposed to 145) as well as a number
of fossil taxa as terminals (see also Caldwell [1999a], who expanded
on the Caldwell and Lee [1997] matrix). Lee's (1997b) analysis was
restricted to platynotans, snakes, and "pythonomorphs" and did not
test global squamate relationships. Lee and Caldwell (1998) used the
data set of Caldwell and Lee (1997), which also provided the basis for
the analysis by Caldwell (1999a). The resulting phylogenetic hypothesis (Fig. 7.15A) framed the current debate by placing *Pachyrhachis*
as basal to all extant snakes, including the scolecophidians, in a clear
three-taxon statement. Whereas Caldwell and Lee's (1997) result was
consistent with that of Janensch (1906) and Nopcsa (1923), it
strongly contrasted with the views of researchers such as Walls (1940,
1942), Underwood (1967), Haas (1979), and Rieppel (1988), all of
whom had concluded that Scolecophidia (=Angiostomata) were the
most basal snakes. This significant implication of the phylogenetic hypothesis by Caldwell and Lee (1997) (Fig. 7.15), that the Scolecophidia are not basal snakes, was largely ignored because the critical
focus of most of the subsequent debate in the literature (e.g., Zaher,
1998; Zaher and Rieppel, 1999b) was on arguing that *Pachyrhachis*
was not a basal snake outside of Alethinophidia, but in fact a derived
macrostomatan snake nested within that larger clade.

At this stage in the developing debate concerning the ingroup
relationships of snakes, Cohn and Tickle (1999) presented the re-

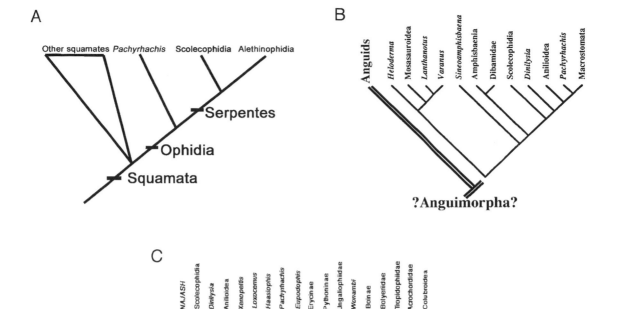

A

Other squamates *Pachyrhachis* Scolecophidia Alethinophidia

Serpentes

Ophidia

Squamata

B

Anguids *Heloderma* Mosasauroidea *Lanthanotus* *Varanus* *Sineoamphisbaena* Amphisbaenia Dibamidae Scolecophidia *Dinilysia* Anilioidea *Pachyrhachis* Macrostomata

?Anguimorpha?

C

NAJASH Scolecophidia *Dinilysia* Anilioidea *Xenopeltis* *Loxocemus* *Haasiophis* *Pachyrhachis* *Eupodophis* Erycinae Pythoninae Ungaliophiidae *Wonambi* Boinae Bolyeriidae Tropidophiidae Acrochordidae Colubroidea

Macrostomata
Alethinophidia
Serpentes

sults of an extremely important study on the molecular genetics of limb development in *Python*. Their conclusions on limb evolution in snakes were based on the squamate sister-group relationships of snakes (not on their ingroup relationships) proposed by Caldwell and Lee (1997) and Lee and Caldwell (1998) (Fig. 7.15A), including *Pachyrhachis* and mosasaurs.

Cohn and Tickle's (1999) molecular genetic research found that *Hox* gene expression domains were expanded anteriorly along the body axis in *Python* and that *HoxC8*, normally expressed axially just posterior to the pectoral girdle in limbed tetrapods, was expressed anteriorly to the first or second precloacal vertebra. Cohn and Tickle concluded, on the basis of their observation of ribs on anterior vertebrae in snakes, that these ribs suggested that thoracic identity had been overprinted onto the snake "neck" and that the trunk resembles an elongated thorax. Furthermore, they noted that far posterior to the tetrapodal pectoral girdle, snakes possess ventral hypapophyses on the vertebrae (a neck character). However, as was discussed by Caldwell (2003), the presence of ribs on anterior vertebrae (cervical ribs and vertebrae) is a primitive character of almost all tetrapods with fully developed pectoral girdles. Caldwell (2003), in contrast to Cohn and Tickle (1999), concluded that snake developmental evolution appears to have "cervicalized" the anterior part of the body rather than the reverse.

An interesting interpretation of Cohn and Tickle's (1999) conclusions was that given by Tchernov et al. (2000:2012) in their description of *Haasiophis* as a second snake with legs nested well within Alethinophidia: "With *Haasiophis* and *Pachyrhachis* related to basal macrostomatan, the conclusion based on parsimony must be that these limbs redeveloped from rudiments such as those present in *Python*." Redevelopment of any evolutionarily lost feature would indeed be a new and exciting finding; however, this idea was not pursued further by Rieppel et al. (2003) in their detailed description of *Haasiophis*. Cohn and Tickle's (1999) groundbreaking work still remains as a stand-alone evo-devo study of snake limb and body axis development and evolution, although it has lately been criticized by several studies for its macroevolutionary conclusions (Wiens and Slingluff, 2001; Shapiro, 2002).

Rieppel and Zaher (2000a) followed with a critique and reanalysis of Lee's (1998) data set; their study was neither an original taxon-character analysis nor a complete reassessment of global squamate relations (Fig. 7.15B), but instead analyzed only a subset of Lee's (1998) matrix. Oddly, Rieppel and Zaher (2000a: fig. 17) illustrated only a portion of one of their results (run 8), which itself was an incomplete version of the bracketed phylogeny (p. 60) given in their text: (Anguidae (Xenosauridae ((*Heloderma* (Mosasauroidea (*Lanthanotus, Varanus*)) ((*Sineoamphisbaenia*, Dibamidae) (Scolecophidia (*Dinilysia* (Anilioidea (*Pachyrhachis*, Macrostomata))))))))). Both phylogenies indicate that the sister-group of snakes is a clade of burrowing lizards, while mosasaurs are derived varanoid lizards. However, a review of the bracketed phylogeny indicates that Rieppel and Zaher's (2000a) exclusion of phylogenetic results from their illustrated phylogeny, i.e., the sister-group relations of anguids and xenosaurids, masked a major phylogenetic hypothesis produced from run 8: snakes + *Sineoamphisbaenia* + Amphisbaenia/Dibamidae are nested within Anguimorpha as the sister-taxon to Varanoidea, and *Sineoamphisbaenia* (now shown to be a macrocephalosaurian teioid; Kearney, 2003), Amphisbaenia, and Dibamidae are anguimorphs. This is an interesting if rather unorthodox relationship (e.g., Greer, 1985) for both Amphisbaenia and Dibamidae, i.e., to be derived anguimorphs, that was not explored by Zaher and Rieppel (2000) in their text.

Coates and Ruta (2000) reviewed the state of the snake phylogeny and origins debate, concluding that "it is worth remembering that phylogeny is an ongoing research programme and that large parts of the evolutionary tree remain unwritten, unexplored and deeply uncertain." This was certainly true at that time, and it remains true now. However, Coates and Ruta drew no major conclusions about which hypothesis was better supported, although they did review aspects of the debate concerning character delimitation.

Meanwhile, Zaher and Rieppel (2002) continued to support the substance of Zaher (1998), who examined three contentious characters from Caldwell and Lee (1997), added six new characters considered critical to his thesis, and criticized Caldwell and Lee (1997) for not including the Cretaceous snake *Dinilysia patagonica* in their analysis of squamate phylogeny. Caldwell's (2000a) critique of Zaher (1998) rebutted these points and in turn criticized Zaher's figure-caption description of his new characters and the absence of a data matrix. Zaher and Rieppel (2002) followed Zaher (1998) with an ingroup analysis of snakes (where the inclusion of *Dinilysia* could be argued as relevant) rather than a sister-group analysis of all squamates consistent with the necessary method for testing the hypothesis by Caldwell and Lee (1997).

Whereas most of the snake fossils at the center of the post-1997 debate have been nearly 100 million years old, the snake species highlighted by the studies of Scanlon and Lee (2000), Rieppel et al. (2003), and Scanlon (2005) are much younger geologically—the giant Pleistocene madtsoiid snakes from Australia typified by *Wonambi naracoortensis* Smith, 1976. Scanlon and Lee's (2000) description of *Wonambi* included a phylogenetic analysis of snake ingroup relationships using an early version of the matrix they would expand upon in Lee and Scanlon (2002; with 212 osteological and 48 soft-tissue characters). The Pachyophiidae were found to be the most basal snakes, with Madtsoiidae and *Dinilysia* at the base of a large clade containing all crown-group snakes. The ingroup relationships of crown-group snakes followed the framework of phylogenies such as that presented by Tchernov et al. (2000).

In response to Scanlon and Lee (2000), Rieppel et al. (2003) published anatomical arguments rebutting Scanlon and Lee's (2000) description of the cranial structure of *Wonambi*, and they included a phylogenetic analysis of 21 terminal taxa (including five nonophidian squamates) that used 45 morphological characters. In complete contrast to Scanlon and Lee's (2000) hypothesis, *Wonambi* and the pachyophiids were all nested within an unresolved polytomy of macrostomatan snakes.

For rebuttal to the anatomical identifications and similarity arguments regarding *Wonambi* by Rieppel et al. (2003), Scanlon (2005:176) responded with a very detailed critique. His conclusions were that

> Rieppel et al.'s (2002) "test" of the phylogenetic relationships of *Wonambi* is rejected because of numerous misinterpretations of its morphology (attributable to the limited representation and partly prepared state of the material they examined), neglect of all comparative data on other Madtsoiidae, and distortion of their data matrix by *a priori* hypotheses of relationships (e.g., in coding the laterosphenoid bridge as "present" in *Wonambi*, although it is unambiguously absent in the only known prootic).

Scanlon concluded that close similarities between the crania of *Wonambi* and extant snakes are not with booid snakes, but with anilioids. Scanlon's (2006) recent description of a second Australian (Oligocene-Miocene) madtsoiid, *Yurlungurr*, reports a phylogenetic result similar to that of Scanlon and Lee (2000): *Pachyrhachis* and *Haasiophis* are the most basal snakes, *Yurlungurr* and *Wonambi* form a monophyletic group, Madtsoiidae, followed by *Dinilysia*, and finally by a monophyletic Serpentes composed of Scolecophidia and Alethinophidia. Scanlon (2006:841) stated with respect to the five most basal snakes, "These long-jawed or 'macro-stomate' fossils are thus excluded from Macrostomata (which is diagnosed by numerous characters, not all related to the jaw apparatus), and collectively represent the ancestral morphology of the diverse modern snake radiation." Scanlon's (2006) closing comment echoes the conclusions of the study on the evolution of snake feeding by Lee et al. (1999a). In a recent and surprising new hypothesis-driven turn of events, both studies are cast in a new light by the unexplored implications of Apesteguía and Zaher's (2006) description and phylogenetic analysis of the new Argentine Turonian-age snake, *Najash rionegrina*.

In the current phase of the sister-group debate, the nature of the evidence has expanded to include molecular data blended with morphological data extracted from extant and fossil squamates. Most recently, Vidal and Hedges (2004) concluded on the basis of the following sister-group relationships and clade statements derived from their Bayesian consensus phylogeny that snakes had a terrestrial origin: (1) snakes share a common ancestor with a clade composed of agamids, iguanids, and chamaeleonids; (2) the snake-iguanian clade shares a common ancestor with Anguimorpha; (3) Anguimorpha-snakes-iguanians share a common ancestor with teiids, gymnophthalmids, and a polyphyletic Amphisbaenia; (4) skinks, xantusiids, and cordylids are the sister-taxon to the previous clade; (5) gekkonids are the sister-group to all other squamates with the exception of dibamids; and (6) dibamids are the most basal squamates. Because this Bayesian consensus phylogeny did not include fossil taxa, Vidal and Hedges (2004) constructed a second summary phylogeny that included two fossil lineages: pachyophiid snakes and mosasaurid lizards. They illustrate the snake clade as monophyletic but unresolved. Mosasaurs are inferred to be anguimorph lizards whose phylogenetic relationships are perfectly resolved as the sister-group to extant varanid lizards with extant anguids as basal to that clade. Stated support for these inferred, but not tested, relationships is based on the literature, i.e., pachyophiids are snakes of uncertain relationship (Caldwell and Lee, 1997; versus Tchernov et al., 2000), and mosasaurs are derived aquatic varanoids (Lee, 1997a; Rieppel and Zaher, 2000a). The development of the scenario by Vidal and Hedges (2004) concerning snake origins is explicitly derived from their figure 2 (note the caption and their reference to the arrows as indicative of the phyletic distance between the aquatic lineages) and not from their molecular data or phylogenetic analysis of that data. Therefore, Vidal and Hedges's (2004) use of their summary phylogeny as support for a scenario with a terrestrial origin for snakes is empirically unsupported, as is their claim for falsification of the hypothesis of aquatic origins. The logic applied in reaching this conclusion is inconsistent and cannot be related to the results of their data analysis (an iguanian sister-group relationship and no close relationship between snakes, amphisbaenians, and/or dibamids), and instead relies on their nonfalsifiable inference of the squamate relationships of pachyophiids and mosasaurs as presented in their summary tree.

In contrast, the phylogeny of Lee (2005) represents the only empirically and epistemically justifiable approach to the integration of molecular and morphological characters, and fossil and extant taxa—the use of as much of the data, particularly in terms of taxic diversity, as is possible (it is the first attempt at total evidence analysis as advocated by Eernisse and Kluge, 1993). The result of Lee's (2005) analysis refuted the findings of ad hoc hypotheses by Vidal and Hedges (2004). Yet the most interesting finding reported by Lee (2005:230) was that, in this seemingly unbalanced data set with numerous molecular characters, the morphological signal was actually much stronger than the molecular signal. This is in complete contrast to notions of molecular data swamping morphological data (Eernisse and Kluge, 1993). Lee (2005:230) stated succinctly that "significantly more molecular data will be needed to robustly overturn (i.e., refute) the morphological results. All these results highlight the need to explicitly incorporate fossils in phylogenetic analyses (Gauthier et al., 1988), instead of making assumptions about their position based on published literature (especially when this literature contains disagreements; e.g., Carroll & deBraga, 1992; Caldwell, 1999)."

In the final twist on this debate, at a point when it seemed that no additional new hypotheses might be forthcoming, Apesteguía and Zaher (2006) removed the Scolecophidia from their safe haven as the most basal snakes, and instead reconstructed the large-bodied, large-mouthed *Najash* as the basal snake plesion. They placed *Pachyrhachis* and the other Cenomanian snakes with hind limbs within the Alethinophidia (Fig. 7.15C). This reversal of perspective on the ancestral snake morphology and habit is profound because it completely overturns previous arguments and explanations such as those given by Rieppel (1984, 1988), who invoked miniaturization to explain size and features shared between tiny scolecophidians and burrowing lizards, or those presented by Walls (1940, 1942), who argued that burrowing habits led to the restructuring of the visual system. Apesteguía and Zaher (2006) argued that the topology of their ingroup phylogeny allows them to unequivocally conclude that snakes had a terrestrial origin; whether this is supported or not by the data remains to be tested. However, I would argue that their topology unequivocally supports an origin of snakes from a large-bodied, large-mouthed snake. This kind of scenario is much more consistent with that proposed by Lee et al. (1999a) and Scanlon (2006).

Conclusions

Fossils remain at the center of evolutionary debates for several reasons. First, fossil specimens are tantalizing fragments of entire and often unusual animals; as such, the evidence they provide for hypothesis construction is powerful, but it is also incomplete, contradictory, and confusing, leading quite naturally to uncertainty and argument. Second, extinct taxa represent fragments of an ancient biotic and phylogenetic nexus that is also incompletely preserved and thus tantalizing regarding the discovery of synapomorphies and subsequently homologies; but again, as repeatedly shown, fossils are critical components of higher-level phylogenies. Third, fossil taxa represent depth of time. This factor is more critical than any other in our consideration and conception of first principles: evolution as change over time. Evolution needs time to effect major transitions, and the fossil record is the only place that preserves some evidence of these past patterns and processes. Thus, despite the arguments arising around something as small as the jugal of *Pachyrhachis,* the fossil record remains the only data set capable of illuminating these questions; the unsuspected legs of *Pachyrhachis* are a perfect example. It seems likely, despite the seemingly unlimited resource of neontological data at the molecular level, that the fossil record will remain the arbiter of future phylogenetic hypotheses. The seminal work of Gauthier et al. (1988) made this point most forcefully.

I also think that taxic diversity is a key element, particularly with respect to the sister-group relationships for snakes; a minor sampling of ingroup diversity is inconsequential to hypotheses of more global snake sister-group relationships. In other words, the cladistic analysis must include as many character data for as many lizard terminals as is possible. Otherwise, the lizard candidate for closest snake sister-group can remain elusive. Total evidence (Eernisse and Kluge, 1993) is a reasonable philosophical underpinning, but it is a recognized methodological impossibility when finding comparable character evidence between fossil and extant taxa (see O'Leary [2000:61A], who commented, "The character congruence basis of the total evidence approach is operationally compromised by extinctions"). Although it is true that we will never have a complete data set for fossil taxa, it remains as nothing more than poor methodology to ignore this critical data set. The centrality and importance of fossil taxa in generating the debate on snake origins and phylogeny are historical facts. The ability of new fossils to reignite the debate after nearly 70 years of stagnation is also historical fact. Despite the visible tensions in the recent literature, a great deal of new data and critical thought has been presented; such efforts can only be perceived, in the long term, as of great value to scientific progress.

To conclude, I will offer some biological and evolutionary thoughts. Snakes, like mosasaurs (Russell, 1967), can lay claim to a terrestrial ancestor at some point in their evolutionary history (e.g., as an extreme, the first terrestrial four-legged vertebrate). Whether

or not snakes and mosasauroids share a most recent common ancestor is very much an open question. An even more vexing problem exists for the opposing hypothesis where no sister-group statements yet exist (e.g., Apesteguía and Zaher, 2006). What is clear is
that the fossil record preserves evidence of a number of very ancient snakes that were certainly aquatically adapted (Caldwell and
Lee, 1997; Rage and Escuillié, 2000; Tchernov et al., 2000); it is
also true that the fossil record preserves evidence of geologically
younger but possibly terrestrial snakes such as the Turonian-age
Najash and the Santonian-age *Dinilysia*. The ecology of the oldest-
known snakes, which to date are known only from isolated vertebrae, remains equivocal. As reviewed above, the position of most
basal snake is a debated question leaving hypotheses as to ancestral
ecology problematic.

Janensch (1906) favored a burrowing or grass-slithering origin
for snakes (expanded on by Camp, 1923), but did not consider angiostomatans (his term for scolecophidians) to be primitive snakes. Although Janensch's (1906) conclusions were not congruent with those
of any other researchers of the Cope to Nopcsa era, they did become
popular into the middle part of the twentieth century. As Rieppel
(1988) noted, Janensch-like ideas of a burrowing origin became en-
trenched during the 1940s with the publication of Mahendra's (1938)
work, and shortly thereafter by Walls's (1940, 1942) studies on the
degenerate eye of scolecophidians versus the uniquely specialized and
thus supposedly re-elaborated or re-evolved eye of alethinophidians
(contra Underwood, 1976, and Caprette et al., 2004).

During the first period of thought on the matter, from Cope to
Nopcsa, both the burrowing and aquatic origins scenarios were
broadly and openly debated. However, as noted, it is also true that
the first iteration of the burrowing scenario was not related to the
phylogenetic position of scolecophidians as the most primitive
snakes. The modern form of the burrowing origins hypothesis, and
its place in our cognitive framework, was proposed during what I
see as the second period of thought on this problem—the neontological phase from the 1920s to the 1970s. It is interesting, therefore, and perhaps not at all surprising, that serious challenges to
the basal position of scolecophidian snakes have come not from
neontological studies (List, 1966; Haas, 1968; Rieppel, 1988; Cundall and Rossman, 1993; Cundall et al., 1993), but rather from the
data unearthed during the third period in the origins of snakes debate—a phase of evolutionary origins research that has been focused on interpreting new character data from new fossils (e.g.,
Caldwell and Lee, 1997; Apesteguía and Zaher, 2006). The current
phase of research on snake phylogeny and evolution finds its insights from the collection and analysis of increasingly large and
confusing molecular data sets hunting for phylogenetic signals
within the evidence of proteins and nucleotides. In terms of this
burgeoning data set, it seems likely that future studies in the newest
phase, similar to Lee's (2005) most recent contribution, will come
to dominate the literature as we attempt to reconcile fossils, osteol-

ogy, soft tissues, and molecules in the recovery of phylogenetic pattern and evolutionary process.

Acknowledgments

It is impossible to acknowledge all of the influences that have shaped my perceptions and thoughts on snake evolution. Some are people I know only through their work (Cope, Estes, Kornhuber, Nopcsa, and Owen), while I can acknowledge many others personally or through correspondence: A. Albino, D. Cundall, H. Greene, J. Head, M. Lee, J.-C. Rage, O. Rieppel, J. Scanlon, G. Underwood, and H. Zaher. These individuals have contributed frank discussion on this topic over the last nine years and as such have profoundly altered my views of science, knowledge acquisition, and such sundry things as the evolution of snakes. Writing a review on this enormous topic is difficult for two reasons: brevity and objectivity. I have discovered that both are elusive, and I therefore accept all responsibility for the ideas and interpretations as expressed herein. In this regard, I am indebted to J. Scanlon, J. Head, C. Bell, and an anonymous referee for highlighting my self-indulgence. In particular I must thank J. Anderson and H.-D. Sues for their long suffering as editors of this volume. My thanks go to everyone who assisted me while studying their collections and prospecting for fossils in their countries: J. Bonaparte, B. Butkovic-Cvorovic, J. Calvo, S. Chapman, A. Currant, M. Fernandez, Z. Gasparini, B. Jurkovsek, T. Kolar-Jurkovsek, K. Krizmanic, H. Kuhlmann, R. Mortimore, and J. Radovcic. Translations of Janensch (1906) and Nopcsa (1903, 1908, 1923, 1925) were prepared by J. D. Scanlon and were obtained courtesy of J. D. Scanlon and the Polyglot Paleontologist Web site (http://ravenel.si.edu/paleo/paleoglot/index .cfm). Financial support was provided by NSERC Operating Grant 238458-01 to the author.

References Cited

Apesteguía, S., and H. Zaher. 2006. A Cretaceous terrestrial snake with robust hindlimbs and a sacrum. Nature 440:1037–1040.

Apesteguía, S., S. de Valais, J. A. Gonzalez, P. A. Gallna, and F. L. Agnolin. 2001. The tetrapod fauna of "La Buitrera," new locality from the basal Late Cretaceous of North Patagonia, Argentina. Journal of Vertebrate Paleontology 21(Suppl. to 3):29A.

Baszio, S. 2004. *Messelophis variatus* n. gen. n.sp. from the Eocene of Messel: a tropidopheine snake with affinities to Erycinae (Boidae). Courier Forschungs-Institut Senckenberg 252:47–66.

Baur, G. 1892. On the morphology of the skull in the Mosasauridae. Journal of Morphology 7:1–22.

Baur, G. 1895. Cope on the temporal part of the skull, and on the systematic position of the Mosasauridae—a reply. American Naturalist 29:998–1002.

Baur, G. 1896. The paroccipital of the Squamata and the affinities of the Mosasauridae once more. A rejoinder to Professor E. D. Cope. American Naturalist 30:143–152.

Bellairs, A. d'A. 1972. Comments on the evolution and affinites of snakes; pp. 157–172 in K. A. Joysey and T. S. Kemp (eds.), Studies in Vertebrate Evolution. Oliver and Boyd, Edinburgh.

Bellairs, A. d'A., and G. Underwood. 1951. The origin of snakes. Biological Reviews of the Cambridge Philosophical Society 26:193–237.

Bolkay, S. J. 1925. *Mesophis nopcsai* n. g., n. sp., ein neues, schlangenähnliches Reptil aus der unteren Kreide (Neocom) von Bilek-Selista (Ost-Hercegovina). Glasnik Zemaljskog Mujeza u Bosni i Hercegovini 27:125–126.

Boulenger, G. 1893. On some newly-described Jurassic and Cretaceous lizards and rhynchocephalians. Annals and Magazine of Natural History (6)11:204–210.

Budney, L. A., M. W. Caldwell, and A. Albino. 2006. Unexpected tooth socket histology in the Cretaceous snake *Dinilysia,* with a review of amniote dental attachment tissues. Journal of Vertebrate Paleontology 26:138–145.

Caldwell, M. W. 1999a. Squamate phylogeny and the relationships of snakes and mosasauroids. Zoological Journal of the Linnean Society 125:115–147.

Caldwell, M. W. 1999b. Description and phylogenetic relationships of a new species of *Coniasaurus* Owen, 1850 (Squamata). Journal of Vertebrate Paleontology 19:438–455.

Caldwell, M. W. 2000a. On the phylogenetic relationships of *Pachyrhachis* within snakes: a response to Zaher (1998). Journal of Vertebrate Paleontology 20:181–184.

Caldwell, M. W. 2000b. An aquatic squamate reptile from the English Chalk: *Dolichosaurus longicollis* Owen, 1850. Journal of Vertebrate Paleontology 20:720–735.

Caldwell, M. W. 2003. "Without a leg to stand on": evolution and development of axial elongation and limblessness in tetrapods. Canadian Journal of Earth Sciences 40:573–588.

Caldwell, M. W., and A. Albino. 2001. Palaeoenvironment and palaeoecology of three Cretaceous snakes: *Pachyophis, Pachyrhachis,* and *Dinilysia.* Acta Palaeontologica Polonica 46:203–218.

Caldwell, M. W., and A. Albino. 2002. Exceptionally preserved skeletons of the Cretaceous snake *Dinilysia patagonica* Woodward, 1901. Journal of Vertebrate Paleontology 22:861–866.

Caldwell, M. W., and M. S. Y. Lee. 1997. A snake with legs from the marine Cretaceous of the Middle East. Nature 386:705–709.

Caldwell, M. W., L. A. Budney, and D. Lamoureux. 2003. Histology of thecodont-like tooth attachment tissues in mosasaurian squamates. Journal of Vertebrate Paleontology 23:622–630.

Camp, C. L. 1923. Classification of the lizards. Bulletin of the American Museum of Natural History 48:289–481.

Caprette, C. I., M. S. Y. Lee, R. Shine, A. Mokany, and J. F. Downhower. 2004. The origin of snakes (Serpentes) as seen through eye anatomy. Biological Journal of the Linnean Society 81:469–482.

Carroll, R. L. 1988. Vertebrate Paleontology and Evolution. W. H. Freeman and Company, New York.

Carroll, R. L., and M. deBraga. 1992. Aigialosaurs: mid-Cretaceous varanoid lizards. Journal of Vertebrate Paleontology 12:66–86.

Coates, M., and M. Ruta. 2000. Nice snake, shame about the legs. Trends in Evolution and Ecology 15:503–507.

Cohn, M., and C. Tickle. 1999. Developmental basis of limblessness and axial patterning in snakes. Nature 399:474–470.

Conrad, J. L. 2004. Skull, mandible, and hyoid of *Shinisaurus crocodilurus* Ahl (Squamata, Anguimorpha). Zoological Journal of the Linnean Society 141:399–434.

Cope, E. D. 1869. On the reptilian orders Pythonomorpha and Streptosauria. Proceedings of the Boston Society of Natural History 12:250–266.

Cope, E. D. 1875. The Vertebrates of the Cretaceous Formations of the West. United States Geological Survey of the Territories. Volume 2, Fossil Vertebrates. Government Printing Office, Washington, D.C., 303 pp.

Cope, E. D. 1895a. Baur on the temporal part of the skull, and on the morphology of the skull in the Mosasauridae. American Naturalist 29:855–859.

Cope, E. D. 1895b. Reply to Dr. Baur's critique of my paper on the paroccipital bone of the scaled reptiles and the systematic position of the Pythonomorpha. American Naturalist 29:1003–1005.

Cope, E. D. 1896a. Criticism of Dr. Baur's rejoinder on the homologies of the paroccipital bone, etc. American Naturalist 30:147–149.

Cope, E. D. 1896b. Boulenger on the difference between Lacertilia and Ophidia; and on the Apoda. American Naturalist 30:149–152.

Cundall, D., and D. A. Rossman. 1993. Cephalic anatomy of the rare Indonesian snake *Anomochilus weberi*. Zoological Journal of the Linnean Society 109:235–273.

Cundall, D., V. Wallach, and D. A. Rossman. 1993. The systematic relationships of the snake genus *Anomochilus*. Zoological Journal of the Linnean Society 109:275–299.

Cuny, G., J.-J. Jaeger, M. Mahboubi, and J.-C. Rage. 1990. Le plus ancien Serpents (Reptilia, Squamata) connus. Mise au point sur l'âge géologique des Serpents de la partie moyenne du Crétacé. Comptes Rendus de l'Academie des Sciences, Paris, sér. II, 311:1267–1272.

Cuvier, G. 1824. Recherches sur les Ossemens Fossiles, ou l'on rétablit les Caractères de plusieurs Animaux dont les Révolutions du Globe ont détruit les espèces. Nouvelle Édition. Volume 5, Part 2. G. Dufour and E. D'Ocagne, Paris, 547 pp.

de Pinna, M. 1991. Concepts and tests of homology in the cladistic paradigm. Cladistics 7:367–394.

Dollo, L. 1903. Les ancêtres des mosasauriens. Bulletin Scientifique de la France et de la Belgique 38:137–139.

Eernisse, D. J., and A. G. Kluge. 1993. Taxonomic congruence versus total evidence, and amniote phylogeny inferred from fossils, molecules and morphology. Molecular Biology and Evolution 10:1170–1195.

Estes, R., K. de Quieroz, and J. A. Gauthier. 1988. Phylogenetic relationships within Squamata; pp. 119–281 in R. Estes and G. Pregill (eds.), Phylogenetic Relationships of the Lizard Families. Stanford University Press, Stanford, Calif.

Estes, R., T. H. Frazzetta, and E. E. Williams. 1970. Studies on the fossil snake *Dinilysia patagonica* Woodward: Part 1. Cranial morphology. Bulletin of the Museum of Comparative Zoology, Harvard University 140:25–74.

Fejérváry, G. J. de. 1918. Contributions to a monography on fossil

Varanidae and on Megalanidae. Annales Musei Nationalis Hungarici 16:341–467.

Gardner, J. D., and R. L. Cifelli. 1999. A primitive snake from the Cretaceous of Utah. Special Papers in Palaeontology 60:87–100.

Gauthier, J., A. G. Kluge, and T. Rowe. 1988. Amniote phylogeny and the importance of fossils. Cladistics 4:105–209.

Greene, H. W. 1997. Snakes: The Evolution of Mystery in Nature. University of California Press, Berkeley, 351 pp.

Greer, A. 1985. The relationships of the lizard genera *Anelytropsis* and *Dibamus*. Journal of Herpetology 19:116–156.

Haas, G. 1968. Anatomical observations on the head of *Anomalepis aspinosus* (Typhlopidae, Ophidia). Acta Zoologica 48:63–139.

Haas, G. 1979. On a new snakelike reptile from the Lower Cenomanian of Ein Jabrud, near Jerusalem. Bulletin du Muséum National de l'Histoire Naturelle, Paris, sér. 4, sect. C, 1:51–64.

Haas, G. 1980a. *Pachyrhachis problematicus* Haas, snakelike reptile from the Lower Cenomanian: ventral view of the skull. Bulletin du Muséum National de l'Histoire Naturelle, Paris, sér. 4, sect. C, 2:87–104.

Haas, G. 1980b. Remarks on a new ophiomorph reptile from the lower Cenomanian of Ein Jabrud, Israel; pp. 177–192 in L. L. Jacobs (ed.), Aspects of Vertebrate History. Museum of Northern Arizona Press, Flagstaff.

Hoffstetter, R. 1959. Un Serpent terrestre dans le Crétacé inférieur du Sahara. Bulletin de la Société géologique de France, sér. 7, 1:897–902.

Hoffstetter, R. 1961. Nouveaux restes d'un Serpent boidé (*Madtsoia madagascarensis* nov. sp.) dans le Crétacé supérieur de Madagascar. Bulletin de le Muséum National de l'Histoire Naturelle, Paris, sér. 2, 33:152–160.

Holman, J. A. 2000. Fossil Snakes of North America. Origin, Evolution, Distribution, Palaeoecology. Indiana University Press, Bloomington, 357 pp.

Huxley, T. H. 1868. On the animals which are most nearly intermediate between birds and reptiles. Annals and Magazine of Natural History (4)2:66–75.

Huxley, T. H. 1870. Further evidence of the affinity between the dinosaurian reptiles and birds. Proceedings of the Geological Society of London 26:12–31.

Janensch, W. 1906. Über *Archaeophis proavus* Mass., eine Schlange aus dem Eocän des Monte Bolca. Beiträge zur Paläontologie und Geologie Österreich-Ungarns und des Orients 19:1–33.

Kearney, M. 2003. The phylogenetic position of *Sineoamphisbaena hexatabularis* reexamined. Journal of Vertebrate Paleontology 23:394–403.

Kluge, A. G. 2003. The repugnant and the mature in phylogenetic inference: atemporal similarity and historical identity. Cladistics 19:356–368.

Kornhuber, A. 1873. Über einen neuen fossilen Saurier aus Lesina. Abhandlungen der k. k. geologischen Reichsanstalt (Wien) 5(4):75–90.

Kornhuber, A. 1893. *Carsosaurus Marchesettii*, ein neuer fossiler Lacertilier aus den Kreideschichten des Karstes bei Komen. Abhandlungen der k. k. geologischen Reichsanstalt (Wien) 17(3):1–15.

Kornhuber, A. 1901. *Opetiosaurus Bucchichi*. Eine neue fossile Eidechse aus der unteren Kreide von Lesina in Dalmatien. Abhandlungen der k. k. geologischen Reichsanstalt (Wien) 17(5):1–24.

Kramberger, K. G. 1892. *Aigialosaurus,* eine neue Eidechse aus den Kreideschiefern der Insel Lesina mit Rücksicht auf die bereits beschriebenen Lacertiden von Comen und Lesina. Glasnik hrvatskoga naravoslovnoga drustva (Societas historico-naturalis croatica) u Zagrebu 7:74–106.

Langer, W. 1961. Über das Alter der Fischschiefer von Hvar-Lesina (Dalmatien). Neues Jahrbuch für Geologie und Paläontologie, Monatshefte 1961:329–331.

Leanza, H. A., S. Apesteguía, F. E. Novas, and M. S. de la Fuente. 2004. Cretaceous terrestrial beds from the Neuquén Basin (Argentina) and their tetrapod assemblages. Cretaceous Research 25:61–87.

Lee, M. S. Y. 1997a. On snake-like dentition in mosasaurian lizards. Journal of Natural History 31:303–314.

Lee, M. S. Y. 1997b. The phylogeny of varanoid lizards and the affinities of snakes. Philosophical Transactions of the Royal Society of London B 352:53–91.

Lee, M. S. Y. 1998. Convergent evolution and character correlation in burrowing reptiles: towards a resolution of squamate relationships. Biological Journal of the Linnean Society 65:369–453.

Lee, M. S. Y. 2005. Molecular evidence and marine snake origins. Biology Letters 1:227–230.

Lee, M. S. Y., and M. W. Caldwell. 1998. Anatomy and relationships of *Pachyrhachis,* a primitive snake with hindlimbs. Philosophical Transactions of the Royal Society of London B 353:1521–1552.

Lee, M. S. Y., and M. W. Caldwell. 2000. *Adriosaurus* and the affinities of mosasaurs, dolichosaurs, and snakes. Journal of Paleontology 74:915–937.

Lee, M. S. Y., and J. D. Scanlon. 2001. On the lower jaw and intramandibular septum in snakes and anguimorph lizards. Copeia 2001:531–535.

Lee, M. S. Y., and J. D. Scanlon. 2002. Snake phylogeny based on osteology, soft anatomy and ecology. Biological Reviews of the Cambridge Philosophical Society 77:333–401.

Lee, M. S. Y., G. L. Bell, and M. W. Caldwell. 1999a. The origins of snake feeding. Nature 400:655–659.

Lee, M. S. Y., M. W. Caldwell, and J. D. Scanlon. 1999b. A second primitive marine snake: *Pachyophis woodwardi* Nopcsa from the Cretaceous of Bosnia-Herzegowina. Journal of Zoology, London 248:509–520.

List, J. C. 1966. Comparative osteology of the snake families Typhlopidae and Leptotyphlopidae. Illinois Biological Monographs 36:1–112.

Mahendra, B. C. 1938. Some remarks on the phylogeny of the Ophidia. Anatomischer Anzeiger 86:347–356.

Marsh, O. C. 1871. Notice of some new fossil reptiles from the Cretaceous and Tertiary formations. American Journal of Science (3)1:447–459.

Marsh, O. C. 1880. New characters of mosasauroid reptiles. American Journal of Science (3)19:83–87.

Marsh, O. C. 1892. Notice of new reptiles from the Laramie Formation. American Journal of Science (4)3:449–453.

Massalongo, A. B. 1859. Specimen photographicum animalium quorundam plantarumque fossilium agri Veronensis. Vicentini-Franchini, Verona, 101 pp.

Mattison, C. 1995. The Encyclopedia of Snakes. Checkmark Books Ltd., New York, 257 pp.

McDowell, S. B. 1987. Systematics; pp. 3–50 in R. A. Siegel, J. T. Collins, and S. S. Novak (eds.), Snakes: Ecology and Evolutionary Biology. Macmillan, New York.

McDowell, S. B., Jr., and C. M. Bogert. 1954. The systematic position of *Lanthanotus* and the affinities of the anguinomorphan lizards. Bulletin of the American Museum of Natural History 105:1–142.

Nopcsa, F. 1903. Über die varanusartigen Lacerten Istriens. Beiträge zur Paläontologie und Geologie Österreich-Ungarns und des Orients 15:31–42.

Nopcsa, F. 1908. Zur Kenntnis der fossilen Eidechsen. Beiträge zur Paläontologie und Geologie Österreich-Ungarns und des Orients 21:33–62.

Nopcsa, F. 1923. *Eidolosaurus* und *Pachyophis*. Zwei neue Neocom-Reptilien. Palaeontographica 65:96–154.

Nopcsa, F. 1925. Ergebnisse der Forschungsreisen Prof. E. Stromers in den Wüsten Ägyptens. II. Wirbeltier-Reste der Baharîje-Stufe (unterstes Cenoman). 5. Die *Symoliophis*-Reste. Abhandlungen der Bayerischen Akademie der Wissenschaften, Mathematisch-naturwissenschaftliche Abteilung 30:1–27.

O'Leary, M. 2000. Operational obstacles to total evidence analyses considering that 99% of life is extinct. Journal of Vertebrate Paleontology 20(Suppl. to 3):61A.

Oppel, M. 1811. Die Ordnungen, Familien und Gattungen der Reptilien, als Prodrom einer Naturgeschichte derselben. Joseph Lindauer, Munich, 86 pp.

Osborn, H. F. 1899. A complete mosasaur skeleton, osseous and cartilaginous. Memoirs of the American Museum of Natural History 1:167–188.

Owen, R. 1850. Description of the fossil reptiles of the Chalk Formation; pp. 378–400 and 403–404 in F. Dixon (ed.), The Geology and Fossils of the Tertiary and Cretaceous Formations of Sussex. Longman, Brown, Green, and Longmans, London.

Nopcsa, F. 1877. On the rank and affinities of the reptilian class of Mosasauridae, Gervais. Quarterly Journal of the Geological Society of London 33:682–715.

Pianka, E. R., and L. J. Vitt. 2003. Lizards: Windows to the Evolution of Diversity. University of California Press, Berkeley, 347 pp.

Polcyn, M. J., L. L. Jacobs, and A. Haber. 2005. A morphological model and CT assessment of the skull of *Pachyrhachis problematicus* (Squamata, Serpentes), a 98 million year old snake with legs from the Middle East. Palaeontologia Electronica 8(1); http://palaeo-electronica .org//2005_1/polycyn26/issue1_05.htm.

Rage, J.-C. 1984. Serpentes. Handbuch der Paläoherpetologie, Teil 11. Gustav Fischer Verlag, Stuttgart, 80 pp.

Rage, J.-C. 1988. Un Serpent primitif (Reptilia, Squamata) dans le Cénomanien (base du Crétacé supérieur). Comptes Rendus de l'Academie des Sciences, Paris, sér. II, 307:1027–1032.

Rage, J.-C., and F. Escuillié. 2000. Un nouveau Serpent bipede du Cénomanien (Crétacé). Implications phyletiques. Comptes Rendus de l'Academie des Sciences, Paris, sér. Ia, 330:1–8.

Rage, J.-C., and F. Escuillié. 2001. *Eupodophis,* new name for the genus *Podophis* Rage and Escuillié, 2000, an extinct bipedal snake, preoccupied by *Podophis* Wiegmann, 1834 (Lacertilia, Scincidae). Amphibia-Reptilia 23:232–233.

Rage, J.-C., and F. Escuillié. 2003. The Cenomanian: stage of hindlimbed snakes. Carnets de Géologie 1:1–11.

Rage, J.-C., and A. Richter. 1994. A snake from the Lower Cretaceous (Barremian) of Spain: the oldest known snake. Neues Jahrbuch für Geologie und Paläontologie, Monatshefte 1994:561–565.

Rieppel, O. 1980. The phylogeny of anguinomorph lizards. Denkschriften der Schweizerischen Naturforschenden Gesellschaft 94 :1–86.

Rieppel, O. 1984. The cranial morphology of the fossorial lizard genus *Dibamus* with a consideration of its phylogenetic relationships. Journal of Zoology, London 204:289–327.

Rieppel, O. 1988. A review of the origin of snakes; pp. 37–130 in M. K. Hecht, B. Wallace, and G. T. Prance (eds.), Evolutionary Biology, Volume 22. Plenum Press, New York and London.

Rieppel, O. 1994. The Lepidosauromorpha: an overview with special emphasis on the Squamata; pp. 22–37 in N. C. Fraser and H.-D. Sues (eds.), In the Shadow of the Dinosaurs: Early Mesozoic Tetrapods. Cambridge University Press, Cambridge.

Rieppel, O., and J. J. Head. 2004. New specimens of the fossil snake genus *Eupodophis* Rage & Escuillié, from the Cenomanian (Late Cretaceous) of Lebanon. Memorie della Società Italiana di Scienze Naturali e del Museo Civico di Storia Naturale di Milano 32(2):1–26.

Rieppel, O., and M. Kearney. 2001. The origins of snakes: limits of a scientific debate. Biologist 48(3):110–114.

Rieppel, O., and M. Kearney. 2002. Similarity. Biological Journal of the Linnean Society 75:59–82.

Rieppel, O., and H. Zaher. 2000a. The intramandibular joint in squamates, and the phylogenetic relationships of the fossil snake *Pachyrhachis problematicus* Haas. Fieldiana, Geology, New Series 43:1–69.

Rieppel, O., and H. Zaher. 2000b. The braincases of mosasaurs and *Varanus*, and the relationships of snakes. Zoological Journal of the Linnean Society 129:489–514.

Rieppel, O., and H. Zaher. 2001a. Re-building the bridge between mosasaurs and snakes. Neues Jahrbuch für Geologie und Paläontologie, Abhandlungen 221:111–132.

Rieppel, O., and H. Zaher. 2001b. The development of the skull in *Acrochordus granulatus* (Schneider) (Reptilia: Serpentes), with special consideration of the otico-occipital complex. Journal of Morphology 249:252–266.

Rieppel, O., A. G. Kluge, and H. Zaher. 2002. Testing the phylogenetic relationships of the Pleistocene snake *Wonambi naracoortensis* Smith. Journal of Vertebrate Paleontology 22:812–829.

Rieppel, O., H. Zaher, E. Tchernov, and M. J. Polcyn. 2003. The anatomy and relationships of *Haasiophis terrasanctus*, a fossil snake with well-developed hind limbs from the mid-Cretaceous of the Middle East. Journal of Paleontology 77:336–358.

Russell, D. A. 1967. Systematics and morphology of American mosasaurs. Peabody Museum of Natural History, Yale University, Bulletin 23:1–241.

Scanlon, J. D. 1992. A new large madtsoiid snake from the Miocene of the Northern Territory. The Beagle 9:49–60.

Scanlon, J. D. 1996. Studies in the palaeontology and systematics of Australian snakes. Ph.D. dissertation, University of New South Wales, Sydney, 648 pp.

Scanlon, J. D. 2004. First known axis vertebra of a madtsoiid snake (*Yurlungurr camfieldensis*) and remarks on the neck of snakes. The Beagle 20:207–215.

Scanlon, J. D. 2005. Cranial morphology of the Plio-Pleistocene giant madtsoiid snake *Wonambi naracoortensis*. Acta Palaeontologica Polonica 50:139–180.

Scanlon, J. D. 2006. Skull of the large non-macrostomatan snake *Yurlungurr* from the Australian Oligo-Miocene. Nature 439:839–842.

Scanlon, J. D., and M. S. Y. Lee. 2000. The Pleistocene serpent *Wonambi* and the early evolution of snakes. Nature 403:416–420.

Scanlon, J. D., and M. S. Y. Lee. 2002. On varanoid-like dentition in primitive snakes (Madtsoiidae). Journal of Herpetology 36:100–106.

Scanlon, J. D., M. S. Y. Lee, M. W. Caldwell, and R. Shine. 1999. The palaeoecology of the primitive snake *Pachyrhachis*. Historical Biology 13:127–152.

Schaal, S., and S. Baszio. 2004. *Messelophis ermannorum* n.sp., eine neue Zwergboa (Serpentes: Boidae: Tropidopheinae) aus dem Mittel-Eozän von Messel. Courier Forschungs-Institut Senckenberg 252:67–77.

Schaal, S., and W. Ziegler. 1992. Messel: An Insight into the History of Life and of the Earth. Oxford University Press, Oxford, 322 pp.

Seeley, H. G. 1881. On remains of a small lizard from the Neocomian rocks of Comén, near Trieste, preserved in the Geological Museum of the University of Vienna. Quarterly Journal of the Geological Society of London 37:52–56.

Shapiro, M. 2002. Developmental morphology of limb reduction in *Hemiergis* (Squamata: Scincidae): chondrogenesis, osteogenesis, and heterochrony. Journal of Morphology 254:211–231.

Sliskovic, T. 1970. Die stratigraphische Lage der Schichten mit Pachyophidae aus Seliste bei Bileca (Ostherzegowina). Bulletin Scientifique Yougoslavie, Zagreb 15:389–390.

Smith, J. 1976. Small fossil vertebrates from Victoria Cave, Naracoorte, South Australia. IV. Reptiles. Transactions of the Royal Society of South Australia 100:39–51.

Tchernov, E., O. Rieppel, H. Zaher, M. J. Polcyn, and L. L. Jacobs. 2000. A fossil snake with limbs. Science 287:2010–2012.

Townsend, T. M., A. Larson, E. Louis, and J. R. Macey. 2004. Molecular phylogenetics of Squamata: the position of snakes, amphisbaenians and dibamids, and the root of the squamate tree. Systematic Biology 53:735–757.

Uetz, P. 2004. The EMBL reptile database; http://www.embl-heidelberg.de/~uetz/LivingReptiles.html.

Underwood, G. 1967. A Contribution to the Classification of Snakes. Trustees of the British Museum (Natural History), London, 179 pp.

Underwood, G. 1970. The eye; pp. 1–97 in C. Gans and T. S. Parsons (eds.), Biology of the Reptilia. Volume 2, Morphology B. Academic Press, London.

Underwood, G. 1976. Simplification and degeneration in the course of evolution of squamate reptiles. Colloques internationaux, Centre National de la Recherche Scientifique 266:341–352.

Vidal, N., and B. S. Hedges. 2004. Molecular evidence for a terrestrial origin of snakes. Proceedings of the Royal Society of London, Biology Letters 271:226–229.

Wallach, V. 1984. A new name for *Ophiomorphus colberti* Haas, 1980. Journal of Herpetology 18:329.

Walls, G. L. 1940. Ophthalmological implications for the early history of snakes. Copeia 1940:1–8.

Walls, G. L. 1942. The vertebrate eye and its adaptive radiation. Bulletin of the Cranbrook Institute of Science 19:1–785.

Wiens, J., and W. Slingluff. 2001. How lizards turn into snakes: a phylogenetic analysis of body-form evolution in anguid lizards. Evolution 55:2303–2318.

Williston, S. W. 1898. Mosasaurs; pp. 83–221 in S. W. Williston, The University Geological Survey of Kansas. Volume 4, Paleontology. Part 1, Upper Cretaceous. J. S. Parks, Topeka.

Williston, S. W. 1904. The relationships and habits of the mosasaurs. Journal of Geology 12:43–51.

Woodward, A. S. 1901. On some extinct reptiles from Patagonia of the genera *Miolania, Dinilysia,* and *Genyodectes.* Proceedings of the Zoological Society of London 1901:169–184.

Zaher, H. 1998. The phylogenetic position of *Pachyrhachis* within snakes (Squamata, Lepidosauria). Journal of Vertebrate Paleontology 18:1–3.

Zaher, H., and O. Rieppel. 1999a. Tooth implantation and replacement in Squamates, with special reference to mosasaur lizards and snakes. American Museum Novitates 3271:1–19.

Zaher, H., and O. Rieppel. 1999b. The phylogenetic relationships of *Pachyrhachis problematicus,* and the evolution of limblessness in snakes (Lepidosauria, Squamata). Comptes Rendus de l'Academie des Sciences, Paris, sér. II, 329:831–837.

Zaher, H., and O. Rieppel. 2000. A brief history of snakes. Herpetological Review 31:73–76.

Zaher, H., and O. Rieppel. 2002. On the phylogenetic relationships of the Cretaceous snakes with legs, with special reference to *Pachyrhachis problematicus* (Squamata, Serpentes). Journal of Vertebrate Paleontology 22:104–109.

8. The Beginnings of Birds: Recent Discoveries, Ongoing Arguments, and New Directions

Luis M. Chiappe and Gareth J. Dyke

ABSTRACT

Unraveling stages in the early evolution and diversification of birds seems simple enough, given the fact that large numbers of new and exceptionally well-preserved fossils have become available for study in recent years. Complications still remain, however; the history of this vertebrate lineage has been central to several heated debates since the earliest days of evolutionary biology. Are birds the extant descendants of theropod dinosaurs? How have new fossil discoveries helped to fill in the gap between *Archaeopteryx* and its extant counterparts? In this review, we focus renewed attention on fossil birds and their relatives—specific lines of evidence that can be used to support the theropod hypothesis for avian origins and update some recent areas of contention. Related to this, we provide a general synopsis of recent fossil discoveries from the Mesozoic, especially the Cretaceous.

Introduction: Discovering Fossil Birds and Their Most Immediate Relatives

With more than 10,000 species—roughly twice as many as there are mammals or lizards—birds are by far the most diverse group of extant tetrapods. However, their current diversity is a remnant of a much older evolutionary radiation that can be dated back to 150 million years, as far as the celebrated Late Jurassic bird *Archaeopteryx* from southern Germany.

Research on the origin and early history of birds has been at the forefront of vertebrate paleontology since the advent of evolutionary thought, but only recently has consensus on their origin been reached and an enormous diversity of Mesozoic birds been revealed. Indeed, for most of the last two centuries, the available evidence for studying the ancestry of birds was largely restricted to animals vastly different in morphology, and stratigraphically much younger, than *Archaeopteryx*. Likewise, knowledge of the early radiation of birds was limited to this taxon and a handful of other Mesozoic birds, fossils for the most part restricted to nearshore and marine environments, greatly separated both anatomically and in age. The only other well-known Mesozoic birds were the marine *Hesperornis* and *Ichthyornis,* both from the Late Cretaceous of North America.

In the last decade, an ongoing explosion of new discoveries of Cretaceous fossils has left us with a greatly improved image of both the nearest relatives of birds and the scale of their earliest evolutionary radiation. Although new fossils have been found all over the world (Chiappe, 1995; Padian and Chiappe, 1998; Chiappe and Dyke, 2002; Padian, 2004; Zhou, 2004), the Early Cretaceous deposits of northeastern China, in particular the Yixian and Jiufotang formations of Liaoning Province, have stunned paleontologists by yielding large numbers of exceptionally well-preserved new taxa that span the entire evolutionary transition between nonavian dinosaurs and birds (Chen et al., 1998; Ji et al., 1998; Xu et al., 1999a,b, 2000, 2003; Zhou and Wang, 2000; Zhou et al., 2000; Zhou and Zhang, 2002a,b; Padian, 2004; Xu and Norell, 2004; Zhang and Zhou, 2004; Zhou, 2004; Xu and Zhang, 2005). Most remarkable of all are a number of new nonavian theropod dinosaurs that are well enough preserved to show both feathers (in some cases) and feather-like structures covering their bodies (Chen et al., 1998; Ji et al., 1998; Xu et al., 1999a,b, 2000, 2003; Zhou and Wang, 2000; Zhou et al., 2000). These taxa, above all, have added new and critical information to the study of the origins of feathers (Brush, 2000; Prum and Williamson, 2001) and the beginnings of avian flight (Burgers and Chiappe, 1999; Padian et al., 2001; Xu et al., 2003).

Taking this new evidence into account, most evolutionary biologists would now consider the main issues of the avian ancestry debate to be over. Anatomical uncertainties that led to the long controversy surrounding the evolutionary origin of birds have been largely clarified on the basis of new fossil material in the last decade, even if some debate persists about identifying the most immediate outgroup to birds like *Archaeopteryx* and its kin. Today, there is an overwhelming consensus that the clade Aves (birds in the traditional sense) is nested within the maniraptoran dinosaurs, a subgroup of the carnivorous theropods that includes the famous *Velociraptor*. In evolutionary terms, birds should then be considered as just a small branch of the enormous dinosaur family tree (Fig. 8.1). Note that we use terms such as *bird, Aves, Neornithes,* and *Neognathae* in their traditional, and cladistically defined, meanings (see Chiappe and

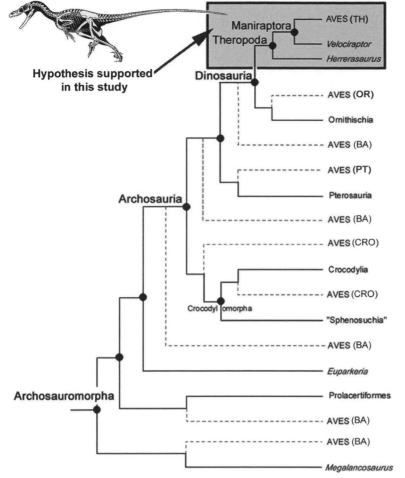

Figure 8.1. Summary cladogram to illustrate diversity of hypotheses proposed over the years to explain origin of birds. **BA**, hypotheses that relate birds to some basal archosaurs or archosauromorphs; **CRO**, hypotheses in favor of a crocodylomorph origin for birds; **OR**, hypotheses that suggest a shared common ancestor between birds and ornithischian dinosaurs; **PT**, origination of birds from within pterosaurs; **TH**, from within theropod dinosaurs. This latter hypothesis is supported by the vast majority of the available data. Modified from Chiappe and Vargas (2004).

Dyke, 2002). Thus "birds," and "Aves" by default, refer to the clade containing *Archaeopteryx* and sparrows, and everything else in between. We do not use the crown clade definitions of these terms.

In light of recent advances in this field, we focus attention in this review on several of the specific lines of evidence that can be used to support the theropod ancestry of birds, revisit and critique some of continuing arguments that are leveled against this hypothesis, and provide a synopsis of the vast number of recent fossil avian discoveries that have filled the gap between *Archaeopteryx* and its relatives, extant birds.

The Theropod Origin of Birds

Heated debates regarding the origin of birds and their relationship to dinosaurs have raged for as long as cohesive evolutionary theories have been in place. After the 1860s, in post-Darwinian times, birds have been hypothesized as being closely related to a wide variety of extinct and extant reptile lineages (Fig. 8.1), including crocodylomorphs (modern crocodylians and their Mesozoic relatives), a diversity of basal archosaurs and archosauromorphs (specifically the Triassic taxa *Euparkeria*, *Longisquama*, and *Megalancosaurus*), pterosaurs (Mesozoic flying reptiles), and various theropod and ornithischian dinosaurs (Witmer, 1991; Padian and Chiappe, 1998; Chiappe, 2001, 2004; Chiappe and Witmer, 2002). In the face of phylogenetic methods, systematists have strongly embraced the view that birds evolved from theropod dinosaurs, although opposition to this widely accepted notion still persists (see reviews in Chatterjee, 1997; Chiappe, 2001; Gauthier and Gall, 2001) and includes hypotheses identifying some primitive archosauromorphs (e.g., Tarsitano and Hecht, 1980; Feduccia and Wild, 1993; Welman, 1995) or crocodylomorphs (Martin et al., 1980; Martin, 1983, 1991; Martin and Stewart, 1999) as the closest relatives of birds. However, couched in a modern systematic framework, these alternative hypotheses are speculative at best; they are not based on repeatable analyses of anatomical or other kinds of data. In contrast, a wealth of recent evidence, including similarities in egg structures, inferable behaviors, and integument, has been accumulated in support of the hypothesis that birds originated within maniraptoran theropods (see reviews in Gauthier, 1986; Chiappe, 2001; Holtz, 2001; Norell et al., 2001; Chiappe and Dyke, 2002; Hwang et al., 2004; Xu et al., 2005), generally gracile dinosaurs that comprise lineages as diverse as alvarezsaurids, dromaeosaurids, oviraptorids, therizinosaurids, and troodontids (for more information about these dinosaurian taxa, see Weishampel et al., 2004) (Fig. 8.2).

Although there are now several competing hypotheses regarding the closest maniraptoran lineage to birds (e.g., Gauthier, 1986; Forster et al., 1998; Elzanowski, 1999; Sereno, 1999; Xu et al., 2000; Holtz, 2001; Norell et al., 2001; Chiappe, 2002a; Clark et al., 2002; Hwang et al., 2003, 2004), this abundant new evidence has met predictions formulated long ago by advocates of the theropod ancestry of birds. We review some of the most salient features of each of these lines of evidence, which, without a doubt, support the maniraptoran origin of birds.

Osteology

For more than a century, anatomical considerations of the similarities between birds and dinosaurs were based only on skeletal features (e.g., Huxley, 1868; Heilmann, 1926; Ostrom, 1973, 1976). It

Figure 8.2. The nonavian maniraptorans *Velociraptor mongoliensis* (left, after Paul, 1988) and *Caudipteryx zoui* (right, after Currie, 2000) are members of the maniraptoran clades Dromaeosauridae and Oviraptorosauria, respectively.

20 cm

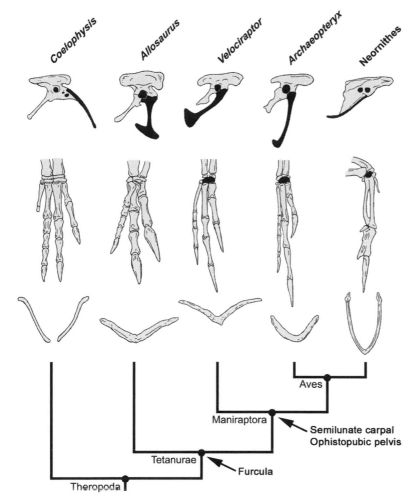

Coelophysis   Allosaurus   Velociraptor   Archaeopteryx   Neornithes

Aves

Maniraptora ———→ **Semilunate carpal**
                  Ophistopubic pelvis

Tetanurae ———→ **Furcula**

Theropoda

Figure 8.3. Examples of osteological synapomorphies (features shared as a result of evolutionary descent) supporting the maniraptoran origin of birds. Modified from Chiappe and Vargas (2004).

was well known shortly after the publication of *The Origin of Species* that many osteological characters are shared by all, or most, nonavian theropods and birds (Huxley, 1868). Osteological comparisons among taxa have been greatly aided in recent years by newly discovered fossils of nonavian maniraptorans and basal birds, which demonstrate increasing levels of similarity in morphology (e.g., corresponding pneumatic sinuses and diverticula, mandibular interdental plates, large epipophyses on cervical vertebrae, semilunate carpal, and posterior rotation and distal boot of pubis) and which fill the morphological gap between the dinosaurs classically used for discussions of bird origins (e.g., *Compsognathus* from the Upper Jurassic Solnhofen Formation of Bavaria) and modern birds (Fig. 8.3). Many osteological features previously thought to be exclusively avian—furcula, laterally facing glenoids, large, ossified sterna, uncinate processes on ribs, and others—are now known to occur among nonavian theropods (Padian and Chiappe, 1998; Chiappe and Dyke, 2002).

Eggs and Eggshells (Oology)

Correlated with improvements in microscopy technology, it has been known for some years that the microstructure of calcified eggshell can be specific to certain groups of extant and extinct reptiles (Mikhailov, 1997). Until recently, however, the precise characteristics of the eggshell microstructure of nonavian theropods remained elusive because of the absence of conclusive embryonic skeletal material preserved inside of eggs. Since the discovery of an embryo of an oviraptorid in the Late Cretaceous strata of the Gobi Desert (Norell et al., 1994; Clark et al., 2001), a number of *in ovo* embryos of nonavian theropods and basal birds have now been discovered and described. Alongside birds (Grellet-Tinner and Norell, 2002; Schweitzer et al., 2002), new dinosaurian discoveries include other species of oviraptorids (Weishampel et al., 2000), therizinosaurids, and troodontids (Varricchio and Jackson, 2004). Furthermore, eggs and egg clutches have been found in association with neonates or adults of dromaeosaurids (Grellet-Tinner and Makovicky, 2000), troodontids (Grellet-Tinner, 2004), oviraptorids (Clark et al., 1999; Sato et al., 2005), and even more basal nonavian theropods (Mateus et al., 1997).

Comparative studies between the eggshell microstructure of all these eggs and those of extant birds have revealed features exclusively common to them (Fig. 8.4; Mikhailov, 1992; Varricchio et al., 1997; Zelenitzky et al., 2002; Grellet-Tinner and Chiappe, 2004). Important among these features is the presence of more than one distinct microstructural layer, most commonly distinguished by differential disposition of the calcitic crystals (Grellet-Tinner and Chiappe, 2003). Furthermore, phylogenetic analyses have revealed a number of other character states uniquely shared by the eggs of nonavian theropods and those of birds: a reduction in the porosity of the shell, a relative increase in the volume of the egg (with respect to the size of the adult), the presence of a distinctly longer axis (eggs that are elongated), and the existence of a heteropolar egg with one pole more blunt than the other (Zelenitzky et al., 2002, 2005; Grellet-Tinner and Chiappe, 2003).

Integument

Although feathers have always seemed to be the quintessential avian character, they too are now well known among the nonavian theropod dinosaurs (Fig. 8.5). The carbonized remains of integumentary structures interpreted as feathers were initially found associated with articulated skeletons of the Early Cretaceous basal coelurosaur *Sinosauropteryx* (which is closely related to *Compsognathus*; Chen et al., 1998; Currie and Chen, 2001), and later with a number of other maniraptorans of similar age from Liaoning Province in northeastern China. These include the therizinosaurid *Beipiaosaurus* (Xu et al., 1999a), the oviraptorosaur *Caudipteryx* (Ji et al., 1998; Zhou and Wang, 2000; Zhou et al., 2000), the dromaeosaurids *Sinornithosaurus* (Xu et al., 1999b) and *Microraptor*

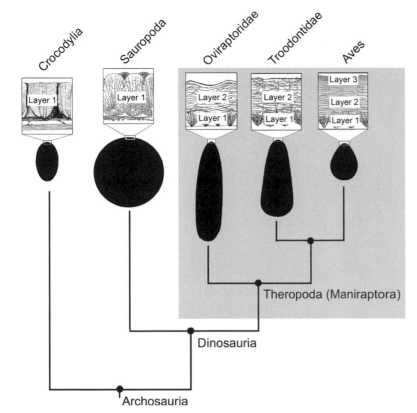

Figure 8.4. Examples of theropod-bird synapomorphies that are seen in eggs and eggshells. Modified from Chiappe and Vargas (2004). Note the presence of two or more layers in the eggshell of maniraptoran theropods (including birds) and the heteropolar shape of the egg in these lineages.

(Xu et al., 2000, 2003), the tyrannosaur *Dilong* (Xu et al., 2004), as well as the long-armed *Protarchaeopteryx* (Ji et al., 1998). Definitive feathers have also been found associated with the incomplete skeleton of *Pedopenna* (Xu and Zhang, 2005), a fossil apparently older than the remaining "feathered dinosaurs"—perhaps dating from the Middle or Late Jurassic—and with features suggesting a greater proximity to the ancestry of birds (although it could be a close relative of the basal dromaeosaurid *Microraptor*). Morphological differentiation in feather type can be seen among these nonavian theropods, just as in present-day birds; although their bodies are covered with filamentous feathers, their tails and forelimbs typically carry pennaceous, undifferentiated feathers. Unquestionable pennaceous feathers, which in *Microraptor* even have asymmetric vanes, have been documented in *Caudipteryx* (Ji et al., 1998; Zhou and Wang, 2000; Zhou et al., 2000), *Protarchaeopteryx* (Ji et al., 1998), *Sinornithosaurus* (Norell et al., 2002), *Pedopenna* (Xu and Zhang, 2005), and *Microraptor* (Xu et al., 2003). The latter two taxa have shown the tantalizing presence of long, vaned feathers attached to their distal hind limbs, structures for which the function remains controversial (Padian, 2003; Xu et al., 2003; Xu and Zhang, 2005). The long pennaceous feathers of the forelimb of *Microraptor* form a wing of essentially modern de-

Figure 8.5. The phylogenetic distribution of the feathered coelurosaurian theropods *Sinosauropteryx prima, Dilong paradoxus, Caudipteryx zoui, Beipiaosaurus inexpectus, Sinornithosaurus millenii, Microraptor gui, Pedopenna daohugouensis,* and *Protarchaeopteryx robusta* with respect to birds (phylogenetic relationships among theropods simplified from the following sources: Chen et al., 1998; Holtz, 1998; Ji et al., 1998; Norell et al., 2001; Xu et al., 2003, 2004; Hwang et al., 2004; Xu and Zhang, 2005). We follow these authors in regarding the undeniably feathered *Caudipteryx* as a nonavian theropod, specifically a member of Oviraptorosauria (see also Chiappe and Dyke, 2002; Dyke and Norell, 2005). Filamentous and vaned feathers are interpreted as coelurosaurian and maniraptoran synapomorphies, respectively.

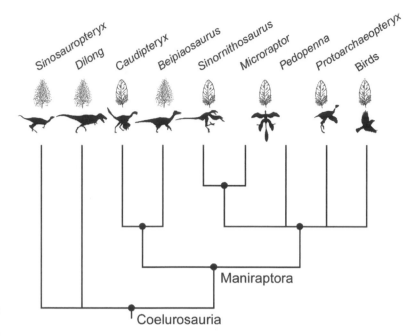

sign, which strongly suggests that this animal (perhaps also *Pedopenna*, although this dinosaur is currently only known from an isolated partial hind limb) was capable of flight, although whether such flight was actively flapping or more gliding remains controversial.

Recent discoveries of feathers in so many nonavian maniraptorans provides undeniable evidence that these integumentary appendages, long considered exclusively avian, evolved much earlier than the last common ancestor of *Archaeopteryx*. The evidence at hand supports the contention that at least simple, filament-like feathers could be considered a potential synapomorphy of Coelurosauria (the basal theropod clade that includes all maniraptoran theropods), whereas the presence of more derived, pennaceous feathers may be interpreted as synapomorphic for Maniraptora (Fig. 8.5). Not only do these new discoveries document the presence of feathers outside birds, but they also suggest that some nonavian theropod dinosaurs may, to some extent, have been able to fly.

Behavior

Documentation of the behavioral attributes of extinct organisms in the fossil record is extremely rare. However, a handful of extraordinarily preserved fossils can cast some light on likely nesting in some of the nonavian maniraptorans. Skeletons of the toothless oviraptorids *Oviraptor* and *Citipati*, for example, from the Upper Cretaceous of the Gobi Desert, have been found sitting atop their egg clutches (Norell et al., 1995; Dong and Currie, 1996; Clark et al., 1999; Fig. 8.6). Similarly, a fragmentary adult specimen of the troodontid *Troodon* was discovered on top of a clutch

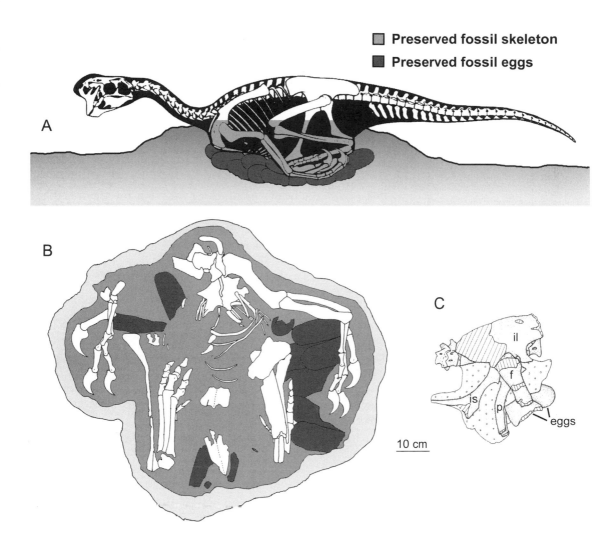

Preserved fossil skeleton
Preserved fossil eggs

A

B

C

10 cm

il

f

is

p

eggs

from the Upper Cretaceous of Montana (Varricchio et al., 1997; Varricchio and Jackson, 2004), and a pair of shelled eggs were recently reported inside an indeterminate oviraptorosaur (Sato et al., 2005) (Fig. 8.6). The best-preserved specimens of the Mongolian oviraptorid *Citipati* demonstrate that these animals used a posture similar to extant birds; their legs were tucked into an open space at the center of the clutch, and their forelimbs surrounded the periphery of the clutch (Clark et al., 1999; Fig. 8.6). Brooding troodontids may have adopted a similar posture, but their clutches lack the empty central space typical of oviraptorid egg clutches (Varricchio and Jackson, 2004), suggesting that these more lightly build animals sat directly on top of their eggs (Grellet-Tinner, 2004). Regardless of their specific functions (e.g., protection, incubation), we know that typical avian nesting behaviors—brooding (sitting atop nests of eggs) and extended nest attention—were present in some nonavian theropods. Furthermore, the recent discovery of a pair of

Figure 8.6. Incomplete skeleton of the maniraptoran *Citipati osmolskae*, an oviraptorid from the Late Cretaceous of the Gobi Desert, brooding a clutch of eggs (**B**), and an interpretation of the posture of this animal in life (**A**; modified from Clark et al., 1999). The recent discovery of a pair of eggs inside the visceral cavity of another oviraptorosaur (**C**) has confirmed previous inferences that these maniraptorans had already evolved the sequential pattern of oviposition that characterizes modern birds (after Sato et al., 2005). **f**, femur; **il**, ilium; **is**, ischium; **p**, pubis.

shelled eggs inside an oviraptorosaur (Fig. 8.6) has confirmed previous inferences that these dinosaurs had evolved an avian type of oviposition (autochronic) in which the two functional ovaries (as opposed to one in birds) produced a single ovule at a time and where it took multiple ovipositions to lay an entire clutch (Sato et al., 2005).

Behavioral inferences highlighting the common descent of birds and certain nonavian theropods have also been documented beyond reproductive traits. Recent fossil discoveries have established the presence of the stereotypical resting posture of present-day birds (with the head tucked under the wing and the legs flexed under the body) in a number of exceptionally well-preserved specimens of the small troodontid *Mei* (Xu et al., 2004), and these findings have suggested similar postures for other troodontid fossils (e.g., *Sinornithoides, Sinovenator*).

Against the Theropod Origin of Birds

Even in the face of the evidence summarized in this chapter and elsewhere over the course of the last 20 years, the theropod hypothesis for avian origins is still not accepted by all evolutionary biologists. Dissenters have highlighted a number of apparent inconsistencies with the known fossil record and the inferred homology of certain structures. Recent work has shown that at least these two areas of contention are nonissues.

Gap in Time? "Temporal Paradox" between Fossils

An apparent inconsistency with the theropod dinosaur hypothesis for avian origins is the presence of a supposed gap in time between *Archaeopteryx* (Late Jurassic; 150 mya) and its putative maniraptoran outgroups (theropods known mostly from the Cretaceous; about 90–65 mya). This argument is referred to as the "temporal paradox" (Feduccia, 1996, 1999), and it asks, "Can it be valid to argue that birds evolved from within a lineage comprising taxa mostly known from much younger rocks?" Although it is true that the bulk of the known fossil record of potential avian sister-taxa are Cretaceous in age, examination of the theoretical basis and evidence used to support the temporal paradox argument shows that it disregards a good deal of the available fossil evidence. It is an artifact caused by failing to consider all the alternative hypotheses of bird origins (Brochu and Norell, 2000). Most importantly, modern formulations of the theropod hypothesis for avian origins do not actually require the sister-taxa of birds to have existed in pre-Jurassic times (the age of *Archaeopteryx*). Rather, this is a cladistic hypothesis that postulates the existence of the most recent common ancestor of these Cretaceous dinosaurs and *Archaeopteryx* before the divergence of the oldest of these taxa (e.g., Gauthier, 1986; Forster et al., 1998; Holtz, 1998; Sereno, 1999; Norell et al., 2001; Clark et al., 2002; Hwang et al., 2004). It is also central to this debate to note that Late Jurassic maniraptoran theropods have, in fact, been well known for decades, even if just from fragmentary remains (Jensen and Padian, 1989). More conclusively, recently described fossils supporting the pre-Cretaceous existence of maniraptoran theropods include a lower jaw of a therizinosaurid from the Lower Jurassic of Yunnan, China (Xu et al., 2001), and the putative dromaeosaurid *Pedopenna* (Xu and Zhang, 2005). The age of this latter taxon appears to be substantially older than any of the other Early Cretaceous feathered maniraptorans known from northeastern China (Zhou, 2004; Xu and Zhang, 2005).

This temporal paradox between the fossil record of early birds and the fossil record of theropods (regarded as closely related to the origin of the latter) appears to exist only if one considers the temporal gap between the 100-million-year-old *Deinonychus,* a commonly cited example of a dromaeosaurid (Ostrom, 1969, 1973, 1976), and the 150-million-year-old *Archaeopteryx.* As demonstrated by Brochu and Norell (2000), when other records of well-known maniraptorans (e.g., the 125-million-year-old dromaeosaurids *Sinornithosaurus* and *Microraptor,* or, more recently, the much older *Pedopenna*) are included in an analysis that compares the goodness of fit between phylogeny and stratigraphy, and when hypotheses of bird origins are compared against each other (e.g., placing birds within groups indicated by other hypotheses like crocodylomorphs or more basal archosaurs), the maniraptoran hypothesis turns out to be the most consistent temporally.

*Embryology of the Avian Hand: "II, III, IV," "I, II, III,"
or Neither*

According to some developmental biologists, the digits of the hand in birds (as documented by experimental embryology) are numbered II, III, and IV when compared to the primitive five-fingered tetrapod condition (e.g., Hinchliffe, 1985; Feduccia, 1999). Embryological investigations of extant birds have identified five precartilaginous condensations, of which only those in positions II, III, and IV develop into the three osseous digits of the adult hand (Feduccia and Nowicki, 2002; Larsson and Wagner, 2002; Fig. 8.7). This has been contrasted to the condition seen in theropod dinosaurs: on the basis of the known fossil evidence, the homologies of the three fingers of *Archaeopteryx* are considered to correspond to digits I, II, and III of the ancestral pentadactyl hand. Fossils show a trend toward reduction of the outermost two digits (VI and V) from the most basal theropods, in which these digits are abbreviated but still present, to maniraptoran theropods with a tridactyl hand (Padian and Chiappe, 1998; Fig. 8.7). It has been argued that if such a discrepancy really does exist, then perhaps the hands of extant birds and theropods are nonhomologous (Feduccia, 1999).

There are two issues at stake in this ongoing debate (but see Vargas and Fallon, 2005): Is there a valid reason to extrapolate the ontogenetic development of modern birds back in time to *Archaeopteryx*? And should the maniraptoran theropod ancestry of

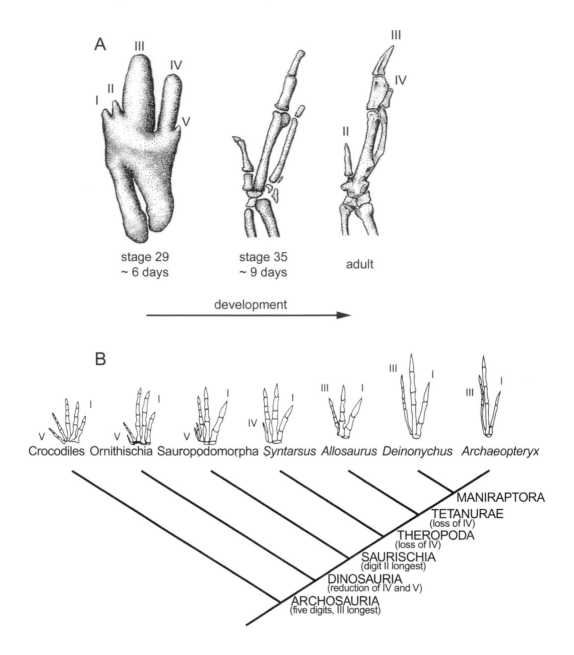

Figure 8.7. Recent embryological research has identified five precartilaginous condensations in the hand of chickens (**A**, modified from Larsson and Wagner, 2002) and other modern birds. These developments have identified the three fingers of adult modern birds as those in positions II, III, and IV. Fossil evidence, however, supports the identity of the fingers of modern birds as digits I, II, and III of the ancestral pentadactyl hand (**B**, modified from Padian and Chiappe, 1998). The hand of nonavian theropods shows a trend toward reduction of the outermost two digits (VI and V) from the most basal taxa to maniraptorans with a tridactyl hand (**B**). Note that there is actually no incongruence in character states throughout the transition between basal archosaurs and birds; changes in proportions and loss or fusion of carpal elements are easily identified.

birds be rejected if nonavian dinosaurs developed their manual digits via a different developmental pathway from that of modern birds? We will argue to the contrary in both cases.

In the first place, extrapolation of the embryogenesis of the extant bird hand back to archaic lineages (including *Archaeopteryx*) is unwarranted because the hand of modern birds is highly transformed, and because embryological evidence is unavailable for *Archaeopteryx* or for any other fossil avian lineage (Padian and Chiappe, 1998; Chiappe and Dyke, 2002; Chiappe, 2004). The second issue is whether embryogenetic differences can take precedence over the enormous volumes of other evidence supporting a phylogenetic relationship between birds and nonavian maniraptorans. We know that the existing embryological evidence for condensation patterns in present-day birds is problematic (Vargas and Fallon, 2005). Can a potential problem with what amounts to a single homology be explained?

Wagner and Gauthier (1999) have argued that homeotic frame shifts can lead to developmental patterns in which digits previously ossified from condensations I–III become ossified from condensations II–IV. Such homeotic frame shifts are relatively common among other vertebrate lineages; examples include hand development in the two-toed earless skink (*Hemiergis quadrilineata*), an Australian scincid lizard (Shapiro, 2002; Fig. 8.8). Other studies have documented the fact that ontogenetic trajectories do evolve and that transformations can occur within lineages without affecting either the morphology or the eventual function of the developing structure (Wagner and Misof, 1993; Mabee, 2000; Hall, 2003). Recent molecular work on the differential expression of *Hox* genes among the digits in developing mouse and chicken embryos actu-

Figure 8.8. The chondrification and ossification patterns of the hand of the scincid lizards *Hemiergis perioni* and *Hemiergis quadrilineata* provide an example of a homeotic frame shift in tetrapods. In the two morphs (three and four fingers) of *Hemiergis perioni* (**A**, **B**), condensations (white circles) of digits II and III develop into the two innermost fingers of the adult hand (gray box), which have three and four phalanges (black circles), respectively. Adults of *Hemiergis quadrilineata* (**C**) have only two fingers with three (innermost finger) and phalanges, respectively. The homeotic frame shift is revealed by the fact that although in *Hemiergis perioni* the two innermost fingers, those with three and four phalanges (gray boxes in **A** and **B**), ossify from condensations II and III, in *Hemiergis quadrilineata*, these digits ossify from condensations III and IV (gray box in C). The morphological similarity between the adult digits of these species is such that the positional identity of the fingers of *Hemiergis quadrilineata* (**C**) can only be verified through developmental investigations. Modified from Shapiro (2002).

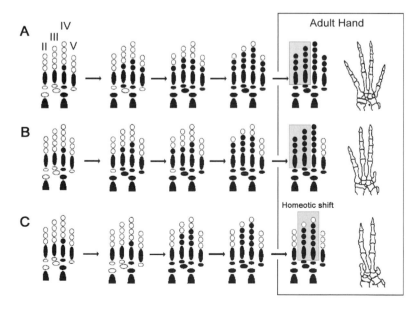

ally identifies the anteriormost digit of the avian wing as digit I (as has been postulated to be the case in *Archaeopteryx,* other basal birds, and theropod dinosaurs; Vargas and Fallon, 2005). Thus, the digit homology problem, so often leveled against the theropod origin of birds, may not even exist in the first place.

New Avian Discoveries from the Mesozoic

As has been discussed elsewhere (Padian and Chiappe, 1998; Chiappe and Dyke, 2002; Zhou, 2004), for more than a century, our knowledge of Mesozoic nonmodern bird diversity (and hence evolution) was greatly limited by the number of known fossils and fossil-bearing sites (Figs. 8.9, 8.10). Just *Archaeopteryx* and a few other taxa from the Cretaceous (mostly the marine birds *Ichthyornis, Hesperornis,* and their kin) formed the bulk of the known diversity of early fossil birds (compare the reviews of Marsh [1880] and Martin [1983], which were separated by a century). This enormous anatomical gap was about all that was evident with regard to the early evolution of the group and the earliest stages of the development of flight (Rayner, 1988, 1991; see also Rayner, 2001). Dramatic discoveries in recent years, however, have served to fill this lacuna in the fossil record (Chiappe, 1995; Feduccia, 1999; Chiappe and Dyke, 2002; Chiappe and Witmer, 2002; Zhou, 2004), turning the study of the early evolution of birds into one of the most dynamic fields of vertebrate paleontology (Fig. 8.11). Many of the new fossils (but by no means all of them) come from Early Cretaceous lacustrine beds in Liaoning Province, northeastern China (Zhou et al., 2003; Zhou, 2004). Over the last five years, more birds from this region have been discovered than from any other area with strata of similar age. Well over a dozen new species (Zhou, 2004) have been described from primarily two lithostratigraphic units: the 128-to-125-million-year-old

Figure 8.9. "Collector curve" for Mesozoic birds. This curve expresses the cumulative number of specimens as a proportion of the total number of Mesozoic bird specimens for which find dates are available, and illustrates the rapid increase in discoveries in recent years.

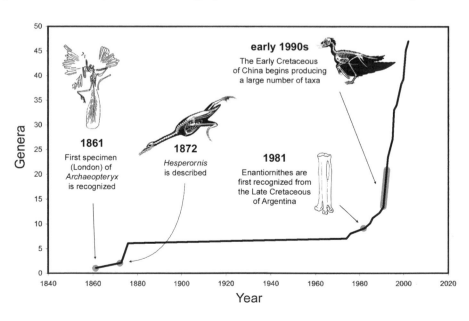

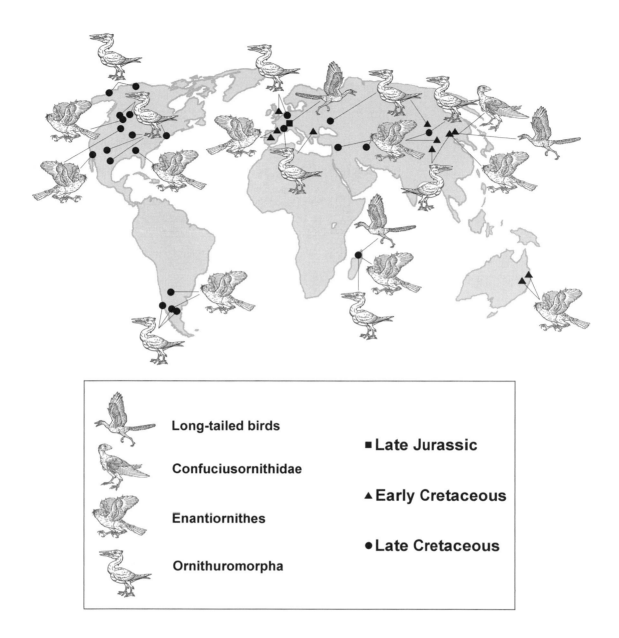

Figure 8.10. Geographical distribution of Mesozoic birds. Note how the Northern Hemisphere is much more extensively sampled than the rest of the world.

Long-tailed birds

Confuciusornithidae

Enantiornithes

Ornithuromorpha

■ Late Jurassic

▲ Early Cretaceous

● Late Cretaceous

Yixian Formation (Swisher et al., 2002) and the overlaying Jiufotang Formation (approximately 120 million years old; Zhou, 2004).

Archaeopteryx *and Long-tailed Birds*

At least four new specimens of *Archaeopteryx* (Wellnhofer, 1992; Elzanowski, 2001) have been discovered and described during the last two decades. The known fossils of this ancient bird now add up to nine skeletal specimens and a feather, all from the same lagoonal Upper Jurassic Solnhofen Limestones of Bavaria, Germany (Elzanowski, 2001). Two other recently discovered taxa are also thought to be basal within Aves; one of them, represented by a

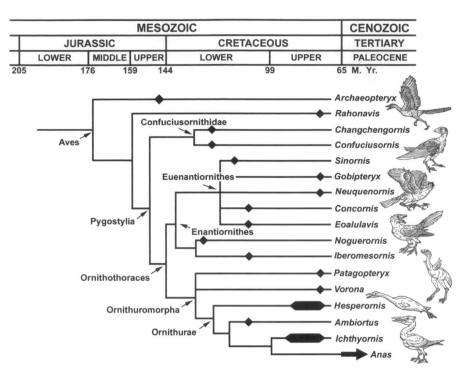

MESOZOIC					CENOZOIC
JURASSIC			CRETACEOUS		TERTIARY
LOWER	MIDDLE	UPPER	LOWER	UPPER	PALEOCENE
205	176	159 144		99	65 M. Yr.

Figure 8.11. Simplified phylogeny of the principal lineages of Mesozoic birds. Most of the taxa (and lineages) illustrated here were discovered during the last two decades. Modified from Chiappe (2002a).

single, mostly disarticulated, and incomplete specimen, is *Rahonavis* from the Upper Cretaceous Maevarano Formation of Madagascar (Forster et al., 1998; Fig. 8.12). Like *Archaeopteryx,* the larger and much younger *Rahonavis* has a forelimb that is significantly longer (in relation to the hind limb) than any other nonavian maniraptoran, and it has a foot with a reversed hallux, a feature thus far undocumented in nonavian theropods. The second recently discovered taxon is *Jinfengopteryx,* a half-meter-long animal from the Lower Cretaceous Qiaotou Formation of northeastern China, a stratigraphic unit that is apparently lower than the renowned Yixian Formation (Ji et al., 2005). *Jinfengopteryx* is known from an articulated, feathered specimen (Fig. 8.12), which has skeletal structure and preserved plumage described as remarkably similar to that in *Archaeopteryx* (Ji et al., 2005). *Jinfengopteryx* may well be the sister-taxon of *Archaeopteryx,* as suggested by Ji et al. (2005), but its abbreviated forelimb and large number of teeth are primitive features absent among other long-tailed birds (and present in some nonavian theropod lineages such as troodontids). A good amount of experimentation characterizes the early evolution of birds, and *Jinfengopteryx* could have reverted to a more primitive condition of wing and dental design. Future studies on this most recent discovery would put these views to test.

Archaeopteryx, Jinfengopteryx, and *Rahonavis* are the most anatomically primitive birds known to date; their skeletal structure only slightly departs from that of nonavian maniraptoran theropods. For example, *Archaeopteryx* and *Jinfengopteryx* have triangular skulls with small tooth crowns lacking serrations, unossified

Figure 8.12. Recent discoveries of long-tailed birds: *Jinfengopteryx elegans* from the Early Cretaceous of China (**A**, after Ji et al., 2005), *Rahonavis ostromi* from the Late Cretaceous of Madagascar (**B**, after Forster et al., 1998), and *Jeholornis prima* from the Early Cretaceous of China (**C**, modified from Zhou, 2004). Until recently, the famed *Archaeopteryx lithographica* from the Late Jurassic of Germany was the only long-tailed bird known. The new diversity of birds with long, bony tails has documented a remarkable pattern of mosaic evolution.

10 cm

sterna (for a reassessment of the bony sternum in *Archaeopteryx,* see Wellnhofer and Tischlinger, 2004), shoulder girdles that lack the specializations seen in more derived birds (e.g., strut-like coracoid, triosseal canal), powerfully clawed hands, pelves with near vertical pubes, and bony tails bearing a frond-like arrangement of pennaceous feathers. As far as can be determined, these primitive features were also present in the more poorly known *Rahonavis* (Forster et al., 1998), although this bird has a number of characteristics (e.g., open glenoid, distally reduced fibula) that suggest it may be more advanced phylogenetically than either *Archaeopteryx* or *Jinfengopteryx*. One other striking feature of *Rahonavis*, and another example of a nonavian theropod specialization absent in both *Archaeopteryx* and apparently *Jinfengopteryx*, is the possession of an enlarged, sickle-shaped foot claw that may have been used for slashing its prey (Forster et al., 1998).

A few other taxa with long, bony tails (a primitive character state for birds), but that have been considered more advanced phylogenetically than *Archaeopteryx*, *Jinfengopteryx*, and *Rahonavis*, have recently been described from the Lower Cretaceous Jiufotang Formation. Two are *Shenzhouraptor* (Ji et al., 2002a, 2003) and *Jeholornis* (Zhou and Zhang, 2002a), which are characterized by their abbreviated dentitions (Fig. 8.12). Like their primitive counterparts, these were large birds (about the size of crows) with long, thin wings. *Jeholornis* and *Shenzhouraptor* are morphologically similar and may represent the same taxon. Although primitive in many respects, the skeletons of *Jeholornis* and *Shenzhouraptor* show significant modifications with respect to *Archaeopteryx*. These birds each have a bony sternum, a much longer coracoid, a mobile glenoid (a shoulder girdle just like that of *Rahonavis*) that extends further upward than in *Archaeopteryx*, and a shorter and more compact hand. Although resembling *Archaeopteryx* in many respects, modifications of the shoulder, sternum, and forelimb indicate that *Jeholornis* and *Shenzhouraptor* had improved their flight performance. Even so, the primitive tails of these birds with vertebral counts exceeding that of *Archaeopteryx*, combined with their feathering (more similar to the dromaeosaurid *Microraptor* than *Archaeopteryx*), highlight the mosaic nature of early avian evolution.

Yet another long-tailed bird also recently described from northeastern China shows features that suggest an even greater flying ability. Although older than *Jeholornis* and *Shenzhouraptor*, the toothless, short-snouted *Jixiangornis* (Ji et al., 2002b) from the Yixian Formation has a shorter and more compact hand—although still with a full set of claws—and a larger sternum with a short and robust ventral keel (Fig. 8.12). Little else is currently known about the anatomy of this very interesting bird, but what has been described suggests that it could be the most advanced of the known long-tailed birds. Evident in the flight apparatus of these primitive taxa is the beginning of a pattern that characterizes eventual evolutionary triumph of birds. Early in their evolutionary history, birds began enhancing their flying abilities long before they developed their characteristic mode of terrestrial locomotion (Chiappe, 1995).

Cretaceous Radiation of Short-tailed Birds

From the same Jiufotang beds as *Jeholornis* and *Shenzhouraptor*, a great diversity of short-tailed taxa—Pygostylia—are also now known. Modern birds are short-tailed, and they have a pygostyle (a stump-like bone) posterior to the remaining free caudal vertebrae; these new Early Cretaceous fossils are genealogically closer to extant birds than to their long-tailed counterparts (Chiappe and Dyke, 2002; Zhou, 2004; Fig. 8.11). Perhaps the most primitive of these new pygostylians is the large *Sapeornis* (Zhou and Zhang, 2002b; Fig. 8.13), an animal about the size of a turkey vulture but with a short skull with robust teeth that are restricted to the tip of its rostrum. *Sapeornis* also has a peculiar flight apparatus (Zhou and Zhang, 2002b): although it has long wings, a strongly reduced outermost digit, and a mobile glenoid, the coracoid is short and hatchet-shaped, and no ossified sternum has been found in the six specimens discovered to date. This suggests that either *Sapeornis* had unexpectedly weak flight musculature, or its sternum could have been cartilaginous, with the large flight muscles needed to power its long wings originating on the transversally expanded coracoids.

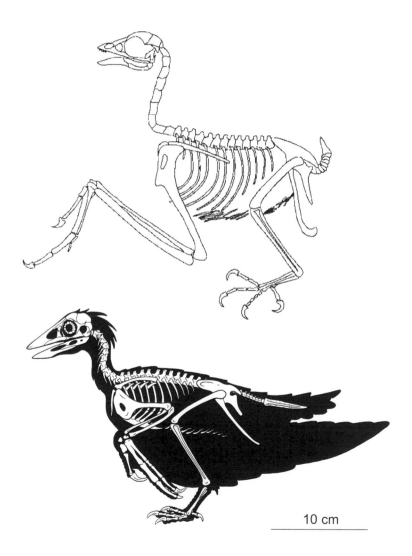

10 cm

Figure 8.13. Basal pygostylians *Sapeornis chaoyangensis* (**A**, after Zhou and Zhang, 2003) and *Confuciusornis sanctus* (**B**, after Chiappe et al., 1999).

Another example of a primitive short-tailed bird is the much better known *Confuciusornis* (Fig. 8.13). This toothless bird is by far the most abundant avian fossil from the Mesozoic anywhere in the world, although it is known only from China (Hou et al., 1995; Chiappe et al., 1999). Hundreds of well-preserved specimens have been collected from the Yixian Formation of Liaoning Province. Because some individual specimens of this bird have long tail streamers and some do not, this taxon has been touted as providing the oldest fossil evidence for sexual dimorphism in birds. Several species of *Confuciusornis* have been named, but most (if not all) of these are likely synonyms (Chiappe et al., 1999). Very similar to *Confuciusornis* is *Changchengornis,* a bird known from a single specimen also from the 125-million-year-old Yixian Formation (Chiappe et al., 1999; Ji et al., 1999), which differs from *Confuciusornis* in its strongly curved beak as well as in a few other details of its skeleton (Fig. 8.13). Confuciusornithids (i.e., *Confuciusornis* and *Changchengornis*) are unique among birds because they have skulls that have retained complete diapsid (upper and lower) temporal fenestrae, an example of a reversal to a more primitive design that would have greatly limited their cranial kinesis (Chiappe et al., 1999). The flight apparatus of these tern-sized birds is also primitive in a number of respects, including the presence of a rigid scapulocoracoid and a forelimb that is proportionally much shorter than that of *Sapeornis*. The confuciusornithid sternum, although ossified, is essentially flat; only a faint caudal ridge is present in some specimens of *Confuciusornis*. These skeletal features of the flight apparatus contrast with the presence of long primary feathers, which make the wings of these birds resemble those of extant terns, and a pair of streamer-like feathers project more than the length of their skeletons.

On the basis of information afforded by the holotype, Zhou and Zhang (2002b) suggested that *Sapeornis* is the phylogenetically most primitive short-tailed bird. However, the presence of features uniquely shared with the much better known *Confuciusornis* (e.g., a large foramen piercing the proximal end of the humerus) and the existence of characteristics that are more derived than in the latter (e.g., a reduced outermost manual digit, a shorter pygostyle) suggest instead that *Sapeornis* may be either the sister-taxon of *Confuciusornis* or a phylogenetically more advanced bird. These alternatives need to be evaluated in light of larger character analyses that include the features these primitive pygostylians share with one another. Regardless of their evolutionary interrelationships, the unusual flight apparatuses of these primitive short-tailed birds hint at aerodynamics and flight mechanisms that were different from those typical of modern avians.

Another basal lineage of short-tailed birds, although clearly more advanced than both *Sapeornis* and *Confuciusornis,* comprises the Enantiornithes (Walker, 1981; Chiappe and Walker, 2002; Figs. 8.11, 8.14). Like no other Mesozoic avian clade, Enantiornithes embody the remarkable number of fossil discoveries that have occurred over the last two decades. Now recorded throughout the Cretaceous and from every continent except Antarctica, nearly half of all Mesozoic birds known to date are members of this group. Yet the existence of Enantiornithes was only recognized a quarter of a century ago (Walker, 1981). Although enantiornithines are most often recorded from inland deposits, they are also known to have inhabited coastal and marine environments, and their range even extended into polar regions (Chiappe, 1995). Having evolved key aerodynamic features (e.g., alula, modern wing proportions, keeled sternum) and sizes similar to those of small songbirds, the earliest members of this group already document a significant departure from the larger and more primitive fliers seen in the Jurassic and

Figure 8.14. Selected taxa of Early Cretaceous Enantiornithes: *Cathayornis yandica* (**A**), *Eoenantiornis buhleri* (**B**, after Zhou et al., 2005), and *Longirostravis hani* (**C**, after Hou et al., 2004). Roughly half of the known diversity of Mesozoic birds can be included within the Enantiornithes.

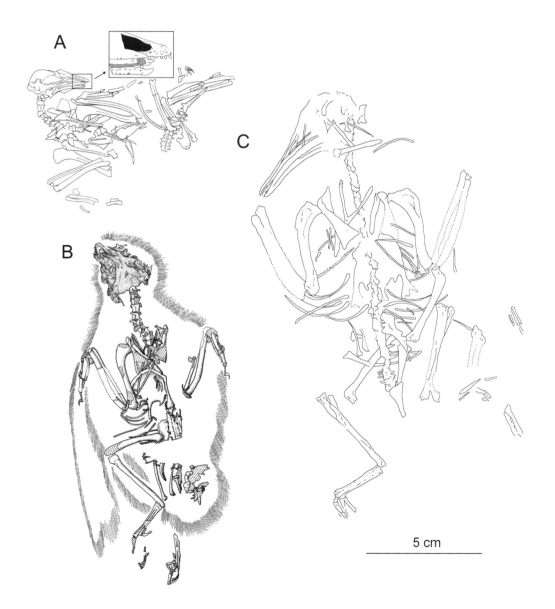

5 cm

Early Cretaceous. These superior flying abilities must have facilitated the occupation of the wide range of environments and the evolution of the diverse feeding specializations seen among these birds (e.g., fish eaters, insect eaters, soft substrate probers; Hou et al., 2004; Zhou, 2004; Fig. 8.14). The later evolution of enantiornithines appears to have been characterized by a distinct increment in size and by the evolution of an even greater diversity of ecomorphological types (e.g., swimmers and waders as well as near flightless birds). Much work remains to be done on these important early birds. For example, we have almost no idea about the details of their interrelationships, which have remained largely unexplored because of the incompleteness of the known material for several taxa and the great similarities in postcranial features among many of these birds (Chiappe and Walker, 2002).

Toward Extant Aves: Ornithuromorpha

In addition to rich accumulations of more basal avians, the Early Cretaceous formations of China in particular have provided critical information for understanding the early evolution of ornithuromorphs—birds even more closely related to those of today. New discoveries include exceptionally well-preserved specimens of two toothed birds, the similar-sized *Yanornis* and *Yixianornis* (Zhou and Zhang, 2001; Zhou et al., 2002; Fig. 8.15). These early ornithuromorphs also retain a number of more primitive avian characteristics as well as their teeth (i.e., gastralia, long fingers that retain claws), while at the same time they have an essentially modern flight apparatus. The pectoral girdles of *Yanornis* and the shorter-snouted *Yixianornis* closely approach the condition seen in present-day birds (i.e., coracoid with a wide base and rounded procoracoid process, curved scapula), and their sterna are large and fully keeled. Both taxa were clearly capable of well-controlled and active flapping flight. Future studies of these birds are likely to clarify further aspects of the anatomical transition toward extant birds.

Equally important evidence of basal ornithuromorphs comes from Late Cretaceous localities in Argentina and Mongolia. Despite being many millions of years younger than *Yanornis* and *Yixianornis*, the hen-sized *Patagopteryx* from the Upper Cretaceous of northwestern Patagonia (Chiappe, 2002b) is possibly the most primitive ornithuromorph known and is the best-represented Mesozoic bird from the Southern Hemisphere to date (Figs. 8.11, 8.15). The short forelimbs of this bird (they are less than half the length of its hind limbs), combined with a flat sternum and reduced furcula, suggest that it was flightless, while the anatomy of its robust pelvis and hind limbs indicate it was an obligated ground dweller (Chiappe, 2002b). Another important basal ornithuromorph is *Apsaravis* from the Upper Cretaceous of Mongolia (Norell and Clarke, 2001; Clarke and Norell, 2002; Fig. 8.15). The single known specimen of this seemingly toothless bird has been considered phylogenetically intermediate between *Patagopteryx* and hesperornithiforms (Clarke and Norell, 2002) and is exquisitely well preserved.

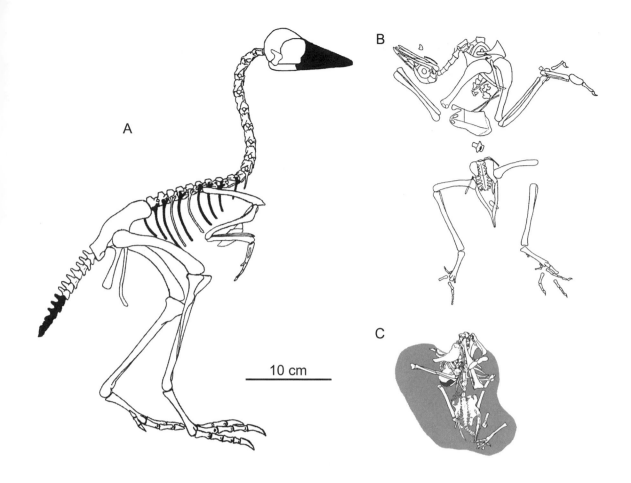

10 cm

Figure 8.15. Selected taxa of basal ornithuromorphs: *Patagopteryx deferrariisi* from the Late Cretaceous of Argentina (**A**, modified from Chiappe, 2002b), *Yanornis martini* from the Early Cretaceous of China (**B**, modified from Zhou and Zhang, 2001), and *Apsaravis ukhaana* from the Late Cretaceous of Mongolia (**C**, modified from Clarke and Norell, 2002).

Many features of this bird's skeleton show similarity to modern avians (e.g., short, fused dentary symphysis, reduced number of trunk vertebrae, abbreviated pygostyle, globose humeral head). An important transformation that is of functional significance is the fact that the innermost metacarpal of *Apsaravis* has a pronounced extensor process; in extant birds, this feature is involved in the automatic extension of the hand by the propatagial ligaments (Clarke and Norell, 2002). The development of this pronounced extensor process in *Apsaravis* indicates that this bird was able to extend its wing automatically (Clarke and Norell, 2002), a functional property that highlights the sophistication already achieved in the flight apparatus of this and more advanced ornithuromorphs by Late Cretaceous times.

For nearly the entire history of paleornithology, the evolutionary transformations leading to the origin of modern birds (i.e., Neornithes) were established on the evidence mostly provided by two lineages of Late Cretaceous seabirds, the specialized diving hesperornithiforms and the less well-known *Ichthyornis* (Marsh, 1880; Fig. 8.11). Varying greatly in size—the largest being comparable in size to an emperor penguin (Fig. 8.16)—hes-

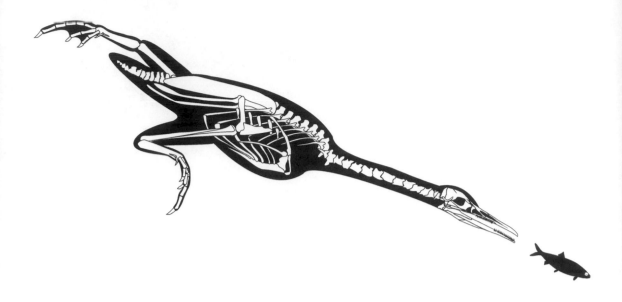

Figure 8.16. With a size comparable to that of an emperor penguin, *Hesperornis regalis* from the Late Cretaceous of North America was one of the largest-known Mesozoic birds. The structure of this formidable fish eater was highly specialized for foot-propelled diving. Modified from Chiappe and Dyke (2002).

perornithiforms show a general trend toward an increase in size, flightlessness, and foot-propelled diving specializations over their 35 million years of recorded evolutionary history. These birds have been recorded over much of the Northern Hemisphere and across a vast latitudinal extent (from the Arctic Circle to the Gulf of Mexico), and from depositional environments ranging from offshore to fluvial (Galton and Martin, 2002). Morphologically, they are characterized by having an elongate skull with a sharp and toothed snout, minute forelimbs (especially in the most advanced forms), a long neck and short trunk, and a robust hind limb specialized for aquatic locomotion (Marsh, 1880; Martin and Tate, 1976; Martin, 1980; Fig. 8.16).

From many of the same environments as the hesperornithiforms comes the historically well-known *Ichthyornis* (Marsh, 1880; Clarke, 2004). Although originally considered to comprise a series of closely related taxa, recent revisionary work has clarified taxonomic problems associated with a large collection of specimens collected in the nineteenth century and has shown that many of these are likely members of just the single species, *Ichthyornis dispar* (Clarke, 2004). As a matter of fact, one of the several original species types (i.e., *I. victor*) has been shown to be a chimera formed by numerous specimens, some even of taxa that may be closer to modern avians (Clarke, 2004). In most aspects of its skeleton, the flying *Ichthyornis* is anatomically modern, although still toothed; this taxon has consistently been regarded as an immediate relative of modern avians (Fig. 8.11). A range of sizes is represented among the known North American specimens (some are 20% larger than others), illustrating a general trend toward larger size over the 15 million years (early Turonian–early Campanian) of the known history of *Ichthyornis*. Whether this large size range is significant taxonomically remains unclear and will to a great extent depend on interpretations of the growth physiology of this bird (Clarke, 2004).

The Beginnings of Modern Birds: New Fossils and New Directions

One of the most heated current debates in ornithology concerns the question of the timing of the modern avian radiation (i.e., Neornithes; Feduccia, 1999; Chiappe and Dyke, 2002; Dyke and van Tuinen, 2004; Clarke et al., 2005). This debate has persisted because of a marked discrepancy between the results of molecular and paleontological data with regard to the timing of the divergence of neornithines: did the modern lineages radiate explosively in the aftermath of the end-Cretaceous extinction event (as hypothesized by some readings of the fossil record; Feduccia, 1995, 1999), or did they originate deep in the Cretaceous (as argued by early proponents of molecular clock divergence dates; Cooper and Penny, 1997)? Although the debate has centered around the end-Cretaceous extinction event (at the K-T boundary), the discrepancy for divergence events is unlike the situation seen, for example, in mammals (Kumar and Hedges, 1997). One serious problem for resolving this debate has been the absence of a good (phylogenetically constrained) modern bird fossil record before the earliest Cenozoic (Paleocene and Eocene; Dyke and van Tuinen, 2004). The fossil record of Neornithes in the early Eocene is excellent (compared with older sediments); a number of well-represented taxa have been included within phylogenetic analyses of the basal modern clades. Some, such as the basal palaeognath clade Lithornithidae (*Lithornis* and its kin), have even proved informative with respect to resolving relationships among extant taxa because *Lithornis* is considered basal among Palaeognathae, close to the divergence of extant Tinamidae (Dyke, 2003).

The actual number of recorded fossil specimens of "neornithine" birds purported to be from the Cretaceous is small (Hope, 2002), and few are represented by more than isolated skeletal elements (Clarke and Chiappe, 2001; Chiappe and Dyke, 2002). Most of these fossils have remained controversial, either because they have not been studied in detail or because their placement within neornithines has been rejected. The incompleteness of most of the specimens referred to as Cretaceous neornithines has rendered few characters useful for phylogenetic analyses (Clarke, 1999; Dyke and Mayr, 1999; Chiappe and Clarke, 2001; Dyke, 2003), and the poor understanding of the higher-level phylogenetic relationships of extant lineages further complicates their systematic consideration. Nevertheless, this scanty fossil record has been used to hypothesize the existence of a number of extant lineages before the end of the Mesozoic (e.g., Pelecaniformes, Charadriiformes, Anseriformes, Gaviiformes, Galliformes, and Psittaciformes; see compilations in Chiappe and Dyke, 2002), either by taking it at face value or by using it for the temporal calibration of morphological or molecular phylogenies.

An exception to this trend of a scanty, incomplete record of supposedly Cretaceous neornithines is *Vegavis iaai* (Clarke et al., 2005) from the Upper Cretaceous of Vega Island, western Antarctica. As has been long suspected on the basis of scrappy fossil material, as well as from phylogenetic extrapolation, representatives of the more basal clades of modern birds were likely present during the latter stages of the Cretaceous (Dyke, 2003). *Vegavis* represents the first skeleton complete enough to be placed unambiguously within a certainly modern clade of birds, Anseriformes (screamers, ducks, and geese; Clarke et al., 2005). Although only one specimen has been found to date, there are certain to be more in the future. More well-preserved fossils and a better understanding of the relationships of modern clades are necessary to unravel the temporal divergence and evolutionary dynamics of neornithines, but the fossil record supports at least a modest divergence during pre-Cenozoic times.

Conclusion

As a direct result of a dramatic increase in fossil material in recent years, new evidence shows that many features previously considered avian hallmarks in fact first evolved within the maniraptoran predecessors of birds. The main aspects of the debate over avian origins are essentially over recent criticisms that relate to the embryology of the bird hand and to the presence of a temporal paradox between the relevant groups of fossils that is both empirically and methodologically misleading; a wealth of evidence indicates that birds are phylogenetically nested within maniraptoran theropod dinosaurs. Even if the notion that the closest relatives of birds must be found among nonavian maniraptorans is indisputable, the debate has shifted toward the search for the identity of the immediate sister-group of birds.

An enormous number of birds are recorded from rocks that are Early Cretaceous in age, but a substantial temporal gap exists between the earliest-known Cretaceous records—125 million years old or perhaps somewhat older—and the Late Jurassic *Archaeopteryx*. Many new discoveries in the last few years have documented an unexpected diversity of long-tailed birds, which show a mosaic of features and not necessarily a progression toward tail reduction. The earliest short-tailed birds appear abruptly in the fossil record and no morphological intermediates are known between these and their more primitive, long-tailed relatives. Early Cretaceous strata—mostly from China and Spain—also document a large diversity of enantiornithines, small birds with a suite of characters suggesting sophisticated flying abilities with respect to their more primitive relatives. New discoveries have also enhanced our understanding of the early history of ornithuromorphs and the earliest evolution of the modern avian body plan, and they have begun to provide reliable information about the pre-Cenozoic radiation of the modern lineages of birds.

Acknowledgments

We thank Jason Anderson and Hans-Dieter Sues for their invitation to participate in this volume, and Chris Brochu and Kevin Padian for helpful comments in review. We also thank the Natural History Museum of Los Angeles County and University College Dublin for support.

References Cited

Brochu, C. A., and M. A. Norell. 2000. Temporal congruence and the origin of birds. Journal of Vertebrate Paleontology 20:197–200.

Brush, A. 2000. Evolving protofeather and feather diversity. American Zoologist 40:631–639.

Burgers, P., and L. M. Chiappe. 1999. The wing of *Archaeopteryx* as a primary thrust generator. Nature 399:60–62.

Chatterjee, S. 1997. The Rise of Birds. 225 Million Years of Evolution. Johns Hopkins University Press, Baltimore, 312 pp.

Chen, P.-J., Z.-M. Dong, and S.-M. Zhen. 1998. An exceptionally well-preserved theropod dinosaur from the Yixian Formation of China. Nature 391:147–152.

Chiappe, L. M. 1995. The first 85 million years of avian evolution. Nature 378:349–355.

Chiappe, L. M. 2001. The rise of birds; pp. 102–106 in D. E. G. Briggs and P. R. Crowther (eds.), Palaeobiology II: A Synthesis. Cambridge University Press, Cambridge.

Chiappe, L. M. 2002a. Basal bird phylogeny: problems and solutions; pp. 448–472 in L. M. Chiappe and L. M. Witmer (eds.), Mesozoic Birds: Above the Heads of Dinosaurs. University of California Press, Berkeley.

Chiappe, L. M. 2002b. Osteology of the flightless *Patagopteryx deferrariisi* from the Late Cretaceous of Patagonia; pp. 281–316 in L. M. Chiappe and L. M. Witmer (eds.), Mesozoic Birds: Above the Heads of Dinosaurs. University of California Press, Berkeley.

Chiappe, L. M. 2004. The closest relatives of birds. Ornitología Neotropical 15:101–116.

Chiappe, L. M., and J. A. Clarke. 2001. A new carinate bird from the Late Cretaceous of Patagonia (Argentina). American Museum Novitates 3323:1–23.

Chiappe, L. M., and G. J. Dyke. 2002. The Mesozoic radiation of birds. Annual Review of Ecology and Systematics 33:91–124.

Chiappe, L. M., and A. Vargas. 2004. Emplumando dinosaurios: la transición evolutiva de terópodos a aves. Hornero 18:1–11.

Chiappe, L. M., and C. A. Walker. 2002. Skeletal morphology and systematics of the Cretaceous Euenantiornithes (Ornithothoraces: Enantiornithes); pp. 240–267 in L. M. Chiappe and L. M. Witmer (eds.), Mesozoic Birds: Above the Heads of Dinosaurs. University of California Press, Berkeley.

Chiappe, L. M., and L. M. Witmer (eds.). 2002. Mesozoic Birds: Above the Heads of Dinosaurs. University of California Press, Berkeley, 520 pp.

Chiappe, L. M., S.-A. Ji, Q. Ji, and M. A. Norell. 1999. Anatomy and systematics of the Confuciusornithidae (Theropoda: Aves) from the Late Mesozoic of northeastern China. Bulletin of the American Museum of Natural History 242:1–89.

Clark, J. M., M. A. Norell, and L. M. Chiappe. 1999. An oviraptorid skeleton from the Late Cretaceous of Ukhaa Tolgod, Mongolia, preserved in an avianlike brooding position over an oviraptorid nest. American Museum Novitates 3265:1–36.

Clarke, J. A. 2004. Morphology, phylogenetic taxonomy, and systematics of *Ichthyornis* and *Apatornis* (Avialae: Ornithurae). Bulletin of the American Museum of Natural History 286:1–179.

Clarke, J. A., and L. M. Chiappe. 2001. A new carinate bird from the Late Cretaceous of Patagonia. American Museum Novitates 3323:1–23.

Clarke, J. A., and M. A. Norell. 2002. The morphology and phylogenetic position of *Apsaravis ukhaana* from the Late Cretaceous of Mongolia. American Museum Novitates 3387:1–46.

Clarke, J. A., C. P. Tambussi, J. I. Noriega, G. M. Erickson, and R. A. Ketchum. 2005. Definitive fossil evidence for the extant avian radiation in the Cretaceous. Nature 433:305–308.

Cooper, A., and D. Penny. 1997. Mass survival of birds across the KT-boundary: molecular evidence. Science 275:1109–1113.

Currie, P. J. 2000. Feathered dinosaurs; pp. 183–189 in G. S. Paul (ed.), The Scientific American Book of Dinosaurs. St. Martin's Press, New York.

Currie, P. J., and P. J. Chen. 2001. Anatomy of *Sinosauropteryx prima* from Liaoning, northeastern China. Canadian Journal of Earth Sciences 38:1705–1727.

Dong, Z.-M., and P. J. Currie. 1996. On the discovery of an oviraptorid skeleton on a nest of eggs at Bayan Mandahu, Inner Mongolia, People's Republic of China. Canadian Journal of Earth Sciences 33:631–636.

Dyke, G. J. 2003. The fossil record and molecular clocks: basal radiations within Neornithes; pp. 263–278 in P. Smith and P. Donoghue (eds.), Telling the Evolutionary Time: Molecular Clocks and the Fossil Record. Taylor and Francis, London.

Dyke, G. J., and G. Mayr. 1999. Did parrots exist in the Cretaceous? Nature 399:317–318.

Dyke, G. J., and M. A. Norell. 2005. *Caudipteryx* as a non-avialian theropod rather than a flightless bird. Acta Palaeontologica Polonica 50:101–106.

Dyke, G. J., and M. van Tuinen. 2004. The evolutionary radiation of modern birds (Neornithes): reconciling molecules, morphology and the fossil record. Zoological Journal of the Linnean Society 141:153–177.

Elzanowski, A. 1999. A comparison of the jaw skeleton in theropods and birds, with a description of the palate in Oviraptoridae; pp. 311–323 in S. L. Olson (ed.), Avian Paleontology at the Close of the 20th Century: Proceedings of the 4th International Meeting of The Society of Avian Paleontology and Evolution, Washington, D.C., 4–7 June 1996. Smithsonian Contributions to Paleobiology 89.

Elzanowski, A. 2001. A new genus and species for the largest specimen of *Archaeopteryx*. Acta Palaeontologica Polonica 41:519–532.

Feduccia, A. 1995. Explosive evolution in Tertiary birds and mammals. Science 267:637–638.

Feduccia, A. 1996. The Origin and Evolution of Birds. Yale University Press, New Haven, 432 pp.

Feduccia, A. 1999. The Origin and Evolution of Birds. Second edition. Yale University Press, New Haven, 480 pp.

Feduccia, A., and J. Nowicki. 2002. The hand of birds revealed by early ostrich embryos. Naturwissenschaften 98:391–393.

Feduccia, A., and R. Wild. 1993. Bird-like characters in the Triassic archosaur *Megalancosaurus*. Naturwissenschaften 80:564–566.

Forster, C., S. D. Sampson, L. M. Chiappe, and D. W. Krause. 1998. The theropodan ancestry of birds: new evidence from the Late Cretaceous of Madagascar. Science 279:1915–1919.

Galton, P. M., and L. D. Martin. 2002. *Enaliornis*, an Early Cretaceous hesperornithiform bird from England, with comments on other Hesperornithiformes; pp. 317–328 in L. M. Chiappe and L. M. Witmer (eds.), Mesozoic Birds: Above the Heads of Dinosaurs. University of California Press, Berkeley.

Gauthier, J. A. 1986. Saurischian monophyly and the origin of birds; pp. 1–55 in K. Padian (ed.), The Origin of Birds and the Evolution of Flight. Memoirs of the California Academy of Sciences 8.

Gauthier, J. A., and L. F. Gall (eds.). 2001. New Perspectives on the Origin and Early Evolution of Birds: Proceedings of the International Symposium in Honor of John H. Ostrom. Peabody Museum of Natural History, Yale University, New Haven, 613 pp.

Grellet-Tinner, G., and L. M. Chiappe. 2004. Dinosaur eggs and nesting: implications for understanding the origin of birds; pp. 185–214 in P. J. Currie, E. B. Koppelhus, M. A. Shugar, and J. L. Wright (eds.), Feathered Dragons: Studies on the Transition from Dinosaurs to Birds. Indiana University Press, Bloomington.

Grellet-Tinner, G., and M. Norell. 2002. An avian egg from the Campanian of Bayn Dzak, Mongolia. Journal of Vertebrate Paleontology 22:719–721.

Hall, B. K. 2003. Descent with modification: the unity underlying homology and homoplasy as seen through an analysis of development and evolution. Biological Reviews of the Cambridge Philosophical Society 78:409–433.

Heilmann, G. 1926. The Origin of Birds. H. F. & G. Witherby, London, 210 pp.

Hinchliffe, J. R. 1985. "One, two, three" or "two, three, four": an embryologist's view of the homologies of the digits and carpus of modern birds; pp. 141–147 in M. K. Hecht, J. H. Ostrom, G. Viohl, and P. Wellnhofer (eds.), The Beginnings of Birds. Proceedings of the International *Archaeopteryx* Conference Eichstätt 1984. Freunde des Jura-Museums Eichstätt, Eichstätt.

Holtz, T. R., Jr. 1998. A new phylogeny of the carnivorous dinosaurs. Gaia 15:5–61.

Holtz, T. R., Jr. 2001. Arctometatarsalia revisited: the problem of homoplasy in reconstructing theropod phylogeny; pp. 99–122 in J. A. Gauthier and L. F. Gall (eds.), New Perspectives on the Origin and Early Evolution of Birds: Proceedings of the International Symposium in Honor of John H. Ostrom. Peabody Museum of Natural History, Yale University, New Haven.

Hope, S. 2002. The Mesozoic radiation of Neornithes; pp. 339–388 in L. M. Chiappe and L. M. Witmer (eds.), Mesozoic Birds: Above the Heads of Dinosaurs. University of California Press, Berkeley.

Hou, L., Z. Zhou, L. D. Martin, and A. Feduccia. 1995. A beaked bird from the Jurassic of China. Nature 377:616–618.

Hou, L., Chiappe, L. M., F. Zhang, and C.-M. Chuong. 2004. New Early Cretaceous fossil from China documents a novel trophic specialization for Mesozoic birds. Naturwissenschaften 91:22–25.

Huxley, T. H. 1868. On the animals which are most nearly intermediate between birds and reptiles. Annals and Magazine of Natural History (4)2:66–75.

Hwang, S. H., M. A. Norell, Q. Ji, and K. Gao. 2004. A large compsognathid from the Early Cretaceous Yixian Formation of China. Journal of Systematic Palaeontology 2:13–30.

Jensen, J. A., and K. Padian. 1989. Small pterosaurs and dinosaurs from the Uncompahgre fauna (Brushy Basin Member, Morrison Formation: ?Tithonian), Late Jurassic, western Colorado. Journal of Paleontology 63:364–373.

Ji, Q., L. M. Chiappe, and S. Ji. 1999. A new late Mesozoic confuciusornithid bird from China. Journal of Vertebrate Paleontology 19:1–7.

Ji, Q., P. J. Currie, M. A. Norell, and S.-A. Ji. 1998. Two feathered dinosaurs from northwestern China. Nature 393:753–761.

Ji, Q., S.-A. Ji, J. C. Lü, H. L. You, W. Chen, Y. Q. Lui, and Y. X. Liu. 2005. First avialan bird from China. Geological Bulletin of China 24:197–210.

Ji, Q., S.-A. Ji, H.-L. You, J.-P. Zhang, N.-J. Zhang, C.-X. Yuan, and X.-X. Ji. 2003. An Early Cretaceous avialan bird *Shenzhouraptor sinensis* from Western Liaoning, China. Acta Geologica Sinica 77:21–26.

Ji, Q., S.-A. Ji, H.-L. You, J.-P. Zhang, C.-X. Yuan, X.-X. Ji, J.-L. Li, and Y.-X. Li. 2002a. [Discovery of an avialan bird, *Shenzhouraptor sinensis* gen. et. sp. nov., from China.] Geological Bulletin of China 21:363–369 [in Chinese with English abstract].

Ji, Q., S.-A. Ji, H.-B. Zhang, H.-L. You, J.-P. Zhang, L.-X. Wang, C.-X. Yuan, and X.-X. Ji. 2002b. A new avialan bird, *Jixiangornis orientalis* gen. et sp. nov. from the Lower Cretaceous of western Liaoning, NE China. Journal of Nanjing University (Natural Sciences) 38:723–736.

Kumar, S., and S. B. Hedges. 1997. A molecular timescale for vertebrate evolution. Nature 392:917–920.

Larsson, H. C. E., and G. P. Wagner. 2002. Pentadactyl ground state of the avian wing. Journal of Experimental Zoology 294:146–151.

Mabee, P. M. 2000. The usefulness of ontogeny in interpreting morphological characters; pp. 84–114 in J. J. Wiens (ed.), Phylogenetic Analysis of Morphological Data. Smithsonian Institution Press, Washington, D.C.

Marsh, O. C. 1880. Odontornithes: A Monograph on the Extinct Toothed Birds of North America. United States Geological Exploration of the 40th Parallel. Government Printing Office, Washington, D.C., 201 pp.

Martin, L. D. 1983. The origin and early radiation of birds; pp. 291–338 in A. H. Brush and G. A. Clark, Jr. (eds.), Perspectives in Ornithology. Cambridge University Press, Cambridge.

Martin, L. D. 1991. Mesozoic birds and the origin of birds; pp. 485–539 in H.-P. Schultze and L. Trueb (eds.), Origins of the Higher Groups of Tetrapods: Controversy and Consensus. Comstock Publishing Associates, Ithaca, N.Y.

Martin, L. D., and J. D. Stewart. 1999. Implantation and replacement of bird teeth; pp. 295–300 in S. L. Olson (ed.), Avian Paleontology at the Close of the 20th Century: Proceedings of the 4th International Meeting of the Society of Avian Paleontology and Evolution, Washington, D.C., June 4–7, 1996. Smithsonian Contributions to Paleobiology 89.

Martin, L. D., and J. Tate, Jr. 1976. The skeleton of *Baptornis advenus* (Aves: Hesperornithiformes); pp. 35–66 in S. L. Olson (ed.), Collected

Papers in Avian Paleontology Honoring the 90th Birthday of Alexander Wetmore. Smithsonian Contributions to Paleobiology 27.

Martin, L. D., J. D. Stewart, and K. N. Whetstone. 1980. The origin of birds: structure of the tarsus and teeth. Auk 97:86–93.

Mateus, I. L., H. Mateus, M. T. Antunes, O. Mateus, P. Taquet, V. Ribeiro, and G. Manuppella. 1997. Couvée, oeufs et embryons d'un Dinosaure Théropode du Jurassique supérieur de Lourinhã (Portugal). Comptes Rendus de l'Académie des Sciences, Sciences de la terre et des planètes 325:71–78.

Mikhailov, K. E. 1992. The microstructure of avian and dinosaurian eggshell: phylogenetic implications; pp. 361–373 in K. E. Campbell, Jr. (ed.), Papers in Avian Paleontology Honoring Pierce Brodkorb. Natural History Museum of Los Angeles County, Science Series 36.

Mikhailov, K. E. 1997. Avian eggshells: an atlas of scanning electron micrographs. Occasional Publications of the British Ornithologists Club 3:1–88.

Norell, M. A., and J. A. Clarke. 2001. Fossil that fills a critical gap in avian evolution. Nature 409:181–184.

Norell, M. A., J. M. Clark, and P. J. Makovicky. 2001. Phylogenetic relationships among coelurosaurian theropods; pp. 49–68 in J. A. Gauthier and L. F. Gall (eds.), New Perspectives on the Origin and Early Evolution of Birds: Proceedings of the International Symposium in Honor of John H. Ostrom. Peabody Museum of Natural History, Yale University, New Haven.

Norell, M. A., J. M. Clark, L. M. Chiappe, and D. Dashzeveg. 1995. A nesting dinosaur. Nature 378:774–776.

Norell, M. A., J. M. Clark, D. Dashzeveg, R. Barsbold, and L. M. Chiappe. 1994. A theropod dinosaur embryo, and the affinities of the Flaming Cliffs dinosaur eggs. Science 266:779–782.

Norell, M. A., Q. Ji, K. Gao, C. Yuan, Y. Zhao, L. Wang. 2002. "Modern" feathers on a non-avian dinosaur. Nature 416:36–37.

Ostrom, J. H. 1969. Osteology of *Deinonychus antirrhopus,* an unusual theropod from the Lower Cretaceous of Montana. Bulletin of the Peabody Museum of Natural History, Yale University 30:1–165.

Ostrom, J. H. 1973. The ancestry of birds. Nature 242:136.

Ostrom, J. H. 1976. *Archaeopteryx* and the origin of birds. Biological Journal of the Linnean Society 8:91–182.

Padian, K. 2003. Four-winged dinosaurs, bird precursors, or neither? Bioscience 53:450–452.

Padian, K. 2004. Basal Avialae; pp. 210–231 in D. Weishampel, P. Dodson, and H. Osmólska (eds.), The Dinosauria. Second edition. University of California Press, Berkeley.

Padian, K., and L. M. Chiappe. 1998. The origin and early evolution of birds. Biological Reviews of the Cambridge Philosophical Society 73:1–42.

Padian, K., Q. Ji, and S.-A. Ji. 2001. Feathered dinosaurs and the origin of flight; pp. 117–135 in K. Carpenter and D. Tanke (eds.), Mesozoic Vertebrate Life. Indiana University Press, Bloomington.

Paul, G. S. 1988. Predatory Dinosaurs of the World. Simon & Schuster, New York, 464 pp.

Prum, R. O. and S. Williamson. 2001. Theory of the growth and evolution of feather shape. Journal of Experimental Zoology, Part B: Molecular and Developmental Evolution 291:30–57.

Rayner, J. M. V. 1988. The evolution of vertebrate flight. Biological Journal of the Linnaean Society 34:269–287.

Rayner, J. M. V. 1991. Avian flight evolution and the problem of *Archaeopteryx*; pp. 183–212 in J. M. V. Rayner and R. J. Wooton (eds.), Biomechanics in Evolution. Cambridge University Press, Cambridge.

Rayner, J. M. V. 2001. On the origin and evolution of flapping flight aerodynamics in birds; pp. 363–385 in J. A. Gauthier and L. F. Gall (eds.), New Perspectives on the Origin and Early Evolution of Birds: Proceedings of the International Symposium in Honor of John H. Ostrom. Peabody Museum of Natural History, Yale University, New Haven.

Sato, T., Y.-N. Cheng, X.-C. Wu, D. K. Zelenitsky, and Y.-F. Hsiao. 2005. A pair of shelled eggs inside a female dinosaur. Science 308:375.

Schweitzer, M. H., F. D. Jackson, L. M. Chiappe, J. O. Calvo, and D. E. Rubilar. 2002. Late Cretaceous avian eggs with embryos from Argentina. Journal of Vertebrate Paleontology 22:191–95.

Sereno, P. C. 1999. The evolution of dinosaurs. Science 284:2137–2147.

Shapiro, M. D. 2002. Developmental morphology of limb reduction in *Hemiergis* (Squamata: Scincidae): chondrogenesis, osteogenesis, and heterochrony. Journal of Morphology 254:211–231.

Swisher, C. C., X.-L. Wang, Z.-H. Zhou, Y.-Q. Wang, F. Jin, J.-Y. Zhang, X. Xu, F.-C. Zhang, and Y. Wang. 2002. Further support for a Cretaceous age for the feathered-dinosaur beds of Liaoning, China: $^{40}Ar/^{39}Ar$ dating of the Yixian and Tuchengzi Formations. Chinese Science Bulletin (English edition) 47:135–138.

Vargas, A. O., and J. F. Fallon. 2005. Birds have dinosaur wings: the molecular evidence. Journal of Experimental Biology, Part B: Molecular and Developmental Evolution 304:86–90.

Varricchio, D. V., and F. Jackson. 2004. Two eggs sunny-side up: reproductive physiology in the dinosaur *Troodon formosus*; pp. 215–233 in P. J. Currie, E. B. Koppelhus, M. A. Shugar, and J. L. Wright (eds.), Feathered Dragons: Studies on the Transition from Dinosaurs to Birds. Indiana University Press, Bloomington.

Varricchio, D. V., F. Jackson, J. J. Borkowski, and J. Horner. 1997. Nest and egg clutches of the dinosaur *Troodon formosus* and the evolution of avian reproductive traits. Nature 385:247–250.

Wagner, G. P., and J. A. Gauthier. 1999. 1, 2, 3 = 2, 3, 4: a solution to the problem of homology of the digits of the avian hand. Proceedings of the National Academy of Sciences USA 96:5111–5116.

Wagner, G. P., and Y. Misof. 1993. How can a character be developmentally constrained despite variation in developmental pathways? Journal of Evolutionary Biology 6:449–455.

Walker, C. A. 1981. New subclass of birds from the Cretaceous of South America. Nature 292:51–53.

Weishampel, D. B., P. Dodson, and H. Osmólska (eds.). 2004. The Dinosauria. Second edition. University of California Press, Berkeley, 861 pp.

Wellnhofer, P. 1992. A new specimen of *Archaeopteryx* from the Solnhofen Limestone; pp. 3–23 in K. E. Campbell, Jr. (ed.), Papers in Avian Paleontology Honoring Pierce Brodkorb. Natural History Museum of Los Angeles County, Science Series 36.

Wellnhofer, P. 1993. Das siebte Exemplar von *Archaeopteryx* aus den Solnhofener Schichten. Archaeopteryx 11:1–48.

Wellnhofer, P., and H. Tischlinger. 2004. Das "Brustbein" von *Archaeopteryx bavarica* Wellnhofer 1993—eine Revision. Archaeopteryx 22:3–15.

Welman, J. 1995. *Euparkeria* and the origin of birds. South African Journal of Science 91:533–537.

Witmer, L. M. 1991. Perspectives on avian origins; pp. 427–466 in H.-P. Schultze and L. Trueb (eds.), Origins of the Higher Groups of Tetrapods: Controversy and Consensus. Comstock Publishing Associates, Ithaca, N.Y.

Xu, X., and M. A. Norell. 2004. A new troodontid dinosaur from China with avian-like sleeping posture. Nature 431:838–841.

Xu, X., and F. Zhang. 2005. A new maniraptoran dinosaur from China with long feathers on the metatarsus. Naturwissenschaften 92:173–177.

Xu, X., Z.-L. Tang, and X.-L. Wang. 1999a. A therizinosauroid dinosaur with integumentary structures from China. Nature 399:350–354.

Xu, X., X.-L. Wang, and X.-C. Wu. 1999b. A dromaeosaurid dinosaur with a filamentous integument from the Yixian Formation of China. Nature 401:262–266.

Xu, X., X. Zhao, and J. M. Clark. 2001. A new therizinosaur from the Lower Jurassic Lower Lufent Formation of Yunnan, China. Journal of Vertebrate Paleontology 21:477–483.

Xu, X., Z. Zhou, and X. Wang. 2000. The smallest known non-avian theropod dinosaur. Nature 408:705–708.

Xu, X., Z. Zhou, X. Wang, X. Kuang, and W. Du. 2003. Four-winged dinosaurs from China. Nature 421:335–340.

Xu, X., M. A. Norell, X. Kuang, X. Wang, Q. Zhao, and C. Jia. 2004. Basal tyrannosauroids from China and evidence for protofeathers in tyrannosauroids. Nature 431:680–684.

Zelenitsky, D. K., S. P. Modesto, and P. J. Currie. 2002. Bird-like characteristics of troodontid theropod eggshell. Cretaceous Research 23:297–305.

Zhang, F.-C., and Z.-H. Zhou. 2004. Leg feathers in an Early Cretaceous bird. Nature 431:925.

Zhang, F.-C., Z. Zhou, and L.-H. Hou. 2003. Birds; pp. 128–149 in M.-M. Chang, P.-J. Chen, Y. Q. Wang, and Y. Wang (eds.), The Jehol Biota: The Emergence of Feathered Dinosaurs, Beaked Birds and Flowering Plants. Shanghai Scientific and Technical Publishers, Shanghai.

Zhou, Z. 2004. The origin and early evolution of birds: discoveries, disputes and perspectives from the fossil record. Naturwissenschaften 91:455–471.

Zhou, Z., and X.-L. Wang. 2000. A new species of *Caudipteryx* from the Yixian Formation of Liaoning, northeast China. Vertebrata PalAsiatica 38:111–127.

Zhou, Z., and F. Zhang. 2001. Two new ornithurine birds from the Early Cretaceous of western Liaoning, China. Chinese Science Bulletin 46:1258–1264.

Zhou, Z., and F. Zhang. 2002a. Largest bird from the Early Cretaceous and its implications for the earliest avian ecological diversification. Naturwissenschaften 89:34–38.

Zhou, Z., and F. Zhang. 2002b. A long-tailed, seed-eating bird from the Early Cretaceous of China. Nature 418:405–409.

Zhou, Z., P. M. Barrett, and J. Hilton. 2003. An exceptionally preserved Lower Cretaceous terrestrial ecosystem. Nature 421:807–814.

Zhou, Z., L. M. Chiappe, and F. Zhang. 2005. Anatomy of the Early Cretaceous bird *Eoenantiornis buhleri* (Aves: Enantiornithes) from China. Canadian Journal of Earth Sciences 42:1331–1338.

Zhou, Z., F. Zhang, and J. A. Clarke. 2002. *Archaeoraptor*'s better half. Nature 420:285.

Zhou Z., X.-L. Wang, F.-C. Zhang, and X. Xu. 2000. Important features of *Caudipteryx*—evidence from two nearly complete new specimens. Vertebrata PalAsiatica 38:241–254.

9. Successive Diversifications in Early Mammalian Evolution

ZHE-XI LUO

ABSTRACT

The evolutionary history of early mammals can be characterized by successive diversifications of relatively short-lived clades, in a series of ra diations, or short episodes of lineage splitting. Most subclades of a major group are clustered into several episodes of diversification. Subclades clustered in a previous episode of diversification tend to be dead-end evolutionary experiments, often with no direct relationship to the emergent clades in the younger mammalian faunas at the order or family level. Most of the ordinal or family-level clades of Mesozoic mammals are without any Cenozoic descendants. Taxonomic turnovers between mammalian faunas of different geological epochs and within major groups are by the successive clusters of emergent clades or subclades, as shown in the transition from Early to Middle Jurassic, from Jurassic to Cretaceous, from Early to Late Cretaceous, and from Cretaceous to Paleogene. A less common macroevolutionary pattern is for a major mammalian group to maintain a relatively long history of low diversity before significant diversification of its subclades, as seen in the faunal transition from Middle to Late Jurassic. Current stratigraphic and phylogenetic data for early mammals show very few examples of rapid splits and diversifications of the subclades of a major group soon after the origination of the major group. The best available dates of the Cretaceous metatherians and eutherians do not contradict the molecular estimates that some present-day mammalian superorders may extend into the Cretaceous, but they fall short of fully corroborating the molecular hypotheses that a number of long-lived extant clades diversified during the Cretaceous Period.

Introduction

"Descent with modification" of mammals from nonmammalian cynodont ancestry is one of the best-known transitions in the evolutionary history of vertebrates and is documented by a good fossil record of nonmammalian cynodonts, mammaliaforms, and stem clades of modern mammals. New discoveries of many well-preserved fossils during the last decade have helped to uncover some new lineages with surprising combinations of features, significantly expanding the range of morphological variation of early mammals. Cladistic analyses during the last two decades have improved the resolution of early mammalian relationships, greatly increasing the number of clade ranks in Mesozoic mammal phylogeny. Thanks to these new discoveries and new analyses, great progress has been made in the understanding of the macroevolution of Mesozoic mammals. The main pattern of Mesozoic mammalian evolution is a vast evolutionary bush with numerous dead-end stem clades clustered in successive episodes of diversification (Figs. 9.1, 9.2). This is a very different picture from the prevailing paradigm of early mammalian evolution of the 1970s and earlier.

Although the mapping of anatomical evolution and documenting the diversity of fossil lineages in early mammalian history have been the domain of morphologists who rely on fossils for their inferences concerning mammalian evolution, studies of timing of early mammalian evolution have also benefited from molecular evolutionary biology and systematics. Recent advances in the study of molecular evolution have made it possible to infer the sequence and timing of divergence of the major extant mammalian lineages on a genomic scale, by using the so-called molecular clock. But inference from molecular data must be corroborated by fossil evidence, and vice versa. Estimating the divergence times of major mammalian clades requires external calibration by the fossil record of the more inclusive amniote groups (Kumar and Hedges, 1998; Benton, 1999; Graur and Martin, 2004; Reisz and Müller, 2004). Estimating the divergence times of modern marsupial and placental groups would require external calibration by the earliest fossils of the marsupial and placental lineages (Ji et al., 2002; Luo et al., 2003; Nilsson et al., 2004). Many recent molecular studies postulated much earlier divergence times of major placental and marsupial groups than can be supported by the currently available fossil records of these respective groups. To some, this discrepancy may indicate the underestimation of the true mammalian history by the poor mammalian fossil record (Kumar and Hedges, 1998; Murphy et al., 2001; Springer et al., 2003, 2004). To others, this may simply reflect incorrect assumptions of molecular clock studies, such as constant rates of molecular evolution (Foote et al., 1999). Some molecular evolutionists have pointed out that molecular estimates of the placental evolution should use multiple calibration points when fossil records are available and appropriate for calibration, to account for the heterogeneity of evolutionary rates among lineages under consideration (Douzery et al., 2003, 2004).

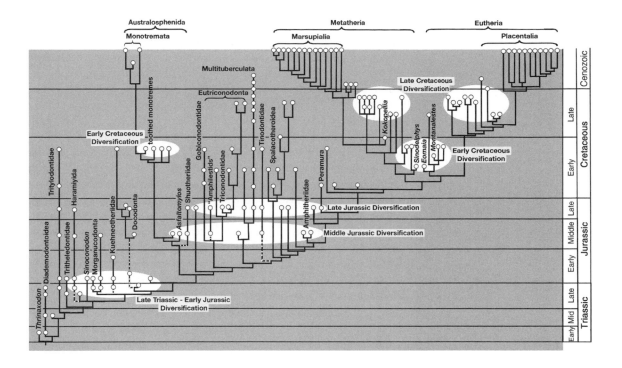

Figure 9.1. Temporal distribution of mammaliamorphs, mammaliaforms, and crown Mammalia and their stem clades, as mapped on the mammaliaform phylogeny (modified from Kielan-Jaworowska et al. [2004], with additional data from the following sources: Luo et al., 2003; Martin and Averianov 2004; Luo and Wible, 2005). The family tree of mammaliaforms is a vast evolutionary bush, enriched by many new fossil lineages from recent discoveries, and many additional clade ranks revealed by the phylogenetic analyses of the last decade. Upon mapping the stratigraphic distributions of Mesozoic mammaliamorph taxa on the current phylogeny, the general pattern of early mammal macroevolution is manifest in several successive episodes of diversification of relatively short-lived clades (Cifelli, 2001; Luo et al., 2003; Kielan-Jaworowska et al., 2004; Luo and Wible, 2005). From this perspective, most clades during a previous episode of diversification are dead-end evolutionary experiments and may not be directly related at the order or family level to the emergent clades of the following episodes of diversification; early mammalian evolution can be characterized as having a series of explosive events with modest diversity. This view is in contrast to the traditional, pre-1980s view that the divergence of prototherians (monotremes and their putative relatives) and therians (marsupials and placentals, and their putative relatives) could be traced to the Late Triassic. It is quite different from the traditional Simpsonian view of the 1940s and 1950s in which multiple long-branch lineages of the Cenozoic mammals, such as multituberculates, the lineage including extant monotremes, and the lineage including extant therians, had originated from separate stocks of therapsids in the Early Triassic. The Simpsonian view is essentially a short-fuse model, postulating that main Cretaceous and Cenozoic mammalian lineages diversified very early in a short time window during the therapsid-mammalian history and subclades of mammals maintained long and separate histories. Diversities of multituberculates, dryolestoids, and basal boreosphenidans are underrepresented because of lack of space in this figure.

New evidence of a much greater diversity of extinct Mesozoic mammalian lineages and the much earlier molecular estimates for the diversification of the superordinal groups of extant mammals have necessitated a reevaluation of the timing and mode of mammalian diversification in the Mesozoic. Recently, Archibald and Deutschman (2001) have proposed three models for the diversification of the placental superordinal groups. (1) The "short-fuse" model from recent molecular studies postulates that placental superordinal groups diversified shortly after the origin of placentals as a group in the Early Cretaceous. Under the short-fuse model, placental superordinal lineages would have long and separate evolutionary histories that can be traced to the origin of the entire placental clade. (2) The "explosive" model favored by some paleontologists postulates that most of the known Late Cretaceous eutherians have no direct relations to Cenozoic placental superordinal groups, which rapidly diversified after the Cretaceous-Cenozoic boundary. Although interrelated as basal members of the eutherian-placental clade, the Late Cretaceous stem eutherians and the early Cenozoic placentals represent different episodes of diversification. Most taxa of the Late Cretaceous episode of diversification represent dead-end evolutionary experiments, and would have neither a direct nor close relationship to the Cenozoic placental superordinal groups. (3) The "long-fuse" model postulates that some Late Cretaceous eutherians are nested within placental superordinal groups; but the placental superordinal lineages have relatively long histories with low diversity, until they have the first chance to diversify after the extinction of dinosaurs at the end of the Cretaceous.

These conceptual models of evolutionary diversification can be extrapolated to the entire Mesozoic mammalian evolution. The short-fuse model posits that the origin of a major clade is immediately followed by the rapid diversification of its subclades, and most subclades would have a very long history that can be traced to the origin of the main clade. This model can be corroborated if many earliest-known stem taxa of a major clade can be reliably placed within each of its subclades. The explosive model posits that the older clades of a previous episode of diversification represent mostly evolutionary dead ends and may not be directly related to the younger clades in the subsequent episode of diversification. This model would be corroborated if the older subclades of an earlier diversification have turned out to be the stem taxa and were not directly related to the subclades of a later diversification in the same lineage. History of a lineage would be best characterized as successive waves of diversifications. The long-fuse model postulates that after the origin of a major clade, its subclades maintained a relatively long history of low diversity, and diversification does not occur until long after each subclade had originated. This model would gain support if early stem taxa are relatively few but can be reliably assigned to each of the subclades with a long history. The timing and mode of diversifications of early mammals can be characterized in terms of these models if their temporal distribution (Fig. 9.1) can be mapped on a well-resolved phylogeny (Fig. 9.2).

Shifting Paradigm of Early Mammalian Evolution

The phylogenetic tree of Mesozoic mammals, as supported by the current evidence, is a vast evolutionary bush (Fig. 9.1). Its richness, with a great many relatively short-lived clades of Mesozoic mammals, represents a conceptual shift from the dominant paradigm before the 1970s, in which a limited number of long-lived lineages extended back in parallel from the Cenozoic into the Triassic (or even the Permian), established by the classic work by G. G. Simpson in the 1920s. Simpson (1928, 1929) and Olson (1944) proposed that mammals of the Cretaceous and Cenozoic were "polyphyletic" and had deep roots in therapsid groups of the Triassic to the Late Permian. For example, the multituberculate mammals of the Late Jurassic to early Cenozoic were linear descendants of the tritylodontid cynodonts of the Late Triassic, and triconodontids of the Late Jurassic were linear descendants of basal cynodonts such as *Thrinaxodon*. Extant monotremes were also derived from an independent stock of therapsid ancestry that is separate from the Triassic ancestor of the extant therians. Thus Cretaceous and Cenozoic mammals as a whole would have at least four parallel lineages extending into the Early Triassic, or even the Permian. This view was embraced and expounded by many subsequent studies (Patterson, 1956; Olson, 1959; Van Valen, 1960). By using the terminology of our current discussion on mammalian origins, the entire therapsid-mammalian evolution would be a gigantic short fuse, in which the major Cretaceous and Cenozoic mammalian groups (i.e., monotremes, triconodonts, multituberculates, and therians) became established relatively quickly in therapsid-mammalian evolution, no later than the Late Triassic and as early as the Permian.

Fossils discovered in the 1950s and 1960s showed that the earliest-known mammals from the Late Triassic and Early Jurassic share several modern mammalian characters, and this led to a new view that mammals form a monophyletic group (Hopson and Crompton, 1969). The origination of Mammalia is neither polyphyletic nor as early as postulated by Simpson (1928) and Olson (1944). Nonetheless, it was widely accepted that the two extant mammalian groups, monotremes and therians, extended their separate lineages into the Late Triassic (Kermack and Mussett, 1958; Hopson, 1964; Hopson and Crompton, 1969; Kermack and Kielan-Jaworowska, 1971; Crompton and Jenkins, 1979). The prototherian lineage (i.e., monotremes) appeared in the Late Triassic with its earliest-known representatives *Morganucodon* and haramiyids. The therian lineage (i.e., marsupials and placentals) appeared in the Late Triassic with its earliest-known representative *Kuehneotherium*. Differing from Simpson's hypothesis of the polyphyletic origination of mammals in regarding Mammalia as monophyletic, the prevailing view in the 1960s and 1970s was that extant mammals consist of two deep lineages (or long branches) that can be traced into the

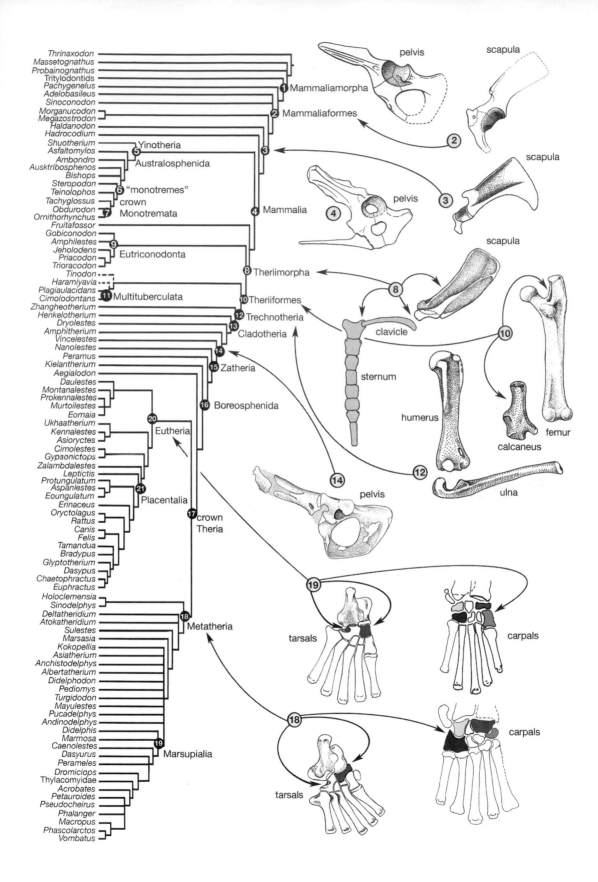

Thrinaxodon
Massetognathus
Probainognathus
Tritylodontids
Pachygenelus
Adelobasileus
Sinoconodon

1 Mammaliamorpha

Morganucodon
Megazostrodon
Haldanodon
Hadrocodium

2 Mammaliaformes

Shuotherium
Asfaltomylos **5** Yinotheria
Ambondro Australosphenida
Ausktribosphenos
Bishops
Steropodon
Teinolophos **6** "monotremes"
Tachyglossus crown
Obdurodon **7** Monotremata
Ornithorhynchus

3

4 Mammalia

Fruitafossor
Gobiconodon
Amphilestes **9**
Jeholodens Eutriconodonta
Priacodon
Trioracodon

8 Theriimorpha

Tinodon
Haramiyavia
Plagiaulacidans
Cimolodontans **11** Multituberculata

10 Theriiformes

Zhangheotherium
Henkelotherium **12** Trechnotheria
Dryolestes
Amphitherium **13** Cladotheria
Vincelestes
Nanolestes
Peramus **14**
Kielantherium **15** Zatheria
Aegialodon

Daulestes
Montanalestes
Prokennalestes
Murtoilestes
Eomaia
Ukhaatherium **20**
Kennalestes Eutheria
Asioryctes
Cimolestes
Gypsonictops
Zalambdalestes
Leptictis
Protungulatum
Aspanlestes **21**
Eoungulatum Placentalia
Erinaceus
Oryctolagus
Rattus
Canis
Felis
Tamandua
Bradypus
Glyptotherium
Dasypus
Chaetophractus
Euphractus

16 Boreosphenida

17 crown Theria

Holoclemensia
Sinodelphys
Deltatheridium
Atokatheridium
Sulestes **18** Metatheria
Marsasia
Kokopellia
Asiatherium
Anchistodelphys
Albertatherium
Didelphodon
Pediomys
Turgidodon
Mayulestes
Pucadelphys
Andinodelphys
Didelphis
Marmosa
Caenolestes **19**
Dasyurus Marsupialia
Perameles
Dromiciops
Thylacomyidae
Acrobates
Petauroides
Pseudocheirus
Phalanger
Macropus
Phascolarctos
Vombatus

pelvis
scapula
2
scapula
3
pelvis
4
Theriimorpha
8
scapula
clavicle
10
humerus
calcaneus
femur
sternum
12
ulna
14 pelvis
19
tarsals carpals
carpals
18
tarsals

Figure 9.2. *(opposite page)* Phylogeny of mammaliaforms with definition of major clades and their postcranial apomorphies (topology from Luo and Wible, 2005). Node (**1**) Mammaliamorpha (Rowe [1988] as modified by Kemp [2005]: common ancestor of tritylodontids, tritheledontids, brasilodontids, possibly also dromatheriids, and mammaliaforms, inclusive of taxa closer to Mammaliaformes than to diademodontoids, chiniquodontoids, and probainognathids): anterior orientation of ilium, as result of elongation of the anterior process and reduction of the posterior process of ilium, posterior rotation of pubis behind the level of acetabulum (reconstructed pelvis of *Oligokyphus* from Kühne, 1956). Node (**2**) Mammaliaformes (Rowe [1988] as modified by McKenna and Bell [1997] and Luo et al. [2001b]: common ancestor of *Sinoconodon* and crown Mammalia, inclusive of taxa closer to crown mammals than to tritylodontids, tritheledontids, brasilodontids, and dromatheriids): saddle-shaped glenoid fossa for the shoulder joint and more elevated scapular spine (partial shoulder girdle of morganucodontans from Jenkins and Parrington, 1976). Node (**3**) unnamed clade (Luo et al., 2002; Martin, 2005: common ancestor of Docodonta and crown Mammalia; to the exclusion of *Sinconodon,* morganucodontans): loss of procoracoid and the coracoid foramen, reduced coracoid process (scapula of *Haldanodon* from Martin, 2005). Node (**4**) Mammalia (Rowe, 1988, 1993: common ancestor of extant mammals and all fossil taxa nested within extant mammals): presence of epipubis, full acetabular rim (reversed in eutriconodontans) (pelvis of *Ornithorhynchus* from Jenkins and Parrington, 1976). Node (**5**) Yinotheria (Kielan-Jaworowska et al., 2004: ancestor of *Shuotherium* and monotremes). Node (**6**) "Monotremes" (Archer et al., 1985; Kielan-Jaworowska et al., 2004: ancestor of *Steropodon* and extant monotremes). Node (**7**) crown Monotremata (ancestor of extant monotremes). Node (**8**) Theriimorpha (Rowe [1988] as modified by Luo and Wible [2005]: common ancestor of eutriconodontans and crown Mammalia, inclusive of fossil taxa closer to crown Theria than to *Fruitafossor*): full scapular spine, broad and laterally facing supraspinous fossa, mobile acromio-clavicle joint, mobile clavicle-interclavicle joint (restoration of shoulder girdle of *Jeholodens* from Ji et al., 1999). Node (**9**) Eutriconodonta (Kielan-Jaworowska et al., 2004: ancestor of *Triconodon* and *Amphilestes,* inclusive of all taxa nested within this clade). Node (**10**) Theriiformes (Rowe, 1988: common ancestors of Multituberculata and crown Mammalia, possibly to the exclusion of haramiyidans): offset femoral neck and spherical femoral head, tall vertical greater trochanter of the femur, absence of broad peroneal shelf and presence of laterally compressed calcaneal tuber (calcaneus and femur of multituberculates from Kielan-Jaworowska et al., 2004; dashed lines represent uncertainties in phylogenetic interpretations of haramiyidans; Jenkins et al., 1997; Luo et al., 2002). Node (**11**) Multituberculata (sensu stricto, Kielan-Jaworowska et al., 2004). Node (**12**) Trechnotheria (McKenna [1975] as modified by Prothero [1981] and Luo et al. [2002]: ancestor of spalacotherioids and crown Theria): incipient ulnar trochlea on humerus and elevated coronal process of ulna (humerus and ulna of *Zhangheotherium,* composite restoration from two specimens). Node (**13**) Cladotheria (McKenna [1975] as modified by Prothero [1981]: ancestor of Dryolestoida and crown Theria). Node (**14**) "Prototribosphenidans" (Rougier et al., 1996: ancestor of *Vincelestes* and crown Theria): gracile pubis and enlarged obdurator foramen (pelvis of *Vincelestes* from Rougier, 1993). Node (**15**) Zatheria (McKenna [1975] as modified by Martin [2002]: ancestor of *Nannolestes* and crown Theria). Node (**17**) crown Theria (ancestor of extant marsupials and placentals). Node (**18**) Metatheria (sensu Luo et al. [2003]: ancestor of *Sinodelphys* and extant marsupials, inclusive of fossil taxa closer to marsupials than to placentals): hypertrophied hamate, triquetrum, and scaphoid in carpals; broad head and laterally spread navicular facet on astragalus, oblique calcaneo-cuboid joint in the tarsals. Node (**19**) Marsupialia (extant marsupials plus fossil forms nested within them; constituent taxa may vary according to different authors). Node (**20**) Eutheria (Ji et al., 2002: ancestor of *Eomaia* and extant placentals, inclusive of fossil taxa closer to placentals than to marsupials): large and oblong trapezium in carpals; broad sustentacular process of calcaneus, restricted astragalar head, tibial crest in upper ankle joint, and narrow navicular in tarsals. Node (**21**) Placentalia (extant placentals, plus fossil forms nested within them; constituent taxa may vary according to different authors). Illustrations are not to scale.

Late Triassic. However, this view shares a common theme with the earlier Simpsonian paradigm of mammalian evolution: that modern mammalian lineages had deep roots extending back into the Late Triassic, if not farther.

The hypothesis that parallel and long-lived lineages including extant mammals could be extended into the Triassic was primarily based on similarities in molars between the respective mammalian groups of the Cretaceous and Cenozoic and their putative Triassic ancestors. But these similarities were later shown to be primitive and therefore phylogenetically uninformative (Kemp, 1983). Because the dental characters are in conflict with nondental morphological features, they must be interpreted as homoplasies in parsimony context of all morphological evidence (Jenkins et al., 1997; Luo et al., 2002; Kielan-Jaworowska et al., 2004).

One of the suggested long lineages is that of triconodontid-*Morganucodon-Thrinaxodon*. The aligned tricuspid cusps on the teeth of Cretaceous triconodontids are similar to those of *Morganucodon;* and the triconodont-like molar pattern of *Morganucodon* bears resemblance to that of the Early Triassic cynodont *Thrinaxodon.* Thus the *Thrinaxodon-Morganucodon*-triconodontid lineage was regarded to be successive stages in a linear evolutionary descent defined by this type of teeth (Hopson and Crompton, 1969; Crompton, 1974). Kemp (1983) first pointed out that straight alignment of the three main cusps on the molar in *Morganucodon* is primitive, and therefore not informative concerning the relationship of *Morganucodon* to other cynodont or mammalian groups with similar teeth. When skeletons of eutriconodontans were discovered (Jenkins and Schaff, 1988; Ji et al., 1999; Hu et al., 2005), it was obvious that their postcranial skeletons were so much more derived than those of morganucodontans (see Fig. 9.2, nodes 2 and 3) that eutriconodontans and morganucodontans must be placed on different hierarchical ranks of the mammalian phylogeny and can no longer be classified into the same group (Luo et al., 2002; Kielan-Jaworowska et al., 2004).

The second of the postulated long lineages is that of multituberculates—tritylodontids. Before the 1930s, when multituberculates and tritylodontids were known primarily from their teeth, both were grouped into one lineage because of their postcanine teeth with multiple rows of cusps (Simpson, 1928). But again, when the cranial and postcranial remains of multituberculates and tritylodontids were discovered, it became abundantly clear that these two groups are not closely related (Simpson, 1937; Young, 1947; Kühne, 1956; Hopson, 1964; Krause and Jenkins, 1983; Sues, 1983, 1986; Sues and Jenkins, 2006). Multituberculates sensu stricto are now considered part of the mammalian crown group by most authors (Rowe, 1988; Luo et al., 2002; Kielan-Jaworowska et al., 2004), whereas tritylodontids are now either a part of the mammaliamorphs (Kemp, 1982, 1983, 2005; Rowe, 1988, 1993; Luo et al., 2002), or part of the cynodont group Gomphodontia (Sues, 1985; Hopson and Barghusen, 1986; Hopson and Kitching, 2001). In either case, tritylodontids would be outside mammaliaforms. Tritylodontids are no more closely related to mammals than other derived cynodonts, such as tritheledontids, brasilodontids (Bonaparte et al., 2003), and dromatheriids (Hahn et al., 1994; Sues, 2001).

The triangular cusp pattern of molars in the Late Triassic *Kuehneotherium* is a key feature for linking it to the more derived therians, including marsupials and placentals (Kermack et al., 1965; Crompton, 1971; Hopson, 1994). This emphasis on the cusp pattern in *Kuehneotherium* and its resemblance to that of extant therians was a key argument in maintaining a fundamental division of "prototherians" versus "therians," beginning in the Late Triassic. But *Kuehneotherium* has a very primitive mandible with many plesiomorphic features of nonmammalian cynodonts and other mammaliaforms, and it lacks many derived mandibular features of the more derived therian mammals. The triangular molar cusp pattern in *Kuehneotherium* is now considered a homoplastic feature because it is in conflict with too many nondental characters (Rougier et al., 1996; Luo et al., 2002; Kielan-Jaworowska et al., 2004).

This paradigm of long-lived extant mammalian lineages with direct Triassic ancestors was abandoned as a growing body of evidence overwhelmingly rejected the Triassic mammaliaforms or mammaliamorphs as the respective precursors to the more derived eutriconodontans, multituberculates, and therians. These "dental" lineages with Triassic ancestors, commonly accepted before the 1980s, have collapsed with the advent of phylogenetic analyses drawing on the large amount of morphological data from the skulls and postcranial skeletons that would become available after the 1980s (Cifelli, 2001; Luo et al., 2002; Kielan-Jaworowska et al., 2004). Instead of a limited number of extant mammalian lineages with long history into the Triassic, today's phylogeny of Mesozoic mammals is an enormous and exuberant bush of many shorter and dead-end branches, grouped in successive clusters in chronological sequence. Most ordinal-level clades in the Mesozoic do not have a linear ancestor-descendant relationship to Cenozoic mammalian orders or families (Fig. 9.1).

This shift in the topology of the early mammalian tree is necessitated by new discoveries of "order-" or "family-level" lineages that simply cannot be accommodated by the classic, pre-1970s morphological schemes and classifications of Mesozoic mammals (e.g., Simpson, 1945; Patterson, 1956). Several lineages discovered after the 1980s are entirely new and do not fit into the dental morphotypes of Mesozoic mammals previously known to science, including *Shuotherium* (Chow and Rich, 1982; Sigogneau-Russell, 1998; Wang et al., 1998), gondwanatherians (Scillato-Yané and Pascual, 1985; Bonaparte, 1986, 1990; Krause et al., 1997; Pascual et al., 1999), *Kollikodon* (Flannery et al., 1995; Kielan-Jaworowska et al., 2004), theroteinidans (Hahn et al., 1989; Kermack et al., 1998), *Tikitherium* (Datta, 2005), and *Fruitafossor* (Luo and Wible, 2005). Other newly discovered taxa have shown some surprising combinations of morphological features. The Early Jurassic *Hadrocodium* is similar to many morganucodontans and *Sinoconodon* in molar morphology, but is far more advanced in the structure of the cranium and the mandible (Luo et al., 2001b). The Cretaceous toothed monotreme *Teinolophos* retains a postdentary trough that can accommodate the middle ear bones on the mandible, but its molars are far more derived than those of any other mammals with a postdentary trough (Rich et al., 2001b, 2005) and bear resemblance to the Cenozoic toothed ornithorhynchid monotremes, in which the mandible has no connection to the middle ear bones (Musser and Archer, 1998). The enrichment of diversity also occurred at lower taxonomic levels. As the result of decades of field exploration, many new genera and species within existing families have been discovered; by 1980, there were about 100 genera of Mesozoic mammals known to science (Lillegraven et al., 1979), but by 2004 there were more than 280 recognized genera (Kielan-Jaworowska et al., 2004).

Ranks of clades of Mesozoic mammals have also multiplied by the increasingly fine resolution of mammalian genealogy. With the progress in phylogenetics in the last two decades, relationships of Mesozoic mammals have become much better resolved (McKenna, 1975; Prothero, 1981; Kemp, 1983; Rowe, 1988, 1993; Wible and Hopson, 1993; Luo, 1994; Rougier et al., 1996, 1998; Luo et al., 2001a,b, 2002, 2003; Ji et al., 2002; Luo and Wible, 2005). These analyses have transformed several traditional groups, formerly held together by nothing more than poorly defined and primitive features, into a succession of resolved clades. For example, symmetrodonts and eupantotheres in Simpson's (1928, 1929, 1945) and Patterson's (1956) classifications have now been broken into multiple ranks of clades by the pioneering studies of McKenna (1975) and Prothero (1981). Another example of a heterogeneous traditional group that has now been resolved into multiple ranks of clades is "Triconodonta" (sensu Kermack et al., 1973). This broadly defined grouping used to include *Sinoconodon*, morganucodontans, docodontans, amphilestids, and triconodontids. "Triconodonts" are now separated into *Sinoconodon*, morganucodontans, and docodontans (mammaliaforms outside crown Mammalia), versus amphilestids and triconodontids (now placed within crown Mammalia). The amphilestids and triconodontids form a monophyletic group of eutriconodontans (sensu Kielan-Jaworowska et al., 2004). Finer resolution of Mesozoic mammal phylogeny contributed to a proliferation of numerous smaller clades at the family level.

Contrary to a limited number of extant branches with long histories extending back into the Triassic, today's much improved phylogeny of Mesozoic mammals is an enormous and exuberant bush of many relatively short-lived branches in successive clusters. Most of the short-lived branches of an early episode of diversification would represent some dead-end evolutionary experiments, without direct ancestor-descendant relationship to the newly emergent branches of a later episode of diversification. In fact, most Mesozoic mammalian clades have no direct descendants among Cenozoic mammals at the familial, ordinal, and superordinal levels (Fig. 9.1). This view of evolutionary diversity of Mesozoic mammals postulates that the main feature of mammaliamorph macroevolution is successive episodes of diversification, or a series of explosive events with modest taxonomic diversity. Taxonomic diversification tends to appear in clusters of many small and basal branches in the history of each major clade (Cifelli, 2001; Luo et al., 2002; Kielan-Jaworowska et al., 2004).

Mammalia (crown mammals) is a monophyletic group—here defined as the common ancestor of extant monotremes (*Ornithorhynchus anatinus* and *Tachyglossus aculeatus*), extant Marsupialia (*Didelphis virginiana* and *Macropus giganteus*), and extant Placentalia (*Homo sapiens* and *Macroscelides proboscideus*), plus all Mesozoic and Cenozoic fossil mammals that are nested within the combined group of monotremes, marsupials, and placentals by the consensus of phylogenetic analyses (e.g., Rowe, 1993; Luo et al., 2002; Kielan-Jaworowska et al., 2004; Luo and Wible, 2005). To the best of our knowledge, all of the Late Triassic to Early Jurassic relatives of mammals are placed outside Mammalia; these taxa are thus considered to be premammalian mammaliaforms (taxa below node 4 of Fig. 9.2; see Rowe, 1993).

On the bases of the fully resolved phylogenies (Fig. 9.2) and on the improved record of the temporal distribution of families and genera (Fig. 9.1; see Kielan-Jaworowska et al., 2004), Mesozoic mammal lineages clearly show several successive episodes of diversifications. The first such episode is the Late Triassic–Early Jurassic diversification of mammaliaforms.

Late Triassic–Early Jurassic Diversification of Mammaliaforms

Mammaliaformes is defined as the common ancestor of *Sinoconodon* and extant mammals (Rowe, 1988; McKenna and Bell, 1997). The clade so defined is synonymous with Mammalia as traditionally defined (Crompton and Sun, 1985; Hopson, 1994; Luo et al., 2002; Kielan-Jaworowska et al., 2004; Kemp, 2005). The basal clades of mammaliaforms are *Sinoconodon*, morganucodontans, kuehneotheriids, haramiyidans, and docodontans if including *Woutersia, Delsatia,* and an unnamed form from the Early Jurassic of India (Butler, 1997; Sigogneau-Russell and Godefroit, 1997; Prasad and Manhas, 2001; Sigogneau-Russell, 2003a). *Adelobasileus* may also belong to this cluster of clades if its mammaliaform affinities suggested by the derived braincase characters (Lucas and Luo, 1993) can be corroborated by additional evidence. The Early Jurassic *Hadrocodium* is a derived mammaliaform with a relatively enlarged braincase, absence of the primitive feature of the pila antotica, and loss of the primitive postdentary trough for the attachment of the middle ear bones (Luo et al., 2001b). The emergent lineages in the diversification during the Late Triassic to Early Jurassic are not closely related among themselves (Fig. 9.1). Although their mandibles are more or less similar in structure, their dentitions are strikingly different from one another because they are specialized for different feeding adaptations. In recent phylogenies, they are resolved into a series of groups that are related to the crown Mammalia in the successive order of *Sinoconodon,* Morganucodonta, Docodonta, and possibly kuehneotheriids. *Hadrocodium* is the closest to the crown mammals among all Late Triassic–Early Jurassic lineages. Haramiyidans are not closely related to the crown mammalian group Multituberculata according to some authors (Jenkins et al., 1997; Luo et al., 2002), but this is not universally accepted (Butler, 2000; Butler and Hooker, 2005).

The evolution of mammaliaforms is an example of an explosive model. Their diversification bursts in a relatively short window from Late Triassic through Early Jurassic, and no individual clades within mammaliaforms can be related to an individual family of the nonmammalian cynodonts even though mammaliaforms as a whole are rooted in the advanced cynodonts (epicynodonts of Hopson and Kitching, 2001; Kemp, 1982, 2005). The emergent lineages of these earliest-known mammaliaforms were also relatively short-lived and represented some dead-end evolutionary experiments, which had no descendants in later mammalian faunas. All these earliest mammaliaforms lacked many derived mandibular features and retained a fully functional jaw joint between the articular and quadrate as in nonmammalian cynodonts and extant nonmammalian amniotes (Kermack et al., 1973; Luo and Crompton, 1994). The earliest mammaliaforms also lacked some derived petrosal features of Late Jurassic mammals (Rougier et al., 1996). In the absence of many of the derived features of the crown mammals, they cannot be placed within them. Even though these mammaliaforms are related to crown mammals as a whole, they cannot be linked, respectively, to any orders or families within the crown Mammalia. With exception of docodontans, these mammaliaforms did not extend beyond the Middle Jurassic, and none of them gave rise to new ordinal level or family-level clades of the Late Jurassic or Cretaceous mammals.

The emergence of the earliest mammaliaforms was accompanied by the evolution of several osteological and dental apomorphies. The most important apomorphy of basal mammaliaforms is the temporomandibular joint consisting of a fully developed dentary condyle in articulation with a fully developed squamosal glenoid, for a more powerful and better-controlled mastication. Other apomorphies include formation of the entire bony housing for the inner ear by a single bone, the petrosal, and development of a petrosal promontorium to accommodate an elongate cochlear canal for hearing. The braincase shows enlargement in the frontal and parietal region and has an orbitosphenoid wall, providing a complete bony enclosure for the enlarged brain (Kermack et al., 1981; Crompton and Luo, 1993). Another major diagnostic feature is the enlarged anterior lamina of the petrosal that forms the lateral wall of the braincase and encloses the exit of the trigeminal nerve. Some, but not all, of these mammaliaforms have also developed diphyodont dental replacement accompanied by a determinate pattern of skull growth (Luo, 1994; Kielan-Jaworowska et al., 2004; Luo et al., 2004). Most mammaliaforms also have precise molar occlusion with wear facets on individual molar cusps, despite markedly diverse patterns of cusps in various mammaliaform groups (reviewed by Kermack and Kermack, 1984; Sigogneau-Russell, 1989a; Jenkins et al., 1997; Butler, 2000; Kielan-Jaworowska et al., 2004; Kemp, 2005).

Records of the most derived nonmammalian cynodonts and basal mammaliaforms are now good enough to show the pattern of their morphological evolution. The diagnostic features for mammaliaforms did not appear together in the fossil record (Luo, 1994; Sidor and Hopson, 1998; Luo et al., 2001b; Kielan-Jaworowska et al., 2004). It is now clear that the evolution of the temporomandibular joint formed by a dentary condyle and squamosal glenoid preceded development of the diphyodont dentition.

Many diagnostic characters of mammaliaforms have precursor conditions in the nonmammalian cynodonts. For example, a fully ossified pars cochlearis, the petrosal component enclosing the cochlear canal of the inner ear and forming the promontorium, was already present in tritylodontids and tritheledontids, although it was not yet fully exposed on the ventral side of the basicranium (Luo, 2001). The tritylodontid and tritheledontid condition represents an intermediate condition between basal cynodonts on the one hand, and the fully enlarged and exposed promontorium in more derived mammaliaforms on the other. Among tritylodontids, *Sinoconodon*, *Morganucodon*, and *Hadrocodium*, there is clearly a morphological gradient toward a longer cochlear canal of the inner ear in the more derived taxa. The mammalian temporomandibular joint also has a precursor condition in nonmammalian cynodonts. The contact between dentary and squamosal is present in tritheledontids, a group considered the sister-taxon to mammaliaforms by many authors, even though the dentary does not have a proper condyle and the squamosal has no distinctive glenoid for this joint. The fully formed orbital wall for the braincase, a derived condition otherwise only known in mammaliaforms, is also present in tritylodontids (Sues, 1986), but not in tritheledontids (Luo, 1994). Kielan-Jaworowska et al. (2004: figs. 3.1–3.25) suggest that most diagnostic features of crown Mammalia have some precursor conditions in mammaliaforms and nonmammalian cynodonts. Their evolution can be shown to have stepwise transformation on the phylogenetic tree (Luo et al., 2001b; Kielan-Jaworowska et al., 2004; Kemp, 2005).

The minimum age of Mammaliaformes is the late Carnian, some 225 million years ago (Table 9.1), as based on *Adelobasileus* (Lucas and Luo, 1993), the earliest-known morganucodontan *Gondwanadon* (Datta and Das, 1996), and *Tikitherium*, an enigmatic mammaliaform from India (Datta, 2005). The earliest mammaliaforms are known from just a few fossils, and their records are not widespread enough for paleobiogeographic inference. For example, *Sinoconodon* and *Hadrocodium* are only known from China

TABLE 9.1

Hierarchical ranks and minimal ages of emergent clades during the successive diversifications of the Mesozoic mammals.*

Geological period		Emergent clade†	Minimal date	Earliest-known taxon	Stage†
Cretaceous	Late	Crown Marsupialia (Didelphis+Phascolarctos)	≥70 mya?	?Glasbius‡	?Maastrichtian
		Crown Placentalia (Elephantulus+Dasypus+Homo)	≥90 mya?	?Kulbeckia/Zhelestes§	?Turonian
	Early	Monotremata (Teinolophos+Tachyglossus)	≥110 mya	Teinolophos/Steropodon	Albian
		Metatheria (Sinodelphys+Phascolarctos)	≥125 mya	Sinodelphys	Middle Barremian
		Eutheria (Eomaia+Homo)	≥125 mya	Eomaia	Middle Barremian
		Crown Theria (Phascolarctos+Homo)	≥125 mya	Eomaia/Sinodelphys	Middle Barremian
		Boreosphenida (Kielantherium+crown Theria)	≥140 mya	Aegialodon/Tribactonodon	Berrasian
Jurassic	Late	Zatheria (Nanolestes/Peramus+crown Theria)	≥151 mya	Nanolestes	Kimmeridgian
	Middle	Cladotheria (Dryolestes+crown Theria)	≥164 mya	Amphitherium/Palaeoxonodon	Late Bathonian
		Trechnotheria (Spalacotherium+crown Theria)	≥164 mya	Ghost lineage range	Late Bathonian
		Theriiformes (Multituberculata+crown Theria)	≥164 mya	Hahnotherium/Kermackodon	Late Bathonian
		Theriimorpha (Eutriconodonta+crown Theria)	≥167 mya	Amphilestes/Phascolotherium	Early Bathonian
		Australosphenida (Asfaltomylos+Monotremata)	≥164 mya	Ambondro/Asfaltomylos	≥late Bathonian
		Yinotheria (Shuotherium+Monotremata)	≥164 mya	Shuotherium kermacki	≥late Bathonian
		Mammalia (Monotremata+crown Theria)	≥167 mya	Amphilestes/Phascolotherium	Early Bathonian
	Early	Unnamed clade (Hadrocodium+Mammalia)	≥195 mya	Hadrocodium	Hettangian

Triassic	Late	Unnamed clade (Docodonta + Mammalia)	≥220 mya	*Delsatia*	Early Rhaetian
		Unnamed clade (*Morganucodon* + Mammalia)	≥220 mya	*Morganucodon/Brachyzostrodon*	Early Rhaetian
		Mammaliaformes (*Sinoconodon* + Mammalia)	≥225 mya	*Adelobasileus*	Late Carnian
		Mammaliamorpha	≥225 mya	*Riograndia/Chaliminia*	Early Carnian
	Middle				

* Ages of the earliest-known taxa of emergent clades are from Kielan-Jaworowska et al. (2004).

† Major clades are also defined by nodes in Figure 9.2.

‡ The putative Late Cretaceous age of crown marsupials is contingent on the placement of several Late Cretaceous North American metatherian genera within crown marsupials (e.g., Marshall et al., 1990). More recent cladistic analyses including these taxa have excluded most of the North American Late Cretaceous taxa from the crown marsupials (Rougier et al., 1998; Luo et al., 2003), although the interrelationship of *Glasbius* (Clemens, 1966) to the marsupial lineage Paucituberculata has not been tested (Marshall, 1987).

§ The putative Late Cretaceous age of crown placentals is contingent on placement of the zalambdalestid *Kulbeckia* and the zhelestid *Zhelestes* within the superordinal placental clades of euarchontoglires and laurasiatherians within the crown placentals (Archibald et al., 2001; Averianov and Archibald, 2005). But in larger analyses including more relevant clades and morphological features, the zalambdalestids and zhelestids are excluded from the crown Placentalia (Asher et al., 2005) but partially included in other studies (e.g., Luo and Wible, 2005). Thus the Cretaceous age of crown placentals is tentative.

(Crompton and Sun, 1985; Crompton and Luo, 1993; Luo et al., 2001b). A relatively common mammaliaform group is Haramiyida, known from Europe and Greenland (Sigogneau-Russell, 1989a; Jenkins et al., 1997; Butler, 2000). Eleutherodontids, a younger group that is related to haramiyidans, achieved a wider distribution in Eurasia during the Middle Jurassic and in Africa in the Late Jurassic. But the distribution of haramiyidans themselves is limited to Europe and Greenland during the Late Triassic and Early Jurassic (Kielan-Jaworowska et al., 2004). Nevertheless, morganucodontans achieved a wide distribution on several continents during the Late Triassic to Early Jurassic (Crompton, 1974; Clemens, 1980; Jenkins et al., 1983, 1994; Sigogneau-Russell and Hahn, 1994; Luo and Wu, 1995; Kielan-Jaworowska et al., 2004). This is consistent with cosmopolitan distribution of other vertebrates that indicates a considerable interchange of terrestrial faunas across Pangaea (Shubin and Sues, 1991; Sues and Reisz, 1995; Datta et al., 2004).

Middle Jurassic Diversification of Basal Clades of Crown Mammalia

At least five new family-level clades appeared during the Middle Jurassic diversification: the basal eutriconodontans or "amphilestids," the earliest-known multituberculates, the amphitheriid cladotherians, the basal mammalian lineage of shuotheriids, and the basal australosphenidans. Originating with this new mammalian diversification are three major morphological innovations: (1) middle ear bones partially separated from the mandible; (2) a "modern" shoulder girdle in which the scapula, clavicle, and reduced interclavicle form mobile joints (Fig. 9.2: node 8); and (3) complex molars with mortar-pestle crushing and grinding in addition to the shearing function in shuotheriids and australosphenidans, although convergent multifunctional teeth also evolved in the Middle Jurassic members of the docodontan lineage.

The emergence of new mammalian clades during the Middle Jurassic represents a case of short-fuse diversification. Eutriconodontans differ from (and are more derived than) Late Triassic and Early Jurassic mammaliaforms with triconodont-like molars in their mandibular structure. Shuotheriids and australosphenidans independently developed some dental apomorphies with grinding function (the pseudotalonid in shuotheriids and the convergent talonid in australosphenidans), in addition to the plesiomorphic shearing structure of the trigonid on the molars in these groups. Therefore, most but not all emergent lineages of the Middle Jurassic are so distinctive that they are not direct descendants (at the family level) of the mammaliaform groups of the Late Triassic and Early Jurassic. Several Late Triassic–Early Jurassic mammaliaform groups did not survive into the Middle Jurassic. However, Morganucodontidae, a common Triassic-Jurassic family, is represented by one taxon, *Wareolestes*, in the Middle Jurassic; another putative morganucodontan represented by a femur is known from Russia (Freeman, 1979; Gambaryan and

Averianov, 2001; Kielan-Jaworowska et al., 2004). Haramiyidans are represented in the Middle Jurassic by eleutherodontids. This group differs from theroteinids and haramiyidans in the wear facets of the molars (Butler and Hooker, 2005) and represents a lineage distinct from theroteinids and haramiyids of the Late Triassic and Early Jurassic. Eleutherodontids had a wide geographic distribution across Eurasia (Kermack et al., 1998; Maisch et al., 2005). There are taxonomic replacements within the haramiyidan lineage between the Early and Middle Jurassic, and then between the Middle and Late Jurassic. Haramiyids and theroteinids are the main groups among haramiyidans during the Late Triassic and Early Jurassic, although two taxa appeared to have extended their range into the Late Jurassic (Heinrich, 1999) and the Cretaceous (Anantharaman et al., 2006). Eleutherodontids are the main haramiyidans of the Middle Jurassic, and were largely replaced by the dominant multituberculates in the Late Jurassic. Contingent on the relationship of the Late Triassic *Delsatia* and *Woutersia* and a taxon from Early Jurassic of India to docodontans (sensu stricto; Prasad and Manhas, 2001; Sigogneau-Russell, 2003a), the docodontan lineage is perhaps the only lineage from the Late Triassic to achieve a good measure of diversity during the Middle Jurassic and later geological times (Kermack et al., 1987; Martin and Krebs, 2000; Maschenko et al., 2002; Martin and Averianov, 2004).

"Amphilestids" are basal eutriconodontans. They share with triconodontids and gobiconodontids several derived mandibular features that are absent in *Sinoconodon* and morganucodontans. This justifies their placement among eutriconodontans (Kielan-Jaworowska et al., 2004), although the monophyly of all eutriconodontans is only weakly supported by the currently available evidence (see Cifelli, 2001; Rougier et al., 2001; Luo et al., 2002). In recent phylogenetic analyses, the amphilestids have been collapsed into a series of successive clades toward the more derived triconodontids (e.g., Ji et al., 1999; Rougier et al., 2001). The 165-million-year-old *Amphilestes* and *Phascolotherium*, the earliest "amphilestids," represent a basal cluster of subclades within the eutriconodontan clade. Because basal eutriconodontans are currently known only from jaws and teeth, inferences concerning cranial and postcranial characteristics of eutriconodontans and the morphological evolution of the theriiform clade including eutriconodontans will have to be based on Cretaceous eutriconodontans documented from skulls and skeletons.

Eutriconodontans lack the postdentary trough retained by mammaliaforms, australosphenidans, and toothed monotremes. The angular region of the mandible is rounded, lacking an angular (or pseudangular) process. The medial side of the mandible has a pronounced fossa for the pterygoideus muscle. In some Cretaceous eutriconodontans, the middle ear bones are separated mediolaterally from the posterior (angular) region of the mandible, but are still connected anteriorly via the Meckel cartilage to the Meckel groove along the ventral margin of the mandible (Wang et al.,

2001; Li et al., 2003; Meng et al., 2003, 2005; Hu et al., 2005). This partial separation of the middle ear from the mandible can be interpreted as an intermediate morphological condition. It is less derived than the completely separated mandible and the middle ear bones in extant mammals, but more derived than the mammaliaform condition in which the middle ear bones are attached both anteriorly via the Meckel cartilage and by the direct contact of the middle ear bones side by side with the postdentary trough (see discussion in Kielan-Jaworowska et al., 2004). It is inferred here that the mediolateral separation of the middle ear and partial independence of the middle ear bones from the mandible occurred with the emergence of eutriconodontans in the Middle Jurassic, and thus long before the reduction of the Meckel's cartilage that resulted in the complete separation of the middle ear bones from the mandible. The latter transformation appears to have occurred convergently in crown monotremes and crown therians (Martin and Luo, 2005; Rich et al., 2005).

Eutriconodontans are the earliest-known mammals with a mobile clavicle-interclavicle joint, a mobile acromion-clavicle joint, and a derived scapula in the shoulder girdle (Fig. 9.2, node 8), assuming that these apomorphies preserved in Early Cretaceous eutriconodontans are characteristic for the entire eutriconodontan clade (Luo et al., 2002; Kielan-Jaworowska et al., 2004). The primitive shoulder girdle of nonmammalian cynodonts and mammaliaforms is characterized by a large interclavicle with long lateral processes. The latter structure either forms extensive contact with the clavicle or is simply synostosed to it. The clavicle-interclavicle joint is immobile. The glenoid of the shoulder is formed by the scapula and a large coracoid. It is saddle-shaped and faces laterally. The shoulder joint has relatively little range of excursion, and the humerus is oriented horizontally in a sprawling posture (Jenkins, 1970). This type of primitive shoulder girdle is present in most cynodonts (Jenkins, 1971, 1984; Sues, 1983; Sun and Li, 1985; Gow, 2001; Sues and Jenkins, 2006), in morganucodontans (Jenkins and Parrington, 1976), and in *Sinoconodon*. The shoulder girdles of docodontans and *Fruitafossor,* as far as is known, retain the same primitive features (Luo and Wible, 2005; Martin, 2005). Among extant mammals, only monotremes have this type of girdle (Klima, 1973; Pridmore, 1985).

In features of the shoulder girdle, eutriconodontans are more similar to multituberculates, spalacotherioid symmetrodontans, and extant marsupials and placentals than to monotremes and mammaliaforms. The interclavicle is small and lacks prominent lateral processes, so that the clavicle-interclavicle joint is mobile. The glenoid of the shoulder joint is a concavity (rather than a saddle) formed mostly or entirely by the scapula, whereas the coracoid is reduced to a small process. The scapula has a laterally facing, broad supraspinous fossa, presumably developed with a neomorphic membranous scapular spine (Sánchez-Villagra and Maier, 2003). The shoulder joint has a greater range of excursion, and the humerus is capable of a wider range of rotation (Jenkins and Weijs, 1979). The forelimb posture is more parasagittal than sprawling. This suite of derived shoulder-girdle features and their related loco-motory function represent a major innovation in the evolution of the mammalian skeleton. The currently available evidence favors a monophyletic origin of these features with the theriimorph clade (contra Ji et al., 1999).

The second group emerging in the Middle Jurassic are the Multi-tuberculata. As reviewed earlier, the relationship of Late Triassic and Early Jurassic haramiyidans to multituberculates has been disputed (e.g., Jenkins et al., 1997; Luo et al., 2002; Kielan-Jaworowska et al., 2004). At a minimum, the geological history of multituberculates can be extended back to the Middle Jurassic by their ghost-lineage his-tory, which is the range that is not yet documented by known fossils but inferred from the range of their sister lineage (Norell, 1993; Padian et al., 1995). The ghost lineage for multituberculates can be inferred to be as old as that for its sister-group Trechnotheria be-cause the latter lineage extends into the Middle Jurassic (Kielan-Jaworowska et al., 2004). More recently, Butler and Hooker (2005) identified *Kermackodon* and *Hahnotherium* of the Middle Jurassic as the earliest-known multituberculates on the basis of the presence of a multituberculate-like, horizontal wear pattern on their molars. This provides additional corroboration that multituberculates are at least as old as the Middle Jurassic. But it should be pointed out that fossils of multituberculates are scarce in the Middle Jurassic.

The third emerging group in the Middle Jurassic are the am-phitheriids. Their lower molars are derived in having a posterior heel with a neomorphic cusp termed the *hypoconulid*, in addition to the three triangulated cusps bearing two shearing crests, known as the *trigonid* (Butler and Clemens, 2001). Amphitheriids are also derived in the possession of a mandibular angle. This angle is posi-tioned posteriorly below the dentary condyle and is inferred to be separated from the angular (ectotympanic) bone, which would have supported the tympanic membrane (Allin and Hopson, 1992). The mandibular angle in amphitheriids is interpreted by some au-thors as a neomorphic structure (the true mandibular angle) and has been distinguished from the pseudangular process, which is an-teriorly positioned on the mandible, and associated with the angu-lar (ectotympanic) bone in mammaliaforms (see Jenkins, 1984; Sues, 1986; Hopson, 1994). *Palaeoxonodon* from the Middle Jurassic of England was previously considered a more derived per-amuran (Freeman, 1979), but is now regarded as an amphitheriid (Sigogneau-Russell, 2003b).

Shuotheriids and australosphenidans are two new mammalian lineages that appeared by the Middle Jurassic (Sigogneau-Russell, 1998; Flynn et al., 1999; Rauhut et al., 2002; Martin and Rauhut, 2005). The pseudotribosphenic molars of shuotheriids have both the primitive function of shearing (*sphen*) and the derived function of mortar-pestle crushing and grinding (*tribo*). Their lower molars have a derived structure of the pseudotalonid basin anterior to the trigonid for receiving the pseudoprotocone of the upper molar for grinding food (Chow and Rich, 1982; Sigogneau-Russell, 1998; Wang et al., 1998). The posterior premolar is strongly triangular, a feature shared by some australosphenidans (Kielan-Jaworowska et al., 2002). In juxtaposition with its derived molar features, the mandible of *Shuotherium* is primitive, with a postdentary trough accommodating the middle ear bones (Chow and Rich, 1982; Kielan-Jaworowska et al., 1998). It is universally accepted that despite the highly derived molars, shuotheriids form one of the basalmost groups of crown mammals and its pseudoprotocone is a convergent feature to the protocone of the boreosphenidans (Wang et al., 1998; Kielan-Jaworowska et al., 2004).

Australosphenidans from the Gondwanan continents have tribosphenic molars with the primitive trigonid structure for shearing and derived talonid for mortar-pestle crushing and grinding. The talonid in australosphenidans is positioned posterior to the trigonid and thus differs from the pseudotalonid anterior to the trigonid in *Shuotherium*. The earliest australosphenidans are *Ambondro* from the Middle Jurassic of Madagascar (Flynn et al., 1999) and *Asfaltomylos* and related species from the Middle Jurassic of Argentina (Rauhut et al., 2002; Forasiepi et al., 2004; Martin and Rauhut, 2005). Australosphenidans also include at least two additional tribosphenic mammals from the Lower Cretaceous of Australia, *Ausktribosphenos* (Rich et al., 1997, 1999) and *Bishops* (Rich et al., 2001a), as well as other unnamed taxa (Rich et al., 2005).

The tribosphenic mammals on the Gondwanan continents are precociously derived for their age. Their appearance during the Middle Jurassic is 20 million years earlier than the first tribosphenic mammals from the Laurasian continents, and they are more derived in many dental features than the northern tribosphenic mammals. When Rich et al. (1997) discovered the first Gondwanan tribosphenic mammal *Ausktribosphenos*, they suggested that it was an erinaceomorph placental (see also Rich et al., 1997, 2001a). Later, Woodburne, Rich, and their colleagues argued that *Ambondro* and *Asfaltomylos* are also placentals (Woodburne, 2003; Woodburne et al., 2003). A contrasting view is that these southern tribosphenic mammals are not closely related to the northern Mesozoic eutherians, metatherians, and their proximate tribosphenic relatives, and that the development of the talonid basin on lower molars is a convergent feature in the southern tribosphenic mammals, which are extinct relatives of monotremes (Luo et al., 2001a, 2002; Rauhut et al., 2002; Kielan-Jaworowska et al., 2004; Martin and Rauhut, 2005; a modified version of this hypothesis was presented by Sigogneau-Russell et al., 2001).

Aside from their primitive postdentary trough, australosphenidans are similar to mammaliaforms such as *Morganucodon* in that the posteriormost mental foramen on the lateral side of the mandibular ramus is located below the canine and anterior premolars (Rich et al., 2001a; Martin and Rauhut, 2005). By comparison, the posteriormost mental foramen in the northern Mesozoic tribosphenic mammals is placed far more posteriorly on the mandibular ramus, either below the penultimate premolar or the first lower molar (e.g., Kielan-Jaworowska and Dashzeveg, 1989; Cifelli and Muizon, 1997; Muizon, 1998; Rougier et al., 1998; Ji et al., 2002; Luo et al., 2003). In these primitive features, southern tribosphenic mammals are similar to the toothed monotreme *Teinolophos* (Rich et al., 2005). However, not all mandibular features of australosphenidans are primitive. For instance, the angular process is in an elevated position and flares transversely in *Ausktribosphenos*, *Ambondro*, and *Asfaltomylos*, as on the mandible of the toothed monotreme *Teinolophos* (Rich et al., 1999, 2001, 2005; Luo et al., 2002; Martin and Rauhut, 2005). After incorporating all of the morphological features known from the australosphenidans into the large morphological data sets for all Mesozoic mammals and major clades of extant mammals, most recent parsimony analyses have corroborated the australosphenidan clade in which the extinct Mesozoic tribosphenic mammals represent an endemic diversification on the Gondwanan continents, with monotremes being a relict extant lineage from this ancient mammalian diversification.

The australosphenidan hypothesis has three elements. From the most readily to the most difficult to corroborate, these are: (1) the dual morphological evolution of the multifunctional tribosphenic molars; (2) the relationship of the southern tribosphenic mammals to the toothed monotremes and their present-day relatives—the platypus and the echidna; and (3) the extent to which this clade is endemic to the Gondwanan continents.

The convergent evolution of the functionally adaptive talonid basin is the least difficult to demonstrate morphologically and is the easiest to corroborate by phylogenetic analysis. From the traditional perspective of morphological evolution (Patterson, 1956; Crompton, 1971; Fox, 1975; McKenna, 1975), the derived molar structure in the geologically older southern tribosphenic mammals from the Middle Jurassic is incongruent with the well-corroborated transformation series of *Amphitherium–Peramus–Aegialodon–Pappotherium*. All of the Laurasian fossil taxa of this well-supported time-evolution series are geologically younger, but less derived in molar features, than all of the australosphenidans (Luo et al., 2002; Rauhut et al., 2002; Forasiepi et al., 2004). The mandibles of amphitheriids and peramurans are also different from those of australosphenidans (Allin and Hopson, 1992; Martin, 2002).

The hypothesis of southern tribosphenic mammals as crown placentals was largely based on the assumption that their molar structure would correspond well with the traditional evolutionary series of tribosphenic molars (Patterson, 1956; Crompton, 1971; Fox, 1975; McKenna, 1975). But many detailed features of the talonid of australosphenidans do not correspond well with those of boreosphenidans. Martin and Rauhut (2005) offered the insightful observation that *Asfaltomylos* lacks the wear facets in the bottom of the talonid basin typical of the boreosphenidan mammals. Dental wear on the talonid basin concentrates on the peripheral crests of the talonid in *Asfaltomylos* and *Ambondro*. Australosphenidans are only represented by the lower teeth, and there are still some uncertainties in how the talonids of these lower molars would occlude with the upper teeth. Martin and Rauhut (2005) outlined four alternative scenarios on how the peculiar wear patterns of the talonid basin in *Asfaltomylos* could correspond to a range of hypothetical upper molar structures, which would include, but not be limited to, a true protocone. Hunter (2004) noted that the molar talonid of *Ausktribosphenos,* with its several internal crests, is atypical for tribosphenic occlusal relationships, and he hypothesized that in *Ausktribosphenos* there could be correspondence of one protocone on the upper molar to two subdivided and side-by-side basins in the talonid of the lower molar. These interpretations of how the lower molar talonid would occlude with the upper molar structure can be tested by future finds of upper molars of these taxa.

From the phylogenetic perspective, it is easy to corroborate that the southern tribosphenic mammals cannot be placed in the same clade as the placental mammals. The morphological evidence of *Asfaltomylos* and *Ausktribosphenos* has been reexamined and reanalyzed. With the exception of Rich et al. (2002) and Woodburne (2003), no other studies have corroborated the placement of *Asfaltomylos* and *Ausktribosphenos* with placentals (Sigogneau-Russell et al., 2001; Archibald, 2003; Forasiepi et al., 2004; Hunter, 2004; Martin and Rauhut, 2005). For example, Sigogneau-Russell et al. (2001) suggested that *Ausktribosphenos* from the Lower Cretaceous of Australia is the sister-group to the toothed monotremes in the same vertebrate assemblage. Forasiepi et al. (2004) also concluded that *Asfaltomylos* and *Ausktribosphenos* could not be placed in the same group as the Laurasian tribosphenic mammals. If any of the southern tribosphenic mammal taxa are placed outside the boreosphenidan clade, then it is inevitable that the tribosphenic molar, as an adaptive functional complex, must have evolved more than once.

It is more difficult to corroborate the hypothesis that monotremes are nested within the australosphenidan clade than to falsify the hypothesis that *Asfaltomylos* and *Ausktribosphenos* are placentals. One difficulty is that fossils of the Early Cretaceous toothed monotremes are generally less complete than those of australosphenidans. Martin and Rauhut (2005) suggested that the wear on the peripheral crests of the talonid basin in australosphenidans is similar to that of toothed monotremes. So far, the dental characters supporting the placement of toothed monotremes in the australosphenidan clade (Luo et al., 2001b; 2002; Kielan-Jaworowska et al., 2004) have been corroborated and expanded by a separate study (Martin and Rauhut, 2005), although questioned by some (Woodburne, 2003). The mandible of *Teinolophos* shows that toothed monotremes retained the postdentary trough for the middle ear bones (Rich et al., 2005). Interestingly, most characteristics in the mandible of *Teinolophos* are very similar to those of *Ausktribosphenos* (Martin and Luo, 2005). Incorporation of these mandibular characteristics of toothed monotremes has helped to consolidate the placement of toothed monotremes within the australosphenidan clade in a recent phylogenetic analysis (Luo and Wible, 2005).

The paleobiogeographic element of the australosphenidan hypothesis is the easiest to falsify and the most difficult to corroborate. Discovery of a single Laurasian mammal that is demonstrably related to australosphenidans would be sufficient to falsify the paleobiogeographic aspect of the australosphenidan hypothesis (although not dual morphological evolution). All currently known australosphenidans are from Gondwana and of Middle Jurassic to Early Cretaceous age. However, the sister-group to australosphenidans is Shuotheriidae, and shuotheriids are from Laurasia (Sigogneau-Russell, 1998; Wang et al., 1998; Kielan-Jaworowska et al., 2002). By inference, the common ancestor of shuotheriids and australosphenidans (the yinotherian clade; Kielan-Jaworowska et al., 2004; Fig. 9.2: node 5) could be either Pangaean or Laurasian. This clade must be rooted in the mammaliaforms of the Early Jurassic, during which time terrestrial vertebrate faunas have many elements with Pangaean distribution, including mammaliaforms.

Late Jurassic Diversification in Mammalian Ecomorphology

The major mammalian groups of the Late Jurassic fit well with the long-fuse model in which a lineage diversifies only after having a relatively long and modest history. Most Late Jurassic mammalian lineages had plesiomorphic relatives in the Middle Jurassic, if not earlier. Although the dominant mammals in the Late Jurassic are rare in the Middle Jurassic (Martin and Krebs, 2000; Kielan-Jaworowska et al., 2004), the taxonomic diversification occurred mostly within the previously established order- or family-level lineages. More specifically, the Late Jurassic triconodontids are related to amphilestids of the Middle Jurassic. Docodontans, which are common in the Late Jurassic, were established by the Middle Jurassic (Kermack et al., 1987; Sigogneau-Russell, 2003b; Martin and Averianov, 2004). Multituberculates are relatively abundant during the Late Jurassic (Kielan-Jaworowska and Hurum, 2001; Hahn and Hahn, 2004; Kielan-Jaworowska et al., 2004), but this group can be traced to some multituberculate-like taxa of the Middle Jurassic, at least according to some workers (Butler and Hooker, 2005). Although no dryolestoids are known from the Middle Jurassic, they have a long ghost-lineage history, extending to the Middle Jurassic, as inferred from the Middle Jurassic age of amphitheriids, the oldest taxon of the sister-group of dryolestoids.

For the first time in the Mesozoic, mammals became numerically abundant in faunas from the Guimarota coal mine in Portugal and the Morrison Formation of the Western Interior of the United States (Krusat, 1980; Prothero, 1981; Hahn, 1988, 2001; Chure et al., 1998; Engelmann and Callison, 1998; Martin, 1999; Martin and Krebs,

2000). The Late Jurassic is the first time when multituberculates appeared in great diversity and large numbers (Hahn, 1988, 1993, 2001; Kielan-Jaworowska and Hurum, 2001), and triconodontids and dryolestoids appeared in relatively high numerical abundance (Engelmann and Callison, 1998; Martin, 1999). Docodontans were also relatively common (Simpson, 1929; Krusat, 1980; Lillegraven and Krusat, 1991; Martin, 2005).

The most important feature of Late Jurassic mammalian evolution is the great scope of ecomorphological diversification, as reflected in a greater variety of feeding adaptations and locomotory specializations. Late Jurassic mammals surpassed previous mammals and mammaliaforms in spreading into some niches inaccessible to their predecessors, developing diverse feeding adaptations outside the primitive and generalized insectivorous feeding for the first time in their evolutionary history. For example, herbivorous multituberculates are far more common and abundant in the faunas than haramiyidans were in the Late Triassic and Early Jurassic. Docodontans developed teeth capable of both shearing and grinding and have been interpreted as insectivorous to omnivorous. Multituberculates and docodontans represent the first major evolutionary experiment of mammals in herbivorous feeding guilds on a significant taxonomic scale and in large numbers. The Late Jurassic *Fruitafossor* is the earliest-known mammal with specialization for feeding on colonial insects (Luo and Wible, 2005). It has open-rooted, tubular teeth that are convergent to the teeth of extant armadillos and is therefore inferred to have a very similar dietary specialization for feeding on colonial insects such as termites.

Late Jurassic mammals developed diverse locomotory functions that may have facilitated their spread into more diverse feeding niches. Early Jurassic mammaliaforms (e.g., *Megazostrodon*) are generalized and terrestrial (Jenkins and Parrington, 1976). By Late Jurassic, at least two mammals have independently developed fully fossorial adaptations: *Haldanodon* (Martin, 2005) and *Fruitafossor* (Luo and Wible, 2005). Several nonmammalian cynodonts are burrowing animals (Sues, 1983; Damiani et al., 2003; Kemp, 2005). Many phylogenetically primitive skeletal features of cynodonts happened to be useful for fossoriality. As shown by the versatile locomotory adaptations of the monotreme *Ornithorhynchus*, skeletal features that contributed to its primitive sprawling posture and gait are also effective for both digging and swimming. Similarly, postcranial features of *Haldanodon* and *Fruitafossor* are phylogenetically primitive, but nonetheless useful for digging. In this regard, both taxa are similar to extant monotremes with digging specializations; their humeri have hypertrophied deltopectoral crests and prominent greater and lesser tubercles, as well as strong crests for M. teres major. The distal end of the humerus is very broad with hypertrophied ectepicondylar and entepicondylar processes. The ulna has an elongate, robust olecranon process. These features are otherwise known from extant mammals with fossorial adaptation (Krusat, 1991; Martin, 2005). Martin and Nowotny (2000) suggested that

Haldanodon has fossorial and semiaquatic habits. In additional to the fossorial features in the scapula, humerus, ulna, and radius, *Fruitafossor* also has a robust manus with some very thick and much shortened metacarpals, phalanges, and very large terminal phalanx (Luo and Wible, 2005). It is now apparent that even small mammals like docodontans and *Fruitafossor* (20 to 40 g) can develop highly fossorial adaptations.

At least one mammal, *Henkelotherium*, developed scansorial or climbing adaptation (Krebs, 1991; Vázquez-Molinero et al., 2001). *Henkelotherium* differs from generalized terrestrial mammals in having many distinctively scansorial features, such as a gracile limb skeleton and relatively elongate phalanges (Krebs, 1991; Vázquez-Molinero et al., 2001). The elongate penultimate phalanx of each pedal digit in *Henkelotherium* is typical of phalangeal proportions of small climbing mammals (Krebs, 1991). Its terminal claw has a high-arched profile and very large flexor tubercle, characteristic of scansorial mammals (MacLeod and Rose, 1993; Vázquez-Molinero et al., 2001). Krebs (1991) argued that the long tail of *Henkelotherium* was indicative of its specialization for climbing, but the inferred function to this feature is questionable (Kielan-Jaworowska and Gambaryan, 1994), and it is uncertain whether it is an arboreal mammal, as previously suggested (Krebs, 1991; Vázquez-Molinero et al., 2001; Kielan-Jaworowska et al., 2004).

Peramurans first appeared in the fossil record during the Late Jurassic with *Nanolestes* (Martin, 2002). The stratigraphic range of peramurans can only be extended back to the Late Jurassic since the Middle Jurassic *Palaeoxonodon* has been reassigned to amphitheriids (Sigogneau-Russell, 2003b), but this group is more common in the earliest Cretaceous. Peramurans are fairly similar in mandibular features to amphitheriids and dryolestoids, but they represent an important stage in the evolution of molar structure. Their molars have a more expanded talonid heel on the lower molar with two cusps (the hypoconulid and the hypoconid), and are more derived than amphitheriids, in which the talonid has only one cusp (Clemens and Mills, 1971; Crompton, 1971; Sigogneau-Russell, 1999; Butler and Clemens, 2001; Martin, 2002). The talonid structure in peramurans is widely regarded as a precursor condition that, with further expansion, would lead to a full talonid basin that can accommodate the upper molar protocone in the true tribosphenic mammals on the Laurasian continents (Crompton, 1971; Fox, 1975), known also as the boreosphenidans (Luo et al., 2001a, 2002; Kielan-Jaworowska et al., 2004).

Early Cretaceous

Emergence of Prototribosphenic and Boreosphenidan Mammals

The most important feature of mammalian faunal evolution during the Early Cretaceous is the emergence of modern mammalian lineages. This epoch is a defining stage in mammalian history because this is the first time when derived anatomical features of some mammals have made it possible for them to be placed unequivocally as the stem taxa to the extant marsupial, placental, and monotreme lineages. The Early Cretaceous marks the beginning of the anatomically modern mammals. The three emergent clades of this epoch are still living today, and eutherians (including placentals) and metatherians (including marsupials) have dominated the terrestrial vertebrate faunas of the Cenozoic.

The new, or newly diversified, lineages of the Early Cretaceous achieved a greater (although still modest) diversity than their phylogenetic relatives (or precursors) of the Late Jurassic. This is especially the case for prototribosphenidans, peramurans, and stem boreosphenidans that represent the successive ranks of phyletic relatives to metatherians and eutherians. The prototribosphenic mammals (sensu Rougier et al., 1996) and the peramuran zatherians were newly diversified and became more common in the Early Cretaceous, in contrast to their scarce records in the Late Jurassic (Dashzeveg, 1994; McKenna and Bell, 1997; Sigogneau-Russell, 1999; Martin, 2002). These taxa do not form a monophyletic group; they are instead resolved into a series of clades that are successively more closely related to boreosphenidan mammals (Sigogneau-Russell, 1999; Martin, 2002; Kielan-Jaworowska et al., 2004). The South American mammal *Vincelestes* has developed some derived cranial and postcranial skeletal structures of crown therians (Bonaparte and Rougier, 1987; Rougier, 1993).

Stem boreosphenidans (northern mammals with tribosphenic molars) are the proximate relatives of the clade comprising marsupial and placental mammals. According to the hypothesis of dual evolution of tribosphenic mammals, the tribosphenic molars of boreosphenidans are a second and convergent evolution of this type of molar, and it occurred on the northern continents some 20 million years later than in the tribosphenic mammals of the Gondwanan continents (Luo et al., 2001a, 2002; Rauhut et al., 2002; Forasiepi et al., 2004; Kielan-Jaworowska et al., 2004; Martin and Rauhut, 2005). Studies in the last two decades have demonstrated that most of these boreosphenidan mammals cannot be reliably placed within either the metatherian or eutherian clades on the basis of their isolated teeth (e.g., Butler, 1978; Dashzeveg and Kielan-Jaworowska, 1984; Clemens and Lillegraven, 1986; Cifelli, 1993). Rather, these stem boreosphenidans represent a burst of diversification of what became blind evolutionary lineages that diverged from the base of crown Theria (Cifelli, 1993; Kielan-Jaworowska et al., 2004). Boreosphenidans gained a modest taxonomic diversity and numerical abundance during the Early Cretaceous, and a wide distribution on the Laurasian continents. By the early 2000s, more than 10 tribosphenic taxa were recognized from the Lower Cretaceous of Eurasia and North America (although some have not been named; Patterson, 1956; Kermack et al., 1965; Slaughter, 1971, 1981; Lillegraven, 1974; Butler, 1978; Dashzeveg and Kielan-

Jaworowska, 1984; Clemens and Lillegraven, 1986; Sigogneau-Russell, 1991; Cifelli, 1993; Kielan-Jaworowska and Cifelli, 2001).

In addition to the evolutionarily derived new lineages, several other preexisting mammalian groups diversified with many new genera during the Early Cretaceous. These include tinodontids (obtuse-triangled symmetrodontans) and spalacotherioids (acute-triangled symmetrodontans). The obtuse-triangled symmetrodonts have sparse records in the Middle and Late Jurassic but gained modest diversity in the Berriasian stage of the earliest Cretaceous, with four documented genera from the Purbeck Group of Great Britain and the Synclinal d'Anoual in Morocco (Sigogneau-Russell, 1989b, 1991; Sigogneau-Russell and Ensom, 1998; Ensom and Sigogneau-Russell, 2000; Kielan-Jaworowska et al., 2004). The youngest known taxon, *Gobitheriodon,* extends the range of obtuse-triangled symmetrodonts to the Albian (Averianov, 2002). These Early Cretaceous taxa are unequivocally related to the Late Jurassic tinodontids and possibly related to the Early Jurassic *Trishulotherium* (Yadagiri, 1984; Prasad and Manhas, 1997; Sigogneau-Russell and Ensom, 1998; Ensom and Sigogneau-Russell, 2000; Kielan-Jaworowska et al., 2004) on the basis of the currently available evidence. The obtuse-triangled symmetrodonts may not be a monophyletic group, but they are clearly the basal taxa of the trechnotherian clade of both spalacotherioids and crown Theria (Kielan-Jaworowska et al., 2004).

Spalacotherioids, or acute-triangled symmetrodontans, are a robust monophyletic group. This group first appeared in the earliest Cretaceous, and achieved modest diversity during the Berriasian through Barremian stages. They ranged into the Late Cretaceous (Campanian; Fox, 1976; Hu et al., 1997, 1998; Cifelli and Madsen, 1999; Averianov, 2002; Gill, 2004; Kielan-Jaworowska et al., 2004; Tsubamoto et al., 2004). On the mammalian phylogeny (Fig. 9.2), spalacotherioids are nested between the more derived *Amphitherium* of the Middle Jurassic and the less derived obtuse-triangled symmetrodontans that extended into the late Early Jurassic (Kielan-Jaworowska et al., 2004). By inference from these related groups, spalacotherioids are at least as old as Middle Jurassic. All known taxa of this group appear to be specialized for feeding on soft-bodied small invertebrates (Kielan-Jaworowska et al., 2004). As far as is known, the postcranial skeletal features of this group indicate a generalized terrestrial habit (Hu et al., 1998; Rougier et al., 2003). The relatively common spalacotherioids in the Early Cretaceous certainly had a longer history in the Jurassic. In this sense, the histories of tinodontids and spalacotherioids can be characterized as examples of the long-fuse model.

Rise of Monotremes, Metatherians, and Eutherians

The earliest-known Mesozoic mammals with definitive and diagnostic dental feature of monotremes are *Teinolophos* and *Steropodon* from the Early Cretaceous of Australia (Archer et al., 1985; Rich et al., 1999, 2001a, 2005); in recent phylogenetic analyses they are placed closer to extant monotremes than to any other Mesozoic mammals. Aside from *Steropodon* and *Teinolophos,* it is generally assumed that the toothed monotremes had a much wider distribution during the Cretaceous, as can be inferred indirectly from the presence of *Monotrematum* in the Paleocene of Argentina (Pascual et al., 1992, 2002; Forasiepi and Martinelli, 2003). The history of the present-day monotreme family Ornithorhynchidae can be extended back to the Miocene on the basis of the well-preserved skull of *Obdurodon dicksoni* (Archer et al., 1992, 1993; Musser and Archer, 1998) and to the Oligocene by teeth of *Obdurodon insignis* (Woodburne and Tedford, 1975). It is beyond question that *Monotrematum* can be referred to the ornithorhynchid lineage on the basis of its dental characteristics (Pascual et al., 1992, 2002; Musser, 1999; Woodburne, 2003).

The Early Cretaceous toothed monotremes cannot be placed within the crown Monotremata. Despite their strong dental resemblance to the toothed ornithorhynchids, toothed monotremes are far more primitive than the late Cenozoic and extant ornithorhynchids in mandibular structure (Rich et al., 2005). At least in *Teinolophos,* the middle ear bones are inferred to have still been connected with the mandible (Martin and Luo, 2005; Rich et al., 2005; but see Bever et al., 2005; Rougier et al., 2005). The latest comprehensive cladistic analysis has placed them close to the crown group of extant monotremes, although not within the crown Monotremata (Luo and Wible, 2005). *Teinolophos* and *Steropodon* would be best regarded as stem clades of earlier diversifications prior to the appearance of modern monotremes, as indicated by the Paleocene record of *Monotrematum* (Pascual et al., 1992, 2002).

The Australian Early Cretaceous mammal *Kollikodon* has bunodont premolars and molars characterized by low and rounded cusps. It was hypothesized that *Kollikodon* is closely related to monotremes because it resembles ornithorhynchids in having a large mandibular canal (Flannery et al., 1995). The monotreme affinities of *Kollikodon* have been questioned by McKenna and Bell (1997), who considered it to be a stem taxon to crown mammals; it may be related to mammals as a whole but not closely related to monotremes. In Cenozoic placental mammals, bunodont molars of more omnivorous mammals usually evolved from the tribosphenic molars of insectivorous ancestors. It cannot be ruled out that *Kollikodon* evolved from an ancestor with tribosphenic molars, by convergence, to the well-established placental molar evolution. In this scenario, *Kollikodon* could be interrelated to monotremes (*Steropodon, Teinolophos,* and extant monotremes) by a common ancestry within the australosphenidan clade, although not necessarily nested within the toothed monotremes.

Placentals and marsupials have dominated the world's terrestrial vertebrate communities for the entire Cenozoic. Today, placentals comprise 5,080 species and marsupials have 331 species; together, they make up 99.9% of all present-day mammalian species (Wilson and Reeder, 2005). But these two lineages are no older than Barremian (Early Cretaceous) and are phylogenetically rooted in a cluster of basal boreosphenidan mammals from the earliest Cretaceous (Fig. 9.3). The earliest fossils that can be reliably referred to the placental lineage (eutherians) and the marsupial lineage (metatherians) are from the Yixian Formation of northeastern China, dated at 125 million years old (Swisher et al., 1999; Chang et al., 2003; Zhou et al., 2003; Ji et al., 2004). The earliest-known eutherians and metatherians can be recognized by many characteristics of the molars, anterior dentition, ear region, and the ankle and wrist bones.

The oldest-known fossil mammal that is more closely related to extant marsupials than to modern placentals is *Sinodelphys szalayi*. Other metatherians, especially the deltatheroidans, are known from the Early Cretaceous of Laurasia (Averianov and Kielan-Jaworowska, 1999; Kielan-Jaworowska and Cifelli, 2001). Metatherians, such as *Sinodelphys, Kokopellia,* and *Asiatherium,* have an approximation of the entoconid and hypoconulid cusps on the talonid basin of the lower molars; this is a precursor condition to the full twinning of these two cusps in the dentally more derived Late Cretaceous metatherians. *Sinodelphys* also shares with the Cenozoic didelphoid marsupials the lanceolate outlines of the posterior upper incisors. All undisputed metatherians have three upper premolars and four upper molars for a total of seven postcanine positions, and a single replacement of the third premolar of the entire postcanine series (Cifelli et al., 1996; Rougier et al., 1998; Luo et al., 2003). Metatherians are also diagnosed by an inflected angular process and a vertical posterior shelf of the masseteric fossa of the mandible (Clemens, 1966; Lillegraven, 1969; Fox, 1987; Szalay and Trofimov, 1996; Sánchez-Villagra and Smith, 1997; Muizon, 1995, 1998). These dental and mandibular apomorphies of metatherians are supplemented by other diagnostic features of the ankle and the wrist (Fig. 9.2, node: 18). Metatherians are characterized by enlargement of three carpal bones in the wrist: the hypertrophied hamate and scaphoid, and an enlarged triquetrum. The more robust structure would enable a more forceful grip of the hands (Szalay, 1994; Argot, 2001). Metatherians also differ from eutherians and stem trechnotherians (such as *Zhangheotherium*) in derived features of the talus (astragalus) and heel bone (calcaneus) of the ankle joint. The upper ankle joint between the astragalus and the tibia and fibula lacks a definitive pulley-like structure; this is phylogenetically primitive but nevertheless is correlated with a wider range of movement for this joint (Szalay, 1994; Szalay and Sargis, 2001; Argot, 2002). The astragalar head is broad and asymmetrical; and the contact facet between the calcaneus and the cuboid is oblique. The trans-tarsal joint (sensu Szalay, 1994), between the astragalus and calcaneus on the one hand and the navicular and cuboid on the other, is oriented more obliquely than in outgroup symmetrodontans and eutherians (Szalay, 1994; Luo et al., 2003). Most of the Cenozoic marsupials are distinguishable from eutherians in having a medial process of the squamosal (Muizon, 1994; Muizon et

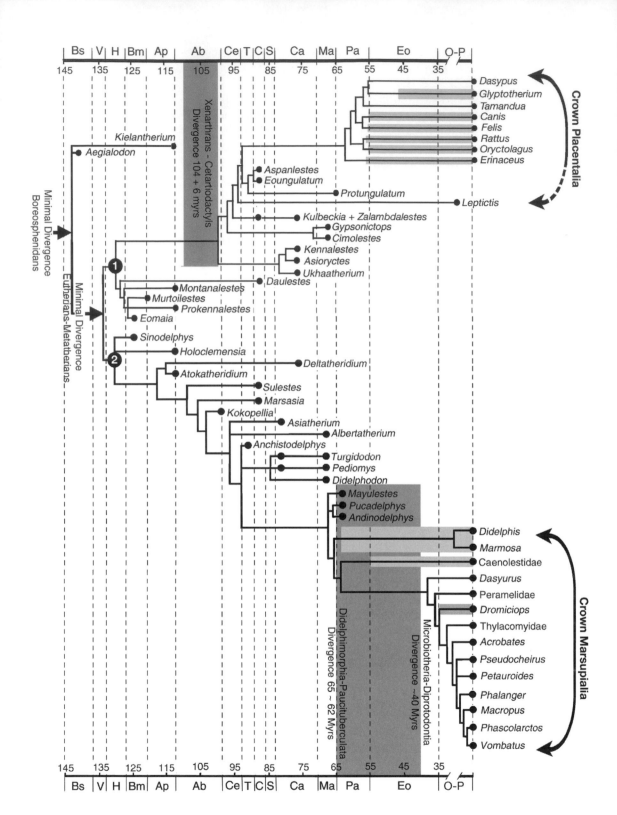

al., 1997) or a large component of the tympanic bulla from the alisphenoid, among some other apomorphies of the ear region (Wible, 1990; Novacek, 1993; Szalay, 1994).

The oldest-known fossil mammal that is more closely related to modern placentals than to modern marsupials is *Eomaia scansoria* (Fig. 9.3). Also known from the Barremian to Albian stages of the Early Cretaceous are *Prokennalestes* (Kielan-Jaworowska and Dashzeveg, 1989; Sigogneau-Russell et al., 1992), *Montanalestes* (Cifelli, 1999), and *Murtoilestes* (Averianov and Skustchas, 2000, 2001). Basalmost eutherians including *Eomaia* are characterized by five premolars and three molars for a total of eight postcanine positions, and replacement of most premolars except the first (Kielan-Jaworowska and Dashzeveg, 1989; Sigogneau-Russell et al., 1992; Luckett, 1993; Cifelli, 1999, 2000; Ji et al., 2002). Most Late Cretaceous and Cenozoic eutherians have three molars but fewer than five premolars (Novacek et al., 1997; Archibald et al., 2001; Kielan-Jaworowska et al., 2004). The earliest-known eutherians are distinguishable by their derived tarsal features (Fig. 9.2, node 19). *Eomaia* and Late Cretaceous eutherians have a relatively narrow navicular facet on the astragalar head (Kielan-Jaworowska, 1977, 1978; Horovitz, 2000; Ji et al., 2002; Luo et al., 2003). The upper ankle joint has developed a medial crest, a precursor condition to the pulley-shaped joint of some Late Cretaceous and all Cenozoic eutherians. Cretaceous eutherians are generalized and

Figure 9.3. *(opposite page)* Temporal distribution of boreosphenidan mammals and divergence of eutherians and metatherians. Topology of eutherians from Luo and Wible (2005); topology of metatherians from Luo et al. (2003). Nodes (1) and (2) are Eutheria and Metatheria. Dashed bracket of crown Placentalia indicates that more Cenozoic taxa related to the Oligocene *Leptictis* (Novacek, 1986) need to be sampled in order to corroborate the traditional hypothesis that *Leptictis* is a part of the crown Placentalia, and that zhelestids are in the laurasiatherian clade (Archibald et al., 2001; Averianov and Archibald, 2003). Minimum age for *Sinodelphys* and *Eomaia* after Swisher et al. (1999); ages of basal boreosphenidans after Kielan-Jaworowska et al. (2004); ages for Cretaceous eutherians from Nessov et al. (1998), Archibald et al. (2001), Averianov and Archibald (2003), and Kielan-Jaworowska et al. (2004); ages of Cretaceous and Paleocene metatherians from Clemens (1966), Fox (1987), Szalay and Trofimov (1996), Cifelli et al. (1997), Muizon (1995, 1998), Muizon et al. (1997), Averianov and Kielan-Jaworowska (1999), Kielan-Jaworowska and Cifelli (2001). Horizontal bars (light gray) representing the fossil ages of extant placental and marsupial orders. Molecular estimate (vertical band in dark gray) of the divergence of placental superordinal clades (xenarthrans versus cetartiodactyls) after Murphy et al. (2001; see also Springer et al., 2003: table 1). Molecular estimates for marsupial superordinal clades (Didelphimorphia vs. Paucituberculata; Microbiotheria vs. Diprotodontia) after Springer (1997; also see Nilsson et al., 2004). Cretaceous and Cenozoic stages: **Ab**, Albian; **Ap**, Aptian; **Bm**, Barremian; **Bs**, Berriasian; **Ca**, Campanian; **Ce**, Cenomanian; **C**, Coniacian; **Eo**, Eocene; **H**, Hauterivian; **Ma**, Maastrichtian; **Pa**, Paleocene; **O-P**, Oligocene through Pleistocene (not to scale); **S**, Santonian; **T**, Turonian; **V**, Valanginian.

more similar to spalacotherioids than metatherians in the structure of their wrists. Early eutherians are also distinguishable from noneutherians by a combination of derived and primitive features, such as the presence of the transpromontorial sulcus of the internal carotid artery, presence of a groove for the stapedial artery branched off from the internal carotid artery, and presence of a notched posttympanic process in the petrosal bone (Wible et al., 1995, 2001, 2004).

The exceptionally complete skeletal remains of various mammals in the Lower Cretaceous Yixian Formation have revealed much ecomorphological information about these mammals (Chang et al., 2003; Zhou et al., 2003; Ji et al., 2004). Mammals in the Yixian biota show a greater range of body sizes than any mammalian faunas in the Jurassic (Hu et al., 2005), and they developed diverse locomotory adaptations. Most eutriconodontans have terrestrial habits (Jenkins and Schaff, 1988; Ji et al., 1999; Hu et al., 2005) but some could have been fossorial (Z.-X. Luo, pers. obs.). Most Cretaceous multituberculates are either generalized terrestrial animals or burrowers (Kielan-Jaworowska and Gambaryan, 1994; Hu and Wang, 2002). The spalacotherioid symmetrodontans from the Yixian Formation are generalized terrestrial animals (Hu et al., 1997, 1998; Rougier et al., 2003; Luo and Ji, 2005; Li and Luo, 2006). By comparison, the metatherian *Sinodelphys* and the eutherian *Eomaia* have many derived features suitable for climbing (Ji et al., 2002; Luo et al., 2003). Coupled with the fact that the Late Jurassic dryolestoid *Henkelotherium* also has scansorial adaptations, *Sinodelphys* and *Eomaia* suggest that the phylogenetic diversification of the derived crown therians and their proximal relatives was accompanied by more frequent development of scansorial skeletal adaptations, which may have facilitated the spreading of the derived groups to other niches that would have been inaccessible to plesiomorphic terrestrial mammals (Weil, 2002; Luo et al., 2003).

Late Cretaceous Diversification of Stem Metatherians and Stem Eutherians

The only new lineage of Mesozoic mammals to appear during the Late Cretaceous is Gondwanatheria, a group characterized by precociously derived and coalesced molar cusps that form transverse ridges and furrows on the tooth crown, and by very thick enamel with continuous, interrow sheets of dentine. Some derived gondwanatheriids have hypsodont (high-crowned) molars. Gondwanatherians were previously considered either placentals (Scillato-Yané and Pascual, 1985; Bonaparte, 1986, 1990) or multituberculates (Krause and Bonaparte, 1993; Kielan-Jaworowska and Bonaparte, 1996). Most recently, however, these mammals have been interpreted as an entirely independent lineage (Pascual et al., 1999; Kielan-Jaworowska et al., 2004). Gondwanatherians had a modest diversity but wide geographic distribution in South America, Madagascar, Africa, and India (Krause et al., 1997, 2003).

The best example of successive episodes of diversification within a major mammalian lineage is found among the Late Cretaceous cimolodontan multituberculates. These herbivorous mammals are taxonomically diverse and numerically abundant, dominating some Late Cretaceous mammalian communities in Asia and North America (Clemens, 1963; Clemens and Kielan-Jaworowska, 1979; Kielan-Jaworowska and Hurum, 1997; Weil, 1999). The faunal successions of the Late Jurassic, Early Cretaceous, and Late Cretaceous multituberculates roughly correspond to the polarity of phylogenetic evolution. The more derived Late Cretaceous multituberculates are the faunal replacements of the more primitive multituberculate groups that thrived during the Late Jurassic and Early Cretaceous (Simmons, 1993; Kielan-Jaworowska and Hurum, 2001). It is most noteworthy that the Late Cretaceous multituberculate families and genera are not linear descendants from the plagiaulacidan multituberculates dominating the Late Jurassic or the arginbaatarids that are common in the Early Cretaceous. Rather, these multituberculates represent a new episode of diversification of clustered subclades within the monophyletic group Cimolodonta, almost all of which are Late Cretaceous in age (Clemens and Kielan-Jaworowska, 1979; Kielan-Jaworowska and Hurum, 2001; Weil, 1999).

There is a consensus (Fig. 9.3) that the Asiatic and North American metatherians of the Early Cretaceous, such as *Sinodelphys,* deltatheroidans, and *Asiatherium* are the basalmost metatherians, and are placed outside the crown Marsupialia (Rougier et al., 1998; Luo et al., 2003; Asher et al., 2004). The earliest-known taxon with derived dental characters of crown Marsupialia is *Kokopellia* from the Cedar Mountain Formation that straddles the Early–Late Cretaceous transition in North America (Cifelli and Muizon, 1997; Cifelli et al., 1997). Many metatherians from the Late Cretaceous of North America are fairly similar to extant marsupials in dental characters, but it is uncertain whether these taxa can be placed within the extant marsupials in cladistic phylogeny, after incorporating all morphological evidence. Determining the relationship (or the lack thereof) of North American Late Cretaceous metatherians to present-day South American marsupials is crucial for rooting the marsupial tree.

Extant didelphoids, a predominantly South American group, are the basalmost clade within Marsupialia, a finding based on most molecular and morphological studies in the last decade (reviewed by Retief et al., 1995; Springer et al., 1997; Amrine-Madsen et al., 2003; Horovitz and Sánchez-Villagra, 2003; Asher et al., 2004; but see Hershkovitz, 1992; Kirsch et al., 1997). Modern didelphoids and their putative relatives are grouped together as Didelphimorphia (e.g., Szalay, 1982; McKenna and Bell, 1997). In traditional classifications, some Late Cretaceous metatherians of North America were placed within the "Didelphimorphia" as a part of the marsupial crown group (Aplin and Archer, 1987; Marshall, 1987). More recent comprehensive cladistic analyses have excluded many of these Late Cretaceous metatherians from the crown

Marsupialia and regarded these to be proximate relatives to Marsupialia as a whole, but not closely related to didelphoids (Rougier et al., 1998; Luo et al., 2003).

In traditional classifications, the North American Late Cretaceous pediomyids, stagodontids, and *Glasbius* (Clemens, 1966; Lillegraven, 1969; Fox, 1987) were related, respectively, to the South American Cenozoic didelphoids, borhyaenoids, and caroloameghinoideans (e.g., Marshall, 1987; Reig et al., 1987). Didelphoids and borhyaenoids are placed with extant American opossums (Didelphidae) in the Didelphimorphia (e.g., McKenna and Bell, 1997), and the Caroloameghinoidea with extant shrew opossums (Caenolestoidea) under Paucituberculata (Aplin and Archer, 1987; Kielan-Jaworowska et al., 2004).

These suggested affinities of North American Late Cretaceous metatherians to the subgroups of Cenozoic marsupials are not supported by recent cladistic studies. Recent parsimony analyses have unequivocally excluded *Pediomys* (a pediomyid), *Didelphodon* (a stagodontid), *Mayulestes* (a borhyaenid), and *Pucadelphys* (a putative didelphoid) from didelphimorphians and placed these taxa outside crown group Marsupialia (Rougier et al., 1998; Horovitz and Sánchez-Villagra, 2003; Luo et al., 2003; Asher et al., 2004). Removal of North American Late Cretaceous metatherians from the didelphimorphian and the paucituberculatan clades has significant implications for the early biogeographic evolution of marsupials and the timing of divergence of the basal marsupial clades (Marshall et al., 1990; Szalay, 1994; Muizon et al., 1997). Association of some North American Late Cretaceous taxa with didelphimorphians and paucituberculatans would imply that these two extant clades had initially diversified in North America during the Late Cretaceous (earlier than Campanian age, or 80 mya or more); their respective lineages dispersed separately to South America in the Early Cenozoic. A corollary of exclusion of the Late Cretaceous North American metatherians from the crown Marsupialia (e.g., Rougier et al., 1998; Luo et al., 2003) would imply that the former represent an episode of dead-end evolutionary experiments before the origin of the crown Marsupialia. They are related to the crown Marsupialia as a whole, but not closely related to the subclades of Marsupialia. Didelphoids and paucituberculatans would largely represent a separate, later diversification in South America during the early Cenozoic.

Ideally, the molecular time estimates of divergence of superordinal clades of extant marsupials should be tested by the Cretaceous fossil record of metatherians. The Didelphimorphia-Paucituberculata divergence was estimated to be 65 to 62 mya by study of 12S rRNA (Springer, 1997), or ~69 mya by analyses of whole mitochondrial genomes (Nilsson et al., 2004). If the putative relationships of North American Late Cretaceous metatherians with either didelphimorphians or paucituberculatans would be corroborated after incorporating these taxa and all relevant morphological features in the analyses, then these molecular time estimates for the origins of marsupial lineages would gain support. Current morphological analyses suggest that none of the Late Cretaceous fossil taxa can be placed within didelphimorphians or paucituberculatans (Fig. 9.3). The crown group of a clade must be younger than stem subclades. The age of the Late Cretaceous fossil metatherians outside the crown group is older than the molecular time estimates for the divergence within the crown group. In this sense, the fossil record (as summarized by Rougier et al., 1998; Luo et al., 2003) does not contradict the molecular time estimates for the latest Cretaceous diversifications of extant marsupial superorders. However, the current fossil evidence falls short of fully corroborating the molecular estimates because no fossil of the same age as the molecular estimates can be placed in any of the extant superorders.

A key issue in interpreting early eutherian evolution is whether some Asiatic Late Cretaceous eutherians can be placed in the crown group Placentalia. Novacek et al. (1997) argued that several Mongolian Late Cretaceous eutherians form an asioryctitherian group, which is an endemic clade of Asia and is outside crown placentals. This has been supported by two subsequent studies (Ji et al., 2002; Wible et al., 2004). Zalambdalestids and zhelestids, two Late Cretaceous eutherian families, were also placed with the superordinal clades of Euarchontoglires and Laurasiatheria within crown placentals (Archibald, 1996; Nessov et al., 1998; Archibald et al., 2001; Averianov and Archibald, 2003, 2005; Ekdale et al., 2004). If their phylogenetic positions can be corroborated, then at least some superordinal lineages of placentals must have originated during the Late Cretaceous. This would be consistent with the long-fuse model that some superordinal clades of placentals have a long history in the Cretaceous but that their great evolutionary diversification did not occur until the beginning of the Cenozoic (Archibald and Deutschman, 2001). A recent analysis has excluded both zalambdalestids and zhelestids from crown Placentalia (Asher et al., 2005). Another comprehensive analysis including all controversial clades has placed zhelestids in crown Placentalia, but excluded zalambdalestids (Luo and Wible, 2005; see also Fig. 9.3). Given the uncertainty in placing any Cretaceous taxa within the superordinal clades of placentals, the Cretaceous age of crown placentals is tentative.

Notwithstanding these uncertainties, it is safe to conclude that all known Early Cretaceous eutherians (*Prokennalestes, Eomaia, Murtoilestes, Montanalestes*) and many Late Cretaceous eutherians (asioryctitherians and several other taxa) represent two successive episodes of diversification before the origin of crown Placentalia and cannot be directly related to any superorders of placentals (Fig. 9.3). Early Cretaceous eutherians are evolutionary dead ends with respect to most of the Late Cretaceous eutherians, and many Late Cretaceous eutherians (with the exception of zalambdalestids and zhelestids) are dead ends with respect to crown placentals.

Discussion

All lineages of Mesozoic mammals, mammaliaforms, and mammaliamorphs have significant shorter longevity than what was generally accepted before the 1970s. Although there is one report on a putative Paleocene therapsid, implying survival of a Mesozoic therapsid lineage into the early Cenozoic (Fox et al., 1992), this claim has been questioned (Sues, 1992). At the grandest scale, the split of the monotreme lineage from the therian lineage occurred no later than the Middle Jurassic, before 165 million years (Rowe, 1993). All subclades within crown Mammalia and all ordinal or family clades of mammaliamorphs have shorter temporal ranges (Table 9.1).

The mammaliamorph group with the longest-known stratigraphic range is Tritylodontidae. It extended from the Rhaetian stage of the Triassic to the Albian stage of the Cretaceous, some 100 million years. The most long-lived mammaliaform orders are the Docodonta and the Haramiyida. The youngest uncontested docodontan is from the Albian stage (about 105 mya; Maschenko et al., 2002), whereas the oldest and uncontested docodontan is from Middle Jurassic (about 165 mya; Kermack et al., 1987). This gives docodontans a stratigraphic range of more than 60 million years. If the docodontan affinities of the Late Triassic *Delsatia* (Sigogneau-Russell and Godefroit, 1997), *Woutersia* (Butler, 1997), or an unnamed, putative docodont from Early Jurassic of India (Prasad and Manhas, 2001) can be further corroborated, then docodontans could have existed for 80 to 100 million years (dashed line of the docodontan lineage in Fig. 9.1). With the earliest record in the Late Triassic and the youngest record in the Late Jurassic, Haramiyida ranged for about 70 million years. Their affinities to multituberculates have been questioned (Rowe, 1988; Jenkins et al., 1997; but see Butler, 2000). The mammalian order with the longest range is Multituberculata (sensu stricto; Kielan-Jaworowska et al., 2004). The earliest uncontested multituberculates are from the Late Jurassic; the youngest uncontested multituberculates are from the late Eocene (Kielan-Jaworowska et al., 2004). These records indicate that multituberculates ranged more than 120 million years. The second-longest lineage of Mesozoic mammalian order is Eutriconodonta with a range from the Middle Jurassic to the Late Cretaceous, spanning 100 million years. The eutriconodontan family Triconodontidae extended from the Late Jurassic to Late Cretaceous for about 80 million years. Another long-lived Mesozoic mammalian order is Dryolestida. The earliest dryolestoidans are from the Late Jurassic and the youngest date from the Paleocene, for a range of about 90 million years (Simpson, 1928, 1929; Prothero, 1981; Martin, 1999; Gelfo and Pascual, 2001).

Mammaliamorph phylogeny comprises more than 30 distinctive clades, and in the traditional Linnean classification, these are all order-level or family-level taxa (Benton, 2004; Kielan-Jaworowska et al., 2004; Kemp, 2005). With the exception of the aforementioned six groups, other orders and families of mammaliamorphs, mammaliaforms, and Mesozoic mammals did not last more than 50 million years on the basis of their current stratigraphic records and best-resolved cladistic relationships (for a summary, see Kielan-Jaworowska et al., 2004).

The temporal ranges of the mammaliamorph clades are not nearly as extensive as postulated before the 1960s (e.g., Simpson, 1945; Patterson, 1956; Olson, 1959). The age for the separation of monotremes from therians is also significantly younger than that inferred from the fundamental dichotomy of Mesozoic mammals in the 1970s (Rowe, 1993). In general, it is very rare for the duration of order- and family-level mammaliamorph clades to exceed 100 million years. This prevalent pattern tends not to favor the hypothesis of multiple long-lived Mesozoic mammalian lineages, as is required by the short-fuse model of mammaliamorph diversification.

After several rounds of revision on the molecular timescales for extant mammalian clades (Kumar and Hedges, 1998; Hedges and Kumar, 1999; Murphy et al., 2001), the latest molecular estimates (Springer et al., 2003: table 1) suggested 65 or 69 mya for didelphimorphian and paucituberculatan marsupials (Springer, 1997; Nilsson et al., 2004), 98–117 mya for the divergence of afrotherians, 94–111 mya for xenarthrans, 80–90 mya for laurasiatherians, and 81–94 mya for euarchontoglirans. These range estimates are close to or exceed the upper ranges for the known durations of ordinal clades in mammaliamorph macroevolution. Some more recent molecular estimates of placental superordinal divergences have used multiple calibrations and taken into account variation in the molecular evolutionary rates (Douzery et al., 2003, 2004), and have reduced the discrepancy between the molecular timescales and the fossil record. Nevertheless, even these latest molecular studies still placed the crucial diversification of the placental superordinal lineages before the Cretaceous-Cenozoic transition.

Conclusions

(1) The Mesozoic mammalian family tree is a vast evolutionary bush. The phylogenetic diversity of mammaliamorphs, mammaliaforms, and mammals has been enriched by new and distinct lineages discovered in recent years, and by addition of numerous clade ranks, as resolved by recent phylogenetic analyses.

(2) The most common and prominent macroevolutionary pattern among mammaliamorphs can be characterized by successive episodes of diversification of relatively short-lived clades, or a series of explosive models with modest diversities. In a major clade, most subclades are clustered into several episodes of diversification. Subclades clustered in a previous episode of diversification tend to be stem clades with regard to the next such cluster of subclades, and they are dead-end evolutionary experiments, often with no direct relationship to the emergent clades in the younger mammalian assemblages at the order or family level. Taxonomic turnovers be-

tween faunas of different geological epochs and within the high-level clades are by the successive clusters of emergent clades. This macroevolutionary pattern of successive diversifications is manifest in the mammalian faunal transitions from the Early to the Middle Jurassic, from the Jurassic to the Cretaceous, from the Early to the Late Cretaceous, and from the Cretaceous to the Cenozoic. The successive clusters of relatively short-lived clades also dominated the evolutionary pattern with the autralosphenidan-monotreme lineage, the metatherian-marsupial lineage, the eutherian-placental lineage, and the successive ranks of clades from theriimorphs to crown Theria (Figs. 9.1, 9.2).

(3) Less common in the macroevolutionary history of mammaliamorphs is the pattern of the long-fuse model in which subclades of a major group maintained a relatively long history of low diversity, and diversification of these respective subclades did not occur until long after they became independent. This pattern is obvious in the faunal transition from the Middle to the Late Jurassic, and in the evolutionary turnover of taxa within docodontans, eutriconodontans, multituberculates, and obtuse-triangled and acute-triangled symmetrodontans. It is likely that this also occurred in one of the placental superordinal lineages.

(4) Current stratigraphic and phylogenetic data of mammaliamorph clades failed to show any examples of the short-fuse model, in which the origin of a main group is immediately followed by rapid splits and diversifications of its subclades that are long branches with great longevity. This is a significant change from the pattern of multiple distinctive lineages of Cenozoic mammals with long history dating back to the Permo-Triassic, as predicted by the Simpsonian paradigm. It is also substantially different from the two main divisions of Cenozoic mammals extending back into the Late Triassic, as widely accepted before the 1970s. Current data on Cretaceous metatherians and eutherians do not contradict the molecular estimates that some extant mammalian superorders may extend into the Cretaceous, but they fall short of fully corroborating the molecular hypotheses that multiple long-lived extant clades diversified during the Cretaceous.

Acknowledgments

This paper is a tribute to Professor Robert L. Carroll for his many pioneering studies on evolutionary transitions of major vertebrate groups. This study has benefited from many years of stimulating discussions on early mammalian evolution with K. C. Beard, J. F. Bonaparte, R. L. Cifelli, W. A. Clemens, A. W. Crompton, M. R. Dawson, Q. Ji, Z. Kielan-Jaworowska, J. A. Hopson, T. Martin, T. H. Rich, G. W. Rougier, T. Rowe, H.-D. Sues, A. Weil, and J. R. Wible. I am indebted to Mary Dawson, Zofia Kielan-Jaworowska, and Gregory Wilson for their help in improving the manuscript and Mark Klingler for assistance in graphics. I also thank Jason Anderson and Hans Sues for their encouragement and patience. Recent research support was received from National Science Foundation (grant DEB-0316558), National Natural Science Foundation of China (grant 40328004), National Geographic Society (grant 7042-01), and the Carnegie Museum of Natural History.

References Cited

Allin, E. F., and J. A. Hopson. 1992. Evolution of the auditory system in Synapsida ("mammal-like reptiles" and primitive mammals) as seen in the fossil record; pp. 587–614 in D. B. Webster, R. R. Fay, and A. N. Popper (eds.), The Evolutionary Biology of Hearing. Springer-Verlag, New York.

Amrine-Madsen, H., M. Scally, M. Westerman, M. J. Stanhope, C. W. Krajewski, and M. S. Springer. 2003. Nuclear gene sequences provide evidence for the monophyly of australidelphian marsupials. Molecular Phylogenetics and Evolution 28:186–196.

Anantharaman, S., G. P. Wilson, D. C. Das Sarma, and W. A. Clemens. 2006. A possible Late Cretaceous "haramiyidan" from India. Journal of Vertebrate Paleontology 26:488–490.

Aplin, K., and M. Archer. 1987. Recent advances in marsupial sysytematics with a new syncretic classification; pp. xv–lxxii in M. Archer (ed.), Possum and Opossums—Studies in Evolution. Surrey Beatty & Sons and Royal Zoological Society of New South Wales, Sydney.

Archer, M., T. F. Flannery, A. Ritchie, and R. Molnar. 1985. First Mesozoic mammal from Australia—an Early Cretaceous monotreme. Nature 318:363–366.

Archer, M., P. Murray, S. J. Hand, and H. Godthelp. 1993. Reconsideration of monotreme relationships based on the skull and dentition of the Miocene *Obdurodon dicksoni;* pp. 75–94 in F. S. Szalay, M. J. Novacek, and M. C. McKenna (eds.), Mammal Phylogeny: Mesozoic Differentiation, Multituberculates, Monotremes, Early Therians, and Marsupials. Springer-Verlag, New York.

Archer, M., F. A. Jenkins Jr., S. J. Hand, P. Murray, and H. Godthelp. 1992. Description of the skull and non-vestigial dentition of a Miocene platypus (*Obdurodon dicksoni* n. sp.) from Riversleigh, Australia, and the problem of monotreme origins; pp. 15–27 in M. L. Augee (ed.), Platypus and Echidnas. Royal Zoological Society of New South Wales, Sydney.

Archibald, J. D. 1996. Fossil evidence for a Late Cretaceous origin of "hoofed" mammals. Science 272:1150–1153.

Archibald, J. D. 2003. Timing and biogeography of the eutherian radiation: fossils and molecules compared. Molecular Phylogenetics and Evolution 28:350–259.

Archibald, J. D., and D. H. Deutschman. 2001. Quantitative analysis of the timing of the origin and diversification of extant placental orders. Journal of Mammalian Evolution 8:107–124.

Archibald, J. D., A. O. Averianov, and E. G. Ekdale. 2001. Late Cretaceous relatives of rabbits, rodents, and other extant eutherian mammals. Nature 414:62–65.

Argot, C. 2001. Functional-adaptive anatomy of the forelimb in the Didelphidae, and the paleobiology of the Paleocene marsupials *Mayulestes ferox* and *Pucadelphys andinus*. Journal of Morphology 247:51–79.

Argot, C. 2002. Functional-adaptive analysis of the hindlimb anatomy of extant marsupials and the paleobiology of the Paleocene marsupials *Mayulestes ferox* and *Pucadelphys andinus.* Journal of Morphology 253:76–108.

Asher, R. J., I. Horovitz, and M. R. Sanchez-Villagra. 2004. First combined cladistic analysis of marsupial mammal interrelationships. Molecular Phylogenetics and Evolution 33:240–250.

Asher, R. J., J. Meng, J. R. Wible, M. C. McKenna, G. W. Rougier, D. Dashzeveg, and M. J. Novacek. 2005. Stem Lagomorpha and the antiquity of Glires. Science 307:1091–1094.

Averianov, A. O. 2002. Early Cretaceous "symmetrodont" mammal *Gobiotheriodon* from Mongolia and the classification of "Symmetrodonta." Acta Palaeontologica Polonica 47:705–716.

Averianov, A. O., and J. D. Archibald. 2003. Mammals from the Upper Cretaceous Aitym Formation, Kyzylkum Desert, Uzbekistan. Cretaceous Research 24:171–191.

Averianov, A. O., and J. D. Archibald. 2005. Mammals from the mid-Cretaceous Khodzhakul Formation, Kyzylkum Desert, Uzbekistan. Cretaceous Research 26:593–608.

Averianov, A. O., and Z. Kielan-Jaworowska. 1999. Marsupials from the Late Cretaceous of Uzbekistan. Acta Palaeontologica Polonica 44:71–81.

Averianov, A. O., and P. P. Skutschas. 2000. A eutherian mammal from the Early Cretaceous of Russia and biostratigraphy of the Asian Early Cretaceous vertebrate assemblages. Lethaia 33:330–340.

Averianov, A. O., and P. P. Skutschas. 2001. A new genus of eutherian mammal from the Early Cretaceous of Transbaikalia, Russia. Acta Palaeontologica Polonica 46:431–436.

Benton, M. J. 1999. Early origins of modern birds and mammals: molecules vs. morphology. BioEssays 21:1043–1051.

Benton, M. J. 2004. Vertebrate Palaeontology. Third edition. Blackwell, Oxford, 455 pp.

Bever, G., T. Rowe, E. G. Ekdale, T. E. Macrini, M. W. Colbert, and A. Balanoff. 2005. Comment on "Independent origins of middle ear bones in monotremes and therians" (I). Science 309:1492a.

Bonaparte, J. F. 1986. A new and unusual Late Cretaceous mammal from Patagonia. Journal of Vertebrate Paleontology 6:264–270.

Bonaparte, J. F. 1990. New Late Cretaceous mammals from the Los Alamitos Formation, northern Patagonia. National Geographic Research 6:63–93.

Bonaparte, J. F., and G. W. Rougier. 1987. Mamíferos del Cretácico Inferior de Patagonia. Actas IV Congreso Latinamericano de Paleontología 1:343–359.

Bonaparte, J. F., A. G. Martinelli, C. L. Schultz, and R. Rubert. 2003. The sister group of mammals: small cynodonts from the Late Triassic of southern Brazil. Revista Brasileira de Paleontologia 5:5–27.

Butler, P. M. 1978. A new interpretation of the mammalian teeth of tribosphenic pattern from the Albian of Texas. Breviora 446:1–27.

Butler, P. M. 1997. An alternative hypothesis on the origin of docodont molar teeth. Journal of Vertebrate Paleontology 17:435–439.

Butler, P. M. 2000. Review of the early allotherian mammals. Acta Palaeontologica Polonica 45:317–342.

Butler, P. M., and W. A. Clemens. 2001. Dental morphology of the Jurassic holotherian mammal *Amphitherium,* with a discussion of the evolution of mammalian post-canine dental formulae. Palaeontology 44:1–20.

Butler, P. M., and J. J. Hooker. 2005. New teeth of allotherian mammals from the English Bathonian, including the earliest multituberculates. Acta Palaeontologica Polonica 50:185–207.

Chang, M.-M., P.-J. Chen, Y.-Q. Wang, and Y. Wang (eds.). 2003. The Jehol Biota. The Emergence of Feathered Dinosaurs, Beaked Birds and Flowering Plants. Shanghai Scientific & Technical Publishers, Shanghai, 208 pp.

Chow, M., and T. H. Rich. 1982. *Shuotherium dongi,* n. gen. and sp., a therian with pseudo-tribosphenic molars from the Jurassic of Sichuan, China. Australian Mammalogy 5:127–142.

Chure, D. J., K. Carpenter, R. Litwin, S. Hasiotis, and E. Evanoff. 1998. The flora and fauna of the Morrison Formation. Modern Geology 23:507–537.

Cifelli, R. L. 1993. Theria of metatherian-eutherian grade and the origin of marsupials; pp. 205–215 in F. S. Szalay, M. J. Novacek, and M. C. McKenna (eds.), Mammal Phylogeny: Mesozoic Differentiation, Multituberculates, Monotremes, Early Therians, and Marsupials. Springer-Verlag, New York.

Cifelli, R. L. 1999. Tribosphenic mammal from the North American Early Cretaceous. Nature 401:363–366.

Cifelli, R. L. 2000. Counting premolars in early eutherian mammals. Acta Palaeontologica Polonica 45:195–198.

Cifelli, R. L. 2001. Early mammalian radiations. Journal of Paleontology 75:1214–1226.

Cifelli, R. L., and C. de Muizon. 1997. Dentition and jaw of *Kokopellia juddi,* a primitive marsupial or near marsupial from the medial Cretaceous of Utah. Journal of Mammalian Evolution 4:241–258.

Cifelli, R. L., and S. K. Madsen. 1999. Spalacotheriid symmetrodonts (Mammalia) from the medial Cretaceous (upper Albian or lower Cenomanian) Mussentuchit local fauna, Cedar Mountain Formation, Utah, USA. Geodiversitas 21:167–214.

Cifelli, R. L., J. I. Kirkland, A. Weil, A. L. Deino, and B. J. Kowallis. 1997. High-precision $^{40}Ar/^{39}Ar$ geochronology and the advent of North America's Late Cretaceous terrestrial fauna. Proceedings of the National Academy of Sciences USA 94:11163–11167.

Cifelli, R. L., T. B. Rowe, W. P. Luckett, J. Banta, R. Reyes, and R. I. Howes. 1996. Origins of marsupial pattern of dental replacement: fossil evidence revealed by high-resolution X-ray CT. Nature 379:715–718.

Clemens, W. A. 1963. Fossil mammals of the type Lance Formation, Wyoming. Part I. Introduction and Multituberculata. University of California Publications in Geological Sciences 48:1–105.

Clemens, W. A. 1966. Fossil mammals from the type Lance Formation, Wyoming. Part II. Marsupialia. University of California Publications in Geological Sciences 62:1–122.

Clemens, W. A. 1980. Rhaeto-Liassic mammals from Switzerland and West Germany. Zitteliana 5:51–92.

Clemens, W. A., and Z. Kielan-Jaworowska. 1979. Multituberculata; pp. 99–149 in J. A. Lillegraven, Z. Kielan-Jaworowska, and W. A. Clemens (eds.), Mesozoic Mammals: The First Two-thirds of Mammalian History. University of California Press, Berkeley.

Clemens, W. A., and J. A. Lillegraven. 1986. New Late Cretaceous, North American advanced therian mammals that fit neither the marsupial nor eutherian molds; pp. 55–85 in K. M. Flanagan and J. A. Lillegraven (eds.), Vertebrates, Phylogeny, and Philosophy. Contributions to Geology, University of Wyoming, Special Paper 3.

Clemens, W. A., and J. R. E. Mills. 1971. Review of *Peramus tenuirostris.* Bulletin of the British Museum (Natural History), Geology 20:89–113.

Crompton, A. W. 1971. The origin of the tribosphenic molar; pp. 65–87 in D. M. Kermack and K. A. Kermack (eds.), Early Mammals. Zoological Journal of the Linnean Society 50, Supplement 1. Academic Press, London.

Crompton, A. W. 1974. The dentition and relationships of the southern African Triassic mammals, *Erythrotherium parringtoni* and *Megazostrodon rudnerae.* Bulletin of the British Museum (Natural History), Geology 24:397–437.

Crompton, A. W., and F. A. Jenkins Jr. 1979. Origin of mammals; pp. 59–73 in J. A. Lillegraven, Z. Kielan-Jaworowska, and W. A. Clemens (eds.), Mesozoic Mammals: The First Two-thirds of Mammalian History. University of California Press, Berkeley.

Crompton, A. W., and Z.-X. Luo. 1993. Relationships of the Liassic mammals *Sinoconodon, Morganucodon,* and *Dinnetherium;* pp. 30–44 in F. S. Szalay, M. J. Novacek, and M. C. McKenna (eds.), Mammal Phylogeny: Mesozoic Differentiation, Multituberculates, Monotremes, Early Therians, and Marsupials. Springer-Verlag, New York.

Crompton, A. W., and A.-L. Sun. 1985. Cranial structure and relationships of the Liassic mammal *Sinoconodon.* Zoological Journal of the Linnean Society 85:99–119.

Damiani, R., S. P. Modesto, A. Yates, and J. Neveling. 2003. Earliest evidence of cynodont burrowing. Proceedings of Royal Society of London, B 270:1747–1751.

Dashzeveg, D. 1994. Two previously unknown eupantotheres (Mammalia, Eupantotheria). American Museum Novitates 3107:1–11.

Dashzeveg, D., and Z. Kielan-Jaworowska. 1984. The lower jaw of an aegialodontid mammal from the Early Cretaceous of Mongolia. Zoological Journal of the Linnean Society 82:217–227.

Datta, P. M. 2005. Earliest mammal with transversely expanded upper molar from the Late Triassic (Carnian) Tiki Formation, South Rewa Gondwana Basin, India. Journal of Vertebrate Paleontology 25:200–207.

Datta, P. M., and D. P. Das. 1996. Discovery of the oldest fossil mammal from India. India Minerals 50:217–222.

Datta, P. M., D. P. Das, and Z.-X. Luo. 2004. A Late Triassic dromatheriid (Synapsida: Cynodontia) from India. Annals of Carnegie Museum 73:72–84.

Douzery, E. J. P., F. Delsuc, M. J. Stanhope, and D. Huchon. 2003. Local molecular clocks in three nuclear genes: divergence times for rodents and other mammals and incompatibility among fossil calibrations. Journal of Molecular Evolution 57:S201–S213.

Douzery, E. J. P., E. A. Snell, E. Baptestes, F. Delsuc, and H. Philippe. 2004. The timing of eukaryotic evolution: does a relaxed molecular clock reconcile proteins and fossils? Proceedings of National Academy of Sciences USA 101:15386–15391.

Ekdale, E. G., J. D. Archibald, and A. O. Averianov. 2004. Late Cretaceous petrosal bones of zhelestid and zalambdalestid eutherian mammals from Uzbekistan. Acta Palaeontologica Polonica 49:161–176.

Engelmann, G. F., and G. Callison. 1998. Mammalian faunas of the Morrison Formation. Modern Geology 23:343–379.

Ensom, P. C., and D. Sigogneau-Russell. 2000. New symmetrodonts (Mammalia, Theria) from the Purbeck Limestone Group, Early Cretaceous of southern England. Cretaceous Research 21:767–779.

Flannery, T. F., M. Archer, T. H. Rich, and R. Jones. 1995. A new family of monotremes from the Cretaceous of Australia. Nature 377:418–420.

Flynn, J. J., J. M. Parrish, B. Rakotosamimanana, W. F. Simpson, and A. Wyss. 1999. A Middle Jurassic mammal from Madagascar. Nature 401:57–60.

Foote, M., J. P. Hunter, C. M. Janis, and J. J. Sepkoski. 1999. Evolutionary and preservational constraints on origins of biologic groups: divergence times of eutherian mammals. Science 283:1310–1314.

Forasiepi, A., and A. Martinelli. 2003. Femur of a monotreme (Mammalia, Monotremata) from the Early Paleocene Salamanca Formation of Patagonia, Argentina. Ameghiniana 40:625–630.

Forasiepi, A., G. W. Rougier, and A. Martinelli. 2004. A new mammal from the Jurassic Cañadon Asfalto Formation, Chubut Province (Argentina). Journal of Vertebrate Paleontology 24(Suppl. to 3):59A.

Fox, R. C. 1975. Molar structure and function in the Early Cretaceous mammal *Pappotherium*: evolutionary implications for Mesozoic Theria. Canadian Journal of Earth Sciences 12:412–442.

Fox, R. C. 1976. Additions to the mammalian local fauna from the upper Milk River Formation (Upper Cretaceous), Alberta. Canadian Journal of Earth Sciences 13:1105–1118.

Fox, R. C. 1987. Palaeontology and the early evolution of marsupials; pp. 161–169 in M. Archer (ed.), Possums and Opossums: Studies in Evolution. Surrey Beatty & Sons and the Royal Zoological Society of New South Wales, Sydney.

Fox, R. C., G. P. Youzwyshyn, and D. W. Krause. 1992. Post-Jurassic mammal-like reptile from the Palaeocene. Nature 358:233–235.

Freeman, E. F. 1979. A Middle Jurassic mammal bed from Oxfordshire. Palaeontology 22:135–166.

Gambaryan, P. P. and A. O. Averianov. 2001. Femur of a morganucodontid mammal from the Middle Jurassic of central Russia. Acta Palaeontologica Polonica 46:99–112.

Gelfo, J. N., and R. Pascual. 2001. *Peligrotherium tropicalis* (Mammalia, Dryolestida) from the early Paleocene of Patagonia, a survival from a Mesozoic Gondwanan radiation. Geodiversitas 23:369–379.

Gill, P. 2004. A new symmetrodont from the Early Cretaceous of England. Journal of Vertebrate Paleontology 24:748–752.

Gow, C. E. 2001. A partial skeleton of the tritheledontid *Pachygenelus* (Therapsida, Cynodontia). Palaeontologica Africana 37:93–97.

Graur, D., and W. Martin. 2004. Reading the entrails of chickens: molecular timescales of evolution and the illusion of precision. Trends in Genetics 20:80–86.

Hahn, G. 1988. Die Ohr-Region der Paulchoffatiidae (Multituberculata, Ober-Jura). Palaeovertebrata 18:155–185.

Hahn, G. 1993. The systematic arrangement of the Paulchoffatiidae (Multituberculata) revisited. Geologica et Palaeontologica 27:201–214.

Hahn, G. 2001. Neue Beobachtungen an Schädel-Resten von Paulchoffatiidae (Multituberculata; Ober-Jura). Geologica et Palaeontologica 35:121–143.

Hahn, G., and R. Hahn. 2004. The dentition of the Plagiaulacida (Multituberculata, Late Jurassic to Early Cretaceous). Geologica et Palaeontologica 38:119–159.

Hahn, G., R. Hahn, and P. Godefroit. 1994. Zur Stellung der Dromatheriidae (Ober-Trias) zwischen den Cynodontia und den Mammalia. Geologica et Palaeontologica 28:141–159.

Hahn, G., D. Sigogneau-Russell, and G. Wouters. 1989. New data on Theroteinidae—their relations with Paulchoffatiidae and Haramiyidae. Geologica et Palaeontologica 23:205–215.

Hedges, S. B., and S. Kumar. 1999. Technical comments: divergence times of eutherian mammals. Science 285:2031a.

Heinrich, W.-D. 1999. First haramiyid (Mammalia, Allotheria) from the Mesozoic of Gondwana. Mitteilungen aus dem Museum für Naturkunde Berlin, Geowissenschaftliche Reihe 2:159–170.

Hershkovitz, P. 1992. Ankle bones: the Chilean opossum *Dromiciops gliroides* Thomas, and marsupial phylogeny. Bonner Zoologische Beiträge 43:181–213.

Hopson, J. A. 1964. The braincase of the advanced mammal-like reptile *Bienotherium*. Postilla 87:1–30.

Hopson, J. A. 1994. Synapsid evolution and the radiation of non-eutherian mammals; pp. 190–219 in R. S. Spencer (ed.), Major Features of Vertebrate Evolution. Short Courses in Paleontology, No. 7. Paleontological Society, University of Tennessee, Knoxville.

Hopson, J. A., and H. R. Barghusen. 1986. An analysis of therapsid relationships; pp. 83–106 in N. Hotton III, P. D. MacLean, J. J. Roth, and E. C. Roth (eds.), The Ecology and Biology of Mammal-like Reptiles. Smithsonian Institution Press, Washington, D.C.

Hopson, J. A., and A. W. Crompton. 1969. Origin of mammals; pp. 15–72 in T. Dobzhansky, M. K. Hecht, and W. C. Steere (eds.), Evolutionary Biology, Volume 3. Appleton-Century-Crofts, New York.

Hopson, J. A., and J. W. Kitching. 2001. A probainognathian cynodont from South Africa and the phylogeny of nonmammalian cynodonts. Bulletin of the Museum of Comparative Zoology, Harvard University 156:5–35.

Horovitz, I. 2000. The tarsus of *Ukhaatherium nessovi* (Eutheria, Mammalia) from the Late Cretaceous of Mongolia: an appraisal of the evolution of the ankle in basal therians. Journal of Vertebrate Paleontology 20:547–560.

Horovitz, I., and M. R. Sánchez-Villagra. 2003. A morphological analysis of marsupial mammal higher-level phylogenetic relationships. Cladistics 19:181–212.

Hu, Y.-M., and Y. Wang. 2002. *Sinobataar* gen. nov.: first multituberculate from Jehol Biota of Liaoning, Northern China. Chinese Science Bulletin 47:933–938.

Hu, Y.-M., J. Meng, Y.-Q. Wang, and C.-K. Li. 2005. Large Mesozoic mammals fed on young dinosaurs. Nature 433:149–153.

Hu, Y.-M., Y.-Q. Wang, C.-K. Li, and Z.-X. Luo. 1998. Morphology of dentition and forelimb of *Zhangheotherium*. Vertebrata PalAsiatica 36:102–125.

Hu, Y.-M., Y.-Q. Wang, Z.-X. Luo, and C.-K. Li. 1997. A new symmetrodont mammal from China and its implications for mammalian evolution. Nature 390:137–142.

Hunter, J. P. 2004. Alternative interpretation of molar morphology and wear in the Early Cretaceous mammal *Ausktribosphenos*. Journal of Vertebrate Paleontology 24(Suppl. to 3):73A.

Jenkins, F. A., Jr. 1970. Cynodont postcranial anatomy and the "prototherian" level of mammalian organization. Evolution 24:230–252.

Jenkins, F. A., Jr. 1971. The postcranial skeleton of African cynodonts. Bulletin of the Peabody Museum of Natural History, Yale University 36:1–216.

Jenkins, F. A., Jr. 1984. A survey of mammalian origins; pp. 32–47 in P. D. Gingerich and C. E. Badgley (eds.), Mammals: Notes for a Short Course. University of Tennessee, Department of Geological Sciences, Studies in Geology 8. Knoxville.

Jenkins, F. A., Jr., and F. R. Parrington. 1976. The postcranial skeletons of the Triassic mammals *Eozostrodon, Megazostrodon* and *Erythrotherium*. Philosophical Transactions of the Royal Society of London B 273:387–431.

Jenkins, F. A., Jr., and C. R. Schaff. 1988. The Early Cretaceous mammal *Gobiconodon* (Mammalia, Triconodonta) from the Cloverly Formation in Montana. Journal of Vertebrate Paleontology 8:1–24.

Jenkins, F. A., Jr., and W. A. Weijs. 1979. The functional anatomy of the shoulder in the Virginia opossum (*Didelphis virginiana*). Journal of Zoology (London) 188:379–410.

Jenkins, F. A., Jr., A. W. Crompton, and W. R. Downs. 1983. Mesozoic mammals from Arizona: new evidence on mammalian evolution. Science 222:1233–1235.

Jenkins, F. A., Jr., S. M. Gatesy, N. H. Shubin, and W. W. Amaral. 1997. Haramiyids and Triassic mammalian evolution. Nature 385:715–718.

Jenkins, F. A., Jr., N. H. Shubin, W. W. Amaral, S. M. Gatesy, C. R. Schaff, L. B. Clemmensen, W. R. Downs, A. R. Davidson, N. Bonde, and F. Osbaeck. 1994. Late Triassic continental vertebrates and depositional environments of the Fleming Fjord Formation, Jameson Land, East Greenland. Meddelelser om Grønland 32:3–25.

Ji, Q., Z.-X. Luo, and S. Ji. 1999. A Chinese triconodont mammal and mosaic evolution of the mammalian skeleton. Nature 398:326–330.

Ji, Q., Z.-X. Luo, C.-X. Yuan, J. R. Wible, J.-P. Zhang, and J. A. Georgi. 2002. The earliest known eutherian mammal. Nature 416:816–822.

Ji, Q., W. Chen, W.-L. Wang, X.-C. Jin, J.-P. Zhang, Y.-Q. Liu, H. Zhang, Y.-P. Yao, S.-A. Ji, C.-X. Yuan, Y. Zhang, and H.-L. You. 2004. Mesozoic Jehol Biota of Western Liaoning, China. Geological Publishing House, Beijing, 374 pp.

Kemp, T. S. 1982. Mammal-like Reptiles and the Origin of Mammals. Academic Press, London, 363 pp.

Kemp, T. S. 1983. The relationships of mammals. Zoological Journal of the Linnean Society 77:353–384.

Kemp, T. S. 2005. The Origin and Evolution of Mammals. Oxford University Press, Oxford and New York, 339 pp.

Kermack, D. M., and K. A. Kermack. 1984. The Evolution of Mammalian Characters. Croom Helm, London, 149 pp.

Kermack, K. A., and Kielan-Jaworowska, Z. 1971. Therian and non-therian mammals; pp. 103–116 in D. M. Kermack and K. A. Kermack (eds.), Early Mammals. Zoological Journal of the Linnean Society 50, Supplement 1. Academic Press, London.

Kermack, K. A., and F. Mussett. 1958. The jaw articulation of the Docodonta and the classification of Mesozoic mammals. Proceedings of the Royal Society of London B 149:204–215.

Kermack, K. A., P. M. Lees, and F. Mussett. 1965. *Aegialodon dawsoni,* a new trituberculosectorial tooth from the lower Wealden. Proceedings of the Royal Society of London B 162:535–554.

Kermack, K. A., F. Mussett, and H. W. Rigney. 1973. The lower jaw of *Morganucodon*. Zoological Journal of the Linnean Society 53:87–175.

Kermack, K. A., F. Mussett, and H. W. Rigney. 1981. The skull of *Morganucodon*. Zoological Journal of the Linnean Society 71:1–158.

Kermack, K. A., D. M. Kermack, P. M. Lees, and J. R. E. Mills. 1998. New multituberculate-like teeth from the Middle Jurassic of England. Acta Palaeontologica Polonica 43:581–606.

Kermack, K. A., A. J. Lee, P. M. Lees, and F. Mussett. 1987. A new docodont from the Forest Marble. Zoological Journal of the Linnean Society 89:1–39.

Kielan-Jaworowska, Z. 1977. Evolution of the therian mammals in the Late Cretaceous of Asia. Part II. Postcranial skeleton in *Kennalestes* and *Asioryctes*. Palaeontologia Polonica 37:65–83.

Kielan-Jaworowska, Z. 1978. Evolution of the therian mammals in the Late Cretaceous of Asia. Part III. Postcranial skeleton in Zalambdalestidae. Palaeontologia Polonica 38:5–41.

Kielan-Jaworowska, Z., and J. F. Bonaparte. 1996. Partial dentary of a multituberculate mammal from the Late Cretaceous of Argentina and its taxonomic implications. Revista del Museo Argentino de Ciencias Naturales "Bernardino Rivadavia" 145:1–9.

Kielan-Jaworowska, Z., and R. L. Cifelli. 2001. Primitive boreosphenidan mammal (?Deltatheroida) from the Early Cretaceous of Oklahoma. Acta Palaeontologica Polonica 46:377–391.

Kielan-Jaworowska, Z., R. L. Cifelli, and Z.-X. Luo. 1998. Alleged Cretaceous placental from Down Under. Lethaia 31:267–268.

Kielan-Jaworowska, Z., R. L. Cifelli, and Z.-X. Luo. 2002. Dentition and relationships of the Jurassic mammal *Shuotherium*. Acta Palaeontologica Polonica 47:479–486.

Kielan-Jaworowska, Z., R. L. Cifelli, and Z.-X. Luo. 2004. Mammals from the Age of Dinosaurs: Origins, Evolution, and Structure. Columbia University Press, New York, 630 pp.

Kielan-Jaworowska, Z., and D. Dashzeveg. 1989. Eutherian mammals from the Early Cretaceous of Mongolia. Zoologica Scripta 18:347–355.

Kielan-Jaworowska, Z., and P. P. Gambaryan. 1994. Postcranial anatomy and habits of Asian multituberculate mammals. Fossils and Strata 36:1–92.

Kielan-Jaworowska, Z., and J. H. Hurum. 1997. Djadochtatheria—a new suborder of multituberculate mammals. Acta Palaeontologica Polonica 42:201–242.

Kielan-Jaworowska, Z., and J. H. Hurum. 2001. Phylogeny and systematics of multituberculate mammals. Palaeontology 44:389–429.

Kirsch, J. A. W., F. J. Lapointe, and M. S. Springerr. 1997. DNA-hybridization studies of marsupials and their implications for metatherian classification. Australian Journal of Zoology 45:211–280.

Klima, M. 1973. Die Frühentwicklung des Schultergürtels und des Brustbeins bei den Monotremen (Mammalia: Prototheria). Advances in Anatomy, Embryology and Cell Biology 47:1–80.

Krause, D. W., and J. F. Bonaparte. 1993. Superfamily Gondwanatherioidea: a previously unrecognized radiation of multituberculate mammals in South America. Proceedings of the National Academy of Sciences USA 90:9379–9383.

Krause, D. W., and F. A. Jenkins Jr. 1983. The postcranial skeleton of North American multituberculates. Bulletin of the Museum of Comparative Zoology, Harvard University 150:199–246.

Krause, D. W., M. D. Gottfried, P. M. O'Connor, and E. M. Roberts. 2003. A Cretaceous mammal from Tanzania. Acta Palaeontologica Polonica 48:321–330.

Krause, D. W., G. V. R. Prasad, W. von Koenigswald, A. Sahni, and F. E. Grine. 1997. Cosmopolitanism among Gondwanan Late Cretaceous mammals. Nature 390:504–507.

Krebs, B. 1991. Das Skelett von *Henkelotherium guimarotae* gen. et sp. nov. (Eupantotheria, Mammalia) aus dem Oberen Jura von Portugal. Berliner Geowissenschafliche Abhandlungen A 133:1–110.

Krusat, G. 1980. Contribuçao para o conhecimento da fauna do Kimeridgiano da mina de lignito Guimarota (Leiria, Portugal). IV Parte. *Haldanodon exspectatus* Kühne & Krusat 1972 (Mammalia, Docodonta). Memórias dos Serviços Geológicos de Portugal 27:1–79.

Krusat, G. 1991. Functional morphology of *Haldanodon exspectatus* (Mammalia, Docodonta) from the Upper Jurassic of Portugal; pp. 37–38 in Z. Kielan-Jaworowska, N. Heintz, and H. A. Nakrem (eds.), Fifth Symposium on Mesozoic Terrestrial Ecosystems and Biota. Extended Abstracts. Contributions from the Paleontological Museum, University of Oslo, 364.

Kühne, W. G. 1956. The Liassic therapsid *Oligokyphus*. Trustees of the British Museum, London, 149 pp.

Kumar, S., and Hedges, S. B. 1998. A molecular timescale for vertebrate evolution. Nature 392:917–920.

Li, C.-K., Y.-Q. Wang, Y.-M. Hu, and J. Meng. 2003. A new species of *Gobiconodon* (Triconodonta, Mammalia) and its implication for the age of Jehol Biota. Chinese Science Bulletin (English Edition) 48:1129–1134.

Li, G., and Z.-X. Luo. 2006. A Cretaceous symmetrodont with some monotreme-like postcranial features. Nature 439:195–200.

Lillegraven, J. A. 1969. Latest Cretaceous mammals of upper part of Edmonton Formation of Alberta, Canada, and review of marsupial-placental dichotomy in mammalian evolution. University of Kansas Paleontological Contributions 50:1–122.

Lillegraven, J. A. 1974. Biogeographical considerations of the marsupial-placental dichotomy. Annual Review of Ecology and Systematics 5:263–283.

Lillegraven, J. A., and G. Krusat. 1991. Cranio-mandibular anatomy of *Haldanodon exspectatus* (Docodonta; Mammalia) from the Late Jurassic of Portugal and its implications to the evolution of mammalian characters. Contributions to Geology, University of Wyoming 28:39–138.

Lillegraven, J. A., Z. Kielan-Jaworowska, and W. A. Clemens (eds.). 1979. Mesozoic Mammals: The First Two-thirds of Mammalian History. University of California Press, Berkeley, 311 pp.

Lucas, S. G., and Z.-X. Luo. 1993. *Adelobasileus* from the Upper Triassic of western Texas: the oldest mammal. Journal of Vertebrate Paleontology 13:309–334.

Luckett, W. P. 1993. An ontogenetic assessment of dental homologies in therian mammals; pp. 182–204 in F. S. Szalay, M. J. Novacek, and M. C. McKenna (eds.), Mammal Phylogeny: Mesozoic Differentiation, Multituberculates, Monotremes, Early Therians, and Marsupials. Springer-Verlag, New York.

Luo, Z.-X. 1994. Sister-group relationships of mammals and transformations of diagnostic mammalian characters; pp. 98–128 in N. C. Fraser and H.-D. Sues (eds.), In the Shadow of the Dinosaurs: Early Mesozoic Tetrapods. Cambridge University Press, Cambridge and New York.

Luo, Z.-X. 2001. Inner ear and its bony housing in tritylodonts and implications for evolution of mammalian ear. Bulletin of Museum of Comparative Zoology, Harvard University 156:81–97.

Luo, Z.-X., and A. W. Crompton. 1994. Transformations of the quadrate (incus) through the transition from non-mammalian cynodonts to mammals. Journal of Vertebrate Paleontology 14:341–374.

Luo, Z.-X., and Q. Ji. 2005. New study on dental and skeletal features of the Cretaceous mammal *Zhangheotherium*. Journal of Mammalian Evolution 12:337–357.

Luo, Z.-X., and J. R. Wible. 2005. A Late Jurassic digging mammal and early mammalian diversification. Science 308:103–107.

Luo, Z.-X., and X.-C. Wu. 1995. Correlation of vertebrate assemblage of the lower Lufeng Formation, Yunnan, China; pp. 83–85 in A.-L. Sun and Y.-Q. Wang (eds.), Sixth Symposium on Mesozoic Terrestrial Ecosystems. China Ocean Press, Beijing.

Luo, Z.-X., R. L. Cifelli, and Z. Kielan-Jaworowska. 2001a. Dual origin of tribosphenic mammals. Nature 409:53–57.

Luo, Z.-X., A. W. Crompton, and A.-L. Sun. 2001b. A new mammaliaform from the Early Jurassic of China and evolution of mammalian characteristics. Science 292:1535–1540.

Luo, Z.-X., Z. Kielan-Jaworowska, and R. L. Cifelli. 2002. In quest for a phylogeny of Mesozoic mammals. Acta Palaeontologica Polonica 47:1–78.

Luo, Z.-X., Z. Kielan-Jaworowska, and R. L. Cifelli. 2004. Evolution of dental replacement in mammals; pp. 159–175 in M. R. Dawson and J. A. Lillegraven (eds.), Fanfare for an Uncommon Paleontologist: Papers in Honor of Malcolm C. McKenna. Bulletin of Carnegie Museum of Natural History 36.

Luo, Z.-X., Q. Ji, J. R. Wible, and C.-X. Yuan. 2003. An Early Cretaceous tribosphenic mammal and metatherian evolution. Science 302:1934–1940.

Maisch, M. W., A. T. Matzke, F. Grossmann, H. Stöhr, H.-U. Pfretzschner, and G. Sun. 2005. The first haramiyoid mammal from Asia. Naturwissenschaften 92:40–44.

Marshall, L. G. 1987. Systematics of Itaboraian (middle Paleocene) age "opossum-like" marsupials from the limestone quarry at São José de Itaboraí, Brazil; pp. 91–160 in M. Archer (ed.), Possums and Opossums: Studies in Evolution. Surrey Beatty & Sons and the Royal Zoological Society of New South Wales, Sydney.

Marshall, L. G., J. A. Case, and M. O. Woodburne. 1990. Phylogenetic relationships of the families of marsupials. Current Mammalogy 2:433–502.

Martin, T. 1999. Dryolestidae (Dryolestoidea, Mammalia) aus dem Oberen Jura von Portugal. Abhandlungen der Senckenbergischen Naturforschenden Gesellschaft 550:1–119.

Martin, T. 2002. New stem-line representatives of Zatheria (Mammalia) from the Late Jurassic of Portugal. Journal of Vertebrate Paleontology 22:332–348.

Martin, T. 2005. Postcranial anatomy of *Haldanodon exspectatus* (Mammalia, Docodonta) from the Late Jurasssic (Kimmeridgian) of Portugal and its bearing for mammalian evolution. Zoological Journal of the Linnean Society 145:219–248.

Martin, T., and A. O. Averianov. 2004. A new docodont (Mammalia) from the Middle Jurassic of Kyrgyzstan, Central Asia. Journal of Vertebrate Paleontology 24:195–201.

Martin, T., and B. Krebs (eds.). 2000. Guimarota: A Jurassic Ecosystem. Verlag Dr. Friedrich Pfeil, Munich, 155 pp.

Martin, T., and Z.-X. Luo. 2005. Homoplasies in the mammalian ear. Science 307:861–862.

Martin, T., and M. Nowotny. 2000. The docodont *Haldanodon* from the Guimarota Mine; pp. 91–96 in T. Martin and B. Krebs (eds.), Guimarota: A Jurassic Ecosystem. Verlag Dr. Friedrich Pfeil, Munich.

Martin, T., and O. W. M. Rauhut. 2005. Mandible and dentition of *Asfaltomylos patagonicus* (Australosphenida, Mammalia) and the evolution of tribosphenic teeth. Journal of Vertebrate Paleontology 25:414–425.

Maschenko, E. N., A. V. Lopatin, and A. V. Voronkevich. 2002. A new genus of the tegotheriid docodonts (Docodonta, Tegotheriidae) from the Early Cretaceous of West Siberia. Russian Journal of Theriology 1:75–81.

McKenna, M. C. 1975. Toward a phylogenetic classification of the Mammalia; pp. 21–46 in W. P. Luckett and F. S. Szalay (eds.), Phylogeny of the Primates. Plenum Publishing Corporation, New York.

McKenna M. C., and S. K. Bell. 1997. Classification of Mammals above the Species Level. Columbia University Press, New York, 631 pp.

MacLeod, N., and K. D. Rose. 1993. Inferring locomotor behavior in Paleogene mammals via eigenshape analysis. American Journal of Science 293-A:300–355.

Meng, J., Y.-M. Hu, Y.-Q. Wang, and C.-K. Li. 2003. The ossified Meckel's cartilage and internal groove in Mesozoic mammaliaforms: implications to origin of the definitive mammalian middle ear. Zoological Journal of the Linnean Society 138:431–448.

Meng, J., Y.-M. Hu, Y.-Q. Wang, and C.-K. Li. 2005. A new triconodont (Mammalia) from the Early Cretaceous Yixian Formation of Liaoning, China. Vertebrata PalAsiatica 43:1–10.

Muizon, C. de. 1994. A new carnivorous marsupial from the Palaeocene of Bolivia and the problem of marsupial monophyly. Nature 370:208–211.

Muizon, C. de. 1998. *Mayulestes ferox*, a borhyaenoid (Metatheria, Mammalia) from the early Palaeocene of Bolivia. Phylogenetic and palaeobiologic implications. Geodiversitas 20:19–142.

Muizon, C. de (ed.). 1995. *Pucadelphys andinus* (Marsupialia, Mammalia) from the early Paleocene of Bolivia. Mémoires du Muséum National d'Histoire Naturelle (Paris) 165:1–164.

Muizon, C. de, R. L. Cifelli, and R. Céspedes. 1997. The origin of dog-like marsupials and the early evolution of Gondwanian marsupials. Nature 389:486–489.

Murphy, W. J., E. Eizirik, S. J. O'Brien, O. Madsen, M. Scally, C. J. Douday, E. Teeling, O. A. Ryder, M. J. Stanhope, W. W. de Jong, and M. S. Springer. 2001. Resolution of the early placental mammal radiation using Bayesian phylogenetics. Science 294:2348–2351.

Musser, A. M. 1999. Diversity and relationships of living and extinct monotremes. Australian Mammalogy 21:8–9.

Musser, A. M., and M. Archer. 1998. New information about the skull and dentary of the Miocene platypus *Obdurodon dicksoni,* and a discussion of ornithorhynchid relationships. Philosophical Transactions of the Royal Society of London B 353:1063–1079.

Nessov, L. A., J. D. Archibald, and Z. Kielan-Jaworowska. 1998. Ungulate-like mammals from the Late Cretaceous of Uzbekistan and a phylogenetic analysis of Ungulatomorpha; pp. 40–88 in K. C. Beard and M. R. Dawson (eds.), Dawn of the Age of Mammals in Asia. Bulletin of Carnegie Museum of Natural History 34.

Nilsson, M. A., U. Arnason, P. B. S. Spencer, and A. Janke. 2004. Marsupial relationships and a timeline for marsupial radiation in South Gondwana. Gene 340:189–196.

Norell, M. A. 1993. Tree-based approaches to understanding history: comments on ranks, rules, and the quality of the fossil record. American Journal of Science 293A:407–417.

Novacek, M. J. 1986. The skull of leptictid insectivorans and the higher-level classification of eutherian mammals. Bulletin of the American Museum of Natural History 183:1–112.

Novacek, M. J. 1993. Pattern of diversity in the mammalian skull; pp. 438–429 in J. Hanken and B. K. Hall (eds.), The Skull. Volume 2, Patterns of Structural and Systematic Diversity. University of Chicago Press, Chicago.

Novacek, M. J., G. W. Rougier, J. R. Wible, M. C. McKenna, D. Dashzeveg, and I. Horovitz. 1997. Epipubic bones in eutherian mammals from the Late Cretaceous of Mongolia. Nature 389:483–486.

Olson, E. C. 1944. Origin of mammals based upon cranial morphology of the therapsid suborders. Geological Society of America, Special Paper 55:1–136.

Olson, E. C. 1959. The evolution of mammalian characters. Evolution 13:344–353.

Pascual, R., M. Archer, E. O. Ortiz-Jaureguizar, J. L. Prado, H. Godthelp, and S. J. Hand. 1992. First discovery of monotremes in South America. Nature 356:704–705.

Padian, K., D. R. Lindberg, and D. P. Polly. 1995. Cladistics and the fossil record: the use of history. Annual Review of Earth and Planetary Sciences 22:63–91.

Pascual, R., F. J. Goin, L. Balarino, and D. E. Udrizar. 2002. New data on the Paleocene monotreme *Monotrematum sudamericanum,* and the convergent evolution of triangulate molars. Acta Palaeontologica Polonica 47:487–492.

Pascual, R., F. J. Goin, D. W. Krause, E. Ortiz-Jaureguizar, and A. A. Carlini. 1999. The first gnathic remains of *Sudamerica:* implications for gondwanathere relationships. Journal of Vertebrate Paleontology 19:373–382.

Patterson, B. 1956. Early Cretaceous mammals and the evolution of mammalian molar teeth. Fieldiana: Geology 13:1–105.

Prasad, G. V. R., and B. K. Manhas. 1997. A new symmetrodont mammal from the Lower Jurassic Kota Formation, Pranhita-Godavari Valley, India. Géobios 30:563–572.

Prasad, G. V. R., and B. K. Manhas. 2001. The first docodont mammal of Laurasian affinity from India. Current Science 81:1235–1238.

Pridmore, P. A. 1985. Terrestrial locomotion in monotremes (Mammalia: Monotremata). Journal of Zoology (London) 205:53–73.

Prothero, D. R. 1981. New Jurassic mammals from Como Bluff, Wyoming, and the interrelationships of non-tribosphenic Theria. Bulletin of the American Museum of Natural History 167:277–326.

Rauhut, O. W. M., T. Martin, and E. O. Ortiz-Jaureguizar. 2002. The first Jurassic mammal from South America. Nature 416:165–168.

Reig, O. A., J. A. W. Kirsch, and L. G. Marshall. 1987. Systematic relationships of the living and Neocenozoic American "opossum-like" marsupials (Suborder Didelphimorphia), with comments on the classification of these and of the Cretaceous and Paleogene New World and European metatherians; pp. 1–89 in M. Archer (ed.), Possums and Opossums: Studies in Evolution. Surrey Beatty & Sons and the Royal Zoological Society of New South Wales, Sydney.

Reisz, R. R., and J. Müller. 2004. Molecular timescales and the fossil record: a paleontological perspective. Trends in Genetics 20:237–241.

Retief, J. D., C. Krajewski, M. Westerman, R. J. Winkfein, and G. H. Dixon. 1995. Molecular phylogeny and evolution of marsupial protamine P1 genes. Proceedings of Royal Society of London B 259:7–14.

Rich, T. H., J. A. Hopson, A. M. Musser, T. F. Flannery, and P. Vickers-Rich. 2005. Independent origins of middle ear bones in monotremes and therians. Science 307:910–914.

Rich, T. H., P. Vickers-Rich, A. Constantine, T. F. Flannery, L. Kool, and N. van Klaveren. 1997. A tribosphenic mammal from the Mesozoic of Australia. Science 278:1438–1442.

Rich, T. H., P. Vickers-Rich, A. Constantine, T. F. Flannery, L. Kool, and N. van Klaveren. 1999. Early Cretaceous mammals from Flat Rocks, Victoria, Australia. Records of the Queen Victoria Museum 106:1–35.

Rich, T. H., T. F. Flannery, P. Trusler, L. Kool, N. van Klaveren, and P. Vickers-Rich. 2002. Evidence that monotremes and ausktribosphenids are not sistergroups. Journal of Vertebrate Paleontology 22:466–469.

Rich, T. H., T. F. Flannery, P. Trusler, A. Constantine, L. Kool, N. van Klaveren, and P. Vickers-Rich. 2001a. An advanced ausktribosphenid from the Early Cretaceous of Australia. The Records of the Queen Victoria Museum 110:1–9.

Rich, T. H., P. Vickers-Rich, P. Trusler, T. F. Flannery, R. L. Cifelli, A. Constantine, L. Kool, and N. van Klaveren. 2001b. Monotreme nature of the Australian Early Cretaceous mammal *Teinolophos trusleri*. Acta Palaeontologica Polonica 46:113–118.

Rougier, G. W. 1993. *Vincelestes neuquenianus* Bonaparte (Mammalia, Theria), un primitivo mamífero del Cretácico Inferior de la Cuenca Neuquina. Ph.D. dissertation, Universidad Nacional de Buenos Aires, Buenos Aires, 720 pp.

Rougier, G. W., A. M. Forasiepi, and A. G. Martinelli. 2005. Comment on "Independent origins of middle ear bones in monotremes and therians" (II). Science 309:1492b.

Rougier, G. W., Q. Ji, and M. J. Novacek. 2003. A new symmetrodont mammal with fur impressions from the Mesozoic of China. Acta Geologica Sinica 77:7–14.

Rougier, G. W., J. R. Wible, and J. A. Hopson. 1996. Basicranial anatomy of *Priacodon fruitaensis* (Triconodontidae, Mammalia) from the Late Jurassic of Colorado, and a reappraisal of mammaliaform interrelationships. American Museum Novitates 3183:1–38.

Rougier, G. W., J. R. Wible, and M. J. Novacek. 1998. Implications of *Deltatheridium* specimens for early marsupial history. Nature 396:459–463.

Rougier, G. W., M. J. Novacek, M. C. McKenna, and J. R. Wible. 2001. Gobiconodonts from the Early Cretaceous of Oshih (Ashile), Mongolia. American Museum Novitates 3348:1–30.

Rowe, T. B. 1988. Definition, diagnosis, and origin of Mammalia. Journal of Vertebrate Paleontology 8:241–264.

Rowe, T. B. 1993. Phylogenetic systematics and the early history of mammals; pp. 129–145 in F. S. Szalay, M. J. Novacek, and M. C. McKenna (eds.), Mammal Phylogeny: Mesozoic Differentiation, Multituberculates, Monotremes, Early Therians, and Marsupials. Springer-Verlag, New York.

Sánchez-Villagra, M. R., and W. Maier. 2003. Ontogenesis of the scapula in marsupial mammals, with special emphasis on perinatal stages of didelphids and remarks on the origin of the therian scapula. Journal of Morphology 258:115–129.

Sánchez-Villagra, M. R., and K. K. Smith. 1997. Diversity and evolution of the marsupial mandibular angular process. Journal of Mammalian Evolution 4:119–144.

Scillato-Yané, G. R., and R. Pascual. 1985. Un peculiar Xenarthra del Paleoceno medio de Patagonia (Argentina). Su importancia en la sistemática de los Paratheria. Ameghiniana 21:173–176.

Shubin, N. H., and H.-D. Sues. 1991. Biogeography of early Mesozoic continental tetrapods: patterns and implications. Paleobiology 17:214–230.

Sidor, C. A., and J. A. Hopson. 1998. Ghost lineages and "mammalness": assessing the temporal pattern of character acquisition in the Synapsida. Paleobiology 24:254–273.

Sigogneau-Russell, D. 1989a. Haramiyidae (Mammalia, Allotheria) en provenance du Trias supérieur de Lorraine (France). Palaeontographica A 206:137–198.

Sigogneau-Russell, D. 1989b. Découverte du premier Symmétrodonte (Mammalia) du continent africain. Comptes Rendus de l'Académie des Sciences, Paris, sér. II, 309:921–926.

Sigogneau-Russell, D. 1991. Découverte du premier mammifère tribosphénique du Mésozoïque africain. Comptes Rendus de l'Académie des Sciences, Paris, sér. II, 313:1635–1640.

Sigogneau-Russell, D. 1998. Discovery of a Late Jurassic Chinese mammal in the upper Bathonian of England. Comptes Rendus de l'Académie des Sciences, Paris, sér. IIa, 327:571–576.

Sigogneau-Russell, D. 1999. Réévaluation des Peramura (Mammalia, Theria) sur la base de nouveaux spécimens du Crétacé inférieur d'Angleterre et du Maroc. Geodiversitas 21:93–127.

Sigogneau-Russell, D. 2003a. Docodonts from the British Mesozoic. Acta Palaeontologica Polonica 48:357–374.

Sigogneau-Russell, D. 2003b. Holotherian mammals from the Forest Marble (Middle Jurassic of England). Geodiversitas 25:501–537.

Sigogneau-Russell, D., and P. C. Ensom. 1998. *Thereuodon* (Theria, Symmetrodonta) from the Lower Cretaceous of North Africa and Europe, and a brief review of symmetrodonts. Cretaceous Research 19:1–26.

Sigogneau-Russell, D., and P. Godefroit. 1997. A primitive docodont (Mammalia) from the Upper Triassic of France and the possible therian affinities of the order. Comptes Rendus de l'Académie des Sciences, Paris, sér. IIa, 324:135–140.

Sigogneau-Russell, D., and G. Hahn. 1994. Late Triassic microvertebrates from central Europe; pp. 197–213 in N. C. Fraser and H.-D. Sues (eds.), In the Shadow of the Dinosaurs: Early Mesozoic Tetrapods. Cambridge University Press, Cambridge and New York.

Sigogneau-Russell, D., D. Dashzeveg, and D. E. Russell. 1992. Further data on *Prokennalestes* (Mammalia, Eutheria *inc. sed.*) from the Early Cretaceous of Mongolia. Zoologica Scripta 21:205–209.

Sigogneau-Russell, D., J. J. Hooker, and P. C. Ensom. 2001. The oldest tribosphenic mammal from Laurasia (Purbeck Limestone Group, Berriasian, Cretaceous, UK) and its bearing on the "dual origin" of Tribosphenida. Comptes Rendus de l'Académie des Sciences, Paris, sér. IIa, 333:141–147.

Simmons, N. B. 1993. Phylogeny of Multituberculata; pp. 146–164 in F. S. Szalay, M. J. Novacek, and M. C. McKenna (eds.), Mammal Phylogeny: Mesozoic Differentiation, Multituberculates, Monotremes, Early Therians, and Marsupials. Springer-Verlag, New York.

Simpson, G. G. 1928. A Catalogue of the Mesozoic Mammalia in the Geological Department of the British Museum. Trustees of the British Museum, London, 215 pp.

Simpson, G. G. 1929. American Mesozoic Mammalia. Memoirs of the Peabody Museum of Yale University 3:1–235.

Simpson, G. G. 1937. Skull structure of the Multituberculata. Bulletin of the American Museum of Natural History 73:727–763.

Simpson, G. G. 1945. The principles of classification and a classification of mammals. Bulletin of the American Museum of Natural History 85:1–350.

Slaughter, B. H. 1971. Mid-Cretaceous (Albian) therians of the Butler Farm local fauna, Texas; pp. 131–143 in D. M. Kermack and K. A. Kermack (eds.), Early Mammals. Zoological Society of the Linnean Society 50, Supplement 1. Academic Press, London.

Slaughter, B. H. 1981. The Trinity therians (Albian, mid-Cretaceous) as marsupials and placentals. Journal of Paleontology 55:682–683.

Springer, M. S. 1997. Molecular clocks and the timing of the placental and marsupial radiations in relation to the Cretaceous-Tertiary boundary. Journal of Mammalian Evolution 4:285–302.

Springer, M. S., J. A. W. Kirsch, and J. A. Case. 1997. The chronicle of marsupial evolution; pp. 129–161 in T. J. Givnish and K. J. Sytema (eds.), Molecular Evolution and Adaptive Radiation. Cambridge University Press, Cambridge and New York.

Springer, M. S., W. J. Murphy, E. Erizirik, and S. J. O'Brien. 2003. Placental mammal diversification and the Cretaceous-Tertiary boundary. Proceedings of National Academy of Sciences USA 100:1056–1061.

Springer, M. S., M. J. Stanhope, O. Madsen, and W. W. de Jong. 2004. Molecules consolidate the placental mammal tree. Trends in Ecology and Evolution 19:430–438.

Sun, A.-L., and Y.-H. Li. 1985. The postcranial skeleton of Jurassic tritylodonts from Sichuan Province. Vertebrata PalAsiatica 23:135–151.

Sues, H.-D. 1983. Advanced mammal-like reptiles from the Early Jurassic of Arizona. Ph.D. dissertation, Harvard University, Cambridge, 247 pp.

Sues, H.-D. 1985. The relationships of the Tritylodontidae (Synapsida). Zoological Journal of the Linnean Society 85:205–217.

Sues, H.-D. 1986. The skull and dentition of two tritylodontid synapsids from the Lower Jurassic of western North America. Bulletin of the Museum of Comparative Zoology, Harvard University 151:217–268.

Sues, H.-D. 1992. No Palaeocene "mammal-like reptile." Nature 359:278.

Sues, H.-D. 2001. On *Microconodon,* a Late Triassic cynodont from the Newark Supergroup of eastern North America. Bulletin of the Museum of Comparative Zoology, Harvard University 156:37–48.

Sues, H.-D., and F. A. Jenkins Jr. 2006. The postcranial skeleton of *Kayentatherium wellesi* from the Lower Jurassic Kayenta Formation of Arizona and the phylogenetic significance of postcranial features in tritylodontid cynodonts; pp. 114–152 in M. T. Carrano, T. J. Gaudin, R. W. Blob, and J. R. Wible (eds.), Amniote Paleobiology: Perspectives on the Evolution of Mammals, Birds, and Reptiles. University of Chicago Press, Chicago.

Sues, H.-D., and R. R. Reisz. 1995. First record of the early Mesozoic sphenodontian *Clevosaurus* (Lepidosauria: Rhynchocephalia) from the Southern Hemisphere. Journal of Paleontology 69:123–126.

Swisher, C. C., III, Y.-Q. Wang, X.-L. Wang, X. Xu, and Y. Wang. 1999. Cretaceous age for the feathered dinosaurs of Liaoning, China. Nature 398:58–61.

Szalay, F. S. 1982. A new appraisal of marsupial phylogeny and classification; pp. 621–640 in M. Archer (ed.), Carnivorous Marsupials. Royal Zoological Society of New South Wales, Sydney.

Szalay, F. S. 1994. Evolutionary History of the Marsupials and an Analysis of Osteological Characters. Cambridge University Press, Cambridge and New York, 481 pp.

Szalay, F. S., and E. J. Sargis. 2001. Model-based analysis of postcranial osteology of marsupials from the Palaeocene of Itaboraí (Brazil) and the phylogenetics and biogeography of Metatheria. Geodiversitas 23:139–302.

Szalay, F. S., and B. A. Trofimov. 1996. The Mongolian Late Cretaceous *Asiatherium,* and the early phylogeny and paleobiogeography of Metatheria. Journal of Vertebrate Paleontology 16:474–509.

Tsubamoto, T., G. W. Rougier, S. Isaji, M. Manabe, and A. M. Forasiepi. 2004. New Early Cretaceous spalacotheriid "symmetrodont" mammal from Japan. Acta Palaeontologica Polonica 49:329–346.

Van Valen, L. 1960. Therapsids as mammals. Evolution 14:304–313.

Vázquez-Molinero, R., T. Martin, M. S. Fischer, and R. Frey. 2001. Comparative anatomical investigations of the postcranial skeleton of *Henkelotherium guimarotae* Krebs, 1991 (Eupantotheria, Mammalia) and their implications on its locomotion. Mitteilungen aus dem Museum für Naturkunde in Berlin, Zoologische Reihe 77:207–216.

Wang, Y.-Q., W. A. Clemens, Y.-M. Hu, and C.-K. Li. 1998. A probable pseudo-tribosphenic upper molar from the Late Jurassic of China and the early radiation of the Holotheria. Journal of Vertebrate Paleontology 18:777–787.

Wang, Y.-Q., Y.-M. Hu, J. Meng, and C.-K. Li. 2001. An ossified Meckel's cartilage in two Cretaceous mammals and origin of the mammalian middle ear. Science 294:357–361.

Weil, A. 1999. Multituberculate phylogeny and mammalian biogeography in the Late Cretaceous and earliest Paleocene Western Interior of North America. Ph.D. dissertation, University of California, Berkeley, 243 pp.

Weil, A. 2002. Mammalian evolution: upwards and onwards. Nature 416:798–799.

Wible, J. R. 1990. Petrosals of Late Cretaceous marsupials from North America and a cladistic analysis of the petrosal in therian mammals. Journal of Vertebrate Paleontology 10:183–205.

Wible, J. R., and J. A. Hopson. 1993. Basicranial evidence for early mammal phylogeny; pp. 45–62 in F. S. Szalay, M. J. Novacek, and M. C. McKenna (eds.), Mammal Phylogeny: Mesozoic Differentiation, Multituberculates, Monotremes, Early Therians, and Marsupials. Springer-Verlag, New York.

Wible, J. R., M. J. Novacek, and G. W. Rougier. 2004. New data on the skull and dentition in the Mongolia Late Cretaceous eutherian mammal *Zalambdalestes*. Bulletin of the American Museum of Natural History 281:1–144.

Wible, J. R., G. W. Rougier, M. J. Novacek, and M. C. McKenna. 2001. Earliest eutherian ear region: a petrosal referred to *Prokennalestes* from the Early Cretaceous of Mongolia. American Museum Novitates 3322:1–44.

Wible, J. R., G. W. Rougier, M. J. Novacek, M. C. McKenna, and D. Dashzeveg. 1995. A mammalian petrosal from the Early Cretaceous of Mongolia: implications for the evolution of the ear region and mammaliamorph relationships. American Museum Novitates 3149:1–19.

Wilson, D. E., and D. M. Reeder. 2005. Mammal Species of the World. A Taxonomic and Geographic Reference. Third edition. The Johns Hopkins University Press, Baltimore, 2142 pp.

Woodburne, M. O. 2003. Monotremes as pretribosphenic mammals. Journal of Mammalian Evolution 10:195–248.

Woodburne, M. O., and R. H. Tedford. 1975. The first Tertiary monotreme from Australia. American Museum Novitates 2588:1–11.

Woodburne, M. O., T. H. Rich, and M. S. Springer. 2003. The evolution of tribosphemy and the antiquity of mammalian clades. Molecular Phylogenetics and Evolution 28:360–385.

Yadagiri, P. 1984. New symmetrodonts from the Kota Formation. Journal of the Geological Society of India 25:512–521.

Young, C. C. 1947. Mammal-like reptiles from Lufeng, Yunnan, China. Proceedings of the Zoological Society of London 117:537–597.

Zhou, Z.-H., P. M. Barrett, and J. Hilton. 2003. An exceptionally preserved Lower Cretaceous ecosystem. Nature 421:807–814.

10. The Terrestrial to Aquatic Transition in Cetacea

Mark D. Uhen

ABSTRACT

Many cetacean synapomorphies are related to their exclusively aquatic lifestyle. Some of these characteristics were convergently acquired by sirenians. Most of the features shared by cetaceans and sirenians involve locomotion, whereas those that differentiate among cetaceans and distinguish them from sirenians involve feeding and sensory systems. The extensive and growing fossil record of cetaceans shows that the shared derived characteristics of extant cetaceans arose in a mosaic pattern over the course of the first 10 million years of cetacean evolution.

Introduction

What makes a whale a whale? Cetaceans (whales, dolphins, and porpoises) are certainly easy to distinguish from other mammals today. Cetaceans share many characteristics that are related to their exclusively aquatic lifestyle. Many, but by no means all, of these features are shared with sirenians (sea cows and dugongs), which are thought to have evolved these characteristics independently. Most of the features that all cetaceans and sirenians share deal with locomotion, whereas those that differentiate among cetaceans and distinguish them from sirenians deal with feeding and sensory systems. The shared derived characteristics of modern cetaceans arose in a mosaic pattern over the course of the first 10 million years of whale evolution.

Reconstructing Evolutionary Scenarios

Reconstruction of the pattern of the acquisition of morphological features in a group of organisms, and by extension the evolutionary scenarios that govern the acquisition of these features, requires an accurate phylogeny on which the features in question can be mapped. Only recently has anything resembling a consensus opinion emerged regarding the origin of cetaceans (Geisler and Uhen, 2003, 2005). Until very recently, paleontologists supported the view that the extinct Mesonychia were the sister-taxon to Cetacea (Geisler, 2001), whereas molecular systematists concluded that Cetacea was nested within a traditionally delimited Artiodactyla (Gatesy, 1998). Currently, most researchers agree that cetaceans are nested within Artiodactyla, or are at least the sister-group to Artiodactyla to the exclusion of Mesonychia or other fossil groups (but see O'Leary, 2001). The consensus opinion arose when new fossil evidence showed that early cetaceans shared a suite of uniquely derived ankle features with artiodactyls (Gingerich et al., 2001; Thewissen et al., 2001). This opinion has been further supported by combined analyses of molecular, morphological, and stratigraphic data (Geisler and Uhen, 2003, in press).

The adoption of the hypothesis that Cetacea is the sister-taxon of Artiodactyla (either mono- or paraphyletic) rather than Mesonychia has profound implications for the origin of Cetacea. Mesonychians are thought to have been carnivorous on the basis of dental structure and dental wear (O'Leary and Rose, 1995; O'Leary and Uhen, 1999). Early artiodactyls, on the other hand, are thought to have been herbivorous or perhaps omnivorous. The nature of the shift in feeding mode from terrestrial herbivore to aquatic carnivore is much more substantive than that of terrestrial carnivore to aquatic carnivore and thus presents a much more complicated evolutionary scenario than if Mesonychia were the sister-taxon to Cetacea. It is also interesting to note that much of the evidence for a mesonychian-cetacean phylogenetic relationship was derived from the dentition, and that these data have been suggested to be misleading when used in phylogenetic analyses (Naylor and Adams, 2001). Although the accusation that the data are misleading has been soundly deflected (O'Leary et al., 2003), nonetheless, the similarities in the dentitions (O'Leary, 1998) and dental wear patterns (O'Leary and Uhen, 1999) between mesonychians and early cetaceans must now be considered convergences, possibly due to similarities in diet and oral processing, rather than synapomorphies.

Cetacean Synapomorphies

Regardless of the identity of the sister-group of cetaceans, the question of what makes a whale a whale is still unanswered. Presumably what transpired at the origin of the cetacean clade was that some population of animals stopped interbreeding with other conspecifics due to physical or behavioral separation of some sort. Over time, the morphology and/or behavior of this isolated population became so different that they could not interbreed with the rest of the group from which they originally separated, creating a new species. As morphological changes accumulated, this new lineage became more readily distinguishable from their ancestors. Unfortunately, it would be difficult to recognize this new lineage as Cetacea until at least one of the shared derived features of later cetaceans arose. Under this scenario, it would be difficult, if not impossible, to identify any fossils from the lineages in the area between the cladogenetic origin of Cetacea and the origin of recognizable cetacean synapomorphies as cetaceans.

The earliest-arising features of cetaceans that are recognized in the fossil record include the following: an inflated tympanic bulla with a well-developed involucrum and sigmoid process (Kellogg, 1936; Gingerich and Russell, 1981); pachyosteosclerotic auditory ossicles (Thewissen, 1994); a partially rotated auditory ossicular chain (Thewissen and Hussain, 1993); an elongate snout with pointed incisors arranged anteroposteriorly in line with the cheek teeth (Thewissen, 1994); high trigonids with corresponding embrasure pits on the palate (Kellogg, 1936; Thewissen, 1994); and elongate, vertical wear facets on the lower molars (O'Leary and Uhen, 1999). It is worth noting that all of the earliest cetacean synapomorphies deal with the auditory and feeding apparatus rather than the locomotor system. Figure 10.1 shows a composite phylogeny of Cetacea on which cetacean features of the various subgroups of Cetacea will be mapped in the following sections. These earliest synapomorphies of all Cetacea map to node 1.

Cranial telescoping, the tendency of the bones of the skull to slip over one another in the postrostral portion of the skull, is characteristic of some, but not all, cetaceans (Miller, 1923). Telescoping results in a shortening of the posterior splanchnocranium and neurocranium concomitant with the lengthening of the rostrum of cetaceans. Archaeocetes lack cranial telescoping altogether. Odontocetes are characterized by having the proximal portion of the maxilla overlap the frontal, posteriorly approaching or meeting the supraoccipital, often posterior to the orbit. In mysticetes, the outer part of the posterior border of the maxilla projects ventrally and posteriorly under the anterior border of the supraorbital wing of the frontal, and the occipital extends anteriorly toward the nasals. Because both forms of telescoping are completely lacking in even the most derived archaeocetes (Uhen, 2004), and neither is well developed in early odontocetes or early mysticetes, it appears that each type of telescoping evolved independently, and they should be considered separately in any kind of evolutionary study. Furthermore, the conformation of the cranial bones in cetaceans is extremely complex, and the evolution of the relative positions of these bones will most likely be best understood on a much finer taxonomic scale than Odontoceti vs. Mysticeti.

Resource acquisition systems, sensory systems, locomotory sys-

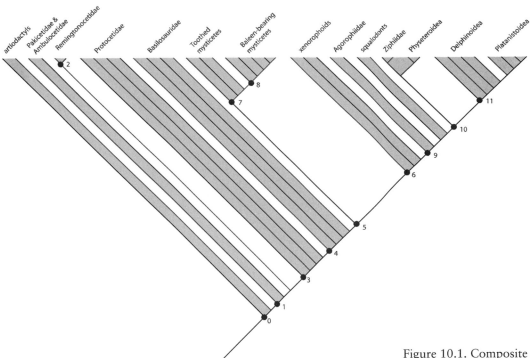

Figure 10.1. Composite phylogenetic tree of Cetacea (based on Geisler and Uhen, 2003; Geisler and Sanders, 2003; Uhen, 2004). Both monophyletic and paraphyletic groups are labeled across the top and are indicated by the shaded areas. Numbered nodes indicate monophyletic groups. Morphological characteristics of these groups are discussed in the text. Nodes are as follows: **1**, Cetacea; **2**, Remingtonocetidae; **3**, unnamed clade (Protocetidae + Basilosauridae + Neoceti); **4**, unnamed clade (Basilosauridae + Neoceti); **5**, Neoceti; **6**, Odontoceti; **7**, Mysticeti; **8**, unnamed clade (Eomysticetidae + "cetotheriids" + Balaenopteridae + Balaenidae + Eschrichtiidae + Neobalaenidae); **9**, unnamed clade (Agorophiidae + "squalodonts" + other fossil groups + Delphinoidea + Platanistoidea (sensu Geisler and Sanders, 2003); **10**, crown group Odontoceti; **11**, unnamed clade (Delphinida + Eurhinodelphinida + Platanistida).

tems, and development have all changed dramatically during the transition of cetaceans from life on land to life in the water. Changes in each of these systems are discussed below with an exploration of the life habits that are suggested by the observed morphological changes. Additional changes in these systems that characterize the origin of the modern suborders Odontoceti and Mysticeti are also explored.

Resource Acquisition

An elongate snout is characteristic of many piscivorous tetrapods, or perhaps better put, carnivorous aquatic tetrapods because many of these animals consume all manner of aquatic prey items (Massare, 1987; Fig. 10.1, node 1). Proportional elongation of the tooth-bearing part of the premaxilla and the anterior part of the dentary has two additional effects on the structure of the cetacean face. First, it aligns the incisors in a row with the cheek teeth. Second, it effectively moves the external nares from the tip of the snout farther posterior on the face. Thus, in addition to changing the shape of the cetacean rostrum to a more piscivorous form, elongation of the anterior portion of the premaxilla began the process of migration of the nares toward the top of the head, which facilitates breathing in many aquatic tetrapods. This example demonstrates the complexity of interpretation of morphological change in an evolutionary scenario because this single morphological change had profound effects on both feeding and breathing.

The teeth and jaws of early cetaceans can tell us a great deal about the diet of these animals, both from the morphology of the structures themselves and from indicators of use, such as the wear facets that are present on the teeth of many early whales. Unlike extant toothed whales, archaeocetes (as well as early mysticetes and odontocetes) retain a differentiated (heterodont) dentition (Fig. 10.1, nodes 1–8, 10). The anterior teeth are elongate cones; the cheek teeth (premolars and molars) are triangular in buccal view. This profound difference in shape is indicative of a profound difference in function. Extant aquatic carnivores use the anterior ends of their snouts mainly for prey acquisition (Werth, 1992). The anterior teeth show mainly apical wear, with little or no wear along the buccal or lingual margins (Uhen, 2004). The structure of the mandibular condyle in archaeocetes indicates that the jaw could be slid forward for occlusion of the anterior teeth separate from the occlusion of the cheek teeth. The shape of the anterior teeth, the wear facets, the snout, and the mandibular condyle indicate that these teeth were used for prey acquisition rather than oral processing of prey (Uhen, 2004).

Early archaeocete dentitions show a greatly reduced, but present, trigonid basin on the lower molars, which indicates that grinding was not an important chewing mode in these animals (Thewissen, 1994). These early archaeocetes do, however, retain a protocone on the upper cheek teeth, which forms a trigon basin between it and the paracone and metacone. This creates two areas for crushing or puncturing: one in the trigon basin itself, and one on the palate between adjacent cheek teeth (Thewissen, 1994; Fig. 10.1, nodes 1–4). It is in this second area on the palate where the embrasure pits form in response to chewing (Uhen, 2004). A significant shearing component develops between the buccal surface of the lower cheek teeth and the lingual surface of the upper cheek teeth (the lingual side of the paracone and metacone), as indicated by the presence of elongate wear facets on these tooth surfaces (O'Leary and Uhen, 1999; Fig. 10.1, nodes 1–5).

More derived archaeocetes (protocetids and basilosaurids; Fig. 10.1, nodes 3–5) reduce the size of the protocone, but retain the embrasure pits between adjacent teeth. In basilosaurids, this trend is completed by the complete elimination of the protocone and the lingual root that supported it on the upper cheek teeth (Uhen, 1998a). The trigonid basin, and thus the grinding component of chewing, is also eliminated completely in protocetids and basilosaurids. Basilosaurids also develop accessory denticles on the mesial and distal edges of the cheek tooth crowns (Fig. 10.1, nodes 4–8, 10). These denticles may help to retain prey while the lower teeth shear past the upper teeth while chewing, but this hypothesis has yet to be evaluated. Swift and Barnes (1996) identified a mass of fish bones as the stomach contents of an individual of *Basilosaurus cetoides* from the Yazoo Clay of Mississippi, whereas Uhen (2004) also identified a mass of fish bones from a skeleton of *Dorudon atrox* from the Birket Qarun Formation of Egypt as the stomach contents. Both observations of stomach contents are at least consistent with the hypothesis of individual prey capture and oral processing suggested by the teeth themselves and the wear facets.

The accessory denticles on the cheek teeth of archaeocetes are retained by early members of both the Odontoceti and Mysticeti, collectively known as Neoceti (Fordyce and de Muizon, 2001; Fig. 10.1, nodes 5–8, 10). Although the teeth of early neocetes are similar, they are thought to have been used in very different fashions. Extant odontocetes use their teeth only for acquisition of individual prey items, but not for chewing. This behavior can be mapped with confidence onto the crown group Odontoceti (Fig. 10.1, node 10), although many odontocetes have greatly reduced the number of teeth, and some have developed suction feeding (Werth, 1992). This entire group also happens to be exclusively heterodont, with all taxa bearing similar single-rooted teeth. Adult modern mysticetes lack teeth altogether, and instead use baleen to filter prey items from the water (Fig. 10.1, node 8). Thus, the feeding modes of all of the toothed mysticetes and early odontocetes between nodes 8 and 10 (Fig. 10.1) remain unclear.

Thus far, there has been no evidence of elongate wear facets on the teeth of the members of either clade that would indicate a use of the teeth in a manner similar to that of archaeocetes. Early neocetes also lose the embrasure pits between adjacent cheek teeth. These features indicate that both clades of early neocetes shifted their oral processing away from shearing toward other modes of feeding. It has been suggested that early mysticetes used their denticulate teeth much like crab-eating seals to filter prey items from the water (Mitchell, 1989). This hypothesis needs to be further evaluated by carefully searching for any evidence of wear facets on the teeth of mysticetes. One unusual early mysticete, *Mammalodon,* has severely worn teeth (Pritchard, 1939), which, when combined with its very small size and blunt, flat snout, suggests an unusual form of oral processing that has yet to be elucidated. Early odontocetes may have adopted the grab-and-swallow mode of oral processing used by many extant odontocetes, but this idea remains to be tested as well.

Sensory Systems: Hearing

The evolution of cetacean hearing has been well documented with a great deal of evidence from both fossil and modern morphology (Nummela et al., 2004; Lancaster, 1990; Oelschläger, 1990; Pilleri et al., 1989; Fleischer, 1976). The hearing system involves bones of the mandible, cranium, tympanic bulla, periotic, auditory ossicles, and basicranium. Each of these skeletal elements contributes to the special aquatic adaptations of cetaceans for hearing underwater. Odontocetes in particular have further specialized hearing systems that perceive high-frequency sound as part of their echolocation systems.

As mentioned above, some of the earliest evidence for the origin of cetaceans comes from changes in the ear region (Gingerich and Russell, 1981). Cetaceans all possess dense, inflated tympanic bullae, with large involucra (Fig. 10.1, node 1). These bullae also possess a sigmoid process to which the malleus is fused. The changes in the ear help to match the ear to the impedance of sound waves in water as opposed to the impedance of sound waves in air. The earliest cetaceans have the tympanic bulla connected to the basicranium and periotic in at least four separate locations (Luo, 1998; Luo and Gingerich, 1999).

The configuration of the tympanic bulla and auditory ossicles of the earliest whales (pakicetids) retained a fully developed tympanic membrane (Luo, 1998; Luo and Gingerich, 1999), in contrast to later archaeocetes (Lancaster, 1990) and extant cetaceans that have an elongate tympanic conus and tympanic ligament in its place, with a rotation of the ossicular chain (Fraser and Purves, 1960). Nummela et al. (2004) noted that the scaling relationships (in relationship to both area and mass) of the pakicetid ear region matched those of terrestrial rather than aquatic mammals. These observations indicate that pakicetids were quite capable of perceiving airborne sounds despite some of the early adaptations for underwater hearing.

This firm connection of the bulla to the skull emphasizes an additional challenge to hearing underwater. Sound waves readily propagate through flesh and bone at about the same speed as through water. In air, sound waves are forced to go through the external auditory meatus (EAM) to reach each ear. Directionality of sound is perceived by sensing the difference in time that the sound waves take to reach each ear. Underwater, when the ear bones are firmly attached to the skull, sound waves can hit the skull at any point and propagate to each ear without being forced through the EAM or any alternative sound path. This method of hearing does not permit the perception of directionality of sound. Pakicetids, remingtonocetids, and protocetids all have tympanic bullae and periotics that are firmly attached to the skull, and thus they probably had difficulty in determining the directionality of sound underwater.

Modern whales isolate the ear from the skull by only loosely attaching the tympanic bulla to the periotic via the posterior process of the bulla and by loosening the attachment of the periotic to the rest of the basicranium. The basilosaurids have a relatively advanced arrangement of the bulla, which is primarily attached to the basicranium through the posterior process (Luo and Gingerich, 1999). Extant whales further isolate the ear region with a system of air-filled sinuses. The earliest cetaceans to exhibit these sinuses are basilosaurids (Uhen, 2004), and they are found in all more modern cetaceans (but not in *Ambulocetus,* contra Thewissen et al., 1996; Fig. 10.1, node 4).

All modern cetaceans retain a patent EAM. Despite being patent, it is not particularly functional because the EAM has virtually no diameter in odontocetes, and it is filled with a wax plug in mysticetes. Luo and Gingerich (1999) noted that the bony EAM is short or absent in outgroup ungulates, early artiodactyls, and odontocetes more derived than *Xenorophus* and its kin. Nonbasilosaurid archaeocetes, early odontocetes including *Xenorophus,* and derived mysticetes have a deep and narrow EAM, whereas basilosaurid archaeocetes and perhaps early mysticetes have a broad EAM. Thus, in those taxa lacking a functional EAM, sound must be channeled to the ears by an alternative path.

Extant cetaceans have a very large mandibular foramen, with a thin plate of bone forming the buccal wall of the posterior end of the dentary (often called the pan bone) opposite the foramen, which is located on the lingual side. This foramen and part of the mandibular canal are filled with a fat pad that extends posteriorly to the ear region. This fat pad has been shown to possess acoustic properties that are significantly different from normal body fat (Norris, 1968). Thus, it appears that sound is transmitted from the lower jaw via the fat pad to the ears. Pakicetids lack the enlarged mandibular foramen of modern cetaceans, suggesting that they retained the sound pathway through the EAM (Thewissen and Hussain, 1993). Ambulocetids have a moderately enlarged foramen (Thewissen et al., 1996). Protocetids and basilosaurids all have enlarged mandibular foramina, suggesting that the alternative sound channel though the jaw was established early in cetacean evolution (Fig. 10.1, node 3). The size of the mandibular foramen has not been reported for Remingtonocetidae.

Modern mammals can perceive sounds over different frequency ranges. It has been shown that features of the periotic, particularly the cochlea, can be indicative of the frequency range that can be perceived. In particular, extant mysticete whales can perceive a much lower range of frequencies when compared with modern odontocetes. Odontocetes can perceive high-frequency sounds as part of their echolocation system. When these anatomical features of the periotic are measured in archaeocetes, it is clear that the frequency range they could perceive is similar to that of modern mysticetes, and that archaeocetes could not perceive the high-frequency sounds that are necessary for echolocation (Uhen, 2004).

Locomotion

Extant terrestrial mammals use a combination of limb and vertebral column movements to generate locomotion (Gambaryan, 1974), whereas present-day cetaceans and sirenians exclusively use vertebral column movements for this purpose (Howell, 1930). The forelimbs of cetaceans and sirenians are used primarily for steering. Thewissen and Fish (1997) proposed a model for the origin and evolution of swimming in cetaceans on the basis of the kinematics of extant mammal swimming modes. They concluded that early whales, such as *Ambulocetus,* went through a locomotor stage like that of modern otters. Gingerich (2003) compared skeletal measurements of a wide range of modern mammals of varying affinities for an aquatic habitat with those of early cetaceans. He concluded that early whales like *Rodhocetus* were not otter-like, but rather desman-like in locomotor mode.

The earliest cetaceans (pakicetids) show few, if any, changes in the gross anatomy of their skeletons that would indicate a shift toward aquatic locomotion (Thewissen et al., 2001:278; Fig. 10.2B). In fact, the limbs of pakicetids indicate that they were running and jumping animals, and the revolute zygapophyses of the vertebrae of pakicetids prevent the dorsal and ventral flexion and extension of the vertebral column that would be necessary for a modern cetacean mode of locomotion (Thewissen et al., 2001). These authors conclude, "Pakicetids were terrestrial mammals, no more amphibious than a tapir."

Ambulocetids, as exemplified by *Ambulocetus natans,* show several postcranial adaptations to locomotion in water (Fig. 10.2C). Robust transverse processes on the trunk vertebrae indicate an increase in axial musculature, and the flat zygapophyses indicate a greater range of possible motions between adjacent trunk vertebrae than in pakicetids (Thewissen et al., 1996). Ambulocetids also have large hind limbs when compared with the forelimbs. Thewissen et al. (1996) hypothesized that *Ambulocetus* may have moved much like a sea lion on land, externally rotating the feet while walking. They suggested that in the water, *Ambulocetus* extended the feet toward the tail, and then used dorsal and ventral flexion and extension of the vertebral column to move the entire pelvis, including the hind limbs and tail, up and down through the water to generate locomotion.

Early protocetids continue to increase the flexibility of the back by reducing the interconnections between adjacent sacral vertebrae (Fig. 10.1, node 3). *Rodhocetus* has four sacral vertebrae, but only the first sacral is attached to the pelvis, and the four sacrals are not ankylosed (Gingerich et al., 1994, 2001; Fig. 10.2D). These authors considered *Rodhocetus* capable of a plantigrade posture while walking on land, and pelvic paddling or quadrupedal paddling in the water, similar to the locomotor capabilities of *Ambulocetus.*

Remingtonocetids are often neglected in discussions of the terrestrial to aquatic transition in cetaceans, probably because they are thought to be a monophyletic group that shows many specializations of their own. Remingtonocetids are characterized by their narrow, elongate cranium and mandible, with long mandibular symphyses (Kumar and Sahni, 1986; Fig. 10.1, node 2). Like protocetids of similar geologic age, remingtonocetids retain a solidly ankylosed sacrum consisting of four vertebrae, and well-developed hind limbs with powerful musculature (Gingerich et al., 1995). Bajpai and Thewissen (2000) suggested that vertebral proportions indicate an otter-like mode of swimming in remingtonocetids, and they also made comparisons to gavials. Perhaps the development of a long, well-ankylosed jaw along with powerful limb musculature indicates a mammalian crocodile analog rather than an otter-like

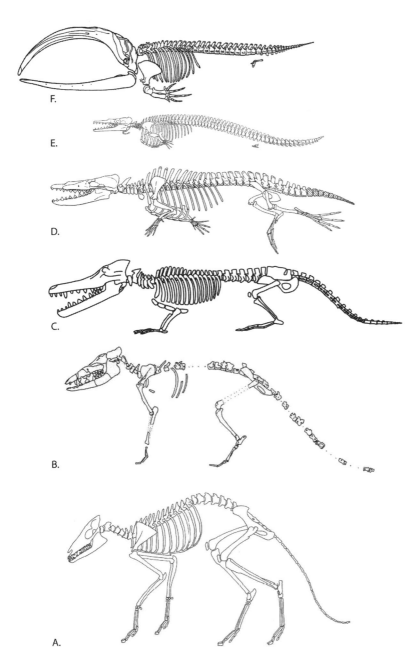

F.

E.

D.

C.

B.

A.

Figure 10.2. Skeletons of various artiodactyls and cetaceans showing the changing proportions of the body and the loss of the hind limbs. **A**, *Diacodexis;* **B**, *Pakicetus;* **C**, *Ambulocetus;* **D**, *Rodhocetus;* **E**, *Dorudon;* **F**, *Balaena*. Note that the earliest cetaceans (*Pakicetus* and *Ambulocetus*) and even the later *Rodhocetus* retain many of the features of the terrestrial *Diacodexis*. Many of the characteristics of the hind limb and vertebral column that characterize modern cetaceans first arose in the basilosaurids, represented here by *Dorudon*. Not to scale.

mode of life. Certainly the gross morphology of the skull is much more like that of a crocodile than that of an otter.

Later protocetids reduced the sacral structure even further, with the number of sacrals reduced to only one (*Protocetus, Natchitochia;* Uhen, 1998b) or none (*Georgiacetus;* Hulbert, 1998). Because *Georgiacetus* has no attachment of the vertebral column to the pelvis, it could not have effectively used the hind limbs for walking. This suggests either that *Georgiacetus* did not move on land, or that *Georgiacetus* used some form of forelimb and/or undulatory locomotion on land. On the basis of the size of the innominate, *Georgiacetus* retained large hind limbs that could have been used for pelvic undulatory locomotion in the water. Presence of tail flukes in all pakicetid, remingtonocetid, and protocetid archaeocetes remains equivocal because no posterior caudal vertebrae are known from any of these groups that could indicate the presence or absence of tail flukes.

The basilosaurids have many features that indicate a fully aquatic existence (Fig. 10.1, node 4; Fig. 10.2E), including great reduction of the hind limb and rotation of the pelvis (see discussion in Uhen and Gingerich, 2001); a significant increase in the number of lumbar vertebrae, trunk vertebrae of similar length, shortened cervical vertebrae, broadening of the scapula and reduction of the possible motions of the forelimb; and a clear indication of the presence of tail flukes in the distal caudal vertebrae. These aquatic adaptations are discussed in detail by Uhen (1998a) and characterize not only the family Basilosauridae, but also the group Basilosauridae + Neoceti (Fig. 10.1, node 4).

Very little has been described of the postcrania of early Neoceti (Fordyce et al., 2000), but the retention of at least pelvic remnants in modern odontocetes, and often pelvic and femoral remnants in modern mysticetes, indicates that although external hind limbs may have been lost at about the time of the origin of Neoceti, both groups never lost the hind limb completely.

Development

Extant cetaceans display a significant delay in skeletal maturation when compared with terrestrial mammals. Fusion of bone epiphyses is not completed before sexual maturation, and in some it is never completed, even in sexually mature adults. This delay in skeletal maturation is first observed in basilosaurid archaeocetes (Uhen and Gingerich, 2001; Uhen, 2004; Fig. 10.1, node 4), although skeletons of immature pakicetids, remingtonocetids, ambulocetids, and protocetids are only poorly known from the fossil record.

Dental eruption has often been used as an indicator of life history in mammals of all kinds. This line of life history evidence is not applicable to modern cetaceans and their fossil relatives because mysticetes lack teeth altogether, and odontocetes erupt only a single set of undifferentiated teeth. Archaeocetes, however, are diphyodont and heterodont, so life history data can be derived from archaeocete dental eruption sequences. The dental eruption sequence for *Dorudon atrox* in particular (Uhen, 2004) and dorudontine archaeocetes in general (*Dorudon* and *Zygorhiza;* Uhen, 2000) is as follows (DD = deciduous dentition; M, m = molar; dP, dp = deciduous premolar; P, p = premolar; I, i = incisor; C, c = canine):

upper dentition—DD; M1; dP1; M2; P4; (P3, I1); (P2, I2); (P1, I3); C1 lower dentition—dd; m1; dp1; m2; m3; p4; (p3, i1); (p2, i2); (p1, i3); c1

As noted by Uhen (2000), dental maturity (eruption of all of the teeth) in extant cetaceans is achieved long before sexual maturity, and skeletal maturity is greatly delayed (if ever achieved). Because archaeocetes retain a more plesiomorphic dental replacement pattern, sexual maturity may be more closely approximated by dental maturity in archaeocetes when compared with modern cetaceans.

Smith (1992) presented a range of life history strategies for mammals ranging from those that "live fast and die young" to those that "live slow and die old." She noted that those mammals that live fast tend to achieve dental maturity first, skeletal maturity second, and sexual maturity last, whereas those that live slow tend to achieve sexual maturity first, dental maturity second, and skeletal maturity last. Extant cetaceans have very long life spans (e.g., Evans and Hindell, 2004), which, in part, are probably the result of large body size (Promislow and Harvey, 1990). Cetaceans achieve dental maturity first, sexual maturity second, and skeletal maturity last. This should, at best, put them in the middle of the life history spectrum, but they are clearly on the "live slow, die old" end.

One partial explanation for this discrepancy could be that because cetaceans live in the water, they are relieved of physical constraints from gravity on the skeletal system. Thus, skeletal maturation can be delayed indefinitely without any resulting loss of skeletal function.

Although it is true that dental eruption in extant odontocetes is complete before the onset of sexual maturity, dental development, in a sense, is never quite complete because odontocete teeth continue to deposit dentin throughout the life of the animal (Uhen, 2002), and in this way is similar to skeletal maturation in following sexual maturity. Thus, if indeed both skeletal and dental maturation are indefinitely delayed in extant cetaceans, they are then in the proper place at the "live slow, die old" end of the life history spectrum.

Reproduction

Recent cetaceans and sirenians never leave the water, and they thus both breed and give birth in the water. All extant pinnipeds, however, haul out on land to breed and give birth. Thus, pinnipeds still retain many features of the skeleton that are necessary for locomotion and support on land that are not found in cetaceans. As mentioned above, the earliest cetaceans that are certainly fully aquatic are the basilosaurids. Many protocetids are thought to be mostly aquatic (Gingerich et al., 2001), but they also still retain features of the vertebral column and limbs that would allow them to support themselves and move on land, in a similar manner as do modern pinnipeds. If the analogy with pinnipeds holds true, perhaps protocetids, like pinnipeds, spent most of their time living and

TABLE 10.1
Secondarily aquatic tetrapods and the distribution of anatomical and behavioral characteristics relating to life in the water

Taxon	Habitus	Body form	Forelimbs	Hind limbs	Tail	Sound channel	Prey detection	Feeding system	Reproduction
Hydrophiidae (sea snakes)	Amphibious/ fully aquatic	Streamlined	Absent	Absent	Long	External tympanum	Taste and vision	Individual prey	Lay eggs on land/live birth at sea
Mosasauroidea	Fully aquatic	Streamlined	Flippers	Flippers	Long	External tympanum	Vision	Individual prey	Live birth at sea?
Cetacea (whales, dolphins, porpoises)	Fully aquatic	Streamlined	Flippers	Absent	Horizontal flukes	Mandibular fat pat	Echolocation/?	Individual prey or bulk feeding	Live birth at sea
Sirenia (sea cows, manatees)	Fully aquatic	Streamlined	Flippers	Absent	Horizontal flukes	EAM (no pinnae)	Vision/tactile	Herbivorous	Live birth at sea
Ichthyosauria	Fully aquatic	Streamlined	Flippers	Flippers	Vertical flukes	External tympanum?	Vision?	Individual prey	Live birth at sea
Sphenisciformes (penguins)	Amphibious	Streamlined	Flippers	Feet	Absent	EAM	Vision	Individual prey	Lay eggs on land
Pinnipedia (seals, sea lions, walruses)	Amphibious	Streamlined	Flippers	Flippers	Absent	EAM (with pinnae)	Vision/tactile	Individual prey or bulk feeding	Live birth on land
Plesiosauria	Fully aquatic	Streamlined	Flippers	Flippers	Long	External tympanum?	Vision?	Individual prey	Live birth at sea
Cheloniidae (sea turtles)	Amphibious	Streamlined	Flippers	Flippers	Short	External tympanum	Smell and vision	Herbivorous to carnivorous	Lay eggs on land

feeding in the water and returned to land to breed and/or give birth. This hypothesis is supported by the observations noted above regarding the adaptations for aquatic feeding and hearing found in protocetids.

Summary and Conclusions

Let us return to the original question asked: What makes a whale a whale? Many of the features listed by workers who focus on extant Cetacea actually map onto the cetacean phylogeny at many different points within Cetacea, broadly construed. This, in retrospect, is not surprising. The features that characterize this highly distinctive mammalian order were not acquired all at once, at least in part because the habit of living on land probably changed gradually to the habit of living in the water.

Different groups of tetrapods have adopted an aquatic existence on numerous evolutionary occasions. Table 10.1 lists a sampling of those groups and lists several features of each. It is illustrative to note that those earliest-arising features that characterize Cetacea are involved with feeding and sensory apparatus. These are also the features that are not generally shared among secondarily aquatic tetrapods. The features that are shared more broadly among these groups (streamlined body form and flipper-shaped forelimbs) deal with locomotion in an aquatic environment. A broad generalization that can be drawn from this comparison is that although there are many ways to sense the aquatic environment, feed in the water, and reproduce in the water, there are few ways to shape a body with minimal drag and develop efficient control surfaces.

Acknowledgements

I thank Jason Anderson and Hans-Dieter Sues for inviting me to contribute to this volume. I also thank my colleagues Lawrence G. Barnes, David J. Bohaska, R. Ewan Fordyce, Jonathan Geisler, Philip D. Gingerich, Sandy Madar, James G. Mead, Christian de Muizon, Albert Sanders, and Hans Thewissen. Without their work, and that of many others, in finding, preparing, describing, and interpreting fossil and extant cetaceans, such a synthetic work as this would not be possible.

References Cited

Bajpai, S., and J. G. M. Thewissen. 2000. A new, diminutive Eocene whale from Kachchh (Gujarat, India) and its implications for locomotor evolution of cetaceans. Current Science 79:1478–1482.

Evans, K., and M. A. Hindell. 2004. The age structure and growth of female sperm whales (*Physeter macrocephalus*) in southern Australian waters. Journal of Zoology (London) 263:237–250.

Fleischer, G. 1976. Hearing in extinct cetaceans as determined by cochlear structure. Journal of Paleontology 50:133–152.

Fordyce, R. E., and C. de Muizon. 2001. Evolutionary history of cetaceans: a review; pp. 169–223 in J.-M. Mazin and V. de Buffrénil (eds.), Secondary Adaptation of Tetrapods to Life in Water. Verlag Dr. Friedrich Pfeil, Munich.

Fordyce, R. E., J. L. Goedert, L. G. Barnes, and B. J. Crowley. 2000. Pelvic girdle elements of Oligocene and Miocene Mysticeti: whale hind legs in transition. Journal of Vertebrate Paleontology 20(Suppl. to 3):41A.

Fraser, F. C., and P. E. Purves. 1960. Anatomy and function of the cetacean ear. Proceedings of the Royal Society of London B 152:62–77.

Gatesy, J. 1998. Molecular evidence for the phylogenetic affinities of Cetacea; pp. 63–112 in J. G. M. Thewissen (ed.), The Emergence of Whales. Evolutionary Patterns in the Origin of Cetacea. Plenum Press, New York and London.

Gambaryan, P. P. 1974. How Mammals Run: Anatomical Adaptations. John Wiley & Sons, New York, 367 pp.

Geisler, J. H. 2001. New morphological evidence for the phylogeny of Artiodactyla, Cetacea, and Mesonychidae. American Museum Novitates 3344:1–53.

Geisler, J. H., and A. E. Sanders. 2003. Morphological evidence for the phylogeny of Cetacea. Journal of Mammalian Evolution 10:23–129.

Geisler, J. H., and M. D. Uhen. 2003. Morphological support for a close relationship between hippos and whales. Journal of Vertebrate Paleontology 23:991–996.

Geisler, J. H., and M. D. Uhen. 2005. Phylogenetic relationships of extinct cetartiodactyls: results of simultaneous analyses of molecular, morphological, and stratigraphic data. Journal of Mammalian Evolution 12:145–160.

Gingerich, P. D. 2003. Land-to-sea transition in early whales: evolution of Eocene Archaeoceti (Cetacea) in relation to skeletal proportions and locomotion of living semiaquatic mammals. Paleobiology 29:429–454.

Gingerich, P. D., and D. E. Russell. 1981. *Pakicetus inachus,* a new archaeocete (Mammalia, Cetacea) from the early-middle Eocene Kuldana Formation of Kohat (Pakistan). Contributions from the Museum of Paleontology, University of Michigan 25:235–246.

Gingerich, P. D., M. Arif, and W. C. Clyde. 1995. New archaeocetes (Mammalia, Cetacea) from the middle Eocene Domanda Formation of the Sulaiman Range, Punjab (Pakistan). Contributions from the Museum of Paleontology, University of Michigan 29:291–330.

Gingerich, P. D., M. U. Haq, I. S. Zalmout, I. H. Khan, and M. S. Malakani. 2001. Origin of whales from early artiodactyls: hands and feet of Eocene Protocetidae from Pakistan. Science 293:2239–2242.

Gingerich, P. D., S. M. Raza, M. Arif, M. Anwar, and X. Zhou. 1994. New whale from the Eocene of Pakistan and the origin of cetacean swimming. Nature 368:844–847.

Howell, A. B. 1930. Aquatic Mammals: Their Adaptations to Life in the Water. Charles C. Thomas, Springfield, Ill., 338 pp.

Hulbert, R. C., Jr. 1998. Postcranial osteology of the North American middle Eocene protocetid *Georgiacetus;* pp. 235–268 in J. G. M. Thewissen (ed.), The Emergence of Whales. Evolutionary Patterns in the Origin of Cetacea. Plenum Press, New York and London.

Kellogg, R. 1936. A review of the Archaeoceti. Carnegie Institution of Washington Special Publication 482:1–366.

Kumar, K., and A. Sahni. 1986. *Remingtonocetus harudiensis,* new combination, a Middle Eocene archaeocete (Mammalia, Cetacea) from western Kutch, India. Journal of Vertebrate Paleontology 6:326–349.

Lancaster, W. C. 1990. The middle ear of the Archaeoceti. Journal of Vertebrate Paleontology 10:117–127.

Luo, Z.-X. 1998. Homology and transformation of cetacean ectotympanic structures; pp. 269–302 in J. G. M. Thewissen (ed.), The Emergence of Whales. Evolutionary Patterns in the Origin of Cetacea. Plenum Press, New York and London.

Luo, Z.-X., and P. D. Gingerich. 1999. Terrestrial Mesonychia to aquatic Cetacea: transformation of the basicranium and evolution of hearing in whales. University of Michigan Papers on Paleontology 31:1–98.

Massare, J. A. 1987. Tooth morphology and prey preference of Mesozoic marine reptiles. Journal of Vertebrate Paleontology 7:121–137.

Miller, G. S., Jr. 1923. The telescoping of the cetacean skull. Smithsonian Miscellaneous Collections 76:1–71.

Mitchell, E. D. 1989. A new cetacean from the late Eocene La Meseta Formation, Seymour Island, Antarctic Peninsula. Canadian Journal of Fisheries and Aquatic Science 46:2219–2235.

Naylor, G. J. P., and D. C. Adams. 2001. Are the fossil data really at odds with the molecular data? Morphological evidence for Cetartiodactyla phylogeny reexamined. Systematic Biology 50:444–453.

Norris, K. S. 1968. The evolution of acoustic mechanisms in odontocete cetaceans; pp. 297–224 in E. T. Drake (ed.), Evolution and Environment. Yale University Press, New Haven.

Nummela, S., J. G. M. Thewissen, S. Bajpai, S. T. Hussain, and K. Kumar. 2004. Eocene evolution of whale hearing. Nature 430:776–778.

O'Leary, M. A. 1998. Phylogenetic and morphometric reassessment of the dental evidence for a mesonychian and cetacean clade; pp. 133–162 in J. G. M. Thewissen (ed.), The Emergence of Whales. Evolutionary Patterns in the Origin of Cetacea. Plenum Press, New York and London.

O'Leary, M. A. 2001. The phylogenetic position of cetaceans: further combined data analyses comparisons with the stratigraphic record and a discussion of character optimization. American Zoologist 41:487–506.

O'Leary, M. A., and K. D. Rose. 1995. New mesonychian dentitions from the Paleocene and Eocene of the Bighorn Basin, Wyoming. Annals of the Carnegie Museum 64:147–172.

O'Leary, M. A., and M. D. Uhen. 1999. The time of origin of whales and the role of behavioral changes in the terrestrial-aquatic transition. Paleobiology 25:534–556.

O'Leary, M. A., J. Gatesy, and M. J. Novacek. 2003. Are the dental data really at odds with the molecular data? Morphological evidence for whole phylogeny (re)reexamined. Systematic Biology 52:853–864.

Oelschläger, H. A. 1990. Evolutionary morphology and acoustics in the dolphin skull; pp. 137–162 in J. Thomas and R. Kastelein (eds.), Sensory Abilities of Cetaceans. Plenum Press, New York and London.

Pilleri, G., M. Gihr, and C. Kraus. 1989. The organ of hearing in Cetacea. II. Paleobiological evolution. Investigations on Cetacea 22:5–185.

Pritchard, B. G. 1939. On the discovery of a fossil whale in the older tertiaries of Torquay, Victoria. Victorian Naturalist 55:151–159.

Promislow, D. E. L., and P. H. Harvey. 1990. Living fast and dying young: a comparative analysis of life-history variation among mammals. Journal of Zoology (London) 220:417–437.

Smith, B. H. 1992. Life history and evolution of human maturation. Evolutionary Anthropology 1:134–142.

Swift, C. C., and L. G. Barnes. 1996. Stomach contents of *Basilosaurus cetoides:* implications for the evolution of cetacean feeding behavior, and evidence for vertebrate fauna of epicontinental Eocene seas. Sixth North American Paleontological Convention. Abstracts of Papers. The Paleontological Society Special Publication 8:380.

Thewissen, J. G. M. 1994. Phylogenetic aspects of cetacean origins: a morphological perspective. Journal of Mammalian Evolution 2:157–184.

Thewissen, J. G. M., and F. E. Fish. 1997. Locomotor evolution in the earliest cetaceans: functional model, modern analogues, and paleontological evidence. Paleobiology 23:482–490.

Thewissen, J. G. M., and S. T. Hussain. 1993. Origin of underwater hearing in whales. Nature 361:444–445.

Thewissen, J. G. M., S. I. Madar, and S. T. Hussain. 1996. *Ambulocetus natans,* an Eocene cetacean (Mammalia) from Pakistan. Courier Forschungsinstitut Senckenberg 191:1–86.

Thewissen, J. G. M., E. M. Williams, L. J. Roe, and S. T. Hussain. 2001. Skeletons of terrestrial cetaceans and the relationship of whales to artiodactyls. Nature 413:277–281.

Uhen, M. D. 1998a. Middle to late Eocene basilosaurines and dorudontines; pp. 29–61 in J. G. M. Thewissen (ed.), The Emergence of Whales. Evolutionary Patterns in the Origin of Cetacea. Plenum Press, New York and London.

Uhen, M. D. 1998b. New protocetid (Mammalia, Cetacea) from the late middle Eocene Cook Mountain Formation of Louisiana. Journal of Vertebrate Paleontology 18:664–668.

Uhen, M. D. 2000. Replacement of deciduous first premolars and dental eruption in archaeocete whales. Journal of Mammalogy 81:123–133.

Uhen, M. D. 2002. Dental morphology (cetacean), evolution of; pp. 316–319 in W. F. Perrin, B. Würsig, and J. G. M. Thewissen (eds.), Encyclopedia of Marine Mammals. Academic Press, San Diego.

Uhen, M. D. 2004. Form, function, and anatomy of *Dorudon atrox* (Mammalia, Cetacea): an archaeocete from the middle to late Eocene of Egypt. University of Michigan Museum of Paleontology Papers on Paleontology 34:1–222.

Uhen, M. D., and P. D. Gingerich. 2001. New genus of dorudontine archaeocete (Cetacea) from the middle-to-late Eocene of South Carolina. Marine Mammal Science 17:1–34.

Werth, A. J. 1992. Anatomy and evolution of odontocete suction feeding. Ph.D. dissertation, Harvard University, Cambridge, 313 pp.

Contributors

Jason S. Anderson, Faculty of Veterinary Medicine, University of Calgary, 3330 Hospital Drive, Calgary, Alberta T2N 4N1, Canada, e-mail: janders@ucalgary.ca

Michael W. Caldwell, Departments of Earth and Atmospheric Sciences and Biological Sciences, University of Alberta, Edmonton, Alberta, T6G 2E3, Canada, e-mail: mw.caldwell@ualberta.ca

Luis M. Chiappe, The Dinosaur Institute, Natural History Museum of Los Angeles County, 900 Exposition Boulevard, Los Angeles, CA 90007, USA, e-mail: chiappe@nhm.org

Gareth J. Dyke, Department of Zoology, University College Dublin, Belfield, Dublin 4, Ireland, e-mail: gareth.dyke@ucd.ie

Brian K. Hall, Department of Biology, Dalhousie University, Halifax, Nova Scotia B3H 4J1, Canada, e-mail: bkh@dal.ca

Gavin F. Hanke, Royal British Columbia Museum, 675 Belleville Street, Victoria, British Columbia V8W 9W2, Canada, e-mail: ghanke@royalbcmuseum.bc.ca

Philippe Janvier, UMR 5143 du CNRS, Département Histoire de la Terre, Muséum National d'Histoire Naturelle, CP 38, 57 rue Cuvier, 75231 Paris Cedex 05, France, e-mail: janvier@mnhn.fr; and The Natural History Museum, Cromwell Road, London SW7 5BD, United Kingdom

Hans C. E. Larsson, Redpath Museum, McGill University, 859 Sherbrooke St. W., Montréal, Québec H3A 2K6, Canada, e-mail: hans.ce.larsson@mcgill.ca

Zhe-Xi Luo, Section of Vertebrate Paleontology, Carnegie Museum of Natural History, 4400 Forbes Avenue, Pittsburgh, PA 15213, USA, e-mail: LuoZ@CarnegieMNH.org

Tiiu Märss, Institute of Geology, Tallinn University of Technology, Estonia Avenue 7, Tallinn 10143, Estonia, e-mail: marss@gi.ee

Robert R. Reisz, Department of Biology, University of Toronto at Mississauga, 3359 Mississauga Road N., Mississauga, Ontario L5L 1C6, Canada, e-mail: rreisz@utm.utoronto.ca

Hans-Dieter Sues, National Museum of Natural History, Smithsonian Institution, MRC 106, P.O. Box 37012, Washington, DC 20013-7012, USA, e-mail: suesh@si.edu

Mark D. Uhen, Department of Paleobiology, National Museum of Natural History, Smithsonian Institution, MRC 106, P.O. Box 37012, Washington, DC 20013-7012, USA, e-mail: uhenm@si.edu

Mark V. H. Wilson, Department of Biological Sciences and Laboratory for Vertebrate Paleontology, University of Alberta, Edmonton, Alberta T6G2E9, Canada, e-mail: mark.wilson@ualberta.ca

P. Eckhard Witten, AKVAFORSK, Institute of Aquaculture Research, N-6600 Sunndalsøra, Norway, e-mail: eckhard.witten@akvaforsk.no

monorhinal condition, 75
Monotremata, 364
Monotrematum, 364
Montanalestes, 367, 371
Morganucodon, 341, 356
morganucodontans (Morganucodonta), 346, 347, 352
Morganucodontidae, 352
mortar-pestle crushing, 355
Mosasauroidea, 288
mosasaurs (Mosasauridae), 9, 256, 258, 290
mucocartilage, 35
multituberculates (Multituberculata), 341, 344, 355, 372
Murtoilestes, 367, 371
Myllokunmingia, 69–72
myllokunmingiids (Myllokunmingiida), 7, 65, 69–72, 108–109
myomeres, 70, 72
Myomerozoa, 61
Myxineides, 76
Myxinekela, 76–77
Myxinidae, 77
mysticetes (Mysticeti), 394, 396, 397

Najash, 256, 266, 272, 290–292
Nannolestes, 361
nasohypophysial complex/opening, 7, 77–81, 99, 107
Natchitochia, 401
natural kinds, 159
Nectridea, 184
Neoceratodus, 30–31, 161, 170
Neoceti, 397, 402
Neognathae, 304
Neornithes, 304, 325, 327
neoteny, 204
neural crest–derived tissues, 70, 74, 102–106
Norselaspis, 92
notochord, 94–95

Obdurodon, 364
Obtusacanthus, 139
odontoblast, 18
odontocetes (Odontoceti), 394, 396–399, 403
odontoclast, 29, 32
odontode, 22–23
odontoid process, 187
Oestocephalus, 189
olfactory placode(s), 107
Onychodontiformes, 141
operculum (of lissamphibians), 199
Ophiacodon, 235
Ophidia, 256, 258
Ophiomorphus, 263–264, 266
oral hood (of lampreys), 71, 78, 104–105

ornithorhynchids (Ornithorhynchidae), 364
Ornithorhynchus, 347, 360, 364
Ornithuromorpha, 324–325
Orobates, 241
orthodentine, 18, 23
osteichthyans (Osteichthyes), 30, 32, 139, 140–141
osteoblast, 16, 31, 32
osteocalcin, 25
osteoclast, 29, 31
osteocyte, 16, 31
osteocytic bone, 19
osteodentine, 18, 23
osteoid, 16, 19
osteonectin, 16–17
osteopontin, 16
osteostracans (Osteostraci), 7, 81–99, 124, 134–136, 141
"ostracoderms" (Ostracodermi), 7, 27–28, 57, 60, 65, 74–81, 84–99, 109
"otic notch," 186
Oviraptor, 310
oviraptorids (Oviraptoridae), 306, 310
oviraptorosaurs (Oviraptorosauria), 308, 311

Pachyophiidae, 261–263
Pachyophis, 260, 263, 268
Pachyrhachis, 9, 255, 264–292
paedomorphosis, 201
pakicetids (Pakicetidae), 398–400, 402
Palaeobranchiostoma, 73
Palaeognathae, 327
Palaeophiidae, 256, 261
Palaeoxonodon, 355, 361
Panderichthys, 60, 162, 171
Pantylus, 206, 229
Pappotherium, 357
Patagopteryx, 324
Paucicanthus, 140
Paucituberculata, 370–371
pedicellate teeth, 206
Pediomys, 370
Pedopenna, 309, 313
Pelecaniformes, 327
pentadactyl hand, 313
peramurans (Peramura), 361
Perca, 37
Pezopallichthys, 132
Phanerosaurus, 241
pharyngeal jaws, 39
Pharyngolepis, 127–128
Phascolotherium, 353
Phlebolepis, 129–130, 141
Phlegethontia, 189, 192, 199
phosphorus metabolism, 29
Phyllolepida, 136
Phyllospondyli, 203